Pharmaceutical Particulate Carriers

DRUGS AND THE PHARMACEUTICAL SCIENCES

A Series of Textbooks and Monographs

edited by

James Swarbrick
Applied Analytical Industries, Inc.
Wilmington, North Carolina

1. Pharmacokinetics, *Milo Gibaldi and Donald Perrier*
2. Good Manufacturing Practices for Pharmaceuticals: A Plan for Total Quality Control, *Sidney H. Willig, Murray M. Tuckerman, and William S. Hitchings IV*
3. Microencapsulation, *edited by J. R. Nixon*
4. Drug Metabolism: Chemical and Biochemical Aspects, *Bernard Testa and Peter Jenner*
5. New Drugs: Discovery and Development, *edited by Alan A. Rubin*
6. Sustained and Controlled Release Drug Delivery Systems, *edited by Joseph R. Robinson*
7. Modern Pharmaceutics, *edited by Gilbert S. Banker and Christopher T. Rhodes*
8. Prescription Drugs in Short Supply: Case Histories, *Michael A. Schwartz*
9. Activated Charcoal: Antidotal and Other Medical Uses, *David O. Cooney*
10. Concepts in Drug Metabolism (in two parts), *edited by Peter Jenner and Bernard Testa*
11. Pharmaceutical Analysis: Modern Methods (in two parts), *edited by James W. Munson*
12. Techniques of Solubilization of Drugs, *edited by Samuel H. Yalkowsky*
13. Orphan Drugs, *edited by Fred E. Karch*
14. Novel Drug Delivery Systems: Fundamentals, Developmental Concepts, Biomedical Assessments, *Yie W. Chien*
15. Pharmacokinetics: Second Edition, Revised and Expanded, *Milo Gibaldi and Donald Perrier*
16. Good Manufacturing Practices for Pharmaceuticals: A Plan for Total Quality Control, Second Edition, Revised and Expanded, *Sidney H. Willig, Murray M. Tuckerman, and William S. Hitchings IV*
17. Formulation of Veterinary Dosage Forms, *edited by Jack Blodinger*
18. Dermatological Formulations: Percutaneous Absorption, *Brian W. Barry*
19. The Clinical Research Process in the Pharmaceutical Industry, *edited by Gary M. Matoren*

20. Microencapsulation and Related Drug Processes, *Patrick B. Deasy*
21. Drugs and Nutrients: The Interactive Effects, *edited by Daphne A. Roe and T. Colin Campbell*
22. Biotechnology of Industrial Antibiotics, *Erick J. Vandamme*
23. Pharmaceutical Process Validation, *edited by Bernard T. Loftus and Robert A. Nash*
24. Anticancer and Interferon Agents: Synthesis and Properties, *edited by Raphael M. Ottenbrite and George B. Butler*
25. Pharmaceutical Statistics: Practical and Clinical Applications, *Sanford Bolton*
26. Drug Dynamics for Analytical, Clinical, and Biological Chemists, *Benjamin J. Gudzinowicz, Burrows T. Younkin, Jr., and Michael J. Gudzinowicz*
27. Modern Analysis of Antibiotics, *edited by Adjoran Aszalos*
28. Solubility and Related Properties, *Kenneth C. James*
29. Controlled Drug Delivery: Fundamentals and Applications, Second Edition, Revised and Expanded, *edited by Joseph R. Robinson and Vincent H. Lee*
30. New Drug Approval Process: Clinical and Regulatory Management, *edited by Richard A. Guarino*
31. Transdermal Controlled Systemic Medications, *edited by Yie W. Chien*
32. Drug Delivery Devices: Fundamentals and Applications, *edited by Praveen Tyle*
33. Pharmacokinetics: Regulatory · Industrial · Academic Perspectives, *edited by Peter G. Welling and Francis L. S. Tse*
34. Clinical Drug Trials and Tribulations, *edited by Allen E. Cato*
35. Transdermal Drug Delivery: Developmental Issues and Research Initiatives, *edited by Jonathan Hadgraft and Richard H. Guy*
36. Aqueous Polymeric Coatings for Pharmaceutical Dosage Forms, *edited by James W. McGinity*
37. Pharmaceutical Pelletization Technology, *edited by Isaac Ghebre-Sellassie*
38. Good Laboratory Practice Regulations, *edited by Allen F. Hirsch*
39. Nasal Systemic Drug Delivery, *Yie W. Chien, Kenneth S. E. Su, and Shyi-Feu Chang*
40. Modern Pharmaceutics: Second Edition, Revised and Expanded, *edited by Gilbert S. Banker and Christopher T. Rhodes*
41. Specialized Drug Delivery Systems: Manufacturing and Production Technology, *edited by Praveen Tyle*
42. Topical Drug Delivery Formulations, *edited by David W. Osborne and Anton H. Amann*
43. Drug Stability: Principles and Practices, *Jens T. Carstensen*
44. Pharmaceutical Statistics: Practical and Clinical Applications, Second Edition, Revised and Expanded, *Sanford Bolton*
45. Biodegradable Polymers as Drug Delivery Systems, *edited by Mark Chasin and Robert Langer*

46. Preclinical Drug Disposition: A Laboratory Handbook, *Francis L. S. Tse and James J. Jaffe*
47. HPLC in the Pharmaceutical Industry, *edited by Godwin W. Fong and Stanley K. Lam*
48. Pharmaceutical Bioequivalence, *edited by Peter G. Welling, Francis L. S. Tse, and Shrikant V. Dinghe*
49. Pharmaceutical Dissolution Testing, *Umesh V. Banakar*
50. Novel Drug Delivery Systems: Second Edition, Revised and Expanded, *Yie W. Chien*
51. Managing the Clinical Drug Development Process, *David M. Cocchetto and Ronald V. Nardi*
52. Good Manufacturing Practices for Pharmaceuticals: A Plan for Total Quality Control, Third Edition, *edited by Sidney H. Willig and James R. Stoker*
53. Prodrugs: Topical and Ocular Drug Delivery, *edited by Kenneth B. Sloan*
54. Pharmaceutical Inhalation Aerosol Technology, *edited by Anthony J. Hickey*
55. Radiopharmaceuticals: Chemistry and Pharmacology, *edited by Adrian D. Nunn*
56. New Drug Approval Process: Second Edition, Revised and Expanded, *edited by Richard A. Guarino*
57. Pharmaceutical Process Validation: Second Edition, Revised and Expanded, *edited by Ira R. Berry and Robert A. Nash*
58. Ophthalmic Drug Delivery Systems, *edited by Ashim K. Mitra*
59. Pharmaceutical Skin Penetration Enhancement, *edited by Kenneth A. Walters and Jonathan Hadgraft*
60. Colonic Drug Absorption and Metabolism, *edited by Peter R. Bieck*
61. Pharmaceutical Particulate Carriers: Therapeutic Applications, *edited by Alain Rolland*

ADDITIONAL VOLUMES IN PREPARATION

Drug Permeation Enhancement: Theory and Applications, *edited by Dean S. Hsieh*

Pharmaceutical Particulate Carriers

Therapeutic Applications

edited by
Alain Rolland
GeneMedicine, Inc.
Houston, Texas

Marcel Dekker, Inc. **New York • Basel • Hong Kong**

Library of Congress Cataloging-in-Publication Data

Pharmaceutical particulate carriers : therapeutic applications /
edited by Alain Rolland.
p. cm. -- (Drugs and the pharmaceutical sciences ; 61)
Includes bibliographical references and index.
ISBN 0-8247-9016-2 (alk. paper)
1. Liposomes--Therapeutic use. 2. Microspheres (Pharmacy)
3. Drugs--Vehicles. I. Rolland, Alain. II. Series:
Drugs and the pharmaceutical sciences ; v. 61.
[DNLM: 1. Drug Carriers--therapeutic use. W1 DR893B v. 61 1993 /
QV 785 P5353 1993]
RS201.L55P48 1993
615'.6--dc20
DNLM/DLC
for Library of Congress 93-19673
CIP

This book is printed on acid-free paper.

Marcel Dekker, Inc.
270 Madison Avenue, New York, New York 10016

Current printing (last digit):
10 9 8 7 6 5 4 3 2 1

PRINTED IN THE UNITED STATES OF AMERICA

Preface

"Every great advance in Science has issued from a new audacity of imagination"
John Dewey, The Quest for Certainty

Over the past few decades, there has been a growing interest in developing new drug delivery systems in order to improve the benefit/risk ratio associated with all drugs. Key considerations in the design of such new drug delivery systems are the route of administration and the controlled release of a drug at its site of action. Among a variety of site-specific and controlled drug delivery systems that have been devised are many particulate carriers, for example, microspheres, liposomes, nanoparticles, and stabilized micellar systems.

Many symposia and publications have focused on the preparation, characteristics, and biological properties of particulate carriers. The aim of this book is to give a comprehensive overview of the various clinical applications of pharmaceutical particulate carriers. Highlighted are the different routes of administration of particulate carriers with respect to anatomical and (patho)physiological considerations and to carrier- and/or disease-related effects. Therefore, this volume covers recent clinical developments and future therapeutic applications of pharmaceutical particulate carriers according to the routes of administration. Each chapter presents a review of the scientific literature and several examples on therapeutic applications of particulate carriers. This leads the reader into a comprehensive understanding of the rationale for administering particulate carriers as site-specific and controlled drug delivery systems via different routes of administration.

Chapter 1 by Drs. Barber and Shek covers topical ocular drug delivery using liposomal ophthalmic formulations. The rationale for associating drugs with liposomes is mainly to control drug release from this depot system or reservoir and to increase the contact time with the corneal surface. In Chapter 2, Professor Edman and Dr. Björk present the potential application of microspheres as a nasal drug delivery system. The intranasal administration of peptides and proteins looks very attractive and promising, in particular when administered via particulate carriers that may enhance drug absorption across the nasal mucosa. Pulmonary administration of liposome-encapsulated drugs is examined by Dr. Meisner in Chapter 3. The two principal areas of interest are the use of liposomes for the treatment of respiratory distress syndrome, (i.e., lung surfactant replacement therapy) and for modulating the pulmonary absorption of drugs such as sodium chromoglycate, nitroglycerin, oxytocin, atropin, and metoproterenol. Oral administration and transport of drugs across the GI epithelium are illustrated and discussed in Chapter 4 by Professor Florence and Dr. Jani, and in Chapter 5 by Dr. Smith et al. As reported in these two chapters, the oral delivery of vaccines and peptidergic molecules, in particular insulin, represents for pharmaceutical scientists perhaps the most formidable challenge that may be approached using particulate carriers. Intratumoral drug delivery using particulate carriers is reviewed by Dr. Gupta and colleagues in Chapter 6. The aim of this therapeutic approach is to ensure access of drugs such as cytotoxic agents (5-fluorouracil, adriamycin, methotrexate) into the target tissue at a controlled rate and maintain adequate levels of drugs at the site of action over adapted periods of time, consequently reducing or avoiding drug exposure to healthy tissue. Intramuscular and subcutaneous injection of liposomes as a drug delivery system is illustrated in Chapter 7 by Dr. Kadir and coworkers. Biodegradable microparticulate drug delivery systems, including microspheres, nanoparticles, and liposomes, have been, for instance, successfully developed for providing a sustained or controlled release of drugs (cytarabine, human interferon-β and -γ, bleomycin, insulin, chloroquine, gentamicin) in order to improve patient compliance, avoid or reduce side-effects, and maintain an optimal therapeutic drug concentration profile for a long period of time. Preclinical and clinical studies regarding subcutaneous administration of liposomal tumor antigen vaccines are presented in Chapter 8 by Dr. Phillips. Liposomal preparations of autologous tumor-associated antigens are capable of eliciting positive clinical responses in patients with stage-III melanoma that correlate with changes in cellular immunity. Such a vaccination procedure may be very promising in controlling metastatic cancer, which is refractory to conventional treatments with chemotherapy and radiotherapy.

Dr. Montini et al. discuss in Chapter 9 clinical evaluation of intramuscularly administered bromocriptine microspheres in pituitary adenomas. Intramuscular administration of bromocriptine-loaded microspheres induces an inhibition of prolactine secretion for several weeks according to the slow release of the drug from the particulate carriers and produces a marked and rapid tumor-size reduction in patients bearing macroprolactinomas. They are also very effective in acromegalic patients and in some patients with functionless pituitary tumors. In Chapter 10, Dr. Petrak gives an overview of potential medical applications of particulate carriers for the intravascular administration of drugs. Most of the intravascular applications of particulate systems relate to the treatment of infectious diseases by targeting drugs such as Amphotericin B, ampicillin, and 8-aminoquinolines to infected cells and to the delivery of anticancer drugs or immunomodulators (doxorubicin, methotrexate, 5-fluorouracil, muramyldipeptide). The liposomal therapy of retroviral infections, particularly AIDS, via the intravenous route is exemplified in Chapter 11 by Drs. Phillips and Tsoukas. The involvement of macrophages in the pathogenesis of human immunodeficiency virus (HIV) infection presents a unique opportunity for targeting drugs such as 3′-azido-3′-deoxythymidine (AZT) with particulate carriers, since after intravascular administration they are generally rapidly and efficiently cleared from the circulation by the mononuclear phagocyte system. Liposome-encapsulated AZT appears to be effective in reducing bone marrow toxicity and enhancing antiviral activity. In Chapter 12, a comprehensive review and perspectives of clinical applications of microspheres in embolization and chemoembolization are given by Dr. Flandroy and coworkers. Chemoembolization using microspheres containing anticancer drugs represents a promising therapeutic approach as a preoperative or palliative treatment for various malignancies. Embolic particles loaded with an exogenous estrogen may be also of interest in local hormonotherapy. Chapter 13 by Dr. Rolland covers the medical applications of particulate carriers in dermal and transdermal drug delivery and presents potential medical applications of microparticulates for targeting drugs to specific cutaneous structures. Polymeric microspheres are presented for instance as a novel topical site-specific drug delivery system for targeting drugs directly into the hair follicles without penetrating the stratum corneum. Drug follicular targeting using particulate carriers should permit an increase of the therapeutic index of drugs designed for the treatment of cutaneous pathologies, in particular those associated with the pilo-sebaceous structures.

This volume is designed to serve as a reference for the latest advances in the therapeutic applications of particulate carriers for industrial, clinical and research pharmacists; medicinal and pharmaceutical chemists; pharmacologists; biochemists; biotechnologists; biologists; ophthalmologists; pulmonologists and allergists; dermatologists; hematologists; oncologists; infectious disease specialists; immunologists; pharmaceutical R&D managers; upper-level undergraduate and graduate students in these disciplines.

Alain Rolland

Contents

Contributors

R. F. Barber, Ph.D. Defence Scientist, Immunology Group, Operational Medicine Section, Biosciences Division, Defence and Civil Institute of Environmental Medicine, North York, Ontario, Canada

Erik Björk, M.Sc.Pharm. Department of Pharmaceutics, University of Uppsala, and Kabi Pharmacia AB, Uppsala, Sweden

Liana Cortesi, M.D. Post Graduate Student, Division of Endocrinology, Neuroendocrine Unit, Ospedali Riuniti di Bergamo, Bergamo, Italy

Daan J. A. Crommelin, Ph.D. Professor, Department of Pharmaceutics, University of Utrecht, Utrecht, The Netherlands

Peter Edman, Ph.D. Professor, Pharmaceutical Department, Astra Draco AB, Lund, and Department of Pharmaceutics, University of Uppsala, Uppsala, Sweden

P. M. J. Flandroy, M.D., Ph.D. Department of Neuroradiology, University Hospital, University of Liège, Liège, Belgium

Alexander T. Florence, Ph.D., D.Sc. Professor, School of Pharmacy, University of London, London, England

Daniela Gianola, M.D. Assistant, Division of Endocrinology, Neuroendocrine Unit, Ospedali Riuniti di Bergamo, Bergamo, Italy

Christian Grandfils, Ph.D. Assistant, Center for Education and Research on Macromolecules (CERM), University Hospital, University of Liège, Liège, Belgium

P. K. Gupta, Ph.D. Senior Scientist, Pharmaceutical Product Development, Abbott Laboratories, North Chicago, Illinois

C. T. Hung, Ph.D. Director, Zenith Technology Corporation, Ltd., Dunedin, New Zealand

Praful U. Jani, M.Sc., Ph.D., M.R. Pharm. S. Centre for Drug Delivery Research, The School of Pharmacy, University of London, London, England

R. J. Jérôme, Ph.D. Professor, Center for Education and Research on Macromolecules (CERM), University Hospital, University of Liège, Liège, Belgium

Farida Kadir, Ph.D. Department of Clinical Pharmacy, University of Utrecht, Utrecht, The Netherlands

F. C. Lam, Ph.D. Lecturer, Department of Mathematics and Statistics, University of Otago, Dunedin, New Zealand

Ioana Lancranjan, M.D., Ph.D. Medical Expert, Department of Neuroendocrinology, Clinical Research, Sandoz Pharma, Ltd., Basel, Switzerland

Dale Meisner, B.Sc.Pharm., Ph.D. Senior Research Pharmacist, Pharmaceutical Research and Development, Merck Frosst Center for Therapeutic Research, Kirkland, Quebec, Canada

Marcella Montini, M.D. Assistant Chief, Division of Endocrinology, Neuroendocrine Unit, Ospedali Riuniti di Bergamo, Bergamo, Italy

Giorgio Pagani, M.D. Chief, Division of Endocrinology, Neuroendocrine Unit, Ospedali Riuniti di Bergamo, Bergamo, Italy

Marina Desiree Pagani, M.D. Assistant, Division of Endocrinology, Neuroendocrine Unit, Ospedali Riuniti di Bergamo, Bergamo, Italy

Alberto Pedroncelli, M.D. Student, Division of Endocrinology, Neuroendocrine Unit, Ospedali Riuniti di Bergamo, Bergamo, Italy

Karel Petrak, Ph.D. Drug Development Department, CIBA-GEIGY Pharmaceuticals, Suffern, New York

Nigel C. Phillips, Ph.D., C.Biol., M.I.Biol.* Research Associate, Department of Medicine, Montreal General Hospital Research Institute, McGill University, Montreal, Quebec, Canada

Alain Rolland, Pharm.D., Ph.D. Director, Vector Delivery, GeneMedicine, Inc., Houston, Texas

P. N. Shek, Ph.D. Head, Immunology Group, Operational Medicine Section, Biosciences Division, Defence and Civil Institute of Environmental Medicine, North York, Ontario, Canada

Philip L. Smith, B.S., M.S., Ph.D. Assistant Director, Drug Delivery Research and Development, SmithKline Beecham Pharmaceuticals, King of Prussia, Pennsylvania

Francesca Tengattini, M.D. Assistant, Division of Endocrinology, Neuroendocrine Unit, Ospedali Riuniti di Bergamo, Bergamo, Italy

Christos M. Tsoukas, M.D. Director, Immune Deficiency Treatment Center, Montreal General Hospital Research Institute, McGill University, Montreal, Quebec, Canada

Doris A. Wall, Ph.D. Senior Investigator, Drug Delivery, SmithKline Beecham Pharmaceuticals, King of Prussia, Pennsylvania

Glynn Wilson, B.Sc., Ph.D. Director, Drug Delivery Research and Development, SmithKline Beecham Pharmaceuticals, King of Prussia, Pennsylvania

Jan Zuidema, Ph.D. Department of Pharmaceutics, University of Utrecht, Utrecht, The Netherlands

---

**Current affiliation*: Centre D'Immunologie et de Biologie Parasitaire, Institut Pasteur de Lille, Lille, France

1

Liposomes as a Topical Ocular Drug Delivery System

R. F. Barber and P. N. Shek

Defence and Civil Institute of Environmental Medicine
North York, Ontario, Canada

INTRODUCTION

The eyes are among the most readily accessible organs in terms of their location in the body and yet drug delivery to the tissues of the eye is particularly problematic. Conventional eye drops deliver drugs in a pulsatile fashion with a very brief period of overdosing, immediately after instillation, followed by a rapid decline to insignificant levels, usually within minutes (Figure 1). As well, most topically applied drugs are extremely limited in their ability to penetrate the various ocular tissues as a result of various anatomical and physicochemical barriers. Drugs that are ingested or administered by the intravenous route generally become very widely distributed and diluted throughout the body. Penetration of these drugs into the tissues of the eye is extremely limited because, like the brain, the eye is protected by highly restrictive blood–retinal and blood–aqueous barriers.

There is a clear need for effective topical formulations capable of promoting drug penetration and maintaining therapeutic levels with a reasonable frequency of application. Such enhancement could be affected by altering the solubility characteristics of the drug molecule or by manipulation of the delivery system. Increasing the permeability characteristics of a drug will not on its own eliminate the problem of pulsed entry. In fact, this strategy could result in the enhancement of side effects that probably would not be acceptable. The development of appropriate delivery systems could increase

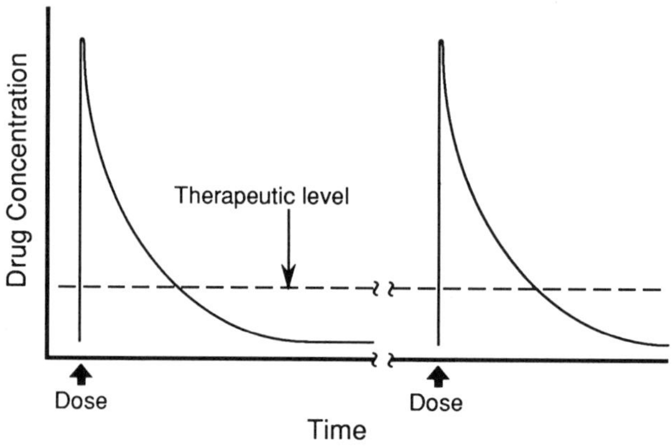

Figure 1 Pulsatile drug delivery achieved by conventional eye drops. The drug level is initially high immediately after instillation and falls below the therapeutic level within minutes resulting in a long period of underdosing.

the contact time of a drug with the eye surface and might promote or facilitate transfer of drug molecules from the tear phase into the eye tissue. The need for controlled and sustained delivery of ophthalmic drugs will become more important with the emergence of new, more potent drugs and biological response modifiers that may also have very short biological half-lives.

Liposomes (lipid vesicles) have been investigated extensively as drug delivery vehicles for over a decade. They offer many potential benefits in this role, including the possibility of controlled, sustained drug release, and protection from metabolic processes while the agent remains sequestered within the vesicles. Our approach has been to associate the drug with liposomes in hope that the vesicles would both increase the contact time with the corneal surface and serve as a slow-release depot. Ours and a number of other animal studies support these contentions and further suggest that liposomes can also enhance the uptake of drug molecules.

PHYSIOLOGICAL AND ANATOMICAL CONSIDERATIONS

Barriers to Drug Bioavailability

Topical ocular drug delivery is particularly complicated by the barriers nature has provided to protect the eyes from the entry of foreign compounds. The tear film is the first barrier and, in conjunction with blinking action, is very effective in rapidly diluting and removing instilled drugs from the precorneal area. As well, the numerous proteins present in tear fluid may bind to, or even metabolize, instilled drugs, thus rendering them unavailable or inactive. Drugs that remain in the tear film long enough to penetrate the sclera are largely carried away by the bloodstream with very little penetrating into other ocular tissues (1,2). The cornea is the primary portal of entry for topically applied drugs and presents the next barrier with its various layers resisting the penetration of either hydrophilic or hydrophobic drugs.

Since tear fluid and the cornea appear to be major obstacles to effective ocular delivery, it is worthwhile to examine some of their key characteristics that must be considered in the design of any ophthalmic drug formulations.

Tear Fluid

The tear fluid itself is divided among three compartments. There is the precorneal film covering the cornea and exposed conjunctiva; the marginal tear strips draining fluid into the lacrimal ducts; and the layer under the lids moistening the conjunctiva. The precorneal film is composed of three layers: an outer oily layer, an aqueous layer, and a mucoid layer. The primary function of the oily layer is to stabilize the underlying aqueous layer and to retard evaporation. The aqueous layer constitutes over 95% of the total tear volume with an average thickness of 7 μm over the cornea. The inner layer is composed of glycoproteins secreted by conjunctival goblet cells and is only 0.02–0.05 μm thick.

The pH of the aqueous layer ranges from 6.5 to 9.3, changing up to 2 units between blinks. The normal osmolarity of tear fluid is approximately 304 mOsm/L. The protein concentration is 4–10 g/L and at least 60 different proteins are present (3). Some of these proteins are filtered from the blood, some are synthesized by the conjunctival epithelial cells, but most are produced by the lacrimal glands. The proteins present in the tear film include albumin, α_1-antichymotrypsin, α_1-antitrypsin, IgA, IgG, lysozyme, α_2-macroglobulin, and transferrin (4). Liotet and Kogbe (3) reported that practically all metabolic and lysosomal enzymes, as well as many hormones, are present and that some enzymes reach concentrations that are 15–100 times higher than those in serum.

The normal tear volume of humans is around 10 μl and the normal flow rate ranges from 0.5 to 2.2 μl/min (this rate decreases with age). Irritants in the precorneal area will increase this flow rate to between 3 and 400 μl/min, thus promoting faster elimination of the irritant from the eye (5). Such reflex tearing often results from the instillation of many ocular pharmaceutical preparations. Nasolacrimal drainage ensures that essentially no soluble drug remains in the precorneal area after several minutes. Lipophilic drugs may leave the precorneal area even faster as they diffuse into the sclera and are then carried away from the eye into the systemic circulation (6).

The Cornea

The cornea is a five-layered avascular structure. However, only three of the layers present any barrier to permeability. The corneal epithelium is five to six cell layers in thickness and has a surface layer of squamous cells joined with tight junctions. Consequently, it is very resistant to the passage of hydrophilic drugs. Lipophilic drugs may diffuse into the cell membranes of the epithelial cells, which can then act as a depot, with the release rate being dependent on the partition coefficient of the drug. It should be obvious at this point that any drug that is going to penetrate any further into the eye must be amphipathic. This is indeed the case as the next barrier layer of the cornea is the stroma with its very high water content. Because of the very open architecture of this layer, it is transparent to water-soluble molecules with molecular weights of less than 500,000 (2). The stroma is the major depot for hydrophilic drugs passing into it from the cornea. Diffusion of drug molecules from the stroma into the anterior chamber is somewhat controlled by the endothelial layer of the cornea. This final layer is one cell thick and is composed of cells with junctional gaps and, as a result, is relatively leaky.

It must be remembered that the above discussion pertains to the healthy eye. In a diseased state, tear protein composition, volume, and flow rate may be altered along with tissue permeability and ocular blood flow. These changes can either reduce or increase drug bioavailability. For example, increased tear protein and flow rate can result in greatly reduced precorneal drug concentrations. Increased corneal permeability, resulting from an inflammatory reaction, could permit easier access to intraocular tissues while increased blood flow would act to remove absorbed drugs more quickly.

STRATEGIES FOR PROMOTING DRUG UPTAKE

Promotion of Corneal Contact and Penetration

Several factors can be manipulated to enhance corneal penetration. For example, with some drugs the contact time with the corneal surface can be lengthened by increasing the viscosity of the preparation through the addition of high molecular weight polymers (5) (e.g., hydroxypropylmethylcellulose). Unfortunately, this approach has had limited success in humans (7). Ointments are a common dosage form employing this strategy but have several drawbacks, including vision impairment, imprecise dosing, and difficult self-administration.

It is possible to modify some drugs to alter their phase solubility. These lipophilic prodrugs could conceivably enter the corneal epithelium and be converted in situ to hydrophilic drugs by native esterases. The drug molecules could then diffuse through the stroma and endothelium into the intraocular tissues. There is also the approach of modifying corneal permeability. Many of the inert ingredients added to ophthalmic preparations to improve shelf life, stability, etc., can increase corneal permeability. Benzalkonium chloride and ethylenediaminetetracetic acid (EDTA) are examples of such additives (8). However, since such alterations in permeability are likely to be nonselective, this is the least acceptable avenue.

Carrier-Mediated Precorneal Drug Release

Another approach is to encapsulate the drug in a particulate carrier. Such vehicles can potentially alter the residence time in the precorneal area, control the rate of drug release, and affect the drug's interaction with the corneal epithelium. At one end of the size spectrum, there are hydrophilic inserts (which may or may not dissolve) and controlled-release devices such as the Ocusert and bioerodable hydrophobic inserts. For example, collagen shields loaded with cyclosporin A can enhance the drug level in rabbit eyes for up to 6 hr (9). However, these devices can elicit a foreign body response in the eye and are not tolerated by all users. They also can be difficult for some users to put in place and then remove. At the other end of the spectrum are aqueous suspensions and microscopic vehicles such as nanoparticles and liposomes. Nanoparticles are colloidal particles ranging in size from 10 nm to 1 μm. Drugs can be entrapped in or absorbed to the polymer matrix. A number of polymers are available, some of which are biodegradable. Nanoparticles are reported to remain in the precorneal area for a longer period than free solution and pilocarpine-loaded particles prolong the drug's miotic effect in rabbit eyes (10). Liposomes are also colloidal particles that are formed from water-insoluble polar lipids, primarily phospholipids. As a drug delivery system, liposomes have shown great potential. This arises out of the immense diversity in liposome size, structure, and composition; their biocompatibility; and their ability to deliver both hydrophilic and hydrophobic agents. The application of liposomes for drug delivery was recently reviewed by Weiner et al. (11). It is the use of liposomes as a topical ocular delivery vehicle to which we will now limit our discussion.

LIPOSOMES

Physicochemical Characteristics

Liposomes are closed vesicles consisting of one or more concentric spheres of lipid bilayers (or lamellae) enclosing an equal number of aqueous compartments. Many lipids that are natural components of biological membranes can be used in the formulation of liposomes. Phosphatidylcholine and cholesterol are the most commonly used lipids but various others are frequently used depending on the application. Because of their structure, liposomes can entrap lipophilic substances in the lipid bilayers and hydrophilic substances within the aqueous compartment (Figure 2).

The permeability of liposomes to entrapped agents is dependent in large part on the partition characteristics of the compound. Agents that are very hydrophilic will be very slowly released because of their insolubility in the lipid phase. Similarly, very hydrophobic (lipophilic) substances will also be very slowly released because of their insolubility in the aqueous phases. Both these types of compounds will only show significant release when the liposomal structure is disrupted. If gradual release is required, then a drug (or

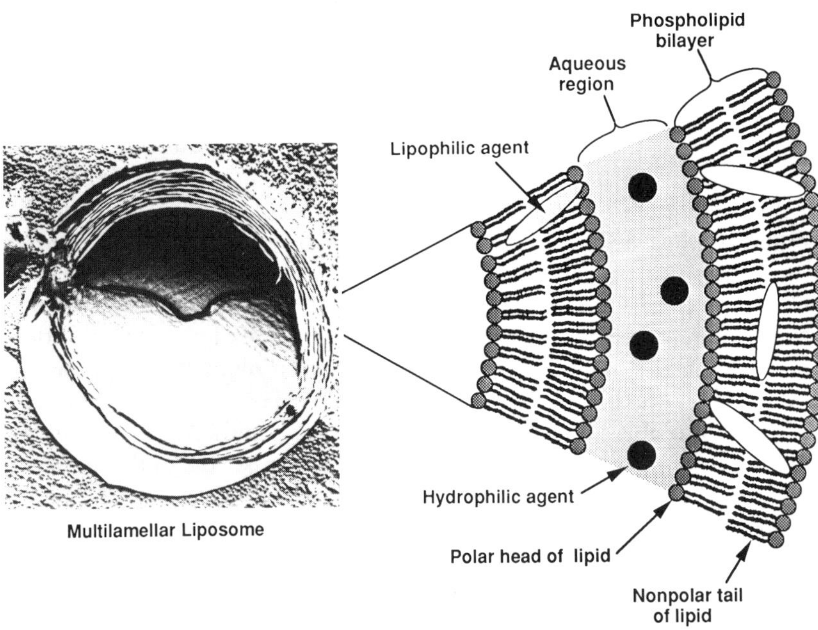

Figure 2 Liposome structure. The multiple concentric lipid bilayers of a multilamellar liposome are clearly indicated by the freeze-fracture electron micrograph illustrated in this figure. The arrangement of the amphipathic phospholipid molecules is shown in the schematic. The polar headgroups situate themselves on the bilayer surface with the nonpolar fatty acid side chains extending into the bilayer core. Amphipathic and nonpolar drugs penetrate the lipid matrix to an extent determined by their inherent hydrophobicity. Water-soluble drugs can be entrapped in the aqueous compartments. (Adapted from Ref. 15.)

modified derivative) must be chosen that has a partition coefficient somewhere between these two extremes.

Liposome permeability is also strongly influenced by the lipid makeup of the bilayers. Both headgroup interactions and acyl chain packing are important. The permeability of vesicles prepared from various phosphatidylcholines increases with increasing temperature, decreasing chain length, and increasing unsaturation (12). Cholesterol is frequently added to liposomal formulations to decrease permeability and increase stability (13).

Preparation

There are numerous methods available for making liposomes and the reader is referred to a recent review by Woodle and Papahadjopoulos (14). Liposomes form spontaneously when most phospholipids are hydrated and, indeed, the simplest method only requires agitation following the addition of an aqueous solution to a dry preparation of lipid. Many different methodologies have been developed because of varying requirements for liposome size, number of lamellae, and because of differences in the solubility and stability of the agents that have been incorporated into the liposomal structure. Liposomes range in size from 25 nm to as much as 10 μm and can be unilamellar or multilamellar. If appropriate lipids are included in the formulation, the surface of the vesicles can be modified by chemically conjugating antibodies, enzymes, glycoproteins, and so forth to the surface.

Liposomes as a Drug Delivery System

The potential advantages that liposomes offer as drug delivery vehicles are clear. They are formulated from naturally occurring membrane components and therefore are biocompatible and biodegradable. They can incorporate both hydrophobic and hydrophilic drugs (or biological response modifiers) within their structure. Their size and the proportion of lipid phase to aqueous phase can be controlled by the preparation method. Their surface can be modified to alter their biodistribution and pharmacokinetics. Liposomes that are stable in vivo can potentially act as slow-release drug depots.

Much of the liposome literature regarding drug delivery pertains to intravenous delivery. There is a large volume of information available on the in vivo stability and disposition of vesicles injected into the vascular and lymphatic systems; the use of liposomes as carriers for chemotherapeutic and antimycotic agents; and their applications as immunological adjuvants (15). The limited success that has been achieved in the use of liposomes for systemic delivery is largely the result of their interactions with serum components, uptake by cells of the reticuloendothelial system (16), and the inability of liposomes to move into extravascular spaces (17). Liposomes cannot extravasate because of their size and therefore targets outside the vascular compartment cannot be accessed by this route. However, some success has been demonstrated in the area of site-specific delivery. For example, liposomes can be injected into joints (18), applied topically to the skin (19), inhaled into the lungs (20), injected intravitreally (21), and applied topically to the eyes.

TOPICAL OCULAR DELIVERY

As indicated above, the use of liposomes as drug carriers has shown some success when they are delivered directly to the anatomical site where the drug is required. The eye is an example of successful site-specific drug delivery by liposomes. For example, intravitreal injection of drug-containing liposomes has demonstrated an improved therapeutic response over the free drug (21,22) and this approach may hold promise for some specific pathological conditions. Of more general applicability is the topical route and liposomal eye drop formulations have been shown to be advantageous in several animal studies.

Animal Studies

The earliest study demonstrating the potential usefulness of liposomes as delivery vehicles for ocular use investigated the efficacy of liposome-entrapped idoxuridine in the treatment of herpes simplex keratitis (23). Rabbits in this study that were treated with liposomes containing idoxuridine suffered a significantly milder disease than those animals treated with free drug. In a later study, it was shown that liposomes actually promote corneal penetration of idoxuridine (24). Other drugs that have shown enhanced ocular uptake when delivered by a liposomal carrier include indoxole (25), penicillin G (25), and triamcinolone acetonide (26). The maximal effect on drug uptake that has been reported was a 10-fold increase in the aqueous humor level using liposomal indoxole (25).

Other studies have revealed that liposome-mediated enhancement does not occur for all drugs. In fact, some drugs are more poorly absorbed when encapsulated in liposomes. For example, epinephrine uptake is reduced by half (27) and dihydrostreptomycin sulfate by at least one-fifth (28).

Meisner et al. (29) prepared neutral, negative, and positively charged multilamellar liposomes containing either atropine base or atropine sulfate. In their study, they compared the pupil dilatory effect of the various preparations in rabbits. Liposomal encapsulation was shown to prolong the action of atropine base from 7 hr for free drug to a maximum of 12 hr with positively charged vesicles. A similar pattern was observed using the water-soluble sulfate salt but with a shorter duration of action. The authors suggest that the enhanced drug response was the result of increased drug loading to the cornea, and not to sustained drug release by the liposomes.

These studies have made it apparent that the bioavailability of drugs is affected by several factors including the physicochemical properties of the drug itself, the stability and permeability characteristics of the lipid vesicles, and the affinity of the liposomes for the corneal surface. The importance of these factors is further illustrated by a comparative study of epinephrine and inulin uptake in rabbits (30). These compounds are similar in that they are water-soluble and penetrate the cornea poorly. Inulin, however, has a high molecular weight and is nonionizable, while epinephrine has a low molecular weight and does ionize. Liposomal delivery enhanced the uptake of inulin by 3- to 15-fold in different tissues while reducing the uptake of epinephrine molecules by 1.5–3 times. In vitro studies revealed that the liposomes were freely permeable to epinephrine and only slightly permeable to inulin. Consequently, when liposomes containing epinephrine were instilled, the rapid drug release stimulated tearing via the drug's β-adrenergic action, thus effecting a rapid clearance of both free drug and liposomes. The residence time of inulin was found to be longer than that of epinephrine. Interestingly, the clearance rate for inulin was the same regardless of entrapment and yet liposomal delivery increased uptake very significantly. The authors were unable to determine the mechanism for this enhancement but it clearly points to the importance of liposomal interaction with both the drug and the corneal surface.

Hydrophilic and Lipophilic Drugs

Molecules that are very hydrophilic cannot penetrate the tight boundary of the corneal epithelium and this unfortunately eliminates many drugs from successful topical use. Drugs that are more lipophilic present a formulation problem; they cannot be delivered in solution and must be prepared in suspensions, gels, or ointments. A significant benefit to using liposomes as a delivery vehicle can be their ability to dissolve lipophilic and amphipathic substances within their lipid bilayers. Anti-inflammatory steroids are commonly used in the treatment of ophthalmic inflammation. These compounds are poorly soluble in water unless modified and then are poorly absorbed by the cornea. Taniguchi et al. (31) synthesized two lipophilic esters of dexamethasone (DM): dexamethasone valerate (DV) and dexamethasone palmitate (DP). All three forms were applied topically to rabbit eyes both in a suspension form and incorporated into liposomes. In the case of DM and DP, liposome-mediated delivery resulted in lower corneal drug levels than the suspension form. However, liposomal delivery of DV achieved the highest corneal drug levels and the authors provide evidence that DV is metabolized by corneal esterases to DM which could then diffuse into the stroma. This study emphasizes the importance of the solubility characteristics of the drug and the affinity of the drug for the liposomal carrier.

FACTORS AFFECTING CORNEAL ADHESION

Surface Charge of Liposomes

Another major determinant of the amount of drug penetrating the cornea is the contact time with the corneal surface. As discussed above, the eye is designed to ensure rapid removal of any foreign material from the precorneal area through the combined action of tear flow and blinking. Liposomes are small enough that, free in the tear fluid, they would be quickly carried away by the tear flow. If liposomes are to remain in the precorneal area for any extended period of time, they must be anchored in some manner to the eye surface. Using uncharged multilamellar liposomes containing [^{3}H]inulin, Lee et al. (32) demonstrated that binding to the corneal and conjunctival surfaces was a prerequisite for enhancement of inulin uptake by the eye. Their study also indicated that the actual number of liposomes binding to the ocular surface, and not the instilled volume or number of instilled vesicles, is an important factor influencing the extent of drug absorption.

The mucin layer overlying the corneal epithelium is negatively charged. Enhanced binding of liposomes to this layer can most easily be accomplished simply by incorporating a positively charged lipid into the bilayer structure. In several studies, the incorporation of such a positive charge has resulted in improved drug transfer (25,26,28,31). Using gamma scintigraphy, neutral and negatively charged liposomes were shown to have a half-life of about 1 min in rabbit eyes versus 3–4 min (increasing with decreasing vesicle size) for positively charged liposomes (33).

Stearylamine is used almost exclusively in liposome research to impart a positive surface charge on vesicles. Despite the fact that several studies have demonstrated that this lipid has cytotoxic effects (34–36), it continues to be used because there is no other readily available alternative. In a recent study by Guo et al. (37), a number of positively charged phospholipid and cholesterol derivatives were synthesized and shown to be noncytotoxic. Using these to formulate liposomes of various sizes, Guo et al. examined retention and clearance characteristics of positively charged vesicles from the precorneal area. These studies revealed that clearance was biphasic. The initial phase was rapid and occurred within the first few minutes following instillation. This was followed by a second phase that was slower and more gradual. They also noted that the ocular surface was saturable, indicating that specific adhesion of positively charged vesicles to the negatively charged mucinous epithelial layer was occurring. When liposomes were prepared with very high charge densities, containing up to 30 mol % positively charged lipid, half-times for clearance of up to 2 hr were measured. Furthermore, they found that liposomes formulated with lipids that form a more rigid bilayer (e.g., by using a high cholesterol content) were retained longer than liposomes with more fluid membranes. This latter observation could be due to steric effects on the orientation of the positively charged moieties on the bilayer surface, or it might reflect some difference in vesicle stability in tear fluid as we will discuss later.

Lectins and Antibodies

Another approach utilized to increase binding to the corneal surface has been to make use of lectins and antibodies. Wheat germ agglutinin (WGA), for example, has a strong affinity for human and rabbit corneal epithelium. Schaeffer et al. (38) pretreated rabbit corneas in vitro with WGA and demonstrated a 2.5-fold increase in the binding of ganglioside-containing liposomes. They further demonstrated a significantly increased corneal uptake and transcorneal flux of vesicle-entrapped carbachol in this same study. Another in vitro study, in which herpes-infected corneal cells were used, demonstrated that the antiviral efficacy of acyclovir-loaded liposomes could be significantly improved by surface-labeling the vesicles with a monoclonal antibody (39). Cells infected with herpes simplex virus express viral glycoproteins to which monoclonal antibodies can be raised. Using a palmitic acid derivative of an antibody specific to the viral marker, acyclovir-loaded liposomes were surface-labeled and then incubated with infected corneal cells. The targeted liposomes were shown to bind preferentially to virus-infected cells with a resultant drop in virus replication. Nontargeted drug-loaded liposomes had no antiviral capability.

The improvement in drug delivery observed in these studies may be brought about in several ways. The liposomes may act as a slow-release depot maintaining a constant "high" drug concentration adjacent to the epithelial layer or, in the case of less water-soluble drugs, they may directly transfer drug molecules between the liposomal membrane and the epithelial membrane as suggested by Schaeffer and Krohn (25) using penicillin G. Regardless of the mechanism, prolonged corneal exposure is necessary and the 2-hr half-time reported by Guo et al. (37) is very promising, remembering that conventional eye drops have half-times of only a few minutes at most.

Tear Fluid Effects on Liposomes

Two major factors in precorneal drug loss are nasolacrimal drainage and binding to tear proteins. As discussed above, normal tear flow ensures rapid clearance of instilled agents from the eye surface. As we have also seen in the case of liposomes, the rate of clearance can be reduced by promotion of corneal adhesion. However, adhesion accomplished by electrostatic interaction is not immune to tear flow and the liposomal preparation should not be irritating to the eye so as not to promote increased tear production.

When any delivery vehicle is employed for the delivery of an entrapped agent, it is important to determine the potential impact of the biological milieu on its stability. Yet despite the variety of liposomal formulations that have been used for ocular delivery, very little information is available on potential interactions between tear fluid and liposomes. Protein binding or, more generally, protein interactions are a very important consideration in liposome-mediated drug delivery. The fate and stability of liposomes in the bloodstream has been extensively examined. These studies have clearly demonstrated the impact of serum proteins on liposome permeability, stability, and ultimate disposition. The interactions are complex and include protein-mediated lipid exchange, protein binding with resultant changes in cellular uptake and vesicle aggregation, and vesicle disruption (40).

Undoubtedly because of the difficulty in obtaining a significant volume of tear fluid, the first report of tear fluid protein effects on liposome permeability utilized an artificial tear solution (30). This solution contained only mucin, serum albumin, α-globulin, and egg white lysozyme. These authors report that the tear proteins had no effect on epinephrine release (to which the vesicles were already freely permeable) and only slightly increased inulin release. Another study, using a homogenate prepared from corneal epithelium, conjunctiva, and tear fluid, noted only a very slow rate of inulin release from liposomes (32). A recent study noted that tear fluid induces the aggregation of positively charged liposomes but did not investigate the effects on permeability (37). Our own research into the use of liposomes for chemical prophylaxis has included an examination of tear fluid–induced effects in terms of drug carrier stability and we will discuss these results below.

A MODEL FOR OCULAR PROPHYLAXIS

In Vivo Studies

Exposure to organophosphate poisons is an occupational risk for several professional groups in both civilian and military populations. The eyes are acutely sensitive to low levels of organophosphates with exposure resulting in a number of side effects, including conjunctival hyperemia, headache, inability to accommodate, blurring of vision, and miosis. While unpleasant in themselves, these side effects can be potentially hazardous for individuals operating equipment and moving vehicles. Protective clothing and masks can, when properly used, protect exposed individuals from serious harm;

however, low-level exposure can easily occur as a result of small leaks and accidents. It is therefore highly desirable to be able to provide a second line of defense to the eye by the instillation of a protective ophthalmic formulation.

Based on the rationale that liposomes can increase the residency time of encapsulated agents in the precorneal area, we tested the efficacy of a liposome-entrapped prophylactic agent in protecting the eye from the miotic effect of an organophosphorous compound, diisopropylfluorophosphate (DFP). DFP is a prototype of a family of organophosphate poisons found in insecticides and chemical warfare agents. It is also used pharmaceutically in the form of an ophthalmic ointment for the treatment of glaucoma. Many of the effects of organophosphate agents are the result of its inhibitory action on acetylcholinesterase (AChE), an enzyme essential for cholinergic synaptic function. Organophosphates bind irreversibly to a serine residue located in the active site of this enzyme rendering it incapable of hydrolyzing its normal substrate, acetylcholine. Since no enzyme that metabolizes DFP was readily available, AChE was used in our study as a prophylactic agent for binding and neutralizing DFP, thereby neutralizing its toxic action (41). Liposomes were employed as the carrier system for the topical delivery of encapsulated AChE to the eye. The idea was that the liposomal enzyme would remove DFP from the tear pool and reduce the level of exposure for the native enzyme in the eye tissues.

A number of methods of liposome preparation were investigated. Initially these methods all involved the exposure of the enzyme to either organic solvents or ultrasonic energy. However, it was found that AChE was rapidly denatured by these methods and ultimately the dehydration-rehydration (freeze-drying) method described by Kirby and Gregoriadis (42) was used instead. Very briefly, this method involves the formation of empty liposomes to which the drug (in our case, AChE) is then added and the resulting mixture lyophilized. In the absence of water, liposomal structure breaks down and the lipid and drug molecules become intermixed. When the preparation is required, the dry powder is rehydrated with a very small volume of water. During the rehydration process, the agent is at a very high concentration as the liposomes reform and a very high entrapment efficiency is usually achieved. Once the preparation has been freeze-dried, it can be stored under inert gas for extended periods. We have determined that such preparations generally retain 75% of the original enzyme activity even after storage at −20°C for 7 months.

For in vivo studies, three different lipid compositions were used to provide either a negative, positive, or neutral surface charge to the liposomal vesicles. Neutral liposomes were prepared using egg phosphatidylcholine (EPC) and cholesterol in an 8:2 molar ratio. Charged liposomes were prepared using EPC, cholesterol, and either phosphatidic acid for a negative charge or stearylamine for a positive charge in a 7:2:1 molar ratio. This preparative method produced multilamellar liposomes with a mean diameter of 1.7 μm. The entrapment efficiency varied from 30% to 70%, depending on whether the bilayers were charged.

An in vivo rabbit model was used to test the efficacy of the liposome-associated AChE in preventing DFP-induced miosis. Prior to DFP challenge, a preparation of either empty liposomes or AChE-containing liposomes was instilled in one eye of the animal and free AChE was applied to the contralateral eye of the same animal. Two hours later, a sufficient quantity of DFP, previously determined to reduce pupil size by 50%, was instilled into both eyes. Pupil diameters were measured at regular intervals under standardized lighting conditions. It was found that only liposome-associated AChE provided any protection against miosis with pupil size being reduced by only 15% using positively charged vesicles. Vesicle surface charge appears to be important since neutral and negatively charged vesicles provided less protection.

The AChE enzyme used to counteract DFP action in our animal model is water-soluble and has been determined to have a molecular weight of between 4×10^5 and 1.15×10^6 daltons (43). A molecule as large as AChE is certainly not able to penetrate the cornea and must exert its protective effect from within the precorneal area. The failure of free AChE to provide any protection at all may be due to its rapid washout from the eye. In contrast, the protection afforded by liposome-associated AChE could be explained by a longer residency time of the liposomal complex in the precorneal area, such prolonged precorneal retention having been shown for other liposomal preparations (37). We have proposed a sponge hypothesis by postulating that the enzyme – liposome complex acts as a sponge that removes DFP from the tear pool before it can penetrate the eye. We have conducted in vitro studies demonstrating that indeed DFP, an amphipathic molecule, has the ability to penetrate and associate with liposomes (41). This observation supports our contention that the liposomal AChE complex can serve as a reservoir for the entrapment of DFP molecules.

Tear-Mediated Effects

Since any substance applied topically to the eye comes into immediate contact with tears, we believe it is important to determine the effects of tear fluid on the stability of ocular liposomal formulations. To this end, tear fluid was collected from the lower lid margins of rabbits with great care taken not to stimulate tearing, which would dilute the tear proteins. The AChE-containing liposome preparation was then incubated in the tear fluid and it was found that tear fluid induced approximately a 20% increase in the amount of AChE released after 15 min of incubation with no further release after 3 hr (41). Any enzyme released in the precorneal area in vivo would of course be rapidly cleared by tear flow. Since a protective effect for at least 2 hr was observed in our miosis model, we concluded that tear-mediated release of AChE involves primarily the release of surface-associated protein. The residual AChE that remained associated with liposomes could serve as active molecules for capturing and neutralizing DFP.

The ability of serum proteins to increase liposome permeability is well documented (44). High-density lipoproteins are primarily responsible for the disruption of liposomes that occurs in the bloodstream and numerous studies have demonstrated that liposomes can be stabilized by the inclusion of lipids such as cholesterol which tighten the molecular packing of the lipid bilayers making them more rigid (45,46). Using unilamellar liposomes containing the fluorescent dye 5-carboxyfluorescein, we have also demonstrated that the incorporation of cholesterol in liposomes has a stabilizing effect in terms of reducing tear fluid-mediated release of the entrapped agent (47). The half-time for carboxyfluorescein release in the presence of tear fluid increased from 0.4 to 6.6 hr as the cholesterol content was increased from 0 to 45 mol %. These values compare with 7.0 and 60.6 hr when the same liposomes were incubated in buffer only.

These data show that rabbit tear fluid can indeed destabilize liposomes in a manner very similar to that observed to occur in the presence of serum and, furthermore, that dye release can be modulated by alterations in the cholesterol content of the liposomal bilayer. However, high-density lipoproteins have never been reported in tear fluid. Nevertheless, the ability of cholesterol to stabilize the liposomes suggests that the putative destabilizing factors require access to the hydrophobic region of the lipid bilayers to achieve disruption of the liposomal membranes.

High-density lipoproteins destabilize liposomes by the removal of phospholipid molecules from the vesicle bilayer. This process is mediated by a transfer protein (48) and is saturable (49). To explore the possibility that tear-induced leakage might be mediated by a similar process, we incubated tear fluid with empty liposomes prior to the addition of freeze-dried liposomes containing either AChE or carboxyfluorescein. In both cases this treatment had no effect, clearly indicating that the process cannot be saturated by an excess of phospholipid (50).

Serum-induced permeability has also been shown to be the result of the activation of the complement system (51). Some other components of the immune system (immunoglobins) have been identified in tear fluid raising the possibility that complement could also be present. However, heat treatment of tear fluid at 56°C, a temperature that inactivates components of the complement system, had no effect on AChE release from liposomes, thus ruling out complement involvement (50).

Based on both our in vivo and in vitro results, we have suggested that tear fluid induces limited liposomal lysis that occurs during the first few minutes of exposure. Subsequent to that initial period, there appears to be at least a partial stabilization of liposome integrity as no further enzyme appears to be released and dye release is slowed significantly. While the exact mechanism and the responsible agents remain to be identified, we have determined that the responsible factors are proteins (unpublished data). Using fast protein electrophoresis, we have isolated an active fraction corresponding to a molecular weight of 30,000. Subsequent analysis by nondenaturing gel chromatography has revealed that this fraction contains 10 proteins with molecular weights ranging from 18,000 to 31,000. Further characterization has not been completed. We have, however, ruled out a number of known tear components. Transferrin, lactoferrin, lysozyme, albumin, and ceruloplasmin have all been evaluated in our laboratory and none have displayed any effect on liposome permeability.

POSSIBLE UNDESIRABLE EFFECTS

One important consideration that has not been included in the discussion so far is that of toxicity and irritability. The primary factor that determines the relative toxicity of liposomes appears to be the lipid composition. In vitro studies on cultured cells have demonstrated that charged liposomes, particularly those containing stearylamine (cationic) or phosphatidylserine (anionic), are growth inhibitory and cytolytic (34–36). Neutral liposomes, on the other hand, are generally nontoxic (52). In a study employing neutral liposomes composed only of phosphatidylcholine, no corneal toxicity was evident by histologic exam following 8 days of treatment with two instillations per day (25). The ocular irritability of neutral and positively charged egg phosphatidylcholine liposomes was assessed by Taniguchi et al. (53). The positively charged vesicles contained a significant proportion of stearylamine with a molar ratio of 8:1 phosphatidylcholine to stearylamine. Draize testing by Taniguchi's group indicated that both preparations were essentially nonirritating and no histological abnormalities were noted following nine instillations over 2 hr. In a blinking test designed to detect weak irritation, the positively charged vesicles appeared to cause some pain or unpleasantness following instillation. These authors concluded that neutral liposomes are safe for ophthalmic application but that a substitute must be found for stearylamine.

The choice of saturated versus unsaturated lipid in the formulation of liposomal products for ocular application could be important. Lipid peroxidation, for example, not only tends to destabilize lipid membrane integrity but may also generate potentially toxic oxidative degradation products (54). Intermediate oxidation products derived from liposomes have been implicated as good possibilities for in vivo toxicity (55). At present, there have been practically no systematic studies designed to refute or substantiate the notion that ocular tissues are susceptible to putative toxicity generated by peroxidation or decomposition of liposomal lipids.

CONCLUDING REMARKS

The major disadvantage of conventional liquid dosage forms for ocular application, as noted previously, is their rapid clearance from the eye often resulting in an immediate pulsatile effect. Liposomal ophthalmic formulations, on the other hand, show promise in improving ocular drug delivery. The attractiveness of using liposomes for delivering encapsulated drugs to the eye stems from the fact that liposomal microvesicles may serve as a depot system or reservoir to achieve a slow, sustained-release effect (15).

The key advantage of liposomal formulations resides in the versatility of custom-designed liposomes to meet specific needs by their variations in size, surface charge, and lipid composition. The pharmacokinetics of ophthalmic drugs entrapped in liposomes are invariably altered when compared to the same but nonentrapped drugs. The extent and pattern of alteration are primarily determined by the method of ocular application and the physicochemical nature of the liposome–drug complex. The administration of the same liposomal drug dosage form to the eye by intravitreal injection versus topical instillation, for example, would undoubtedly produce quite different drug distribution and retention patterns in ocular tissues in each case. Therefore, the implementation of strategies for the development of ophthalmic liposomal drugs must take into consideration the method intended for ocular administration. The method of intravitreal injection allows the liposomal drug to bypass the corneal barrier and therefore the formulational design does not need to address the issue of promoting corneal adhesion. On the other hand, the design of a liposomal drug for topical ocular instillation must tackle the rapid washout problem by having a formulation that favors some form of anchorage/adhesion to corneal and conjunctival tissues.

Recent advances in liposome technology have enabled pharmaceutical industries to test various liposomal drugs in human clinical trials (56). While antineoplastic liposomal drugs always appear in the forefront of clinical trials, some liposomal ophthalmic drugs have also been considered for human studies, e.g., in the treatment of dry eyes and glaucoma. The current rapid progress in the field of liposome research reinforces the likelihood of fruitful exploitations of liposomal pharmaceutical products for applications in therapeutic and prophylactic medicine.

ACKNOWLEDGMENTS

The authors wish to thank Dr. Brian Sabiston for his critical review of this manuscript and Mr. Doug Saunders for excellent graphic preparations.

REFERENCES

1. M. G. Doane, A. D. Jensen, and C. H. Dohlman, *Am. J. Ophthalmol.*, *85*, 383 (1978).
2. M. Peduzzi, A. Debbia, and A. Monzani, in *Ophthalmic Drug Delivery: Biopharmaceutical, Technological and Clinical Aspects* (M. S. Saettone, G. Bucci, and P. Speiser, eds.), Liviana Press, Padova, 1987, p. 1.
3. S. Liotet and O. Kogbe, in *Ophthalmic Drug Delivery: Biopharmaceutical, Technological and Clinical Aspects* (M. S. Saettone, G. Bucci, and P. Speiser, eds.), Liviana Press, Padova, 1987, p. 93.
4. A. M. Gachon, P. Lambin, and B. Dastugne, *Ophthalmic Res.*, *12*, 277 (1980).
5. M. M. Van Ooteghem, in *Ophthalmic Drug Delivery: Biopharmaceutical, Technological and Clinical Aspects* (M. S. Saettone, G. Bucci, and P. Speiser, eds.), Liviana Press, Padova, 1987, p. 7.
6. D. M. Maurice, in *Ophthalmic Drug Delivery: Biopharmaceutical, Technological and Clinical Aspects* (M. S. Saettone, G. Bucci, and P. Speiser, eds.), Liviana Press, Padova, 1987, p. 19.
7. J. W. Shell, in *Drug Delivery Systems: Fundamentals and Techniques* (P. Johnson and J. G. Lloyd-Jones, eds.), Ellis Horwood, New York, 1987, p. 243.
8. V. H. L. Lee and J. R. Robinson, *J. Ocular Pharmacol.*, *2*, 67 (1986).
9. B. Elkins, U. Pleyer, D. Ruckert, and B. J. Mondino, *Invest. Ophthalmol. Vis. Sci.*, *32*, 1292 (1991).
10. J. Kreuter, in *Ophthalmic Drug Delivery: Biopharmaceutical, Technological and Clinical Aspects* (M. S. Saettone, G. Bucci, and P. Speiser, eds.), Liviana Press, Padova, 1987, p. 101.
11. N. Weiner, F. Martin, and M. Riaz, *Drug Dev. Ind. Pharm.*, *15*, 1523 (1989).
12. R. A. Demel, W. S. M. Geurts van Kessel, and L. L. M. van Deenen, *Biochim. Biophys. Acta*, *266*, 26 (1972).
13. K. E. Block, in *Critical Reviews in Biochemistry*, Vol. 14 (G. D. Fasman, ed.), CRC Press, Boca Raton, 1983, p. 47.
14. M. C. Woodle and D. Papahadjopoulos, *Meth. Enzymol.*, *171*, 193 (1989).
15. P. N. Shek and R. F. Barber, *Modern Med. (Canada)*, *41*, 314, (1986).
16. F. H. Roerdink, J. Dijkstra, H. H. Spanjer, and G. L. Scherphof, *Biochem. Soc. Trans.*, *12*, 335 (1984).
17. G. Poste, C. Bucana, A. Raz, P. Bugelski, R. Kirsh, and I. J. Fidler, *Cancer Res.*, *42*, 1412 (1982).

18. J. T. Dingle, J. L. Gordon, and B. L. Hazleman, *Nature*, *271*, 372 (1978).
19. P. S. Uster, in *Drugs and the Pharmaceutical Sciences*, Vol. 42 (D. W. Osborne and A. H. Amann, eds.), Marcel Dekker, New York, 1990, p. 327.
20. P. N. Shek, R. F. Barber, and M. Jurima-Romet, *Modern Med. (Canada)*, *45*, 135 (1990).
21. P. H. Fishman, G. A. Peyman, and T. Lesar, *Invest. Ophthalmol. Vis. Sci.*, *27*, 1103 (1986).
22. C. Tremblay, M. Barza, F. Szoka, M. Lahav, and J. Baum, *Invest. Ophthalmol. Vis. Sci.*, *26*, 711 (1985).
23. G. Smolin, M. Okumoto, S. Feiler, and D. Condon, *Am. J. Ophthalmol.*, *91*, 220 (1981).
24. S. K. Dharma, P. H. Fishman, and G. A. Peyman, *Acta Ophthalmol.*, *64*, 298 (1986).
25. H. E. Schaeffer and D. L. Krohn, *Invest. Ophthalmol. Vis. Sci.*, *22*, 220 (1982).
26. K. Singh and M. Mezei, *Int. J. Pharm.*, *16*, 339 (1983).
27. R. E. Stratford, D. C. Yang, M. A. Redell, and V. H. L. Lee, *Int. J. Pharm.*, *13*, 263 (1983).
28. K. Singh and M. Mezei, *Int. J. Pharm.*, *19*, 263 (1984).
29. D. Meisner, J. Pringle, and M. Mezei, *Int. J. Pharm.*, *55*, 105 (1989).
30. R. E. Stratford, D. C. Yang, M. A. Redell, and V. H. L. Lee, *Curr. Eye Res.*, *2*, 377 (1983).
31. K. Taniguchi, K. Itakura, N. Yamazawa, K. Morisaki, S. Hayashi, and Y. Yamada, *J. Pharmocobio. Dyn.*, *11*, 39 (1988).
32. V. H. L. Lee, K. A. Takemoto, and D. S. Iimoto, *Curr. Eye Res.*, *3*, 585 (1984).
33. P. Fitzgerald, J. Hadgraft, J. Kreuter, and C. G. Wilson, *Int. J. Pharm.*, *40*, 81 (1987).
34. T. M. Allen, L. McAllister, S. Mausolf, and E. Gyorffy, *Biochim. Biophys. Acta*, *643*, 346 (1981).
35. P. I. Campbell, *Cytobios*, *37*, 21 (1983).
36. E. Yoshihara and T. Nakae, *Biochim. Biophys. Acta*, *854*, 93 (1986).
37. L. S. S. Guo, R. Radhakrishnan, and C. T. Redemann, *J. Liposome Res.*, *1*, 319 (1990).
38. H. E. Schaeffer, J. M. Breitfeller, and D. L. Krohn, *Invest. Ophthalmol. Vis. Sci.*, *23*, 530 (1982).
39. S. G. Norley, L. Huang, and B. T. Rouse, *J. Immunol.*, *136*, 681 (1986).
40. F. Bonte and R. L. Juliano, *Chem. Phys. Lipids*, *40*, 359 (1986).
41. P. N. Shek and R. F. Barber, *Biochim. Biophys. Acta*, *902*, 229 (1987).
42. C. Kirby and G. Gregoriadis, in *Liposome Technology*, Vol. 1 (G. Gregoriadis, ed.), CRC Press, Boca Raton, 1984, p. 19.
43. J. Massoulie and S. Bon, *Ann. Rev. Neurosci.*, *5*, 57 (1982).
44. G. Gregoriadis, in *Phospholipids: Biochemical, Pharmaceutical, and Analytical Considerations* (I. Hanin and G. Pepeu, eds.), Plenum Press, New York, 1990, p. 123.
45. M. C. Finkelstein and G. Weissmann, *Biochim. Biophys. Acta*, *587*, 202 (1979).
46. T. M. Allen, *Biochim. Biophys. Acta*, *640*, 385 (1981).
47. R. F. Barber and P. N. Shek, *Biochim. Biophys. Acta*, *879*, 157 (1986).

48. J. Damen, J. Regts, and G. Scherphof, *Biochim. Biophys. Acta*, *712*, 444, (1982).
49. P. I. Lelkes and H. G. Tandeter, *Biochim. Biophys. Acta*, *716*, 410 (1982).
50. R. F. Barber and P. N. Shek, *Int. J. Pharm.*, *60*, 219 (1990).
51. M. L. Shin, W. A. Paznekas, and M. M. Mayer, *J. Immunol.*, *120*, 1996 (1978).
52. E. Mayhew, M. Ito, and R. Lazo, *Exp. Cell Res.*, *171*, 95 (1987).
53. K. Taniguchi, Y. Yamamoto, K. Itakura, H. Miichi, and S. Hayashi, *J. Pharmacobio. Dyn.*, *11*, 607 (1988).
54. B. Halliwell and J. M. C. Gutteridge, in *Free Radicals in Biology and Medicine* (B. Halliwell and J. M. C. Gutteridge, eds.), Clarendon Press, Oxford, 1985, p. 67.
55. C. Vigo-Pelfrey and G. Reza, in *Membrane Lipid Oxidation* (C. Vigo-Pelfrey, ed.), CRC Press, Boca Raton, 1990, p. 1.
56. A. Klausner, *Biotechnology*, *6*, 20 (1988).

2

Microspheres as a Nasal Drug Delivery System

Peter Edman

Astra Draco AB
Lund, Sweden and
University of Uppsala
Uppsala, Sweden

Erik Björk

University of Uppsala and
Kabi Pharmacia AB
Uppsala, Sweden

INTRODUCTION

The administration of drugs to the nose in order to obtain a systemic effect is widely accepted. The intranasal administration of peptides and proteins has attracted particular interest (1). Oxytocin, vasopressin, and vasopressin analog are examples of peptides that have been administered by the nasal route for many years with an acceptable clinical effect (2,3). During recent years, several other peptides and proteins have also been tested nasally. Insulin, one of the most commonly used proteins, has been well studied in different experimental systems (4). The reason for this activity lies in the fact that diabetes mellitus is spread throughout the world and its incidence is growing. Furthermore, the biological effect of insulin is easy to monitor by recording the decrease in blood sugar, and sensitive radioimmunoassays of insulin are available at a relatively low price. These reasons taken together make it possible for almost all researchers worldwide to work with insulin as a model protein, particularly in the field of alternative delivery systems such as nasal administration.

However, recent studies have also been made on several other peptides/proteins of wide clinical interest, e.g., growth hormone (GH) (5), growth hormone–releasing factor (GHRF) (1), calcitonin (6), adrenocorticotropic hormone (ACTH) (7), and luteinizing hormone–releasing hormone (LHRH) (8).

The idea of giving proteins and peptides intranasally is logical. These drugs have a macromolecular nature and are very sensitive to both chemical and enzymatic degradation. Consequently, the only previously available route of administration for peptides was parenteral. The nasal route offers attractive advantages for both protein and peptide delivery such as rapid absorption, avoidance of hepatic first-pass metabolism, and a high degree of self-medication.

Although there are clear advantages with the nasal route, the low bioavailability of larger molecules is a drawback. A search for absorption-promoting substances or systems is therefore a natural step in order to overcome this major limitation of the nasal route.

Attempts have been made and are ongoing to use substances such as bile acids (9), sodium taurodihydrofusidate (STDHF) (10,11), bacitracin (12), etc., as absorption promotors. Other technical solutions have also been proposed, e.g., the use of polymers with different properties that become gels in situ (13).

In this chapter, we focus on solid particulate systems of different materials such as starch, cellulose, and dextran that are given nasally as a dry powder. The chapter is divided into in vitro and in vivo characteristics, biological effects, and toxicological/morphological aspects of particles given intranasally. The anatomy and physiology of the nose will not be discussed in this chapter but the interested reader can go to the references (14,15). A short description of the nasal mucosa follows.

The nasal epithelial membrane consists of squamous, ciliated secretory, and olfactory epithelia lined with microvilli. The epithelium consists of four major cell types. These are nonciliated columnar cells, goblet cells, ciliated cells, and basal cells. The main function of the ciliated cells is to transport mucus toward the pharynx by the movement of the cilia. The goblet cells produce the mucous secretion together with the nasal glands. The submucosa contains a network of fenestrated capillaries that provide the nasal mucosa with a good supply of blood. The total surface area of both nasal cavities is about 160 cm^2.

NASAL POWDER SYSTEMS

Particles of Different Materials

In a pioneer work by Nagai et al. (16), freeze-dried powders of lactose, crystalline cellulose, hydroxypropylcellulose, and carbomer 934, all containing insulin, were prepared. The powders were sieved through a 100- to

200-mesh sieve. When crystalline insulin (5 IU/kg body weight) was given intranasally to beagle dogs, the maximal decrease in blood glucose, approximately 60%, was seen 1–1.5 hr after administration. If the insulin dose was given together with 100 μl purified water, the effect on the blood glucose level was drastically decreased and only a 20% change in glucose level was noted.

Addition of microcrystalline cellulose enhanced the absorption, and 30 min postdosing the glucose level had decreased to 49.1%. The minimum level was 32% of normal, which was reached 1.5 hr after administration. Both hydroxypropylcellulose and carbomer produced a similar enhancing effect on insulin absorption. Lactose was least effective in promoting insulin absorption.

The best effect was obtained when insulin and carbomer 934 were freeze-dried together and the resulting powder was further mixed with cellulose of 100–200 mesh. A rapid effect on the blood glucose level, which decreased to 60% after 30 min and to a minimum level of 20% after 1 hr, was seen with this system. When the amount of carbomer was increased, the duration of the effect on the blood glucose was extended (Figure 1). A minimum glucose level of 24% was reached after 3 hr.

The authors concluded that insulin solutions persist for only a short while in the nose because of rapid clearance. Powders, on the other hand, are in close contact with the mucosal surface, so that the concentration of dissolved insulin on the surface of the mucosa is high, favoring absorption. It was explained that the lack of effect of lactose was due to its solubility in

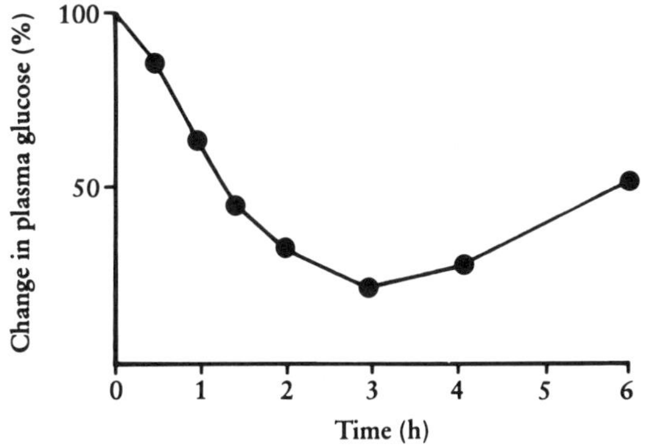

Figure 1 Plasma glucose level after intranasal administration of insulin (10 mg) freeze-dried with carbomer 934 (50 mg) and dry-mixed with crystalline cellulose (40 mg). Total insulin dose given was 3 IU/kg body weight. (Redrawn from Ref. 16 with permission from the copyright owner.)

mucus and short contact time with the nasal mucosa. Carbomer, hydroxypropylcellulose, and cellulose form viscous gels with mucus and therefore stay considerably longer in the nose, creating a favorable situation for insulin absorption.

Starch and Dextran Microspheres

Degradable starch microspheres (DSM) with a mean particle diameter of 45 μm were administered intranasally for the first time in 1987 (17). The spheres consisted of epichlorohydrine-crosslinked starch produced by a traditional emulsion polymerization technique. The spheres were originally developed as a chemoembolization agent for use in cancer treatment to temporarily arrest the blood flow in the liver artery, giving increased local concentrations of coadministered drugs (18). The spheres are degradable by α-amylase. DSM administered nasally to healthy volunteers shows an increased persistance and reduced clearance in the nose (17).

The impact of DSM on drug or peptide absorption across the nasal epithelium has been thoroughly studied with several peptides and drugs during recent years. The first study dealing with peptides/proteins was published in 1988 (19). In that study insulin was administered via DSM to rats. The results indicated that DSM acts as an absorption-promoting system, providing bioavailability of approximately 30% (see Table 1). Maximal peak concentrations of insulin were seen within 8–9 min after administration and a concomitant decrease in blood sugar occurred 30 min later.

In a follow-up dose optimization study of DSM in the rat (20), it was shown that the optimal dose was 3–7 mg DSM per kg bodyweight. However, it is evident that the dose will vary among species.

The degree of crosslinking in the spheres has no significant influence on the in vivo performance of the DSM, though a difference could be noted in vitro with respect to release of insulin. This is probably because the spheres

Table 1 Effect of Degradable Starch Microspheres (5 mg/kg body weight) on Insulin Absorption Across Nasal Mucosa

	Serum insulin			Blood glucose	
Dose (IU/kg)	Peak level (μU/ml)	Peak time (min)	BA (%)	Maximal decrease (%)	Peak time (min)
0.75	146 ± 64	8.0 ± 1.5	30	40 ± 13	36 ± 12
1.70	167 ± 46	7.3 ± 2.6	33	64 ± 8	38 ± 8

Note: BA = bioavailability. All figures are expressed as mean ±SD (n = 6–8).
Source: Adapted from Ref. 19.

are overall very similar, so that they behave in a similar manner in vivo. An interesting finding was seen when comparing starch powders with different molecular weights (MW) of 25,000 and 11,000, respectively (Figure 2). Starch with a MW of 11,000 is water soluble and no effect on the blood glucose level was seen when insulin was given with this type of starch in a dry form. On the other hand, the same amount of insoluble starch with a MW of 25,000 given with equivalent doses of insulin produced a rapid reduction in blood glucose and a kinetic profile showing an effect equivalent to that seen with DSM.

These results with starch powders of different molecular weights are predictable when comparing data from Nagai (16). He showed that crystalline cellulose and its propylhydroxy derivate increased insulin absorption effec-

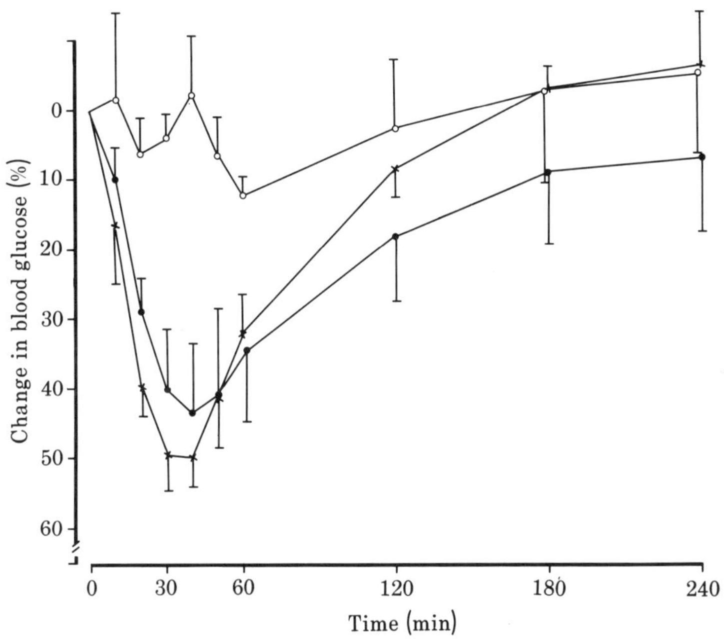

Figure 2 Change in plasma glucose after nasal administration of degradable starch microspheres (DSM) and starch powder together with insulin. The insulin dose was 1 IU/kg body weight. The amount of spheres and starch administered was 4 mg/kg body weight. The following preparations were used: (x) degradable starch microspheres (DSM); (○) water-soluble starch of MW 11,000; and (●) water-insoluble starch of MW 25,000. The points represent the means and standard deviations of the means from six animals. (Reproduced from Ref. 20 with permission.)

tively. However, lactose, which is water-soluble, had a considerably lower effect on insulin absorption than the water-insoluble cellulose.

Dextran spheres crosslinked with epichlorohydrine (Sephadex) were recently tested as a nasal delivery system for insulin in the rat (21). The in vitro release rate of insulin from Sephadex particles is comparable to that obtained with starch microspheres; a more or less instantaneous release. Consequently, the in vivo behavior of Sephadex and insulin should be similar to that seen with DSM and insulin. A rapid decrease in plasma glucose was seen when giving an insulin dose of 1 IU/kg body weight together with Sephadex in dry form. The maximal effect was a 25% decrease of plasma glucose seen 40–60 min after administration. The efficacy of insulin in Sephadex particles in lowering blood glucose is inferior to that of insulin in DSM at equivalent doses. The reasons for this may involve discrepancies in sphere size and the location of the insulin either inside or at the surface of the spheres. The latter parameter is dependent on the cutoff limit of the spheres. DSM is much more porous whereas Sephadex is considerably denser. Use of Sephadex spheres with an ion exchange group (DEAE) containing bound insulin produced no significant biological effect, probably due to the low ionic strength in mucus, which is insufficient to displace insulin from the DEAE groups. A general conclusion that can be drawn from these in vivo studies with DSM, Sephadex, and the studies by Nagai (16) is that there are certain criteria to be fulfilled by the particles or spheres before they can act as absorption enhancers. If the particles are water-absorptive, water-insoluble, and administered in powder form to the nose they will probably promote the absorption of drugs through the nasal epithelia.

That the spheres promote drug absorption has been shown in many studies by Illum and coworkers. They have published several papers on different drugs such as gentamicin (22), insulin (23), growth hormone (24), given nasally together with DSM to rabbits and sheep. The combination of DSM and lysophosphatidylcholine causes a further increase in the enhancing effect of DSM (25).

The mechanism behind the enhancing effect of drug absorption by insoluble microspheres is not clear. A plausible hypothesis launched by Edman and Björk (19,20) is that a dehydration of the mucus and the epithelial cells leads to a temporary widening of tight junctions in the epithelium.

Preliminary studies on cultured Caco-2 epithelial cells in vitro have shown that after exposure to dry starch microspheres for 3 min, the cell layer has a significant number of opened tight junctions (Figure 3). This is followed by an increased permeability to drugs, e.g., insulin. The transepithelial resistance is also decreased. However, the effects seen are reversible 180 min after exposure to the spheres all deviations are nearly normalized. The effects seen can only be induced with dry spheres that take up water and swell. Preswollen

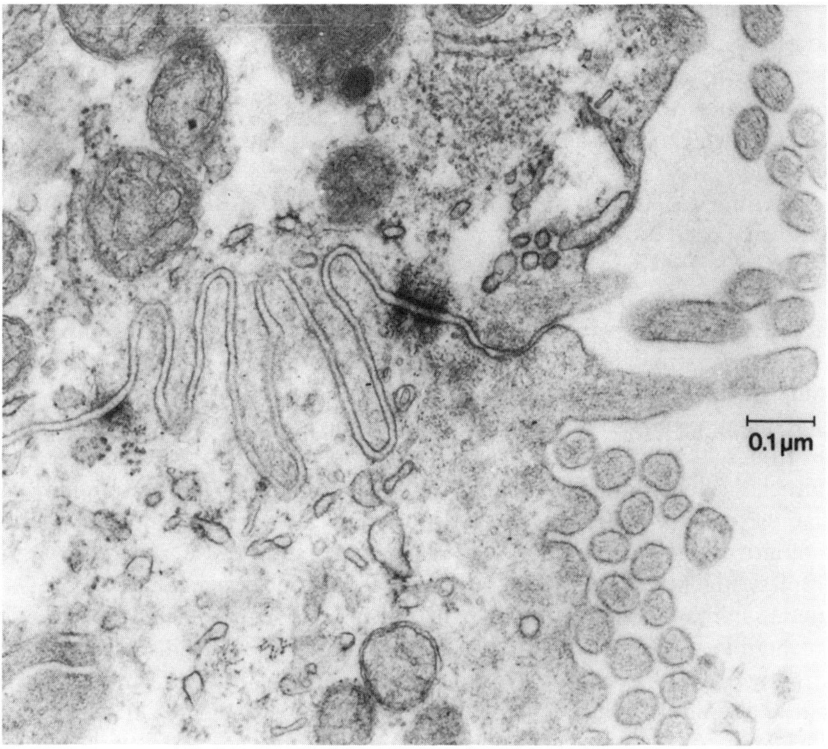

Figure 3 Transmission electron microscopy of an open tight junction in the CaCo-2 cell line model after exposure to degradable starch microspheres (DSM) for 3 min (×63,000).

spheres have no effect on this system. These results support and strengthen the hypothesis that the spheres have to be water-absorptive and water-insoluble to induce the desirable effect on drug absorption in the nose (26).

IRRITATIVE AND TOXICOLOGICAL CONSIDERATIONS

Use of enhancers is often linked to adverse effects on the nasal epithelia. When considering that the aim of enhancers is to increase drug absorption, it is evident that one must disturb the epithelial barrier in order to promote an increased absorption rate. Consequently, one has the classical ratio of risk and benefit to consider.

Surface-active enhancers such as bile acids, saponins, laureth-9, etc., have all been shown to affect the mucosa and the cilia (27–29).

Microspheres of starch have been tested in rabbits for 8 weeks (30). Two doses of spheres were used; 10 and 20 mg given twice daily nasally. At different times up to 8 weeks the rabbits were killed and the nasal cavities were microscopically examined by using light microscopy and scanning electron microscopy. The nasal cavities showed no inflammatory reactions or ulcerations.

In some rabbits that were treated with DSM for 8 weeks, the nasal septum wall had a mild focal hyperplasia of the epithelium. An increased number of goblet cells and mild hyperplasia of the columnar epithelium was seen. A dose–response relationship was noted. The hyperplasia was more developed in the animals given the highest dose of DSM. However, this finding was only seen in animals treated for 8 weeks. No changes in the septum wall could be observed in animals given DSM for 2 or 4 weeks. The cilia were intact in all groups. The distribution pattern of ciliated cells was similar for all groups of treated animals. The reason for the light hyperplasia of the epithelial cells on the nasal septum wall is not clear. An explanation may be that the dose of spheres hits the septum wall at the same spot after each administration, which induces a mechanical irritation causing secondary hyperplasia. Dehydration of the mucosa could be the explanation for the increased number of goblet cells. DSM given as a dry powder to the nose is not harmless, which supports the hypothesis that all systems that interact with the epithelial barrier to increase the permeability rate of drugs will induce alterations in the mucosa, temporarily or permanently. It seems reasonable to conclude that the ideal enhancer has not yet been found, though microspheres are quite promising in this respect. In a recent study by Björk et al. (31), it was shown in healthy volunteers that insufflation of DSM 10 mg twice daily for 7 days does not induce irritation and is well accepted. The mucociliary clearance was not affected, as demonstrated with the saccharin test. This "clinical" trial indicates that DSM may have a future as a nasal delivery system for peptides and drugs.

SUMMARY

The use of dry spheres of different material, which have the ability to take up water and swell, clearly has interesting possibilities for enhancing drug absorption across the nasal mucosa. However, it has to be made clear that all steps taken to promote drug absorption must be linked to the risk–benefit ratio. Disturbance of the epithelial barrier that leads to an increased permeability of drug substances might also induce morphological and cellular alterations and toxic effects in the mucosa. This topic has to be carefully considered, especially if the treatment is to be lifelong.

Presently, there are no ideal enhancers on the market, but several concepts are under evaluation. Microspheres and naturally occurring phospholipids are examples of enhancers that have in vivo behavior that appears to be acceptable clinically. The rapid promotion of insulin absorption by starch microspheres, almost simulating the effects of intravenous injection, is particularly exciting.

Even though there are indications that microspheres might be biologically acceptable when given nasally, there are still extended toxicological studies to be undertaken before they can be used as a general pharmaceutical technique for increased drug absorption.

REFERENCES

1. K. S. E. Su, in *Peptide and Protein Drug Delivery* (Vincent H. L. Lee, ed.), Marcel Dekker, New York, 1991, p. 595.
2. G. Hendricks and R. A. Gabel, *Am. J. Obstet. Gynecol.*, *79*, 789 (1960).
3. A. S. Harris, I. M. Nilsson, Z. G. Wagner, and U. Alkner, *J. Pharm. Sci.*, *75*, 1085 (1986).
4. Y. Hirata, T. Yokosuka, T. Casahara, M. Kikuchi, and K. Ooi, *Proceedings of a Symposium on Proinsulin, Insulin and c-Peptide.* Excerpta Medica, Amsterdam, 1978, p. 319.
5. A. L. Daugherty, H. D. Liggitt, J. G. McCabe, J. A. Moore, and J. S. Patton, *Int. J. Pharm.*, *45*, 197 (1988).
6. A. E. Pontiroli, M. Alberetto, A. Calderara, E. Pajetta, and G. Pozza, *Eur. J. Clin. Pharmacol.*, *37*, 427 (1989).
7. C. R. DeBold, W. R. Sheldon, G. S. DeCherney, R. V. Jackson, W. E. Nicholson, D. P. Island, and D. N. Orth, *J. Clin. Endocrinol. Metab.*, *60*, 836 (1985).
8. G. Fink, G. Geunser, P. Liedholm, J. Thorell, and J. Mulder, *J. Endocrinol.*, *63*, 351 (1974).
9. G. S. M. J. E. Duchateau, J. Zuidema, and F. W. H. M. Merkus, *Int. J. Pharm.*, *31*, 193 (1986).
10. J. P. Longenecker, A. C. Moses, J. S. Flier, R. D. Silver, M. C. Carey, and E. J. Dubovi, *J. Pharm. Sci.*, *76*, 351 (1987).
11. M. J. M. Deurloo, W. A. J. J. Hermens, S. G. Romeyn, J. C. Verhoef, and W. H. M. Merkus, *Pharm. Res.*, *6*, 853 (1989).
12. S. C. Raehs, J. Sandow, K. Wirth, and H. P. Merkle, *Pharm. Res.*, *5*, 689 (1988).
13. K. Morimoto, H. Yamaguchi, Y. Iwakura, K. Morisaka, Y. Ohashi, and Y. Nakai, *Pharm. Res.*, *8*, 471 (1991).
14. D. F. Proctor and I. J. Andersson (eds.), *The Nose: Upper Airway Physiology and the Atmospheric Environment.* Elsevier Press, Amsterdam, 1982.
15. Y. E. Chen (ed.), *Transnasal Systemic Medications: Fundamental Developmental Concepts and Biomedical Assessments*, Elsevier Press, Amsterdam, 1985.
16. T. Nagai, Y. Nishimoto, J. Nambu, Y. Suzuki, and K. Sekine, *J. Controlled Release*, *1*, 15 (1984).

17. L. Illum, H. Jørgensen, H. Bisgaard, O. Krogsgaard, and N. Rossing, *Int. J. Pharm.*, *39*, 189 (1987).
18. B. Lindberg, K. Lote, and H. Teder, in *Microspheres and Drug Therapy: Pharmaceutical, Immunological and Medical Aspects* (S. S. Davis, L. Illum, J. G. McVie, and E. Tomlinson, eds.), Elsevier Press, Amsterdam, 1984, p. 153.
19. E. Björk and P. Edman, *Int. J. Pharm.*, *47*, 233 (1988).
20. E. Björk and P. Edman, *Int. J. Pharm.*, *62*, 187 (1990).
21. L. Rydén and P. Edman, *Int. J. Pharm.*, *83*, 1 (1992).
22. L. Illum, N. Farraj, H. Critchley, and S. S. Davis, *Int. J. Pharm.*, *46*, 261 (1988).
23. N. F. Farraj, B. R. Johansen, S. S. Davis, and L. Illum, *J. Controlled Release*, *13*, 253 (1990).
24. L. Illum, N. F. Farraj, S. S. Davis, B. R. Johansen, and D. T. O. Hagan, *Int. J. Pharm.*, *63*, 209 (1990).
25. L. Illum, N. Farraj, H. Critchley, and S. S. Davis, *Int. J. Pharm.*, *46*, 261 (1988).
26. E. Björk, U. Isaksson, P. Artursson, and P. Edman, manuscript submitted.
27. W. A. J. J. Hermens, P. M. Hooymans, J. C. Verhoef, and W. H. M. Merkus, *Pharm. Res.*, 7, 144 (1990).
28. B. J. Aungst and N. J. Rogers, *Pharm. Res.*, *5*, 305 (1988).
29. S. G. Chandler, L. Illum, and N. W. Thomas, *Int. J. Pharm.*, *70*, 19 (1991).
30. E. Björk, S. Bjurström, and P. Edman, *Int. J. Pharm.*, *75*, 73 (1991).
31. E. Björk, K. Holmberg, and P. Edman, *Rhinology*, (in press).

3

Liposomes as a Pulmonary Drug Delivery System

Dale Meisner

Merck Frosst Centre for Therapeutic Research
Kirkland, Quebec, Canada

INTRODUCTION

Traditional forms of inhalation therapy date back to the earliest records of ancient cultures. Although the delivery of therapeutic compounds directly to the respiratory tract is more sophisticated today, the advantages of inhalation therapy have essentially remained the same (1). First, the onset of action is very rapid. An oral dose of bronchodilator may take 2–3 hr to be fully effective while an inhaled dose usually takes a maximum of 15–30 min. Second, relatively small doses are required for effective therapy. A few hundred micrograms of inhaled bronchodilator is as effective as an oral dose of 5–10 mg. This also reduces exposure of drug to the systemic circulation and potentially minimizes adverse effects. Lower dosage regimens may provide considerable cost savings especially with expensive therapeutic agents.

In the treatment of pulmonary disorders and, in some cases, for systemic action, inhalation is the preferred route of administration. Delivering small doses of active ingredient directly to the lung effectively localizes the drug there by maximizing therapeutic effect while minimizing unwanted side effects. Despite these advantages and the widespread use of therapeutic aerosols, there are several shortcomings associated with drug delivery to the respiratory tract. Although the onset of action is very rapid, the duration is often short-lived as the drug can be quickly removed from the lung through various clearance mechanisms (2–4). The relationship between the functional anatomy of the respiratory tract and aerosol kinetics makes drug delivery to the lung highly inefficient. Typically only 10% of inhaled dose reaches the lower airways (1). Furthermore, pulmonary tissue has a highly developed cytochrome P-450 system capable of inactivating inhaled compounds (5). Regardless of the type of aerosol device employed, i.e., metered dose inhaler (MDI), nebulizer, or insufflation device, most patients will require dosing every 6–8 hr and in some instances more frequently.

Improving drug delivery to the pulmonary system has become an area of increasing interest among several disciplines. There has been extensive effort to define the factors that influence the deposition of aerosols within the respiratory tract, clearance mechanisms from the lung, circulation in the airways, and absorption and metabolism of compounds by the lung (for detailed discussions, see Refs. 3–9). With this information, strategies to improve drug delivery to the respiratory system have developed. Particulate carriers such as liposomes have many attractive features as pulmonary drug delivery systems particularly with respect to controlled delivery. The objective of this chapter is to review the problems and constraints of controlled drug delivery to the respiratory system and take a comprehensive look at the application of particulate systems as a means of improving inhalation therapy. It will deal briefly with the relevant physiology and pharmacology of the airways as well as the physical behavior of particle deposition in order to provide a basis to evaluate and assess the potential role of particulate carriers, particularly liposomes, in pulmonary drug delivery.

CONSTRAINTS OF PULMONARY DRUG DELIVERY

Structure and Function of the Respiratory Tract

The respiratory tract can be divided into upper and lower airways with the line of division being the junction of the larynx and trachea (10,11). As such the upper airways or nasopharyngeal region consists of the nose, mouth, larynx, and pharynx. Below the contours of the nasopharyngeal region, the lower airways resemble a series of tubes undergoing regular dichotomous branching (12) (Figure 1). Successive branching from the trachea to the alveoli reduces the diameter of the tubes but markedly increases the surface area of the airways, a design intrinsic to its role in gas exchange. The lower airways can be divided into three physiological zones: conducting, transitional, and respiratory zone (12). The conducting zone consists of the larger tubes responsible for the bulk movement of air and blood. In the central airways, airflow is rapid and turbulent (13) and, having no alveoli, no gas exchange occurs. The transitional zone plays only a limited role in gas exchange as

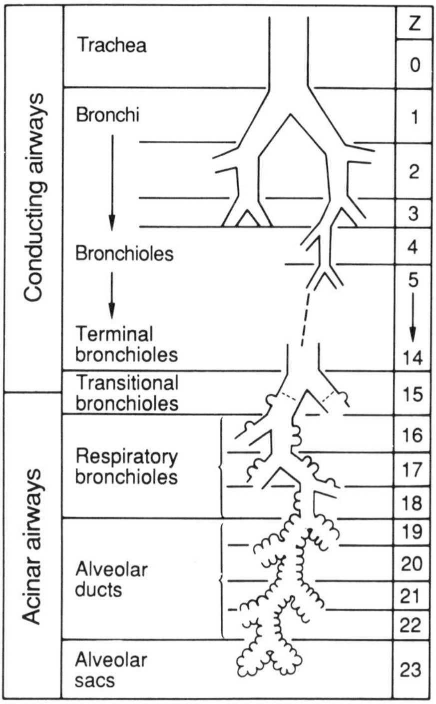

Figure 1 Organization of the airways in the human lung. (Reproduced from Ref. 12.)

compared as the respiratory zone in which airflow is smooth and laminar in nature (13).

The characteristic D shape of the trachea is maintained by longitudinally stacked cartilage supported by smooth muscles fibers. Interspersed throughout the pseudostratified epithelial layer of the trachea and main bronchi are a number of cell types including ciliated, basal, goblet, and Kulchitsky cells that are supported by a fully developed basement membrane. A large number of mucus and serum-producing glands are located in the submucosa. The human lung consists of 5 lobes and 10 bronchopulmonary segments. Distal to each segment are lung lobules, composed of three to five terminal bronchioles. Each bronchiole supplies the smallest structural unit of the lung, the acinus, which consists of alveolar ducts, alveolar sacs, and alveoli. The structural morphology of the main bronchi is similar to that of the trachea except smooth muscle is interspersed between cartilage segments. At the level of small bronchi and bronchioles the amount and organization of cartilage diminishes as the number of bronchial bifurcations increases. In the smaller bronchioles ciliated squamous epithelium replaces ciliated columnar epithelium and another secretory cell, the Clara cell, replaces goblet cells.

The acinus represents a marked change in morphology. The primary cell of the epithelium is the type I pneumocyte. Type I pneumocytes are flattened thin cells which extend to cover about 90% of the entire alveolar surface. Type II pneumocytes are more numerous but have a smaller total volume. These cells are responsible for storage and secretion of lung surfactant that continuously line the alveolar ducts and sacs to maintain surface tension at the air–water interface. Less prevalent cell types include type III pneumocyte (alveolar brush cell) and alveolar macrophages. No glandular structures, goblet or ciliated cells are found at this level although sphincter-like muscles line the alveolar ducts. The cells of the alveolar epithelium are closely jointed to form a continuous lining for alveoli and is rather impermeable to all but small molecules. The endothelial cells of the alveolar capillaries are in many respects similar to the type I pneumocyte. Both are extremely flat and extended, have nuclei close to septal corners, and are tightly apposed to form tight junctions. The alveolar–blood barrier in its simplest form consists of a single epithelial cell, a basement membrane, and a single endothelial cell. The distance between air and blood is less than 1 μm. While this morphological arrangement readily facilitates gas exchange, it can still represent a major barrier to large molecules.

Inhalation Therapy

Principles Governing Aerosol Therapy

Drugs for inhalation therapy are administered in aerosol form. An aerosol is defined as a suspension of liquid or solid in the form of fine particles dispersed in a gas. An obvious prerequisite for therapeutic efficacy is the ability of the aerosolized drug to reach the peripheral airways. Herein lies the fundamental problem of inhalation therapy as the anatomy and physiology of the respiratory tract has evolved to prevent the entry of particulate matter. The extent of aerosol deposition in the lungs is difficult to predict because particles can come in vastly different shapes and sizes. To account for shape and size, aerosol particles are often described by their aerodynamic diameter (see Refs. 6 and 14 for detailed discussions). This diameter is related to particle size and density and is equivalent to the diameter of a perfect sphere of unit density (1 g/cm^3) with the same settling velocity in still air. Two numbers are used to describe the wide range of particles within an aerosol cloud. The mass median aerodynamic diameter (MMAD), derived from a cumulative plot of probability, states that half the aerosol mass is contained in particles smaller than the MMAD and half the mass in particles larger. The geometric standard deviation (GSD) is defined as the ratio of the diameter for 84.1% cumulative probability to that of a 50% probability and is a measure of aerosol polydispersity. Aerosols with a GSD of greater than 1.22 are generally termed *heterodisperse*.

The deposition of inhaled aerosols is governed by the physical principles that apply to particles of different sizes (see Refs. 6 and 14 for detailed discussions). The three main processes of aerosol particle behavior in the respiratory tract are inertial impaction, gravitational sedimentation, and Brownian diffusion. Generally particles greater than 5 μm are unable to follow the abrupt changes in the direction of the air stream during inspiration and impact in the upper airways of the respiratory tract. Smaller particles in the 1- to 5-μm range settle in the lower airways and may reach the alveoli by gravitational sedimentation. Submicrometer particles behave like a gas and can deposit in the alveoli but are also readily exhaled. Although particle size may be the most important determinant in aerosol deposition, the efficiency of deposition can be influenced by a number of factors. Solutions containing salts or chemical substances that exert an osmotic effect can affect the stability and behavior of droplets once aerosolized. Hypertonic particles can attract water molecules, grow in size, and deposit more proximal while particles of low tonicity may evaporate and shrink. Droplets can also collide and coalesce. Patient factors can also influence the efficiency of particle deposition in the lung. Slow deep breathing with end-inspiratory hold favors distal deposition. Compromised or abnormal lung function in disease states such as obstructive lung disease and asthma can restrict the dose of aerosol reaching the peripheral airways.

Drug Absorption in the Lung

Deposition of drug in the respiratory tract does not guarantee successful pharmacotherapy. An important determinant of drug fate following inhalation relates to the permeability characteristics of the lung tissue and its effect on the distribution and absorption of molecules. The thin alveolar–blood barrier, high rate of tissue perfusion, and enormous surface area are physiological properties of the lung that help promote rapid absorption of solutes. The alveolar–capillary barrier generally limits penetration of large molecules such as proteins (15); however, systemic absorption can occur rapidly especially for small lipophilic molecules (16). The rate of absorption can vary widely depending on the physicochemical properties of the compounds and on whether absorption occurs by diffusion or carrier-mediated processes (16–18). Lipophilic compounds appear to be absorbed primarily by membrane diffusion and are transported more rapidly than hydrophilic compounds. The absorption of hydrophilic compounds appears to occur through extracellular junctions and has been shown to be inversely related to their molecular weight (19). Carrier-mediated processes have been identified for several substances including sodium chromoglycate (20) and cycloleucine (21).

Pulmonary Metabolism and Elimination

Pulmonary tissue has a highly developed P450 system capable of inactivating inhaled compounds (4). The pulmonary cytochrome P-450 system is similar to that of the liver and a number of phase I–type oxidations and phase II–type conjugations such as glucuronidation and sulfation have been documented (5). The metabolism of endogenous substances such as prostaglandins and bradykinin occurs within endothelial cells of the lung (22,23) while most of the xenobiotic-metabolizing enzyme systems are located in Clara cells and type II pneumocytes (24,25). Alveolar macrophages have also been shown to have oxidase, *N*-acetylation, reductase, and esterase activities (26–29).

Inhaled drugs can also be cleared from the lung by breathing and the mucociliary elevator (3,30,31). The ciliated epithelium of the tracheobronchial tree is covered by mucoid fluid secreted by bronchial glands and goblet cells. Inhaled particles lodged in mucus are transported to the oral cavity through continuous propulsion of mucus up the respiratory tree and eventually swallowed (32). In nonciliated regions, the viscoelastic properties of mucus allow it to capture inhaled particles and transport them to more central areas of the lung. In the alveolar regions, particles can be actively taken up by macrophages that are responsible for "cleaning" alveolar surfaces. Alveolar macrophages eventually migrate to ciliated regions and are eliminated by the mucociliary elevator. Generally parasympathomimetic agents are known to stimulate mucus secretions while an inhibitory effect is produced by parasympatholytic substances (3). β-Agonists appear to stimulate mucociliary clearance in a dose-dependent manner while anticholinergic drugs have been reported to increase, decrease, or produce no effect depending on the drug and the dose (30). In terms of controlled drug delivery, the importance of mucociliary clearance mechanisms as a means of drug removal may be dependent on the formulation components as well as the active species.

Controlled Drug Delivery to the Lung

Controlled drug delivery to the respiratory tract, whether for local or systemic activity, provides an interesting challenge. The anatomical design and physiological processes of the respiratory tract represent major barriers to efficient drug delivery and effective prolonged pharmacotherapy. Regardless of which type of aerosol delivery device is employed, generally only about 10% of the dose reaches the lung (1). A large percentage of the dose deposits in the oropharyngeal region and is eventually swallowed. The fraction of the dose that actually reaches the lung is subjected to a variety of efficient clearance mechanisms. An obvious strategy to improve aerosol therapy is to increase the fraction of the dose reaching the lungs. To do so one must optimize the formulation and then ensure its proper utilization. To provide a prolonged effect, it is necessary to prevent rapid clearance and metabolism of the compound by the lung. Particulate systems such as liposomes represent a delivery system that can potentially fulfill these objectives. The evaluation of particulate systems for inhalation therapy must critically examine the pharmaceutical aspects of the dosage form, define and establish their clinical application, and verify an improvement over existing treatments.

THE USE OF LIPOSOMES IN PULMONARY DRUG DELIVERY

Composition and Physical Description

Liposomes, as originally described by Bangham (33), are microscopic vesicles composed of phospholipid bilayers surrounding aqueous compartments. The driving force for bilayer assembly is the amphiphilic nature of phospholipid molecules. Phospholipids typically consist of a hydrophilic headgroup attached to two hydrophobic fatty acid chains. When suspended in an excess of aqueous solution, phospholipid molecules orientate themselves in ordered bilayers so that the polar heads are hydrated and hydrophobic tails are excluded from the aqueous environment (Figure 2). Although suspended phospholipids may also assume other geometries such as micelles and tubular aggregates in hexagonal phases, this can be controlled by several factors including lipid composition and method of preparation (see Refs. 34–36 for detailed discussions).

Liposomes can be prepared with a variety of natural and synthetic phospholipids. Most phospholipids and combinations thereof, except phosphatidylethanolamine alone, will form closed vesicles (35). Phosphatidylcholine (lecithin) with or without cholesterol is often used. Cholesterol is added to increase bilayer rigidity and to modify the permeability of liposomes. Charged amphiphiles such as stearylamine and dicetylphosphate are added to alter the net surface charge of vesicles. Entrapment of compounds is highly influenced by their physicochemical properties. Generally hydrophobic molecules are incorporated into the lipid bilayers whereas hydrophilic compounds are entrapped in the internal aqueous volume (37).

Different types of liposomes can be prepared and are classified by size and structure (Figure 2). Multilamellar vesicles (MLVs) consist of numerous concentric bilayers separated by aqueous spaces and range up to 15 μm in diameter. Vesicles consisting of a single bilayer encompassing a central aqueous compartment are referred to as small unilamellar vesicles (SUVs), which range

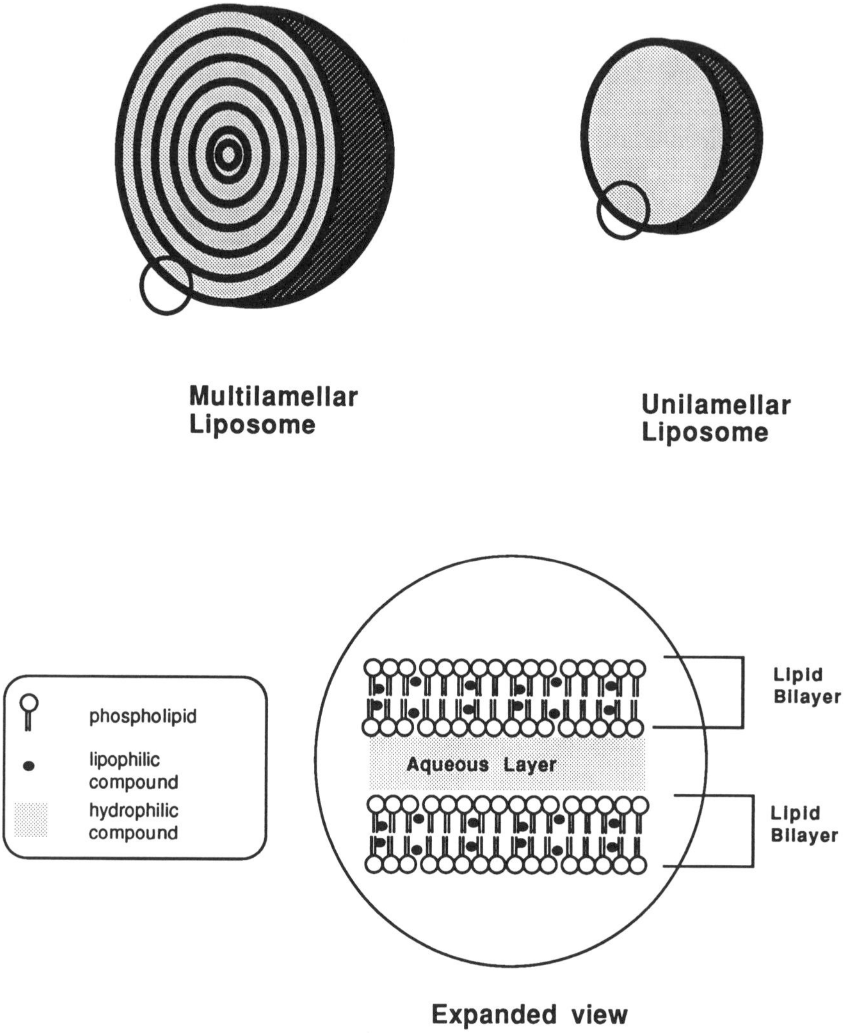

Figure 2 The structure of multilamellar and unilamellar vesicles showing the organization of phospholipid bilayers and the encapsulation of lipophilic and hydrophilic compounds. (Diagram courtesy of W. C. Foong.)

up to 100 nm in diameter and large unilamellar vesicles (LUVs) ranging from 100 to 500 nm in diameter.

Liposomes have been extensively studied as drug carriers. Potential applications in cancer chemotherapy, enzyme therapy, immunomodulation, antimicrobial therapy, metal detoxification, diagnostics, hemoglobin substitution, and topical therapy were recently summarized (38). The cumulative work in this volume has highlighted the potential advantages of liposomes as drug delivery systems. Liposomal encapsulation can alter the pharmacokinetic fate of the entrapped species and thereby increase drug efficacy and reduce toxicity. Protection against metabolism and sustained-release properties are additional benefits advocated. Biodegradability and compatibility are also attractive features of lipid vesicles. Liposomes have been successfully administered by several routes of administration including inhalation to the respiratory tract.

Surfactant Replacement Therapy

Pulmonary Surfactant

One of the more intensely pursued areas involving liposomes as a pulmonary drug delivery system has exploited the composition of lipid vesicles rather than the carrier ability of vesicles. Respiratory distress syndrome is a disease characterized by a deficiency of lung surfactant. From a historical point of view, Clements and colleagues first described the surface tension properties of lung fluid extracts and suggested that dipalmitoylphosphatidylcholine was the main surface-active component (39). Later it was shown that infants suffering from hyaline membrane disease had an impaired ability to lower surface tension in their lung (40). These observations stimulated investigative efforts into the chemical characterization of lung surfactant and its physiological role at the air–water interface. Potential strategies for the use of exogenous phospholipids, particularly in the form of liposomes, in surfactant replacement therapy evolved as the pulmonary surfactant system became better understood.

Several recent reviews provide complete descriptions of the composition and physiological role of lung surfactant (41–44). Briefly, extracellular surfactant consists of lipids (80% by weight), several proteins (10%), and small amounts of carbohydrates. The majority of the surfactant lipids is phospholipid (80–90%) whereas cholesterol constitutes the largest amount of neutral lipid. The major phospholipid components are PC (60–70%), phosphatidylglycerol (PG) (5–10%), and phosphatidylethanolamine (PE) (5–10%), with small quantities of phosphatidylserine (PS) and phosphatidylinositol (PI). Although these phospholipids are not unique to lung surfactant, they contain an unusually high percentage of PG and saturated PC, in particular DPPC.

The protein fraction consists of specific surfactant-associated proteins – SP-A, SP-B and SP-C – essential for secretion, rapid film formation at the air–liquid interface, and reuptake of lung surfactant.

Surfactant lipids are synthesized in alveolar type II epithelial cells and transported intracellularly to lamellar bodies where they are packaged and stored along with cholesterol and SP-A, a regulator of surfactant secretion. Surfactant is secreted by exocytosis of lamellar bodies into alveolar air spaces where SP-A, SP-B, and calcium ions promote the lipids to form densely packed leaflets known as tubular myelin (45). Tubular myelin represents an intermediate form of surfactant for alveolar transport and is believed to be the form that adsorbs rapidly at the air–water interface. Both SP-B and SP-C play a role in enhancing the adsorption and spreading of the surfactant. During the process of adsorption, the monolayer is somewhat "purified" by removal of molecules other than DPPC. Ninety to ninety-five percent of secreted surfactant is recycled, reprocessed, and repackaged for secretion.

In the normal lung, the surfactant system provides a stable, low surface tension within the lung to prevent alveolar collapse at low transpulmonary pressures. Maintaining the functional integrity of the monolayer formed at the air–liquid interface is dependent on the balance between surfactant release and clearance. The rate of phospholipid turnover is rapid. In adult rabbits, PC turnover in the alveolar surfactant pool has been calculated at 10–30%/hr (46,47). Synthesis and release of surfactant are influenced by deep lung inflation, adrenergic stimuli, hormonal effects, and pH of the parenchyma (43,48,49), while multiple pathways may be responsible for the removal of alveolar phospholipids (43,50). These include reuptake by type II epithelial cells, degradation in alveolar macrophages following phagocytosis, extracellular enzymatic degradation in the alveoli, transport in the airways via the mucociliary elevator, and clearance via lymph and blood. Although the extent to which each mechanism contributes to surfactant clearance has yet to be ascertained, surfactant lipids are cleared from the alveolar air space more rapidly than from the whole lung, indicating that these two processes may be regulated by different factors (51). Some studies have reported that synthetic lipids are taken up at the same rate as PC from natural surfactant (52,53) while others suggest that there is a specificity for natural surfactant (54). Surfactant proteins have been shown to stimulate phospholipid uptake by type II cells *in vitro* (55,56), and *in vivo*, following intratracheal instillation (57), and thus have been implemented as an important regulator of phospholipid/surfactant uptake. Clearance is also under physiological control. Increased lung ventilation was reported to enhance clearance possibly through β-adrenergic stimulation (48). Present evidence indicates that exogenous phospholipids associate rapidly with alveolar and enter the intracellular lipid pool without causing any apparent untoward effects. The report of Oyarzun et al. (48) that administration of exogenous phospholipids did not adversely affect arterial blood gases, pH, lung appearance, lavage fluid lactase dehydrogenase activity, protein component, and cell consistency supports the biocompatibility and safety of phospholipid administration.

Studies with Natural and Artificial Substitutes in Respiratory Distress Syndrome

Early attempts to treat respiratory distress syndrome (RDS) with nebulized DPPC did not provide any long-term benefits (58,59). Nebulization of sonicated vesicles was not entirely successful due to an inability of the phospholipids to dissociate readily from the liposomes resulting in slow formation of a monolayer at the air–liquid interface (60). However, in an uncontrolled study, a suspension containing a lipid extract of bovine lung supplemented with DPPC and PG produced beneficial responses in infants with RDS (61). Over the last decade the efficacy of this therapeutic approach has been well established in both animal experiments and clinical trials (see Refs. 42, 43, 62, 63 for reviews). Nonetheless, some fundamental issues concerning the use of exogenous surfactant still exist. These include optimal composition, standardization of the formulation, dosage, detailed description of their distribution and fate, and possible adverse effects.

As listed by Morton (64), the types of surfactant that have emerged for the treatment of RDS can be generally classified as either natural surfactant, modified natural surfactant, artificial surfactant, or synthetic natural surfactant. Natural surfactant is obtained from alveolar lavage or amniotic fluid (42) and can be modified by the selective addition or deletion of lipid or protein components. Modified natural surfactants such as Surfactant TA, Curosurf, and Survanta are formulated to reproduce or improve the quality of isolated natural surfactant, to decrease protein contamination, and to provide sterility (42). Artificial surfactants are generally composed of phospholipids such as DPPC and PG derived from different sources and, in comparison to natural surfactants, often vary in proportion and acyl chain composition. They may contain other components not found in natural surfactant such as hexadecanol and tyloxapol (65). Current preparations under clinical investigation include Exosurf and ALEC (artificial lung-expanding substance) (64).

Comparison of the numerous clinical studies conducted with particulate systems as surfactant replacement therapy in neonatal RDS is difficult because protocols have varied considerably. Although surfactant compositions have been vastly different and not directly compared, studies have suggested that an effective surfactant substitute does not have to be identical to that of the normal alveolar fluid (43). Initial uncontrolled studies that used extracts of bovine alveolar fluid enriched with synthetic DPPC and unsaturated PG or with DPPC, tripalmitin, and palmitic acid (Surfactant TA) indicated that, based on improvement in oxygenation, treatment of RDS in infants was successful (61). Comparison of the physiological activity of four different exogenous surfactants in immature newborn lambs demonstrated that the effect of Surfactant TA on blood gases and lung compliance was not superior to crude natural rabbit or sheep surfactant (66). Consequently, the contamination of surfactant with membrane lipids and other material unrelated to the surfactant system may not be critical as there is selective adsorption of active components to the air–water interface (67). It has been demonstrated that manipulation of phospholipid composition can change significantly surface properties *in vitro* (68) but the benefits of such changes in the treatment of RDS have not been verified (66). The fundamental problem to be resolved with the use of natural surfactants is related to standardization of a formulation and its preparation, i.e., reproducibility of composition and surface properties, sterilization, protein sensitization, and suspension properties.

Artificial surfactants can avoid the potential problems of protein sensitization and sterile administration (69); however, their use has not produced impressive results. Initial animal studies indicated that mixtures of DPPC and other phospholipids were less effective at restoring pressure-volume relationships in the surfactant-deficient rat lung (70). LUVs showed the greatest improvement as compared to MLVs and SUVs indicating that the vesicular form may be an integral part of an artificial surfactant preparation (71–74). The administration of LUVs did not approach the improvement achieved with natural surfactant (71,72). Similarly, a 7:3 mixture of DPPC/PG was not as effective as natural surfactant in improving the compliance of spontaneously breathing rabbits (75) regardless of their possessing similar properties to natural surfactant. Egan and colleagues (76) also reported a poor response in lambs to a 7:3 mixture of DPPC/PG. Furthermore, it was found that a 9:1 mixture of DPPC/PG produced only a modest improvement in oxygenation during ventilation, although it may have augmented the response to natural surfactant (77).

Human clinical studies with artificial surfactant have shown essentially the same results. Morley et al. (69) used a dry powder mixture of DPPC/PG (7:3) and reported only modest reductions in mortality rates in surfactant-deficient infants. In other studies treatment with the same dry mixture showed no significant changes in compliance and blood gas values (78), had no immediate effects on lung mechanics (79), and did not reduce ventilatory requirements, oxygen use, or neonatal mortality in the first 2 years of life (80). It has been suggested that artificial surfactant may not have any significant effects on the surface properties of the lung but provides substrate for surfactant metabolic pathways (42); consequently, no immediate benefits are observed. The amount of DPPC required to form a complete monolayer in the neonatal lung is approximately 5 mg/kg body weight (43). Assuming that exogenous surfactant contains about 40% DPPC, a dose of 12.5 mg/kg would be theoretically required to prevent the development of RDS in infants. Clinical studies have administered doses ranging from 50 to 200 mg phospholipid/kg in a volume of 2–4 ml (61,81–83), roughly the same size as the surfactant pool. This suggests that the large doses of exogenous surfactant are not well distributed or poorly utilized by the lung. Furthermore, a monolayer is probably not enough to stabilize the terminal air spaces for a long time but a reservoir of phospholipid is required to provide continuous adsorption and recycling. Supplementing the phospholipid pool alone may be insufficient for treatment of RDS because mechanisms regulating recycling and utilization may also be defective.

Collectively, it has been demonstrated that surfactant replacement therapy for neonatal RDS reduces mortality and morbidity due to pneumothorax, pulmonary interstitial emphysema, bronchopulmonary dysplasia, and intraventricular hemorrhage (64). Synthetic surfactants such as liposomes appear to be useful in prophylaxis while natural surfactants appear more appropriate for treatment of established RDS (63). Prophylactic use of either type of surfactant have provided a 50% reduction in mortality whereas in established RDS cases the administration of natural surfactant in rescue studies reduced mortality by 40% (63). Improvement in pulmonary function is usually rapid and sometimes sustained after a single dose of surfactant (64). In other cases multiple doses are required (84).

Studies in Adult Respiratory Distress Syndrome

Some evidence supports the potential use of exogenous surfactant to treat adult respiratory distress syndrome (ARDS). It has been demonstrated that the administration of exogenous surfactant can alleviate lung injury in several animal models of ARDS (85–87) and may be helpful in the treatment of inflammatory lung disease (88). It seems particularly beneficial if administered in the early stages of ARDS (64) and prophylactic administration of exogenous surfactant may delay hyperoxic lung injury (89). It has also been demonstrated that liposome encapsulation of corticosteroids may be useful in the prophylactic treatment of early ARDS (90). Using a porcine model, aerosolized liposomes containing a corticosteroid modified and partly counteracted the changes in respiratory mechanics and mean pulmonary arterial pressure induced by endotoxin. The data suggested that the liposomal preparation was of some value as a beneficial modulator without any significant systemic effects. Only limited clinical trials have been conducted in patients suffering from severe ARDS. A single 50 mg/kg dose of Curosurf (porcine surfactant) showed a transient increase in arterial oxygen tension suggesting that repeated doses would be necessary (91,92). Lachmann et al. (93) suggested that a 200–300 mg/kg dose is required to improve pulmonary function.

Concluding Remarks

Phospholipid mixtures, whether in suspension or liposomal form, have demonstrated usefulness in surfactant replacement therapy; however, several issues have yet to be completely resolved. Although surfactant is generally considered safe because it is not distributed to the systemic circulation (94), increases in the incidence of apneas have been reported (95). On the positive side, studies to date have demonstrated that long-term adverse effects including chronic pulmonary disease, neurological abnormalities, developmental delay, and allergy were not significantly different between treatment and control groups (96–98). The only adverse effect noted in treated infants has been an increase in the incidence of patent ductus arteriosus (62). The optimum dosing protocol that includes not only the dose but the timing and method of administration has not been established. Pharmaceutically, a standardized preparation needs to be formulated and tested on a larger scale which may in turn resolve the details of dosing protocols. Indeed, this appears to have come to fruition with the recent FDA approval of Exosurf Paediatric.

Pharmacokinetic and Pharmacological Studies

One of the perceived benefits of liposomes as a drug carrier is based on their ability to alter favorably the pharmacokinetic profile of the encapsulated species and thus provide selective and prolonged pharmacological effects at the site of administration. Administration of liposomes to the respiratory tract is particularly attractive because of the accessibility of the lung as a target organ, the compatibility of liposomes and lung surfactant components, and the need for sustained local therapy following inhalation. Consequently, numerous studies have explored the effect of liposomal encapsulation on the distribution and fate of compounds administered directly to the lung by either intratracheal instillation or inhalation. In some cases, the correlation between altered pharmacokinetics and improved pharmacological responses has been established.

The first study was reported by McCullough and Juliano and compared the effectiveness and distribution of "free" and encapsulated β-cytosine arabinoside (ARA-C) following intratracheal administration to rats (99). MLVs were composed of egg PC/cholesterol/stearylamine (25:7:1 weight ratio). "Free" ARA-C was rapidly cleared from the lung ($t_{1/2}$ = 40 min) and entered the systemic circulation while liposomal ARA-C displayed little redistribution to other tissues and persisted throughout the lung ($t_{1/2}$ = 8 hr). Furthermore, liposome-encapsulated ARA-C selectively suppressed the incorporation of thymidine and DNA synthesis in the lung but not in the gut and bone marrow as observed with the "free" compound (100). The liposomal form favorably altered the distribution of ARA-C and successfully produced a prolonged local pharmacological effect without causing any untoward effects in other tissues.

Negatively charged MLVs containing superoxide dismutase or catalase administered intratracheally prior to hyperoxia exposure showed elevated enzyme levels, improved survival rate, and less lung injury as compared to the control groups (101). The protective effect was attributed to the elevated levels of enzymes maintained in the lungs by liposome encapsulation. *In vitro* study by Sone and coworkers concluded that macrophages treated with liposome-encapsulated muramyl dipeptide had significantly higher tumoricidal activity and maintained its action longer than macrophages treated with muramyl dipeptide (102). Similar success was reported using a lipophilic derivative of the dipeptide (103). Acyl tripeptide and its analogs were 800 times more effective in liposomal form than the unencapsulated form in potentiating the tumoricidal activity of macrophages (104).

Farr and colleagues investigated the pulmonary deposition and clearance of liposomes by labeling MLVs and SUVs composed of DPPC with 99m-technetium (105). Deposition of nebulized vesicles was dependent on the droplet size of the aerosol produced and not the vesicle size. In fact the process of nebulization reduced the apparent mean diameter of MLVs from 5.7 to 2.9 μm while the mean size of SUVs (70 nm) was unchanged. The MMAD of the aerosol as determined by impaction was similar for both products, 3.7 and 3.2 μm respectively, and consequently the regional deposition of MLVs and SUVs was similar in human subjects. For both preparations, over 75% of the vesicles were retained in the lung after 6 hr indicating that the mucociliary elevator was a major mechanism of clearance. Since the removal of particulate matter by the mucociliary elevator becomes progressively faster from the peripheral to central airways, the initial regional deposition of aerosol droplets was considered an important factor in determining the rate of vesicle clearance. From a therapeutic point of view, alveolar deposition would be preferred not only on the basis of providing local but prolonged activity due to the inability of the lung to clear vesicles as quickly from that site.

Woolfrey et al. investigated the pulmonary absorption of liposome-encapsulated 6-carboxyfluoroscein (CF) following intratracheal instillation to anaesthetized rats (106). The extent and rate of CF absorption was influenced by the dose of lipid and the net surface charge of the vesicles. Increasing the lipid dose from 14 to 34 mg/kg resulted in an increase in the rate of absorption but a decrease in the extent of absorption. The fraction of dose available to the systemic circulation decreased from 40% to 14% with the higher dose of lipid. The authors suggested that the higher lipid dose accelerated removal of the liposomal CF from the lung, perhaps by specific pathways capable of removing particulate material. The absorption of CF from negatively charged vesicles was at least twice as fast as from neutral vesicles. The fraction of tracer in the systemic circulation was also greater for negatively charged SUVs. These vesicles were slightly larger than neutral vesicles, approximately 145 nm as compared to about 110 nm in diameter. However, there were no discernible differences between the *in vitro* release of CF from the two vesicle preparations.

In contrast to studies with CF, neither the dose nor the ratio of drug to lipid had a significant effect on the *in vivo* release of bronchodilators (107). However, changes in the lipid composition of vesicles greatly influenced the release rate of encapsulated bronchodilators as well as a lipophilic steroid. The retention of drug in the lung was prolonged by an increase in the chain length and saturation of the fatty acyl side chains and by the inclusion of cholesterol. Consequently, the authors suggested liposomal formulations with a wide range of release rates can be designed by careful manipulation of lipid composition. In pharmacological studies with bronchodilators, it was reported that formulations with a half-life of clearance greater than 17 hr provided no protection against histamine-induced bronchoconstriction while those with half-times less than 2 hr provided essentially the same effect as the free drug (108).

Taylor et al. performed studies in humans that compared the systemic absorption of sodium chromoglycate in solution and liposomal form after inhalation therapy (109). Vesicles were composed of DPPC/cholesterol (1:1 molar ratio) and had a mean diameter of 1.2 μm. Greater than 95% of the sodium chromoglycate molecules were associated with the vesicles. Solution and liposomal formulations were delivered by jet nebulization and a previous study had established that vesicles were stable to the process of nebulization (110). With sodium chromoglycate in solution form, drug levels in plasma reached a maximum concentration of approximately 35 ng/ml within 15 min after the end of inhalation (Figure 3). In comparison, C_{max} and t_{max} for the liposomal preparation was about 5 ng/ml and 20 min, respectively. After 24 hr, no detectable levels of drug were found in the plasma when sodium chromoglycate was administered in solution form while drug was still detectable from the liposomal form. The absorption rate constant for sodium chromoglycate following inhalation was similar for the solution form and for the initial absorption of liposome-encapsulated drug. The liposomal drug showed a second much slower rate of absorption following the initial phase. This suggested that the initial rapid phase was due to unencapsulated drug in the liposomal preparation and the slow phase was attributed to the absorption of liposome-encapsulated drug following release from the vesicles.

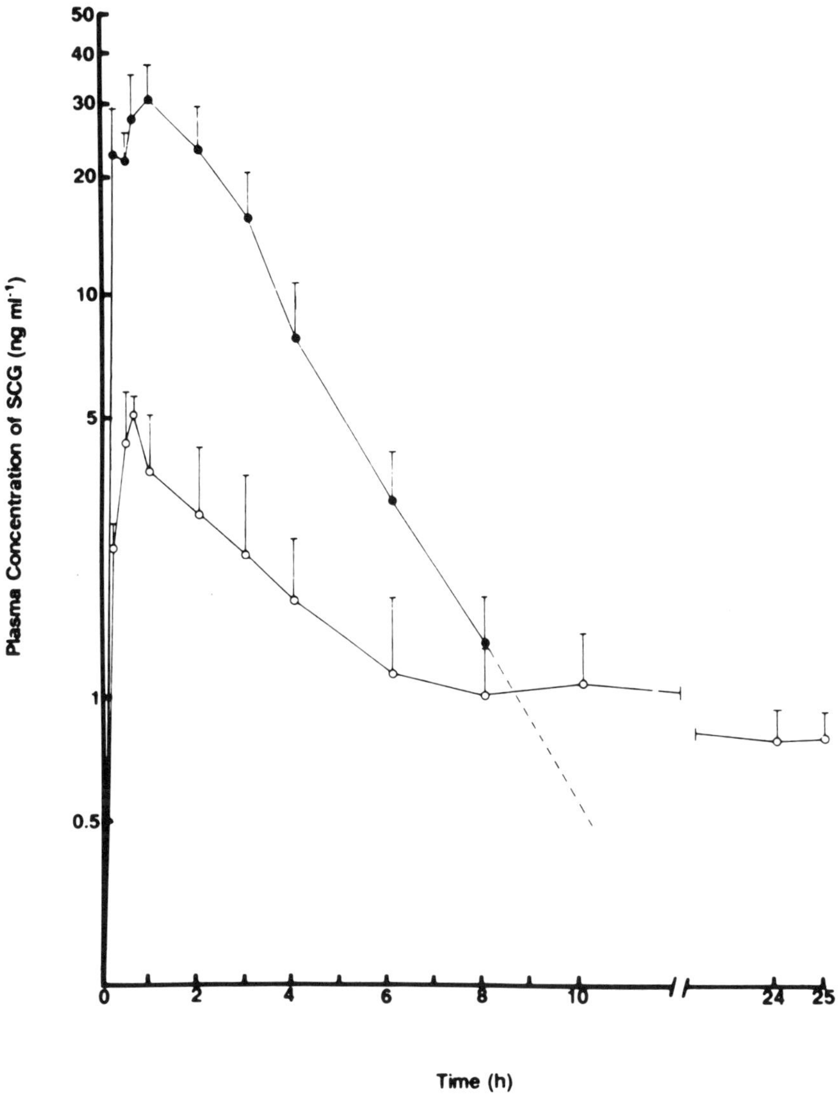

Figure 3 Plasma levels following jet nebulization of 20 mg of sodium chromoglycate to volunteers. Values are mean ± SE at each time point; *n* = 5 for solution (●); *n* = 4 for liposomal form (○). (Reproduced from Ref. 109.)

The pharmacokinetic profile and pharmacological activities of aerosolized pentamidine in solution and liposomal form have been compared by Debs et al. (111). MLVs composed of PC/PS/Chol (8:2:5 molar ratio) were labeled with fluorescent PE and extruded through a 0.2-μm filter prior to nebulization. It was found that greater than 95% of the labeled lipid was removed from the lung by 48 hr as compared to both free and encapsulated pentamidine which were not cleared at all during this time period. The actual clearance of the phospholipid may be misleading because the labeled lipid could possibly have been metabolized and quenched within the lung rather than cleared. Approximately 75% of the encapsulated pentamidine remained associated with liposomes following administration yet there was limited absorption of both the free and encapsulated drug to the systemic circulation. As a result, there was also low distribution to the liver and kidney. Based on these data, the use of liposomes as a carrier for pentamidine may be of little additional benefit because the sequestering of free drug in the lung provides similar clearance to that of encapsulated drug. However, it was interesting that the authors noted a subtle difference in lung deposition between the two preparations of pentamidine. The fraction of the dose recovered in the cell pellet 1 hr after administration was significantly increased for liposome-encapsulated pentamidine and was probably due to uptake by pulmonary macrophages. Assessment of the therapeutic efficacy of aerosolized pentamidine preparations in immunosuppressed rats indicated that both free and encapsulated drug eradicated *Pneumocystis carinii* pneumonia in 75% of the treated animals (112). As in the previous study, no difference in tissue distribution of drug was observed between the two products.

Enviroxime, a lipophilic compound, was incorporated into liposomes and evaluated *in vitro* for rhinoviral activity and cell culture toxicity (113–115). The *in vitro* activity of free drug and drug encapsulated in MLVs composed of egg PC were equivalent; however, it was determined that the liposome preparation was 10- to greater than 50-fold less toxic to strain 1A and 13 culture cells than the free drug. In the ensuing *in vivo* experiments, only liposome-encapsulated enviroxime was administered because free drug could not be delivered as an aerosol due to the insolubility of the compound in solution. Liposome-encapsulated enviroxime delivered by jet nebulization produced significant levels of drug in the lungs and noses of treated mice. Using a fluorescent marker to label the phospholipid component, it was observed that fluorescence became evident in the large and small airways within 10 min and accumulated continually with time particularly in and around the columnar epithelium of the bronchioles. Drug levels were detected in nose washes of these animals immediately after nebulization but declined rapidly, reaching undetectable levels by 2 hr. Levels of enviroxime in the lung were lower than levels detected in the nose and had disappeared 1 hr after cessation of nebulization. *In vitro* antiviral activity of lung lavages obtained from mice treated with liposome-encapsulated enviroxime coincided with drug levels detected in the lungs and ceased 2 hr after aerosolization. The authors stated several advantages of the liposomal drug delivery system including high drug loading of a water-insoluble compound for aerosolization and decreased *in vitro* toxicity. Aerosol exposure of liposome-encapsulated enviroxime to normal volunteers showed that the preparation was well tolerated (115). As in animal experiments, enviroxime collected in nose washes retained their antiviral activity.

Another interesting aspect of these studies was the concentrating effect of drug observed in both the aerosol and reservoir of the nebulizer with the liposomal formulation (114). During a 75-min period of operation, the drug concentration in the aerosol increased from its original concentration of 5 mg/ml to a final concentration of 40 mg/ml. In this regard, liposomes appear to behave similarly to other nebulized products that concentrate during aerosolization due to evaporative losses (116–118). There was also a reduction in vesicle size during nebulization and the MMAD of liposomal enviroxime aerosols was dependent on the type of nebulizer employed and not vesicle size.

Investigations by Mihalko and coworkers examined the effect of liposomal encapsulation on the pharmacokinetic fate of hydrophilic and lipophilic compounds following intratracheal instillation to rats (119). [^{14}C]Benzylpenicillin and [^{125}I]oxytocin were entrapped in the aqueous phase while [^{14}C]nitroglycerin was incorporated into the bilayers of MLVs composed of Egg PC/PG/cholesterol (5:1:4 molar ratio) extruded through a 0.4-μm polycarbonate filter. Although the concentration of benzylpenicillin in plasma following administration in either solution or liposomal form showed a gradual increase and no peak levels after 3 hr, the level produced by free drug at this time point was approximately two-fold greater than that provided by the encapsulated drug. There was a three-fold decrease in the area under the plasma curve for the liposomal product compared to the free drug; however, absorption rate constants from the lung were similar for both dosage forms. A higher percentage of encapsulated drug remained in the lung at 3 hr post-instillation. For oxytocin, peak drug levels in plasma and AUC was about eight times less for the encapsulated drug. For these two hydrophilic compounds, the authors concluded that the lung epithelium represented the major rate-limiting barrier for systemic absorption, not the rate of drug release from vesicles. In contrast, intratracheal administration of nitroglycerin as either free or encapsulated drug produced almost identical plasma levels suggesting that diffusion of the lipophilic compound through vesicle bilayers and lung epithelium was unrestricted.

A similar comparison of hydrophilic and lipophilic compounds was investigated by Mezei and colleagues following endotracheal administration to rabbits (120–122). Liposomal encapsulation of atropine base maintained drug concentrations in the lung approximately three-fold over a 48-hr period while free atropine disappeared rapidly (120) (Figure 4A). After 1 hr only 19% of the total dose administered in solution form was detected in the lung while in liposomal form about 53% was retained. After 48 hr, 21% of liposomal atropine was still associated with lung tissue as compared to 4% when instilled in solution form. Peak levels of drug in plasma occurred rapidly for both preparations, however, were nearly three times higher for free atropine (Figure 4B). This is consistent with the observation that small lipophilic molecules are easily and rapidly absorbed into the systemic circulation following administration to the lung (16). The area under the plasma curve was two times higher for the free drug and appeared in the urine more rapidly. The distribution of atropine base to other tissues corresponded to the rate and extent of systemic absorption. Free atropine, having rapid accessibility to the general circulation, quickly redistributed to other tissues such as liver and spleen. It was concluded that the amount of drug available to lung tissue from liposomes was controlled by the rate of drug release from the MLVs.

Pharmacological studies were performed to ascertain if prolonged localization of atropine in the lung could also provide a sustained therapeutic effect. Guinea pigs were pretreated with either normal saline or atropine in solution and liposomal form and then subjected to a series of methacholine challenges (122). Normal saline aerosols provided no protection against methacholine-induced bronchoconstriction. A protective effect was observed with aerosolized atropine in both solution and liposomal form; however, liposome-encapsulated drug provided a longer duration of protection than the free drug. The prolonged effect provided liposomal atropine was not as superior as was predicted from *in vivo* disposition data. This could have been a result of inadequate release of drug from vesicles and thus inability to maintain therapeutic levels at receptor sites. It was also observed that aerosols of atropine in solution produced an increase in heart rate and dilation of blood vessels in the ears of guinea pigs within 5 min of administration. These effects were absent when treated with liposome-encapsulated atropine and supported *in vivo* disposition data demonstrating that free drug had produced greater systemic exposure. The implication of these findings is that selective localization of drug by liposome encapsulation can reduce or retard its systemic absorption and toxicity.

Endotracheal administration of pyridine-2-aldoxime methiodide (2-PAM) in solution and liposomal form suggested that hydrophilic compounds, as demonstrated by Mihalko et al. (119), behave differently from lipophilic compounds (121). In solution form, there were high levels of 2-PAM in the lung initially but rapidly declined (Figure 5A). This coincided with rapid systemic absorption (Figure 5B) and redistribution to other tissues. In contrast, liposomal encapsulation eliminated the initial rapid absorption of drug

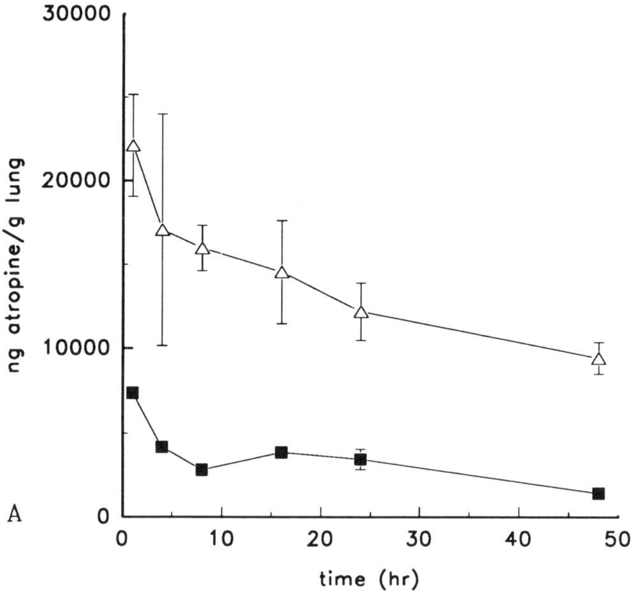

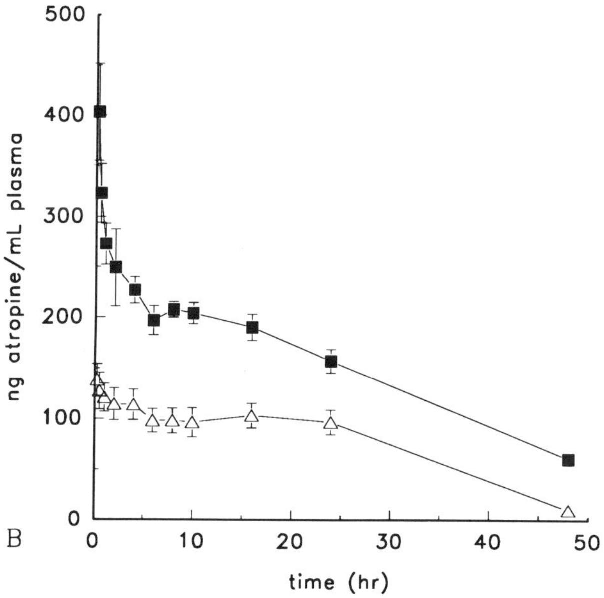

Figure 4 Concentration of atropine in lung (A) and plasma (B) following endotracheal instillation to the rabbit lung in solution (■) and liposomal (△) form. Values are mean ± SE at each time point (n = 3). (Reproduced from Ref. 120.)

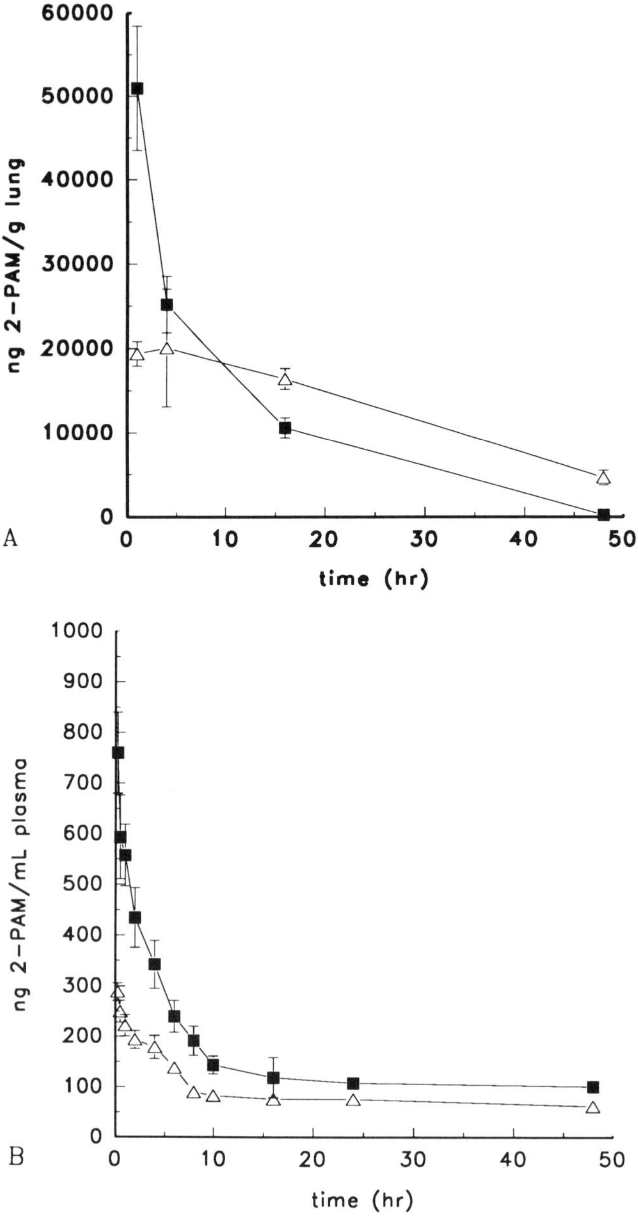

Figure 5 Concentration of 2-PAM in lung (A) and plasma (B) following endotracheal instillation to the rabbit lung in solution (■) and liposomal (△) form. Values are mean ± SE at each time point ($n = 3$).

into the lung and systemic circulation. Collection of lung perfusate showed that, in liposome-encapsulated form, a greater fraction of dose was retained in the lung over time. It was concluded that liposome encapsulation imparted different influences on the deposition of hydrophilic and lipophilic compounds such as 2-PAM and atropine and these may be attributed to the rate of drug release from vesicles.

The use of liposome-encapsulated bronchodilators as a means to extend duration of action and diminish systemic side effects was recently reviewed (123). Original studies in guinea pigs determined that liposome-encapsulated metoproterenol sulfate was superior to saline and the drug in solution following intratracheal instillation (124). Saline offered no protection against histamine-induced changes in airway resistance while in animals treated with metoproterenol in solution the protective effect began to dissipate 45 min after dosing. Instillation of liposome-encapsulated drug showed a more prolonged inhibition of the histamine-induced changes, although the authors stated that the experiments were too short to conclude definitively that the duration of action was significantly longer than that of the free drug (125). Liposome encapsulation reduced systemic side effects of metoproterenol; however, this was product-dependent (124). Of four liposomal preparations evaluated, only one produced a tachycardia similar to that of the free drug. Although all four liposomal formulations elicited a modest drop in diastolic blood pressure, the effect produced by the solution form of the drug was greater. Subsequent experiments demonstrated that liposomes provided sustained delivery of metoproterenol to the systemic circulation (125). Delivered as a solution, the compound produced more significant tachycardia initially but disappeared more quickly than the liposome-encapsulated drug. Liposomal metoproterenol produced only a small increase in heart rate that persisted over 10–12 hr.

In these studies relatively high doses of the bronchodilator agent were administered. When liposomal metoproterenol was delivered to guinea pigs in lower doses, approximately 10 times that delivered to humans, results obtained were unimpressive (125). The bronchodilator response dissipated within the first 2 hr and there was no significant protection against histamine challenge at any time points studied. The authors suggested that the loading dose of free drug and the eventual release of the encapsulated drug was inadequate to provide therapeutic drug concentrations at appropriate receptor sites in the lung.

The effect of liposome encapsulation on the bronchodilator effect of albuterol was similarly evaluated (123). Administration of albuterol in solution form displayed a dose-dependent reduction in the response to histamine with a maximum duration of 2 hr while the liposomal form inhibited the response for at least 4 hr. Since the liposomal formulation was identical to that of metoproterenol tested in previous studies, it was concluded that drug potency was a contributing factor. Albuterol had sufficient potency to achieve and maintain a pharmacological response with the small amount of drug provided initially and during release from vesicles while metoproterenol lacked sufficient potency and required higher dose loading.

A pharmacokinetic model that describes the release of drug from carriers such as liposomes following administration directly into the lung was recently reported (126). It stresses the distinction between the free and "unreleased" drug in relationship to regional deposition and mechanisms of clearance within those regions. In the pulmonary region, both free and encapsulated drug can be equally removed by mucociliary transport mechanisms, whereas in the alveolar region absorption of drug into the systemic circulation or lymph tissue is only significant for free drug molecules. Based on these assumptions the effect of control of release rate from vesicles on the duration of residence of drug in the lung can be predicted. Drugs with high rates of absorption in the alveolar region will display very low levels of activity if the rate of release from vesicles is slow. Studies with atropine and metoproterenol suggest that these compounds may fit this description. There was rapid systemic absorption of drug and although liposomal encapsulation extended the residence time within the lung, the pharmacological effect was not significantly prolonged. Alternatively, substances that are not rapidly absorbed show prolonged residence in the lung even without the carrier. The sequestering of pentamidine in the lung appears to behave in this manner. Applying this pharmacokinetic model to most of the liposome-encapsulated compounds discussed is difficult because there was generally no distinction made between free and unreleased drug as well as no attempt to define the regional deposition following administration. It was also stressed that unless it can be removed by other mechanisms, carrier material can accumulate in pulmonary regions without mucociliary clearance capabilities (126). Since the exogenous phospholipids are reutilized and cleared by the lung, this problem may not arise with liposomal drug delivery.

Development of a Liposomal Dosage Form

Delivery as an Aerosol

Many of the pharmacological and pharmacokinetic studies that have explored the potential of liposomes to act as a selective and prolonged carrier for pulmonary use have administered the formulation by endotracheal instillation (99,100,106,120,125). Although endotracheal administration may be practical to define distribution and elimination of liposome-encapsulated drugs and test the efficacy of the product, it is obviously unacceptable as a routine method for drug delivery. Nebulization is probably the simplest and most effective way to deliver liposomes to the respiratory tract; however, one of the first questions to consider is whether or not vesicles are stable during nebulization. Vesicles must remain intact without the loss of entrapped solutes to exert sustained release of the compound. The energy required by jet and ultrasonic nebulizers to reduce aqueous solutions to a respirable aerosol may have a detrimental effect in this regard.

Investigations involving the stability of liposomes during nebulization have been limited; however, a few studies were recently reported. Using a jet nebulizer, the loss of entrapped drug was related to vesicle size (127). MLVs containing sodium chromoglycate lost 51% of the entrapped drug during nebulization. This was associated with a significant decrease in mean vesicle diameter from 5.4 to 2.7 μm indicating that vesicles were fragmented by the process of nebulization. REVs showed a decrease in size from 3.4 to 2.5 μm during nebulization and lost 32% of the encapsulated drug. Reducing the size of reverse phase evaporation vesicles to 1.2 μm by extrusion reduced drug losses during nebulization to 17% without any significant change in vesicle size. The type of liposome formulation used had no effect on the MMAD of the aerosol produced, which was approximately 2.7 μm for all preparations. This observation supports a previous study demonstrating that deposition of nebulized liposomes was dependent on the droplet size of the aerosol and not on vesicle size (105).

Niven et al. (128) also reported that loss of entrapped solute was dependent on vesicle size. Following an 80-min period of nebulization, the loss of encapsulated carboxyfluoroscein (CF) ranged from 8% for vesicles extruded through a 0.2-μm filter to 76% for vesicles approximately 5 μm in size. Release of CF over a 7-hr period fit a two-compartment kinetic model consisting of a fast initial phase and a slow terminal phase. A cutoff in release was observed at a size near the MMAD of the aerosol produced by the nebulizer suggesting that vesicles smaller than the aerosol droplets are likely to be less exposed to the fragmentation forces produced by nebulization. Consequently, it was suggested that nebulizers that produce aerosols with larger MMADs may have less impact on solute retention. This may hold true only if the vesicle size is smaller than the droplet size. Taylor and coworkers also had smaller losses as long as vesicle size did not exceed droplet size (127).

Niven et al. performed additional studies that investigated the effects of lipid composition on the stability of liposomes during nebulization (129). Although the lipid concentration had no effect on the rate or percentage of CF released, lipid composition of the vesicles dramatically affected their ability to retain CF during nebulization. Vesicles composed of hydrogenated soy phosphatidylcholine (HSPC) and 30 mol % cholesterol had the greatest stability while soy phosphatidylcholine (SPC) with 30 mol % dipalmitoylphosphatidylglycerol (DPPG) had the least. The presence of DPPG had little effect on the release of CF from HSPC liposomes but had significant effect on vesicles composed of unsaturated SPC. Cholesterol effects were concentration-dependent. For vesicles containing only 10% cholesterol or less, there were marked increases in the release of encapsulated CF. Cholesterol content and degree of saturation have previously been shown to be important factors in determining vesicle permeability and appear to influence the rate and extent of loss of entrapped solutes during nebulization (36). In formulating liposome preparations intended for nebulization, lipid composition and vesicle size have been identified as two important factors. Fragmentation of vesicles is the most plausible cause for the release of water-soluble compounds. It has not been determined if hydrophobic compounds that are incorporated into vesicle bilayers exhibit similar losses, although it has been suggested that maintaining vesicle integrity may be less important for intercalated compounds (127). Based on these investigations, to limit solute losses during nebulization the size of the vesicles should not exceed the MMAD of the aerosol; thus the nebulizer must be considered an integral part of the liposome-drug delivery system. It also could complicate the development of a liposomal product for nebulization as a change in the type of nebulizer can alter the release characteristics of the vesicles. Specifically defining which type(s) of nebulizers could be used for the liposomal product may ensure proper drug/liposome delivery but may also limit flexibility and possibly marketability.

A potential alternative approach to the delivery of liposomes to the respiratory tract was described by Farr and coworkers (130). Phospholipids were dissolved in trichlorofluoromethane and pressurized with dichlorodifluoromethane or dichlorotetrafluoroethane, commonly used chlorofluorohydrocarbons for MDI formulation. Phospholipid aerosols produced by pressurized packs were shown *in vitro* to form liposomes spontaneously upon release and complete evaporation of propellant in a water-rich environment. Phospholipid concentration, vapor pressure, and the diameter of the actuator orifice were influential in determining the fraction of aerosol in the respirable range. Using salbutamol and hydrocortisone 21-octanoate as model compounds it was found that solute partitioning into liposomes derived from MDI aerosols was comparable to those prepared by the conventional solvent evaporation technique. Release of hydrocortisone was also comparable suggesting that MLVs produced from phospholipid aerosols were structurally similar as well and that lipophilic drugs may be more suitable candidates for this approach.

Another possible dosage form that may also overcome the problems of long-term stability of liposomes was described (131). The system consists of two compartments. The nonaqueous compartment, comprising a pressurized bottle fitted with a valve, contains a mixture of propellant and phospholipid solution. The aqueous compartment comprises a pressurized container fitted with a metered valve and a mixing chamber. Depending on its solubility, the compound can be added to either phase. Actuation of the valve system allows mixing of the two phases to form an emulsion that is released as a metered dose aerosol.

Pharmaceutical Considerations

The potential of liposomes as a pulmonary drug delivery system may depend on whether a suitable pharmaceutical product can be developed (see Ref. 132 for detailed discussion). Present evidence indicates that the production of liposomal aerosols is feasible; however, several questions pertaining to product stability remain. Retention and chemical stability of the entrapped species is an obvious pharmaceutical requirement. An acceptable shelf-life for most pharmaceutical products is 2 years, a requirement unlikely to be met if liposome preparations are stored in aqueous form. However, there are alternatives. It has been demonstrated that dehydrated liposomes maintain their original chemical and physical properties and retain greater than 90% of the entrapped species upon rehydration (133, 134). In the dehydrated state the stability of the product not only may be enhanced but effectively reduces the aqueous stability problems to the period of use. Polymerized vesicles, composed of synthetic phospholipids capable of producing membranes substantially more stable than conventional bilayers, exhibit increased resistance to leakage; however, their biodegradability and toxicity remain uncertain (135). Oxidation of phospholipids can be minimized by the addition of antioxidants such as tocopherol to the formulation (136).

Sterility of the preparation may also be difficult because most methods of sterilization are unsuitable for liposomal drug products (132). Temperatures required for autoclaving can cause irreparable damage to liposomes. Filtration can guarantee sterility but reduces vesicle size to an average of 0.2 μm and thus is not applicable for MLV preparations. Components of the liposome-drug delivery system, i.e., lipid, drug, and buffer, can be sterilized separately and then combined in a sterile environment; however, it is highly impractical from a manufacturing point of view. The problems of liposome formulation are not insurmountable as illustrated by the market entry of Pevaryl Lipogel, a topical formulation containing 1% econazole. Furthermore, the fact that several intravenous, aerosol, and topical liposome products are presently undergoing clinical trials suggests that formulation issues are being resolved.

SUMMARY

The present evidence suggests that liposomes have potential applications in pulmonary drug delivery. The two areas of primary interest are the use of liposomes in surfactant replacement therapy and the ability of vesicles to modulate the pulmonary absorption of therapeutic compounds. Although it is still not clear which is the best surfactant preparation to use, the instillation of exogenous phospholipids represents a significant advance in the treatment of RDS. An artificial formulation that has a composition closest to natural surfactant may prove to be superior. With the availability of DNA recombinant technology, a highly purified surfactant containing the specific apoproteins can be developed. As surfactant preparations become commercially available, there is hope that they will be beneficial in other lung diseases such as ARDS.

Biodisposition studies have demonstrated that liposome encapsulation of compounds can localize and maintain drug levels in the lung for an extended period of time while decreasing the extent of systemic absorption. Pharmacologically, there is also evidence that liposomal drugs can produce selective and prolonged effects with decreased systemic toxicity. The pulmonary absorption of liposome-encapsulated solutes can be influenced by the physicochemical properties of the entrapped species, the lipid composition of the vesicles, and the site of deposition in the lung. A better understanding of these factors is essential in designing a liposomal formulation that will not only optimize pharmacokinetic and pharmacological profiles but also satisfy pharmaceutical requirements. The technological issues related to the delivery of liposomes as an aerosol have been largely ignored and have to be appropriately addressed. Studies to date have produced some promising results; however, the successful use of liposomes as a pulmonary drug delivery system has yet to be conclusively demonstrated. Future work should concentrate on verifying the mechanism(s) of liposome-drug release, absorption, and clearance; developing a pharmaceutically acceptable product for aerosol delivery; and establishing the biocompatibility and toxicity of chronic administration to the respiratory tract.

REFERENCES

1. S. P. Newman and D. Pavia, in *Aerosols in Medicine* (F. Morén, M. T. Newhouse, and M. B. Dolovich, eds.), Elsevier, New York, 1985, p. 193.
2. G. M. Shenfield, M. E. Evans, and J. W. Paterson, *Br. J. Clin. Pharmacol.*, *3*, 1218 (1976).
3. C. Marriott, *Adv. Drug Deliv. Rev.*, *5*, 19 (1990).
4. G. Taylor, *Adv. Drug Deliv. Rev.*, *5*, 37 (1990).
5. E. A. B. Brown, *Drug Metab. Rev*, *3*, 33 (1974).
6. T. R. Gerrity, in *Respiratory Drug Delivery* (P. R. Byron, ed.), CRC Press, Boca Raton, 1990, p. 1.
7. L. A. Damani and P. A. Crooks, in *Respiratory Drug Delivery* (P. R. Byron, ed.), CRC Press, Boca Raton, 1990, p. 61.

8. A. D. Chediak and A. Wanner, *Adv. Drug Deliv. Rev.*, *5*, 11 (1990).
9. P. R. Byron and E. M. Phillips, in *Respiratory Drug Delivery* (P. R. Byron, ed.), CRC Press, Boca Raton, 1990, p. 107.
10. N. Mygind, in *Aerosols in Medicine* (F. Morén, M. T. Newhouse, and M. B. Dolovich, eds.), Elsevier, New York, 1985, p. 1.
11. J. B. Forrest, in *Aerosols in Medicine* (F. Morén, M. T. Newhouse, and M. B. Dolovich, eds.), Elsevier, New York, 1985, p. 21.
12. E. R. Wiebel, in *The Lung: Scientific Foundations* (R. G. Crystal and J. B. West, eds.), Raven Press, New York, 1991, p. 712.
13. S. W. Clarke, in *Aerosols and the Lung: Clinical and Experimental Aspects* (S. W. Clarke and D. Pavia, eds.), Butterworths, Toronto, 1984, p. 1.
14. J. E. Agnew, in *Aerosols and the Lung. Clinical and Experimental Aspects* (S. W. Clarke and D. Pavia, eds.), Butterworths, Toronto, 1984, p. 49.
15. E. E. Schneeberger, *Fed. Proc.*, *37*, 2471 (1978).
16. L. S. Schanker, *Biochem. Pharmacol.*, *27*, 381 (1978).
17. L. S. Schanker, E. W. Mitchell, and R. A. Brown, Jr., *Pharmacology*, *32*, 176 (1986).
18. L. S. Schanker, E. W. Mitchell, and R. A. Brown, Jr., *Drug Metab. Dispos.*, *14*, 79 (1986).
19. L. S. Schanker and J. A. Hemberger, *Biochem. Pharmacol.*, *32*, 2599 (1983).
20. T. H. Gardiner and L. S. Schanker, *Xenobiotica*, *4*, 725 (1974).
21. Y-J. Lin and L. S. Schanker, *Am. J. Physiol.*, *240*, C215 (1981).
22. J. W. Ryan, *Ann. Rev. Physiol.*, *44*, 241 (1982).
23. S. I. Said, *Ann. Rev. Physiol.*, *44*, 257 (1982).
24. T. R. Devereux and J. R. Fouts, *Meth. Enzymol.*, 77, 147 (1981).
25. C. J. Serabjit-Singh, S. J. Nishio, R. M. Philpot, and C. G. Plopper, *Mol. Pharmacol.*, *33*, 279 (1988).
26. G. E. R. Hook, J. R. Bend, and J. R. Fouts, *Biochem. Pharmacol.*, *21*, 3267 (1972).
27. B. A. Domin, T. R. Devereux, and R. M. Philpot, *Mol. Pharmacol.*, *30*, 296 (1986).
28. W. D. Reid, J. M. Glick, and G. Krishna, *Biochem. Biophys. Res. Commun.*, *49*, 626 (1972).
29. C. Meban, *Acta Anat.*, *108*, 184 (1980).
30. D. Pavia, in *Aerosols and the Lung: Clinical and Experimental Aspects* (S. W. Clarke and D. Pavia, eds.), Butterworths, Toronto, 1984, p. 127.
31. J. D. Brain, P. A. Valberg, and S. Snedden, in *Aerosols in Medicine* (F. Morén, M. T. Newhouse, and M. B. Dolovich, eds.), Elsevier, New York, 1985, p. 123–147.
32. M. A. Sleigh, *Chest*, *80*, 791 (1980).
33. A. D. Bangham, M. M. Standish, and J. C. Watkins, *J. Mol. Biol.*, *13*, 238 (1965).
34. H. K. Kimelberg and E. G. Mayhew, *Crit. Rev. Toxicol.*, *6*, 25 (1978).
35. B. E. Ryman and D. A. Tyrrell, *Front. Biol.*, *48*, 549 (1979).
36. F. Szoka, Jr. and D. Papahadjopoulos, *Ann. Rev. Biophys. Bioeng.*, *9*, 467 (1980).

37. D. Stamp and R. L. Juliano, *Can. J. Physiol. Pharmacol.*, *57*, 535 (1979).
38. G. Gregoriadis, In *Liposomes as Drug Carriers: Recent Trends and Progress* (G. Gregoriadis, ed.), John Wiley and Sons, Toronto, 1988.
39. J. A. Clements, *Proc. Soc. Exp. Biol. Med.*, *95*, 170 (1957).
40. M. E. Avery and J. Mead, *Am. J. Dis. Child.*, *97*, 517 (1959).
41. T. Higenbottam, *Respiration*, *55* (*suppl. 1*), 14 (1989).
42. A. Jobe and M. Ikegami, *Am. Rev. Respir. Dis.*, *136*, 1256 (1987).
43. L. M. G. Van Golde, J. J. Batenburg, and B. Robertson, *Physiol. Rev.*, *68*, 374 (1988).
44. L. G. Dobbs, *Ann. Rev. Med.*, *40*, 431 (1989).
45. M. C. Williams, *J. Cell. Biol.*, *72*, 260 (1977).
46. A. G. Baritussio, M. W. Magoon, J. Goerke, and J. A. Clements, *Biochem. Biophys. Acta*, *666*, 382 (1981).
47. H. C. Jacobs, A. Jobe, M. Ikegami, and S. Jones, *J. Biol. Chem.*, *257*, 1805 (1982).
48. M. J. Oyarzun, J. A. Clements, and A. Baritussio. *Am. Rev. Respir. Dis.*, *121*, 709 (1980).
49. A. B. Fisher, C. Dodia, and A. Chander, *J. Appl. Physiol.*, *59*, 743 (1985).
50. J. R. Wright and J. A. Clements, *Am. Rev. Respir. Dis.*, *135*, 426 (1987).
51. J. R. Wright, *Am. J. Physiol.*, *259*, L1, (1990).
52. H. Jacobs, A. Jobe, M. Ikegami, and D. Conaway, *J. Biol. Chem.*, *258*, 4159 (1983).
53. H. C. Jacobs, A. Jobe, M. Ikegami, S. Jones, and D. Miller, *Biochim. Biophys. Acta*, *752*, 178 (1983).
54. K. Oguchi, M. Ikegami, H. Jacobs, and A. Jobe, *Exp. Lung Res.*, *9*, 221 (1985).
55. W. D. Claypool, D. L. Wang, A. Chander, and A. B. Fisher, *J. Clin. Invest.*, *74*, 677 (1984).
56. W. D. Claypool, D. L. Wang, A. Chander, and A. B. Fisher, *Exp. Lung Res.*, *6*, 215 (1984).
57. J. R. Wright, R. E. Wager, R. L. Hamilton, M. Huang, and J. A. Clements, *J. Appl. Physiol.*, *60*, 817 (1986).
58. J. Chu, J. A. Clements, E. K. Cotton, M. H. Klaus, A. Y. Sweet, and W. H. Tooley, *Pediatrics*, *40*, 709 (1967).
59. D. C. Shannon and J. B. Bunnell, *Pediatr. Res.*, *10*, 467 (1976).
60. H. Ivey, S. Roth, and J. Kattwinkel, *Pediatr. Res.*, *10*, 462 (1976).
61. T. Fujiwara, H. Maeta, S. Chida, T. Morita, Y. Watabe, and T. Abe, *Lancet*, *1*, 55 (1980).
62. H. L. Halliday, *Dev. Pharmacol. Ther.*, *13*, 173 (1989).
63. H. L. Halliday, *Intensive Ther. Clin. Monit.*, *11*, 75 (1990).
64. N. S. Morton, *Thorax*, *45*, 825 (1990).
65. D. J. Durand, R. I. Clyman, M. A. Heymann, J. A. Clements, F. Mauray, J. Kitterman, and P. Ballard, *J. Pediatrics*, *107*, 775 (1985).
66. M. Ikegami, Y. Agata, T. Elkady, M. Hallman, D. Berry, and A. Jobe, *Pediatrics*, *79*, 38 (1987).
67. C. J. Morley and A. Bangham, *Prog. Respir. Res.*, *15*, 188 (1981).

68. K. Nohara, P. Berggren, T. Curstedt, G. Grossmann, R. Nilsson, and B. Robertson, *Eur. J. Respir. Dis.*, *69*, 321 (1986).
69. C. J. Morley, A. D. Bangham, N. Miller, and J. A. Davis, *Lancet*, *1*, 64 (1981).
70. M. Ikegami, J. Silverman, and F. H. Adams, *Pediatr. Res.*, *13*, 777 (1979).
71. M. Obladen, F. Brendlein, B. Krempien, *Eur. J. Pediatrics*, *131*, 219 (1979).
72. M. Obladen, D. Popp, C. Scholl, H. Schwarz, and F. Jahnig, *Biochem. Biophys. Acta*, *735*, 215 (1983).
73. M. Obladen, W. Kampmann, I. Zimmermann, and B. Lachmann, *Eur. J. Pediatrics*, *144*, 195 (1985).
74. O. Ferrer-Roca, M. C. Padilla, and E. Domenech, *Biol. Neonate*, *57*, 61 (1990).
75. C. Morley, B. Robertson, B. Lachmann, R. Nilsson, A. Bangham, G. Grossmann, and N. Miller, *Arch. Dis. Child.*, *55*, 758 (1980).
76. E. A. Egan, R. H. Notter, M. S. Kwong, D. L. Shapiro, *J. Appl. Physiol.*, *55*, 875 (1983).
77. M. Ikegami, A. Jobe, H. Jacobs, and S. J. Jones, *J. Clin. Invest.*, *68*, 491 (1981).
78. A. D. Milner, H. Vyas, and I. E. Hopkin, *Arch. Dis. Child.*, *58*, 458 (1983).
79. A. D. Milner, H. Vyas, and I. E. Hopkin, *Arch. Dis. Child.*, *59*, 369 (1984).
80. A. Wilkinson, P. A. Jenkins, and J. A. Jeffrey, *Lancet*, *2*, 287 (1985).
81. M. Hallman, T. A. Merritt, A.-L. Jarvenpaa, B. Boynton, F. Mannino, L. Gluck, T. Moore, and D. Edwards, *J. Pediatrics*, *106*, 963 (1985).
82. M. Hallman, T. A. Merritt, H. Schneider, B. L. Epstein, F. Mannino, D. K. Edwards, and L. Gluck, *Pediatrics*, *71*, 473 (1983).
83. J. A. Smyth, I. L. Metcalfe, P. Duffty, F. Possmayer, M. H. Bryan, and G. Enhorning, *Pediatrics*, *71*, 913 (1983).
84. N. S. Morton, *Br. J. Hosp. Med.*, *42*, 52 (1989).
85. S. Matalon, B. A. Holm, and R. H. Notter, *J. Appl. Physiol.*, *62*, 756 (1987).
86. B. Lachmann, M. Hallman, and K.-C. Bergmann, *Exp. Lung Res.*, *12*, 163 (1987).
87. D. Berry, M. Ikegami, and A. Jobe, *J. Appl. Physiol.*, *61*, 1741 (1986).
88. J. Floros, *Am. J. Physiol.*, *258*, L238 (1990).
89. G. M. Loewen, B. A. Holm, L. Milanowski, L. M. Wild, R. H. Notter, and S. Matalon, *J. Appl. Physiol.*, *66*, 1087 (1988).
90. P. E. Forsgren, J. A. Modig, C. M. O. Dahlback, and B. I. Axelsson. *Acta Chir. Scand.*, *156*, 423 (1990).
91. P. S. Richman, R. G. Spragg, T. A. Merritt, B. Robertson, and T. Curstedt, *Am. Rev. Respir. Dis.*, *135* (*suppl.*), A5 (1987).
92. P. S. Richman, R. G. Spragg, T. A. Merritt, and T. Curstedt, *Eur. Respir. J.*, *2* (*suppl. 3*), 109 (1989).
93. B. Lachmann, *Eur. Respir. J.*, *2* (*suppl. 3*), 98 (1989).
94. M. S. Reynolds and K. A. Wallander, *J. Pediatr. Health Care*, *4*, 209 (1990).
95. S. L. Nightingale, *JAMA*, *262*, 1296 (1989).
96. M. S. Dunn, A. T. Shennan, E. M. Hoskins, K. Lennox, and G. Enhorning, *Pediatrics*, *82*, 543 (1988).
97. H. L. Halliday, G. McClure, and M. M. Reid, *Early Hum. Dev.*, *13*, 323 (1986).

98. Y. E. Vaucher, T. A. Merritt, M. Hallman, A. V. Jarvenpaa, A. M. Tesley, and B. L. Jones, *Am. J. Dis. Child.*, *142*, 927 (1988).
99. H. N. McCullough and R. L. Juliano, *J. Natl. Cancer Inst.*, *63*, 727 (1979).
100. R. L. Juliano and H. N. McCullough, *J. Pharmacol. Exp. Ther.*, *214*, 381 (1980).
101. R. V. Padmanabhan, R. Gudapaty, I. E. Leiner, B. A. Schwartz, and J. R. Hoidal, *Am. Rev. Respir. Dis.*, *132*, 164 (1985).
102. S. Sone, K. Tachibana, M. Shono, F. Ogushi, and E. Tsubura, *J. Biol. Respir. Mod.*, *3*, 185 (1984).
103. N. C. Phillips, M. L. Moras, L. Chedid, P. LeFrancier, and J. M. Bernard, *Cancer Res.*, *45*, 128 (1985).
104. S. Sone, S. Mutsuura, M. Ogawara, T. Utsugi, and E. Tsubura, *Cancer Immunol. Immunother.*, *18*, 169 (1984).
105. S. J. Farr, I. W. Kellaway, D. R. Parry-Jones, and S. G. Woolfrey, *Int. J. Pharm.*, *26*, 303 (1985).
106. S. G. Woolfrey, G. Taylor, I. W. Kellaway, and A. Smith, *J. Controlled Release*, *5*, 203 (1988).
107. R. M. Fielding, *Proc. West. Pharmacol. Soc.*, *32*, 103 (1989).
108. R. M. Fielding, E. Kerr, A. Wong, T. Tsukamoto, and R. Abra, *Pharm. Res.*, *5*, S-66 (1988).
109. K. M. G. Taylor, G. Taylor, I. W. Kellaway, and J. Stevens, *Pharm. Res.*, *6* 633 (1989).
110. K. M. G. Taylor, G. Taylor, I. W. Kellaway, and J. Stevens, *J. Pharm. Pharmacol.*, *40*, 137 (1988).
111. R. J. Debs, R. M. Straubinger, E. N. Brunette, J. M. Lin, E. J. Lin, A. B. Montgomery, D. S. Friend, and D. P. Papahadjopoulos, *Am. Rev. Respir. Dis.*, *135*, 731 (1987).
112. R. J. Debs, W. Blumenfeld, E. N. Brunette, R. M. Straubinger, A. B. Montgomery, E. Lin, N. Agabian, and D. P. Papahadjopoulos, *Antimicrob. Agents Chemother.*, *31*, 37 (1987).
113. P. R. Wyde, H. R. Six, S. Z. Wilson, B. E. Gilbert, and V. Knight, *Antimicrob. Agents Chemother.*, *32*, 890 (1988).
114. B. E. Gilbert, H. R. Six, S. Z. Wilson, P. R. Wyde, and V. Knight, *Antiviral Res.*, *9*, 355 (1988).
115. V. Knight and B. Gilbert, *Eur. J. Clin. Microbiol. Infect. Dis.*, 7, 721 (1988).
116. C. O'Callaghan, A. R. Clarke, and A. D. Milner, *Eur. J. Pediatrics*, *148*, 473 (1989).
117. D. W. Cockcroft, T. S. Hurst, and B. P. Gore, *Chest*, *96*, 505 (1989).
118. M. H. Schoni and R. Kraemer, *Eur. Respir. J.*, *2*, 887 (1989).
119. P. J. Mihalko, H. Schreier, and R. M. Abra, in *Liposomes as Drug Carriers: Recent Trends and Progress* (G. Gregoriadis, ed.), John Wiley and Sons, Toronto, 1988, p. 679.
120. D. Meisner, J. Pringle, and M. Mezei, *J. Microencapsulation*, *6*, 379 (1989).
121. D. Meisner, J. Pringle, and M. Mezei, Liposomes for pulmonary drug delivery. II. Studies of pyridine-2-aldoxime methiodide, in *Proceedings of the Canadian Federation of Biological Sciences*, Calgary, Canada, 1989.

122. D. Meisner, Ph.D. thesis, Dalhousie University, Halifax, 1989.
123. T. A. McCalden, *Adv. Drug Deliv. Rev.*, *5*, 253 (1990).
124. T. A. McCalden, R. A. Abra, and P. J. Mihalko, *Proc. West. Pharmacol. Soc.*, *31*, 129 (1988).
125. A. Kamarei, A. Wong, and T. McCalden, *Proc. West. Pharmacol. Soc.*, *32*, 33 (1989).
126. I. Gonda, *J. Pharm. Sci.*, 77, 340 (1988).
127. K. M. G. Taylor, G. Taylor, I. W. Kellaway, and J. Stevens, *Int. J. Pharm.*, *58*, 57 (1990).
128. R. W. Niven, M. Speer, and H. Schreier, *Pharm. Res.*, *8*, 217 (1991).
129. R. W. Niven and H. Schreier, *Pharm. Res.*, 7, 1127 (1990).
130. S. J. Farr, I. W. Kellaway, and B. Carman-Meakin, *J. Controlled Release*, *5*, 119 (1987).
131. J. G. McGurk, A. R. Ross, and G. C. Gilroy, *J. Pharm. Pharmacol.*, *38* (*suppl.*), 20 (1985).
132. M. J. Ostro, in *Liposomes as Drug Carriers: Recent Trends and Progress* (G. Gregoriadis, ed.), John Wiley and Sons, Toronto, 1988, p. 855.
133. L. H. Crowe, J. H. Crowe, A. Rudolph, C. Womersley, and L. Appel, *Arch. Biochem. Biophys.*, *242*, 240 (1985).
134. T. D. Madden, M. B. Bally, M. J. Hope, P. R. Cullis, H. P. Schieren, and A. S. Janoff, *Biochim. Biophys. Acta*, *817*, 67 (1985).
135. F. J. Freeman and D. Chapman, in *Liposomes as Drug Carriers: Recent Trends and Progress* (G. Gregoriadis, ed.), John Wiley and Sons, Toronto, 1988, p. 821.
136. Y. Yamamoto, E. Niki, J. Eguchi, Y. Kamiya, and H. Shimasaki, *Biochim. Biophys. Acta*, *819*, 29 (1985).

4

Particulate Delivery: The Challenge of the Oral Route

Alexander T. Florence and Praful U. Jani

University of London
London, England

INTRODUCTION

Much published work on microspheres has been directed to their use as intravenous and intra-arterial targeting and delivery systems. Research on the oral administration of microparticulates has been concerned mainly with their sustained-release potential to achieve both reduced local concentrations in the gut and prolonged plasma levels or limited site direction through enteric coating. A number of years ago it might have been believed that the applications of microspheres for oral use were limited to controlled release. However, with increasing evidence that colloidal particles can pass intact through intestinal membranes, the oral route poses both new opportunities and new challenges. There has been for some time a variety of evidence for the transcellular passage of small particles across the gastrointestinal wall, but there has been considerable controversy about these findings, with the consequence that until recently there have been few attempts to use the potential of oral administration of microparticulate carriers for the systemic delivery of drugs in intact carriers.

This chapter reviews some of the evidence for particulate uptake and the potential, either for drug delivery of nonabsorbed or poorly absorbed drugs by this mechanism or for oral immunization. Viral and bacterial particles can penetrate the mucosal barrier by way of Peyer's patches, which endocytose and transport both macromolecules and microorganisms into the associated lymphoid tissues. It seems not unreasonable that if bacterial particles can be transported from the lumen then so too might some synthetic particles of appropriate size and surface character. Such absorbable particles, loaded with drug, could be used as carriers of labile drug molecules, such as proteins or other poorly absorbed materials, or could be used to enhance the response to oral antigens.

PARTICULATE UPTAKE FROM THE GASTROINTESTINAL TRACT

The oral route is the most convenient and popular route for the administration of medication. Drugs normally are absorbed by passive diffusion or active transport processes from the lumen of the gastrointestinal tract in molecular form. It has been commonly taught that absorption can only occur from solution, as the wall of the mammalian gastrointestinal tract has been assumed to be an impenetrable barrier to the passage of inert particulate matter. O'Mullane and colleagues (1) concluded in a review that "the transport of intact carriers across the gastrointestinal (GI) tract is restricted to exceptional and unusual circumstances." What these circumstances are, and how unusually they occur, has still to be delineated. Although a number of investigators have shown that passage of particles across the intestinal mucosa is possible (2–5), the controversy over the uptake of solid particles in the colloidal size range continues, with some authors claiming, like Volkheimer (6,7), that uptake is the result of paracellular events. Volkheimer claimed that starch granules ranging in diameter from 7 to 10 μm, when administered orally (up to 12 million granules/g), could later be observed in the venous blood. He speculated that the passage of starch and other particulate matter such as carbon across the intestinal barrier resulted from the muscular activity in the *muscularis mucosa*, proposing that this resulted in the "kneading" of the particles between cells at the desquamation zones of the intestinal villi, allowing movement from the intestinal lumen into the subepithelial region. For this type of uptake Volkheimer (6) introduced the term *persorption*, defining it as the paracellular passage of large solid or nondeformable food and other particulate matter, such as polyvinylchloride (PVC), in the micrometer size range (up to 100 μm), through the epithelial cell layer of the intestinal tract.

Before Volkheimer, a number of other investigators had reported the passage of inert particulates across the intestinal mucosa. As early as 1960, uptake was observed of small resin particles, of size 1–5 μm, administered orally in an aqueous suspension to calves (8), the particles accumulating in the tonsils, small intestine, and certain lymph nodes. In 1961, Sanders and Ashworth (3) administered orally to Sprague–Dawley rats a concentrated emulsion of polystyrene latex and were able to detect a large number of particles in the jejunal epithelial cells. Particles between 100 and 200 nm diameter were occasionally observed between the microvilli or in the terminal web and the intracellular spaces of the epithelium, and many were found in the cytoplasm of the jejunal epithelial cells, where they were always enclosed in vesicles.

Probably the most extensive work carried out to date is that by LeFevre and his group (5–12) who showed, perhaps for the first time, that the epithelial *gut-associated lymphoid tissue* (GALT) was the route of entry for 2-μm polystyrene latex particles administered to mice during chronic feeding of up to 61 days duration. LeFevre et al. (11) also showed that chronic administration of particles of diameter 6 and 16 μm for 60 days failed to cause particles to appear in the blood and body tissues. Histological examination of tissues after dosing with a very concentrated suspension of 16-μm particles for 60 days failed to reveal the presence of particles; 6-μm particles were found mostly adsorbed in the submucosal layer of the Peyer's patches, none being found in significant numbers in blood or other body tissues, either after short-term (3 days) or long-term (60 days) administration.

In spite of the evidence that has been published, there has been no universal acceptance of the phenomenon of colloidal particulate absorption. The question still remains of the circumstances and reproducibility of colloidal uptake as a prelude to drug delivery. Several issues need to be resolved, such as the route by which colloidal drug carriers cross the endothelial barrier and whether or not the process of uptake is consistent, whether or not it is saturable, what factors affect the process, and, not least, what are the kinetics of particulate uptake and disposition?

MACROMOLECULAR AND BACTERIAL UPTAKE BY PEYER'S PATCHES

There is, of course, evidence that small quantities of macromolecules can penetrate mucosal surfaces of the small intestine intact. The concept is well accepted of the transference of maternal immunoglobulin via breast milk to the newborn, thus allowing transmission of passive immunity to the young. Several workers have demonstrated uptake of macromolecules across the GI tract using tracer proteins such as horseradish peroxidase (HRP) (13–16), tritiated bovine serum albumin (17), cationized ferritin (18,19), IgA (20), and γ-lactoglobulin (21–23). These substances have been found to be absorbed via the enterocytes as well as the solitary and aggregated lymphoid follicles such as the Peyer's patches. Absorption from the GI tract of the rabbit of Ru 41740 (a glycoprotein extract from *Klebsiella pneumoniae*) was found to be greater from the duodenum than from Peyer's patches (24), a result similar to that obtained by Owen (13) for the glycoprotein HRP. However, most of the Ru 41740 crossing Peyer's patches was still in a high molecular weight form, whereas the fraction absorbed from the duodenum had been degraded into lower molecular weight species. Many workers have demonstrated the uptake of various macromolecules in the specialized regions of the GI tract and through specific and nonspecific sites, e.g., lymphoid tissue or through the pinocytotic vacuoles of the enterocytes, respectively. The extent of uptake has also been found to depend on age (22). The neonatal mammalian small intestine, like the reticuloendothelial (RE) system in the human adult, has a capacity to ingest macromolecules by endocytotic mechanisms in the transference of maternal immunoglobulin to the newborn (14,15,25).

Recently, there has been renewed interest in the role of the failure of the intestinal barrier in the development of systemic infection. Deitch (26) states that, based on current knowledge, bacteria do not translocate from the normal gut because of the presence of an intact epithelial barrier, normal host immune defenses, and a GI flora that prevents bacterial overgrowth. However, bacteria [such as *Vibrio cholerae* (27) and *C. jejuni*] and viruses have been found not only to be associated with the Peyer's patches but to translate across the epithelial barrier via the M cells. Walker and Owen (28) and Grutzkau et al. (29) discuss M-cell uptake and transport mechanisms of virulent organisms, features that might be utilized in the design of delivery systems with specific affinities for Peyer's patches. Newton et al. (30) have, for example, incorporated the cholera toxin epitope into *Salmonella* flagellin. Walker and Owen (28) suggest that improved delivery systems will facilitate the use of lymphokines and immunomodulators, thereby giving the potential to enhance nonspecific resistance to enteric pathogens. The oral route thus offers several challenges: in the design of carriers that are both resistant to the ravages of the GI milieu, and that are taken up quantitatively by the epithelial cells for transport to the systemic circulation where this is required, or to induce cellular or humoral immunity.

POSSIBLE ROUTES OF ENTRY OF PARTICULATES

There are three possible modes of uptake of antigens and colloidal particles, some less efficient than others: transport across intact epithelial membranes (transcellular route) possibly by endocytosis, transport through endothelial tight junctions (the paracellular route), and transport via the M cells of Peyer's patches (Figure 1). Absorption of macromolecules and particulate matter involving the process of endocytosis is complex and involves formation of pinocytotic vacuoles, a process dependent on both the hydrophobicity of the membrane and the molecule or particle, among other factors such as particle

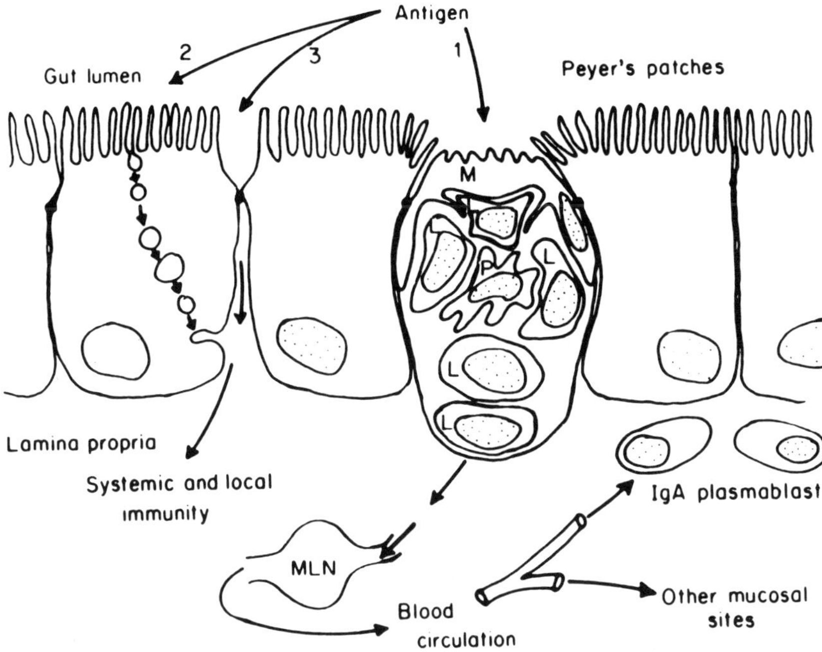

Figure 1 Schematic representation of the mode of entry of food antigens across the intestinal mucosa. The antigens in the gut lumen enter via (1) the specialized microfold epithelial (M) cells of the Peyer's patches in the normal intestine, (2) the enterocytes in the immature intestine, or (3) the interepithelial tight junction in osmotically stressed intestine.

charge. Non-receptor- and receptor-mediated endocytosis also requires the solute in question to make contact with the membrane and form enclosing vacuoles (35). The chemistry of the material under test will play an important role in the presentation and eventual uptake by endocytic processes (32,36). There is no good evidence that particulate matter, particularly in the size range above 50 nm, is taken up by pinocytosis.

The epithelial tight junction will facilitate the transport of some molecules such as lipids and proteins. As demonstrated by Cereijido and colleagues (31), the monolayer of tight junctions at places may behave as a natural "leaky epithelium." Along the GI tract there are fixed and numerous negatively charged sites and this fact coupled with the ubiquitous negative characteristics of cell membranes makes the GI tract prevalently a negatively charged organ (32–34).

PEYER'S PATCHES

In investigation of macromolecular and particulate uptake most attention focuses on the specialized epithelia and structures of the GALT, in particular Peyer's patches, which are aggregates of lymphoid nodules. While solitary lymphoid nodules may occur along the entire intestine, they are more numerous and concentrated in the ileum, where they are recognized in aggregate form as Peyer's patches (Figure 2). The intestinal epithelium overlying Peyer's patches has mainly columnar epithelial cells which are seen to lie upon the basement membrane (37). Beneath the basement membrane, there is a layer of flattened connective tissue cells with elongated processes separated from one another as well as from the epithelial layer. In the interval between these processes and dense lymphoid tissues there are loose reticular connective tissues containing blood and lymphatic vessels, occasional smooth muscle cells and unmyelinated nerve fibers, as well as lymphocytes, plasma cells, eosinophil and neutrophil granulocytes. Sometimes lymphatic blood vessels are related to elongated cell processes and appear to follow the contour of the epithelium of the intestinal gland.

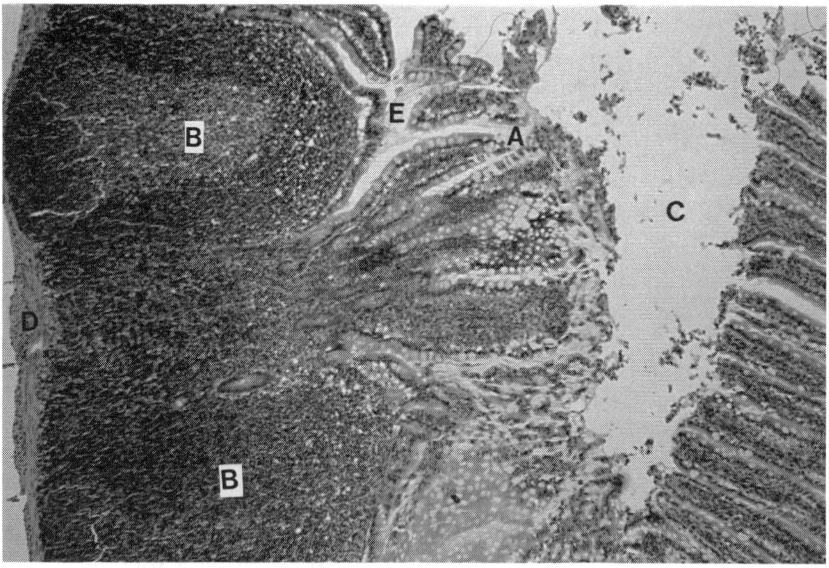

Figure 2 Photomicrograph (×200) of a section of a Peyer's patch of a rat. The overhanging villi (A) nearby the lymphoid follicles (B) are clearly seen. C is the intestinal lumen and D the serosal layer, featured in later figures. E marks the position of M cells, which cannot be identified by light microscopy.

Peyer's patches are generally oval in shape and are usually found on the side opposite to the mesenteric (i.e., the antimesenteric) wall of the intestine. Their number and size is relevant to the extent of colloidal particle uptake. In humans the number of Peyer's patches increases with age from about 45 at 24 weeks of gestation to around 300 at 12–14 years of age, declining after puberty to an average of 200 at 20 years and then further decreasing but more slowly to around 100 at the age of 75 years (38,39). Morphological studies carried out in different species, e.g., mice, rodent, rabbit, and human, have shown that there is no difference in the structure and physiological properties of Peyer's patches (39–42). Although age affects their numbers, it does not appear to affect function. LeFevre et al. (12) administered 1.8 μm rhodamine polystyrene microspheres to young (24 days) and old (18 months) mice. After 25 days oral administration, he found no significant differences in the uptake of particles between the young and the old. So it is debatable if there is a significant change in the *function* of Peyer's patches after puberty.

Peyer's patches occur more frequently in the distal areas of the small intestine and the terminal section of the colon, and there are follicular lymphoid aggregates like Peyer's patches around the appendix. Both light microscopy and electron microscopy (40,43) have shown that each Peyer's patch is divisible into four zones: (a) the germinal center, (b) the small lymphocytic areas, (c) the interfollicular area, and (d) the subepithelial zone (see Figure 2). Intraepithelial lymphocytes (IEL) compose the largest population of lymphocytes in the body and by the virtue of their position have the closest direct relationship with foreign antigens. Thus they have an important role to play in the intestinal mucosa and the GALT (44,45).

The membranous epithelial cell layer or the follicle-associated epithelium (FAE) that lines the lymphoid follicles of Peyer's patches is composed of absorptive cells, goblet cells, M cells, and enteroendocrine cells. The FAE overlying Peyer's patches has two important characteristics, compared with the adjacent villus epithelium, which contribute to its ability to transport macromolecules as well as minimizing macromolecular degradation on passage. The first is that the FAE has fewer goblet cells and consequently less mucus is produced, a circumstance that allows greater contact between luminal contents and the FAE, facilitating, for example, antigen sampling and processing. The second most important difference and feature is the presence of M cells. These cells are specialized antigen-transporting cells that are interspersed among the columnar epithelial cells of the FAE. The derivation of membraneous epithelial (M) cells is unknown (46–49), but it was recently suggested that M cells differentiate from absorptive cells on the follicular epithelium. To examine whether M cells, like other intestinal epithelial cells, derive directly from undifferentiated crypt cells, Bye et al. (50) studied selected functional features, the proliferation and distribution of Peyer's patches, and M-cell structure ranging from mature to immature. Most immature M cells lack the central cytoplasmic hollow found in the mature M cells that contain lymphoid cells, the focus of uptake; hence the immature M cells behave like the other intestinal epithelial cells until maturation takes place.

Both germinal center and lymphocytic zones are permeated by small capillaries, and in the interfollicular zones lymphatic vessels and large venules are located. Postcapillary venules are found in the intervening region. The structure of the postcapillary endothelial venules suggests transportation of lymphocytes and antigenic particles from the Peyer's patches to the lymph nodes takes place by this route (51) (Figure 4). The subepithelial zone immediately below the membrane epithelial layer of the M cells has a large number of macrophages and plasma cells (40,43,52).

To understand more fully the process of uptake and the fate of particles after uptake, it is important to consider the structure and function of all parts of the gut-associated lymphoid tissue (Figure 3).

The Germinal Center

The Peyer's patch has three cell types: (a) lymphocytes, which are subdivided into large (13–15 μm in diameter), medium (8–12 μm in diameter), and small (less than 8 μm in diameter); (b) the "apoptotic (tingible) body" macrophages, which are macrophages laden with phagocytosized lymphocytes, approximately 20–25 μm in diameter, and (c) the dendritic reticular cells, stellate in shape, with attenuated processes embracing the lymphocytes, forming the scaffolding of the germinal center.

The Small Lymphocytic Zone

The small lymphatic zone partially surrounds the germinal center. The great majority of cells are typical small lymphocytes, with only a small number of reticular cells and very few large or medium-sized lymphocytes. A very small number of "apoptotic (tingible) body" macrophages are seen in this zone (39,43); hence in comparison to the germinal center, phagocytosis of degenerated lymphocytes and other foreign bodies is minimal. There is, however, evidence of moderate macro- and micropinocytosis. In some cells, cytoplasmic vesicles and vacuoles have been seen to contain 50-nm viral particles (53,54).

The Interfollicular Zone

The interfollicular zone has cells that are loosely packed with larger intercellular spaces. The majority of the cells in the interfollicular area are small

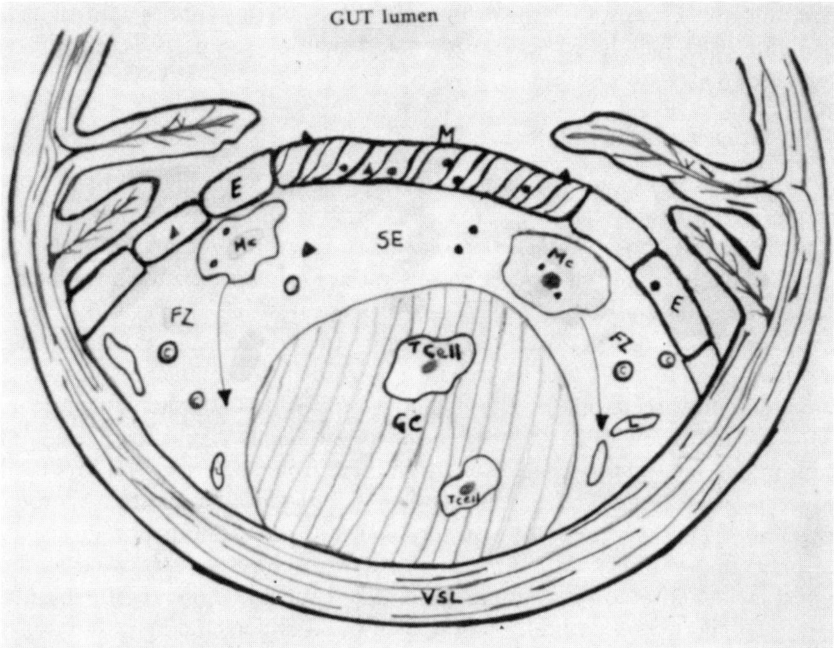

Figure 3 Schematic diagram of a Peyer's patch lymphoid follicle with germinal center surrounded by overhanging villi (V) (with lymphatic vessels); SE, subepithelial zone; FZ, interfollicular zone; E, follicular-associated epithelial surfaces; Mc, macrophage; M, M cell; VSL, vascularized serosal layer (contains mainly lymphatic and blood vessels which form the mesentery supply to the lymph nodes); GC, germinal center with T cells; L, lymphatic vessels; C, endothelial capillary venules; ● = antigen or particulate matter; ▲ = protein or macromolecule. Lymphatic vessels surround the periphery of each follicle and the capillary venules and are mostly present in the interfollicular area; tiny capillaries are scattered throughout the follicles. Antigen or particulate matter is brought into the Peyer's patches through M cells that themselves have phagocytic and pinocytic activity and thence into cells like the Mc. The central shaded area is the germinal center where there are mainly B and T lymphocytes, where an antigen such as a virus which may have been processed (opsonized) in the gut lumen by the secretory IgA antibody is presented to a T cell. The T cells then traverse to the serosal layer via the lymphatic supply, which collects on the antimesenterial wall of the Peyer's patches and thus to the mesentery lymphatic nodes.

lymphocytes. However, as the more central area of this zone is approached, medium-sized lymphocytes, immature and mature plasma cells, and macrophages increase in number (39,40,43).

The Subepithelial Zone

The subepithelial zone is characterized by a large accumulation of macrophages and plasma cells. The cytoplasm contains numerous lysosomal granules similar to those of the apoptotic body macrophages in the germinal center. These cells are phagocytic mainly toward extraneous substances. There is a greater abundance of plasma cells (plasmocytes that are derived from B lymphocytes and are active in the formation of antibodies) in this zone than in other parts of the Peyer's patches. The rest of the cellular population is predominantly medium-sized lymphocytes (39,40,43). In the subepithelial area, accumulation of macrophages and plasma cells occurs as a result of continuous and intense stimulation by antigens from bacterial, viral, and food sources, which enter the Peyer's patches (23,55–58). In addition to their large number, the cytoplasm of macrophages in this zone is primarily rich in lysosomal granules, in contrast to that of macrophages observed in the follicular area. Some cytoplasmic vesicles and vacuoles contain 50- to 100-viral particles (53,54), a finding significant to an understanding of the translocation of colloidal particles.

CELL PROLIFERATION IN THE GI TRACT

Epithelial cell proliferation is of interest because of the possibility of saturation of particulate uptake that might occur in an otherwise static cell population. It is believed that there is controlled cell proliferation and cell replacement processes in the GI tract of both experimental animals, such as rats, and of humans (59). The whole population of macrophages from the center of the Peyer's patches will be replaced entirely after 20 hr, and will mature and be ready for action after 36 hr. If a dose of 0.1 ml polystyrene (about 2×10^{10} particles) is phagocytosized by this macrophage cell population, on chronic administration there will be an increase in cell demand that might reduce the probability of high uptake.

Loss of villous architecture with age may occur but little is known about this (60). It is a well-known fact that the conventional process of drug absorption will be controlled by physiological factors such as intestinal blood flow (61), neurohumoral factors (62), and the presence of food (25,63), but many complex factors and intestinal architecture will affect the absorption of particulates and macromolecules because the movement of particles in the GI tract involves colloidal particulate interactions, obstruction effects in the complex milieu of the lumen, and convection and diffusion.

Role of Mucus

The general lack of susceptibility of the mammalian GI tract to attack by microorganisms and dietary antigens may represent a dilemma when attempting to explain and rationalize the uptake of colloidal carriers. There are along the GI tract significant elements of defense in the mucosal layer, such as layers of mucus and mucin gel, and glycocalyx (composed mainly of carbohydrates) (60,64–66). If so, it might also represent a barrier to particulate uptake. Mucus and its gel-forming glycoprotein component, mucin, overlies the surface of the villous epithelium. Like many gels, mucus has the ability to change its physical state from that of a viscous, mobile solution to a relatively impermeable state with fluid-rejecting properties. The fluidity of the mucus and mucin can be altered, for example, by surfactant (33,67–69), a factor that perhaps can be exploited in attempts to increase the extent or rate of uptake. Reduced viscosity of mucus will result in increased uptake of dietary protein, in high molecular mass form, and particulate matter, along the lamina propria (19) and mucosal surfaces. Experiments with the behavior of mucus and colloidal particle show that mucus can cause aggregation, as can very low pH. If microspheres were administered with an appropriate surfactant, then the extent of aggregation of the microspheres might be reduced and uptake may be increased (11,70), but if any additive results in excess mucus and mucin production this may counteract such an effect (166). It is difficult to define precisely the extent of uptake and the consequences of mucus-microsphere aggregation, but it can be surmised that this must be one vital factor reduces the overall quantitative uptake of the microspheres from the GI tract. In the glycocalyx, the major sources of the carbohydrate, enzymes composed mainly of sucrase, maltase, and aminopeptidase, are attached to the external surface of the membrane and can easily be removed by proteinases and detergents. Goblet cells that secrete mucus to entrap microorganisms and undigested macromolecules and impede their attachment to the epithelium are largely absent over the antigen-trapping epithelium of lymphoid follicles or are in small numbers but are nevertheless present to slow down and decrease uptake.

LAMINA PROPRIA

In addition to the organized lymphoid tissues of the GALT, the small intestinal mucosa contains large numbers of lymphoid cells scattered throughout the wall of the intestine, in both the lamina propria and the epithelial layer. Uptake of microparticles may also take place from these "solitary" regions. The normal lamina propria is heavily infiltrated by a wide variety of lymphoid cells, including lymphocytes, plasma cells, macrophages, eosinophils, basophils, and mast cells (71). Although macrophages are only rarely found in the villous epithelium, they are abundant in the lamina propria, frequently extending cytoplasmic protrusions into the epithelium, enabling the sampling of luminal antigen (71,72).

LYMPHATIC SYSTEM AND TRANSPORT OF ABSORBED PARTICLES

Materials taken up via M cells usually are transported away from the Peyer's patches by the lymphatics (Figure 4). The lymphatic system consists of two elements: (a) an extensive network of vessels containing lymph and (b) scattered masses of lymphoid tissues with which these vessels may be associated. Lymph, the interstitial tissue fluid that enters the lymphatic vessels, comprises mainly plasma proteins, such as albumin, fibrinogen, and globulins, which have initially leaked out of the capillary fenestrations, but also foreign antigenic

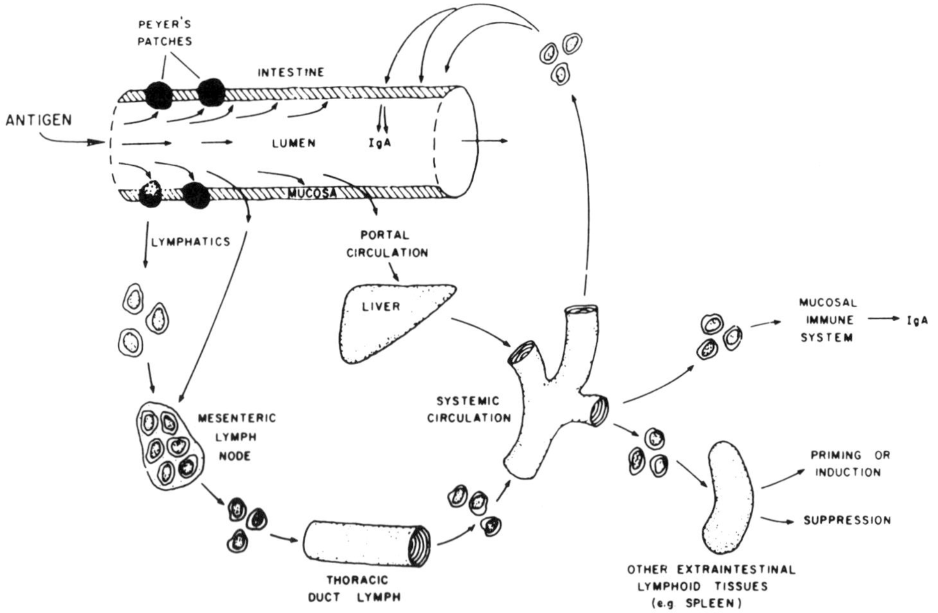

Figure 4 Representation of the course followed by antigen in the gastrointestinal tract which might explain oral tolerance. It is assumed that small colloidal particles taken up by Peyer's patches would follow similar routes, although size constraints will limit the translocation of some colloids. Reproduced with permission from Kagnoff et al. Ann. N.Y. Acad. Sci. 1982, *392* (p 248–65).

substances from the gut. It also contains water-insoluble fats especially those absorbed from the intestinal lumen. Lymph from all regions of the body finally enters the blood from the thoracic duct, into the left subclavian (innominate) vein and the right lymphatic duct into the right subclavian vein (73).

Lymphoid tissue, especially the aggregated nodules of the Peyer's patches as we have discussed, plays an important part in the host defense system and uptake of the particulate matter from the GI tract. The lymph nodes constitute an extensive screening system to prevent the systemic distribution of infectious and particulate matter. In the vertebrate, however, as the concentration of the infectious and particulate matter increases, the flow of the lymph transports it to other parts of the body for further processing in the liver, spleen, and the bone marrow.

Lymphatic Vessels

The network of lymphatic capillary and large vessels originates in connective tissue spaces and is seen to have gaps in their walls, thus making a relatively open system. For example, cell junctions have gaps 15 nm in width; the cell junctions are devoid of desmosomes. There appear to be no intercellular tight junctions (74) as occasionally gaps of 100 nm width have been seen between neighboring cells. Large and medium-sized lymphatic vessels, especially in the mesentery, have an inside diameter of 100–200 μm. Like blood vessels, lymphatic vessels compose a system of endothelium-lined tubes but, unlike blood vessels, they do not form a continuous circular system. originating in the connective tissue spaces and carrying their contents, in one direction, toward the base of the neck which empties into the subclavian vein (74,75).

The GI tract is supplied with lymphatic vessels that arise as blind tubes in the mucosa and submucosa. In the submucosa the larger lymphatic vessels branch freely and have numerous valves. They cross the muscle layer spreading into the intermuscular tissue and serosa and pass through the mesentery network to the mesentery lymph nodes. The flow of the lymph is from the mesentery to the efferent lymphatic, first into cortical or subcapsular sinus, and from there, the medullary sinuses into the efferent lymphatic and then into the thoracic duct. Lymph-containing particles such as microspheres and other matter either pass through the medullary sinuses, which may be clear channels, or traverse to a variable extent through a reticular edifice, consisting of cells that hold up the microspheres and viral particles from the GI tract. They are also actively phagocytocized by the abundant macrophages, free and fixed in the lymph nodes (74,75).

Lymphatic Tissue and Lymph Nodes

Lymphoid tissue is divided into "central" and "peripheral" components. Gut-associated lymphoid tissue such as Peyer's patches and gut-associated lamina propria compose the central lymphoid tissue, while lymph nodes (especially those of the legs, skin, and other parts of the body) are regarded as the peripheral lymphoid tissue. The immunological functions are carried out in the peripheral region, while the mesentery lymph nodes are the bridges between the central and peripheral tissues (73–75).

The outer capsule and trabeculae of lymph nodes are composed of fibrous connective tissue. The subcapsular sinuses and trabecular sinusoids are lined with macrophages, and there are many macrophages and lymphocytes among the interstitial cells that surround the lymphoid follicles. The lymphoid follicles contain densely packed masses of small lymphocytes with macrophages scattered between them. Although all mammals possess lymph nodes, there is a little variation between species. For example, in monkey and in humans, the mesentery lymph nodes appear to be smaller, discreet, and more numerous than in animals such as dog, cat, rabbit, and rat. In lower animals, though, the lymphoid tissue in the mesentery tends to form one large node (75).

THE RETICULOENDOTHELIAL SYSTEM

In order for a drug–carrier complex to successfully reach target cells without the reticuloendothelial system, it must not only be able to exit from the circulation, passing the endothelial and basal lamina barriers, but must also be able to escape the grasp of the reticuloendothelial system (RES). Mononuclear phagocytic cells originate from precursors in the bone marrow, enter the circulation as monocytes, and then pass into various tissues where they differentiate into macrophages, chiefly concerned with phagocytosis. In connective tissue and the lymphatic system they may occur as free or fixed macrophages or form part of the lining of the sinusoids in the liver (Kupffer cells) or spleen.

The cells most closely involved with foreign particle uptake are the Kupffer cells of the liver and the splenic macrophages. The Kupffer cells lie astride the channels of the liver sinusoid and thus are in an ideal position to engulf passing particles. In the spleen, especially the red pulp, sheaths of macrophages border the sinusoids. These cells are in an advantageous position to take up circulating particles.

Macrophages take up not only particles but also proteins capable of interacting with macrophage surface receptors (76). The protein–receptor complex is internalized into a vesicular structure that fuses with lysosomes;

the internalized protein is then usually degraded while the receptors may recycle to the cell surface, resembling the process of receptor-mediated endocytosis seen with less active phagocytic function.

Nonionic surfactant block copolymers have been used to coat microspheres to prevent their sequestration by the RES (77,78). It might be that block copolymers such as the poloxamer Pluronic F-68 can be exploited not only to modify the surface properties of microspheres and perhaps also to decrease the viscosity of the mucus barrier, so that more of the microspheres will be free for presentation to uptake sites such as the M cells. If the surfactant coating was retained it might well also reduce uptake of the translocated particles by the liver, but it is not yet known if such coating will also reduce uptake from the gut.

RECENT EVIDENCE OF MICROPARTICULATE UPTAKE

In the work carried out by the present authors, originally in the University of Strathclyde, Glasgow in collaboration with Dr. G. W. Halbert, and later in London, fluorescent polystyrene nanoparticles and microparticles of various sizes and different surface characteristics were administered orally for 10 days to female Sprague–Dawley rats. The aim was to investigate the uptake, if any, from the GI tract and the possible translocation of the microspheres from the lumen of the experimental animals. The results are discussed here in some detail in the context of those studies reported earlier in the chapter and later discussions of the fate of systems comprising drug-loaded carrier particles.

Polystyrene is not biodegradable and has low toxicity; thus it can be used with assurance that the integrity of the intestinal barriers will not be compromised, avoiding the complication of interpretation that would follow from macromolecular degradation and paracellular uptake. Most importantly, the particles are nonimmunogenic so there is no opportunity for them to enhance their own uptake or influence uptake by lymphoid tissue. Polystyrene is widely used as a diagnostic tracer tool in animal and human investigations (79–85). Monodisperse, negatively charged carboxylated microspheres with covalently linked rhodamine and nonionized microspheres with covalently linked fluorescein were used in the work (86–88), which is described below in some detail.

The Experiments

The microspheres were administered orally by gavage needle. A dose of 12.5 mg/kg body weight (equivalent to 0.1 ml) was administered once daily for 10 days. The animals were given free access to food and water, but food was

Table 1 Uptake in Various Tissues After Oral Feeding of Microspheres

Particle type	Stomach	Peyer's patches	Small intestine	Mesentery node	Colon	Liver	Spleen	Kidney	Heart	Lungs
YG	+ +	+ + + +	+ +	+ + + +	+ + +	+ + +	+ +	○	○	○
YO	+	+ +	+	+	+ +	+ +	○	○	○	○

Note: YG = fluorescein dye nonionized (monodispersed) microspheres of size 0.13 μm and 0.95 μm. YO = rhodamine dye carboxylated (negative) (monodispersed) microspheres of size 0.1 μm and 1.1 μm. ○ = no evidence of latex particles, + = very low uptake, + + = low uptake, + + + = moderate uptake, + + + + = high uptake.

removed overnight (about 8–10 hr) prior to the morning dosing of the microspheres. Food was restored at least 1–2 hr after the dosing to ensure a clear passage of the beads through the GI tract. Daily urine and feces were collected and occasionally, at random, the urine was freeze-dried and examined for the fluorescent beads. The animals were weighed daily and kept in metabolic cages to ease the collection of urine and feces and to prevent coprophagia. On the last day, the final dose was administered and the animals kept for 2 days in a microsphere-free environment to clear the GI tract of any unabsorbed polystyrene microspheres. Before sacrificing, the animals were fasted for 15 hr to clear the gut of food particles, but to further facilitate the removal of the unabsorbed microspheres access to water was allowed. In later experiments, single oral doses have been administered to assess the extent and rate of uptake (Jani and Florence, unpublished). After sacrifice, the rats were dissected avoiding any cross-contamination at all times. The various organs were carefully removed, weighed, and frozen as described by Jani et al. (86–88). Some methods of traditional sectioning and fixing the tissues in absolute alcohol and clearing using chloroform destroy the polystyrene microspheres. Sectioning and mounting using the frozen method was preferred. On the whole this is the most innocuous of all the histological methods. The first evidence of uptake was histological, later confirmed by analysis of tissue for polystyrene using gel permeation chromatography (87).

Data gleaned from histology are summarized in Table 1. The microphotographic evidence (see Figures 5–12) shows clear evidence of uptake in a size-dependent manner. One-micrometer-diameter polystyrene microspheres were taken up less efficiently than the smaller (50-, 100-, 300-, and 500-nm) nonionic fluorescent microspheres. When viewed under the ultraviolet fluorescence microscope the tissues of Peyer's patches, villi, lymph nodes, liver, and spleen of the rats fed on the nonionic microspheres showed evidence of the presence of the microspheres. This visual evidence, summarized in Figures 5–12, shows that the microspheres are distributed in organs but associated with specific structures such as the serosal layer, the mesentery lymph vessels, and liver sinusoids. The animals fed carboxylated (negatively charged) fluorescent microspheres showed evidence of a smaller extent of uptake. Even the smaller, 100-nm particles of carboxylated microspheres appeared not to be absorbed to a significant extent in comparison to virtually identical diameter nonionic microspheres, although the evidence is not unequivocal because of the difficulty in distinguishing the rhodamine fluorescence from the autofluorescence of some of the body tissue components such as red blood cells. The evidence shows that the microspheres are translocated from the submucosal and serosal layer of the Peyer's patches to the lymphatic nodes via the lymphatic and blood vessels of the mesentery network. Most importantly, there is substantial evidence of these micro-

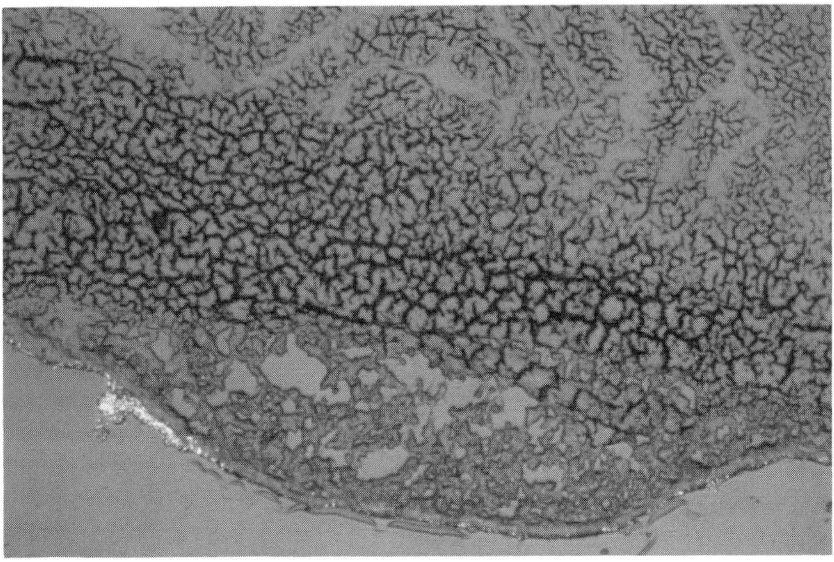

Figure 5 Photomicrograph (×200) of a Peyer's patch (obtained by frozen section) showing a collection of 500-nm nonionic fluorescent microspheres arranged uniformly in the serosal layer, after oral feeding (as described in text) daily for 10 days. Some microspheres are in the mucosal and submucosal layer. Comparison of a similar section of villi shows that more microspheres accrete in the Peyer's patches than in villi; the microspheres in villi occur mainly in granular structures.

spheres in the Peyer's patches (Figures 5–7), confirming LeFevre's findings (10,12) on the main site of uptake. As discussed above, the Peyer's patches are richly supplied with lymphatic vessels that arise as blind tubes in the mucosa, while in the submucosal layer the larger lymphatic vessels branch freely and have various openings.

It has been observed in this work that microspheres will travel from these to the mesentery nodes, evidence supported by Eldridge's data (89). Figure 8a–c shows the frozen sections of the mesentery network and minor lymphatic vessels and veins containing some 500-nm nonionic fluorescent polystyrene spheres, whereas Figure 8b is a magnification of a section of 8a and 8c the corresponding bright field photomicrograph showing the surrounding tissue serosal architecture.

Our experiments showed that no 3-μm microspheres were located in any major target organs, such as mesentery network, lymph nodes, liver, and spleen, even after oral administration for 10 days. None were found in the

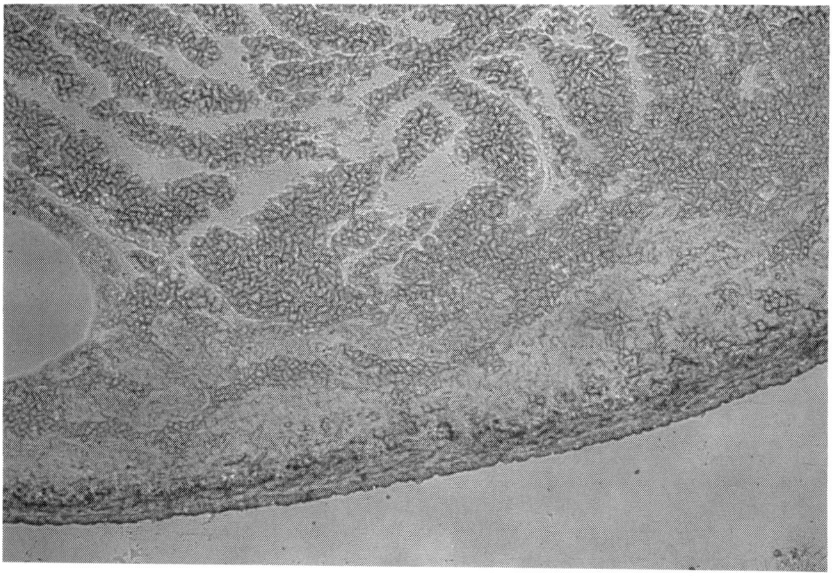

Figure 6 Photomicrograph (×200) showing a few 500-nm nonionic fluorescent microspheres in the serosal layer in a region of gut not associated with Peyer's patches (section prepared by frozen sectioning) after administration of the latex spheres daily for 10 days as described in the text. Latex spheres can also be seen in the submucosal layer and mainly appear in the granular cells.

serosal layer of the Peyer's patches unlike the 1-μm, 500-nm, 300-nm, 100-nm, and 50-nm microspheres. However, the 3-μm microspheres were found to be adsorbed and immobile within the submucosal layers of the thicker mucosa of the stomach and also of the Peyer's patches (see Figure 11). Eldridge et al. (89) and LeFevre et al. (11) also reported similar findings, with particles of about 6 μm in diameter.

The study of the translocation of the microspheres could be taken a stage further by cannulating the hepatic bile duct and jugular vein of an experimental animal (167,168). Recovering the microspheres in the bile and blood selectively would then show that the microspheres follow the RES blood clearance. The microspheres while circulating in the lymph would eventually enter the liver and general circulation (Figure 12). The venous circulation would then take the microspheres to the liver sinusoids where they (Figure 12) would be phagocytosized by the Kupffer cells with some of them ending up in the bile canaliculi and later excreted into the bile duct. Recently, Alpar et al. (70) demonstrated the ability of 1.1-μm albumin particles to reach the

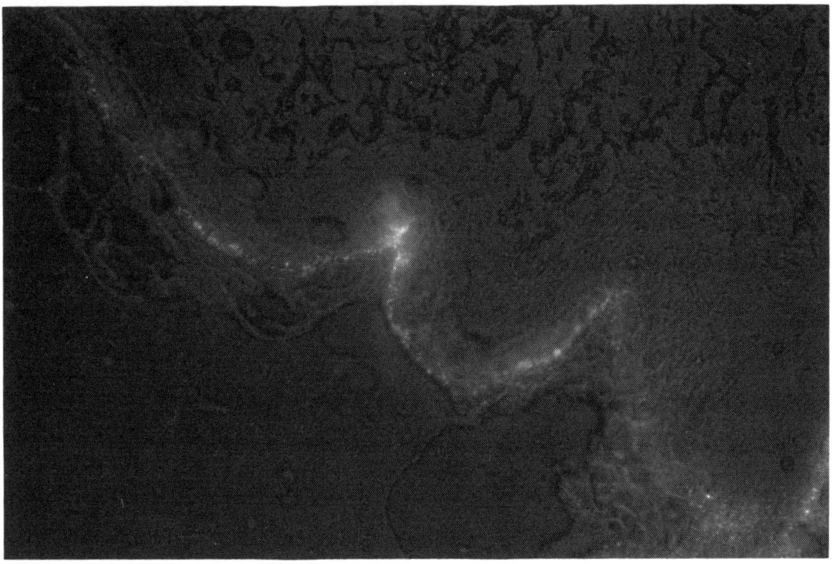

Figure 7 Photomicrograph (×200) of a folded section of the serosal layer of a Peyer's patch with a collection of 300-nm nonionic fluorescent beads, again emphasizing that the major accumulation of the microspheres after oral feeding is specific and consistent in this and similar regions.

general circulation, within 40 min of oral administration. Our data clearly show the uptake and subsequent translocation of the microspheres from the GI tract absorptive surfaces.

IS THE ROUTE A POTENTIAL ROUTE FOR DRUG UPTAKE IN CARRIERS?

Lymphoid organs in all locations are characterized by an epithelium that differs from that of the adjacent areas and that has features, as discussed above, which enhance trapping and uptake of antigenic and particulate materials. This contrasts with the remainder of the epithelium, which forms a barrier to large particulates, organisms, and undigested macromolecules. The area occupied by the specialized uptake regions (which can be considered to be the target for particle association prior to delivery) is relatively small and, so far, no means of targeting carriers to the Peyer's patches has been shown to be effective. Yet particles are taken up, albeit in small quantities. It is not essential that to be useful in therapeutic applications the uptake by this route be a high percentage of dose administered, but obviously this is normally a requirement. If there are few alternative modes of administering a drug or a vaccine, then the potential of this specialized route should be pursued. With vaccines, of course, the dose–response relationship is somewhat complex and does not require absorption of a large percentage of the administered antigens.

What Is the Extent of Uptake?

Crucial to any consideration of oral administration of carriers is the question of the percentage of the administered dose that is absorbed. Possibly equally important is a knowledge of the proportion of the particles absorbed into Peyer's patches that are then translocated to other sites. The results derived in our laboratories from quantitative measurements (87) confirm the presence of significant amounts of polystyrene microspheres in the size range 50 nm to 1 μm in several organs (Figures 13 and 14). The largest microspheres studied (3 μm diameter) were not found in the liver, spleen, and blood. Total polystyrene in the liver, spleen, bone marrow, and blood possibly shows the truest extent of translocation, as it avoids interpretation of analysis of tissue in contact with the lumenal contents which may have both *adsorbed* and *absorbed* material present. The highest level of uptake is around 7% of the administered dose after 10 daily doses for 50-nm particles decreasing to about 1% for 1-μm particles (Figure 16).

No latex particles were found in heart and lungs, but 50-nm latex particles were found in the kidney (Table 2). There was no histological or chromatographic evidence of the microspheres of other sizes in the kidney. The 100-nm latex microspheres seem to represent the upper limit for the transport into the bone marrow.

Although the 3-μm latex microspheres were detected neither in the systemic circulation nor in the liver and spleen, the histological evidence showed that these microspheres were *adsorbed* and *immobile* within the submucosal layer of the thicker mucosa and the mucosa of the Peyer's patches. Other workers (11,89,90) obtained similar results with larger sizes of latex particle. Hence such particles may have value in delivering drug close to the mucosal surface. In general, it may not be essential for absorption of the carrier particles to occur. In the stomach we have evidence of small as well as large microspheres trapped in the gastric glandular pits. These may be unlikely sites for the uptake of the microspheres but may prove to be a powerful area to exploit for storage and release of some therapeutic entities.

Single oral dosing experiments (Jani and Florence, unpublished) show that microspheres in the size range 50 nm to 1 μm are quickly translocated into the serosal layer (Figure 17a, b) and thence into the systemic circulation.

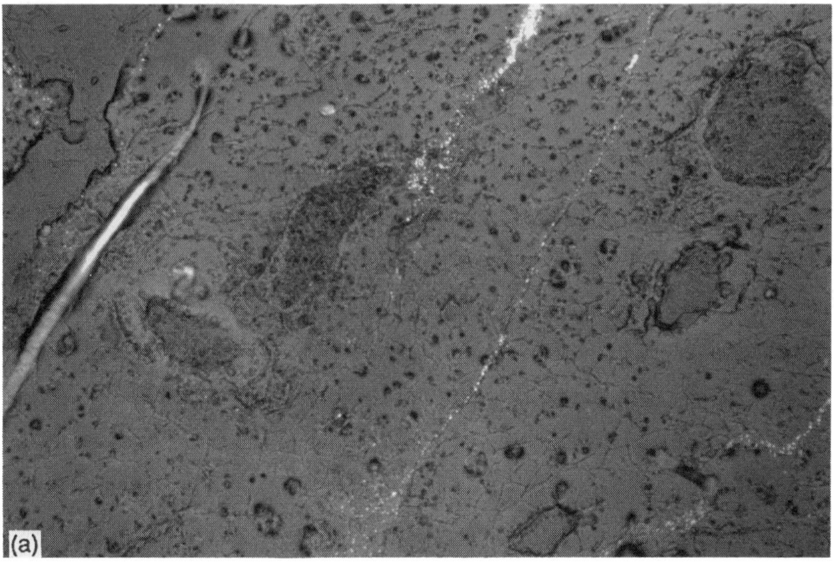

Figure 8 (a) Photomicrograph (×200) of a frozen section of mesentery lymph node connective tissue network, demonstrating lymph vessels and 500-nm nonionic fluorescent microspheres translocating toward the lymph nodes. (b) A magnification (500×) of a portion of Fig. 8a showing mesentery vessels, translocating 500-nm nonionic fluorescent microspheres, cut in the horizontal plane. (c) Photomicrograph of the above at the same magnification with bright field illumination demonstrating the outline of the mesentery vessel among the mesentery connective tissue.

Adsorption of smaller and larger microspheres to the mucosal and submucosal layer of the GI tract is undoubtedly a prelude to their absorption.

Factors Affecting Particulate Uptake and Distribution

There are now several reports in the literature on the size dependence of particle absorption from the small intestine (10,11,70,86–91). No studies exist on the extent of uptake of particles below 50 nm in diameter. There are, however, limited data on the size constraints on subsequent transport via the lymphatic system. We have shown that microparticles can gain access to the mesentery lymph by passing through the Peyer's patches, then travel to the systemic circulation by way of the lymph and thoracic duct, and finally are located in the organs of the RES, which includes the spleen and bone; only the smallest (50-nm) size microspheres find their way to the bone marrow. Paraphrasing Owen's (13) view of the stages involved in the transport of HRP (Figure 18), we can imply that uptake of colloidal particles involves

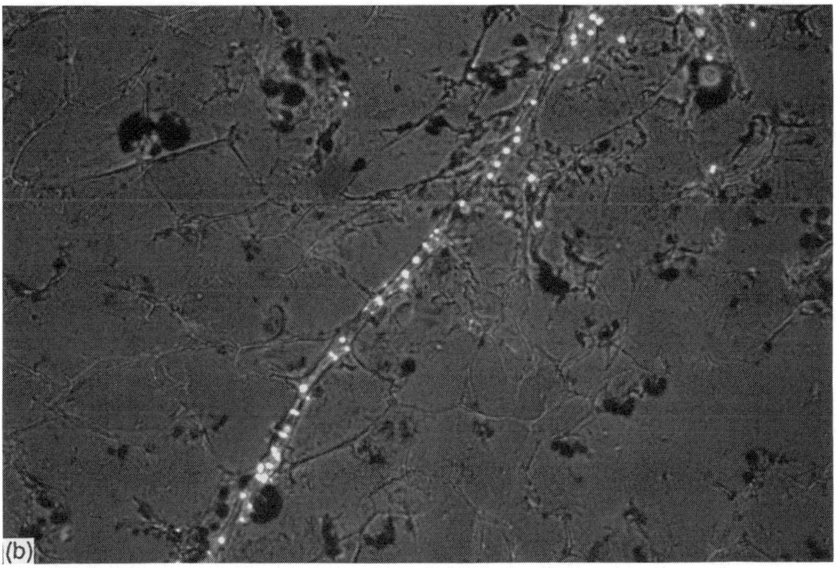
(b)

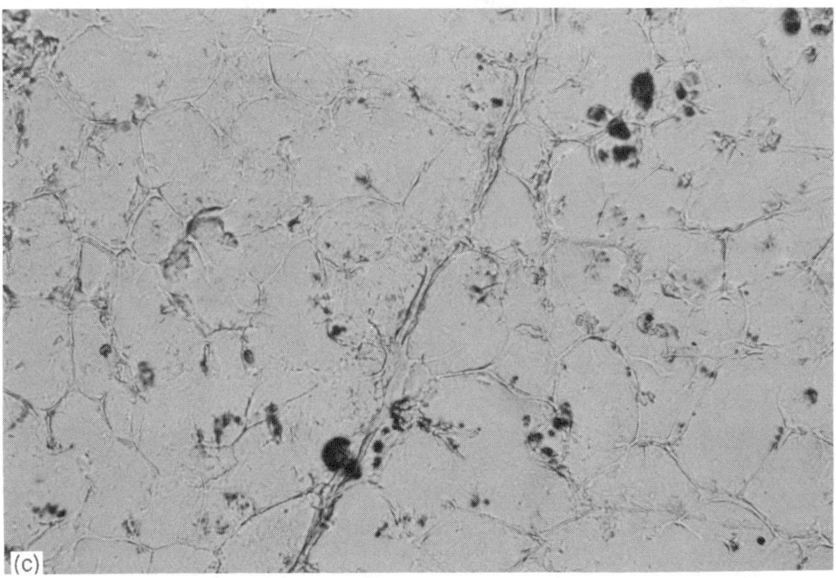
(c)

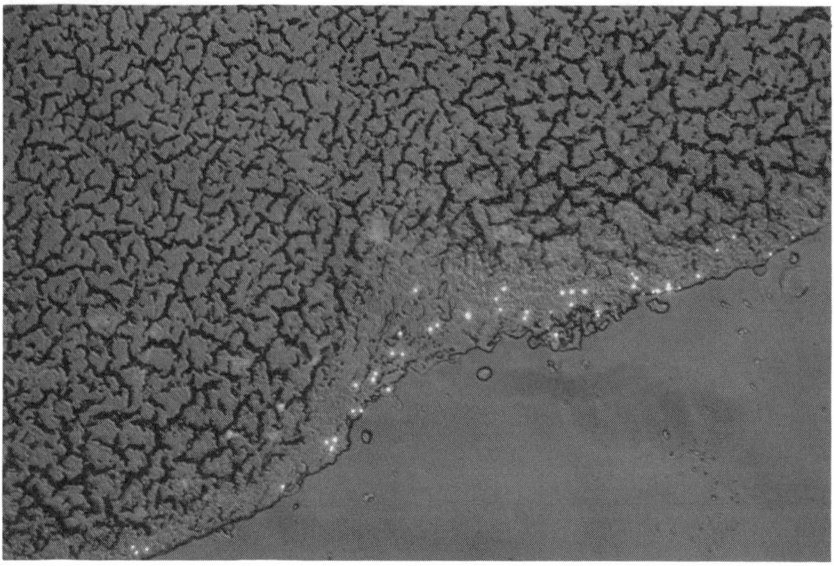

Figure 9 Photomicrograph of a lymph node, where the 500-nm nonionic microspheres are deposited in the lymph capsules and then in the subcapsular sinuses.

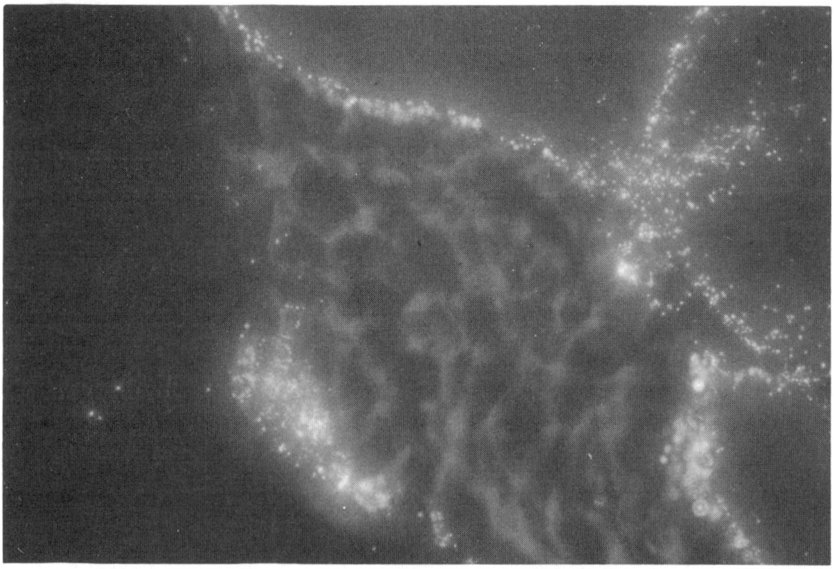

Figure 10 Photomicrograph (×200) of a portion of the mesentery network neighboring the GI tract in which 300-nm nonionic fluorescent microspheres are being transported.

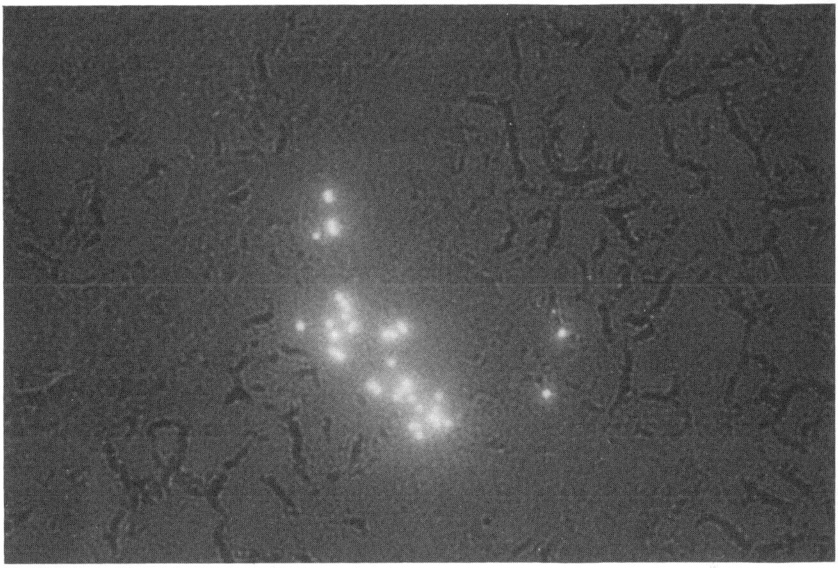

Figure 11 Photomicrograph (×200) of the mucosal/submucosal region of a Peyer's patch with 3-μm nonionic fluorescent microspheres, which appear to be *adsorbed* in this region. None were found to translocate to the serosal layer or to any major target organ.

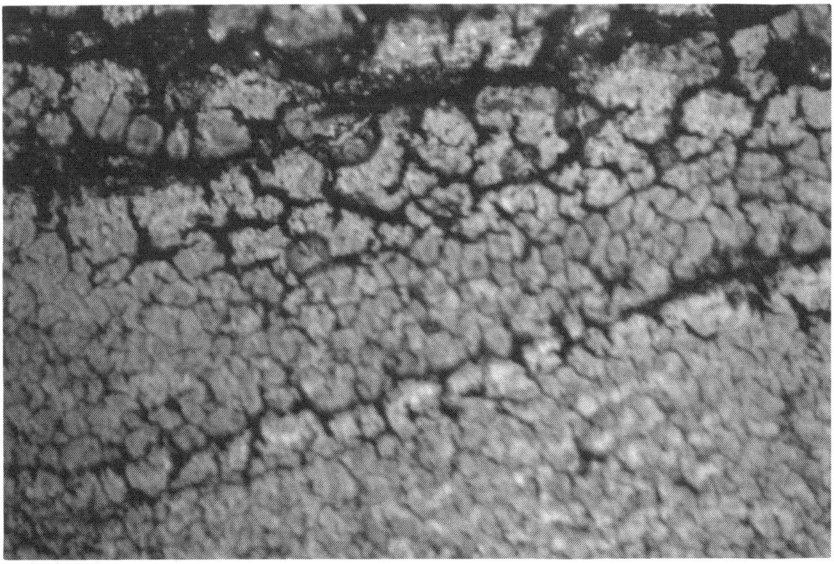

Figure 12 Photomicrograph (×200) of frozen section of liver showing 300-nm nonionic fluorescent microspheres in the sinusoid spaces arranged in a consistent order.

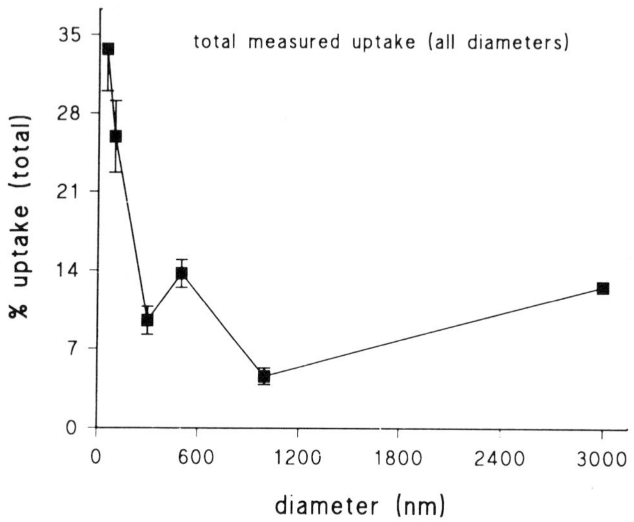

Figure 13 Total uptake (absorbed and adsorbed) of polystyrene microspheres from the gastrointestinal tract of female Sprague–Dawley rats ($n = 3$) following oral administration for 10 days at a dose of 12.5 mg/kg as measured by extraction of polystyrene from the tissues and application of gel permeation chromatography. Data plotted as a function of the particle size of the latex. (From Ref. 87.)

the same initial stages: adherence, pinocytosis by M cells, and transport, either free or in vesicles. This is followed by release into the extracellular space, but whether or not uptake by lymphocytes occurs and subsequent transport by these cells is involved is uncertain.

Rate

In colloidal drug delivery both the *rate* and *extent* of uptake and translocation are key issues. Eldridge's data (89) and our own (unpublished) suggest that the process is relatively rapid with microparticles appearing in the mesentery lymph and beginning to accumulate 2 hr after a single oral dose. Time intervals less than 2 hr have not been investigated.

Surface Characteristics: Charge and Hydrophobicity

Eldridge (89) produced the most comprehensive information on the relationship between uptake and composition of particles (Table 3). Cellulosic particles are not absorbed by Peyer's patches. Our own limited data showed that charged particles of polystyrene were less efficiently absorbed than nonionic particles. Matsuno and colleagues (92) found that nearly spherical polyvinylpyrrolidine-coated silica gel particles with an average diameter of 28–30 nm were taken up by mouse enterocytes and that a small number of the

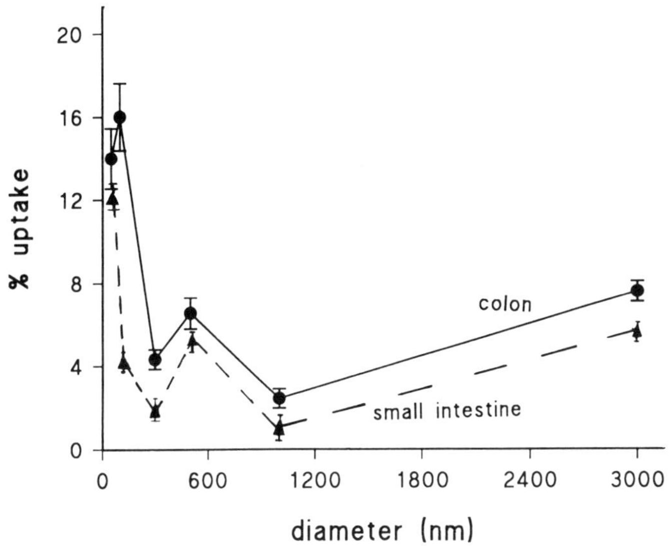

Figure 14 Uptake of orally administered polystyrene latex by the small intestine and colon, determined by gel permeation chromatography, as a function of particle diameter. Histological evidence showed that the microspheres are present mainly in the Peyer's patches and mesentery network of the small intestine and smaller lymphoid aggregates present in the colon. The data for the 3-μm-diameter latex are probably the result of *adsorption* to epithelium of the GI tract, as none was found in the internal organs, as discussed in the text. (From Ref. 87.)

particles enter the general circulation after repeated administration (7 days) per os. We have found 500-nm titanium dioxide particles also to be absorbed and translocated (93), and there is a growing literature on the presence of a variety of environmental airborne pollutants that find their way into Peyer's patches after swallowing (94) aluminum- and magnesium-rich silicates suggestive of atmospheric dust have been found in Peyer's patches and black pigment in the livers and spleens of coal workers has been detected by LeFevre et al. (95) and others (96,169).

Eldridge's study (89) remains the most comprehensive, although mechanisms of interaction of Peyer's patches with M cells still need to be addressed. Materials that might form the basis of drug carriers include polyalkylcyanoacrylates (97) and the terpolymers methyl methacrylate, 2-hydroxyethylmethacrylate, and butylacrylate nanoparticles (98), which have been found to be taken up after administration per os. In the latter case, 7 days after administration of a 1-ml dose to rats, at least 4% of the dose of ^{14}C was absorbed from the GI tract.

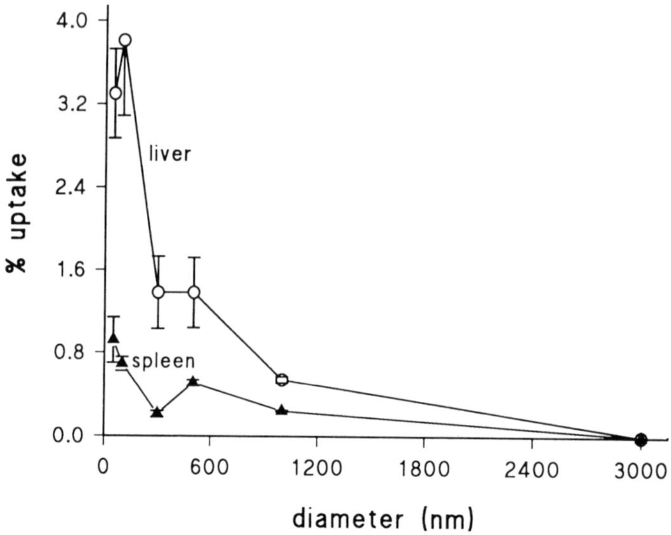

Figure 15 Concentration of orally administered polystyrene latex in liver (○) and spleen (▲) as a percentage of administered dose plotted as a function of particle size. The particles are present mainly in the macrophage cells of the liver and the granular follicles of the spleen. (From Ref. 87.)

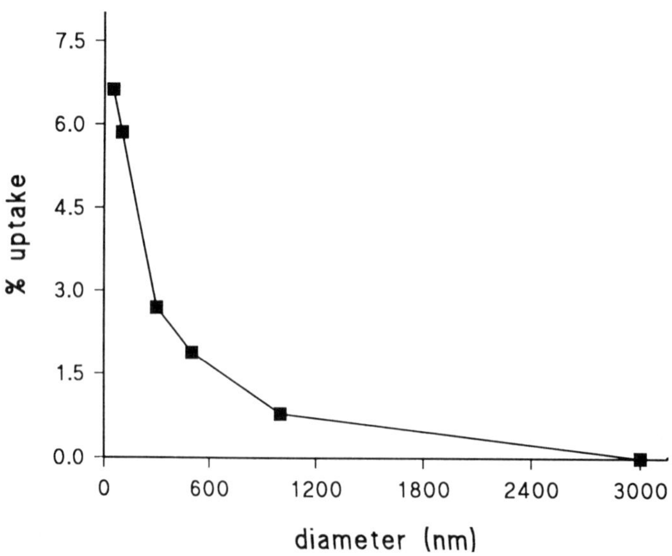

Figure 16 Cumulative uptake of polystyrene, orally administered to female Sprague–Dawley rats for 10 days at a dose of 12.5 mg/kg, as a function of particle diameter in liver, spleen, blood, bone marrow, and kidney. In the case of particles of 500 nm and 1 μm these data refer only to liver and spleen as no microspheres were detected in blood, bone marrow, and kidney. For the 300-nm latex the data refer to liver, spleen, and blood. (From Ref. 87.)

Table 2 Polystyrene Latex Uptake (% ±s.d.) as a Function of Size and Tissue

	Particle size					
Organs (g) (n = 21)	50 nm	100 nm	300 nm	500 nm	1 μm	3.0 μm
Stomach (1.76 ± 0.163)	1.1 ± 0.189	0.65 ± 0.15	0.45 ± 0.07	1.013 ± 0.06	0.27 ± 0.12	1.376 ± 0.08
Small intestine with Peyer's patches and mesentery (5.142 ± 0.386)	12 ± 0.47	3.4 ± 0.21	2.027 ± 0.14	4.28 ± 0.52	1.082 ± 0.12	3.627 ± 0.05
Colon with Peyer's patches, appendix, and mesentery (3.621 ± 0.282)	14 ± 1.46	16 ± 1.61	4.32 ± 0.47	6.53 ± 0.75	2.43 ± 0.46	7.53 ± 0.5
Liver (8.335 ± 0.662)	3.3 ± 0.93	3.8 ± 0.73	1.38 ± 0.35	1.38 ± 0.34	0.54 ± 0.03	ND
Spleen (0.53 ± 0.083)	0.92 ± 0.22	0.69 ± 0.07	0.21 ± 0.03	0.507 ± 0.03	0.24 ± 0.01	ND
Blood 2 ml per animal (n = 3)	2.2 ± 0.39	1.255 ± 0.44	1.1 ± 0.19	ND	ND	ND
Bone marrow 2 ml per animal (n = 3)	NE	0.1 ± 0.011	ND	ND	ND	ND
Kidney (2) (1.7 ± 0.282)	0.2 ± 0.05	ND	ND	ND	ND	ND
Lungs (2) (1.302 ± 0.061)	ND	ND	ND	ND	ND	ND
Heart (0.682 ± 0.063)	ND	ND	ND	ND	ND	ND
Total	33.72 ± 3.71	25.95 ± 3.21	9.487 ± 1.25	13.71 ± 1.2	4.562 ± 0.73	See text

NE = not examined; ND = not detected.

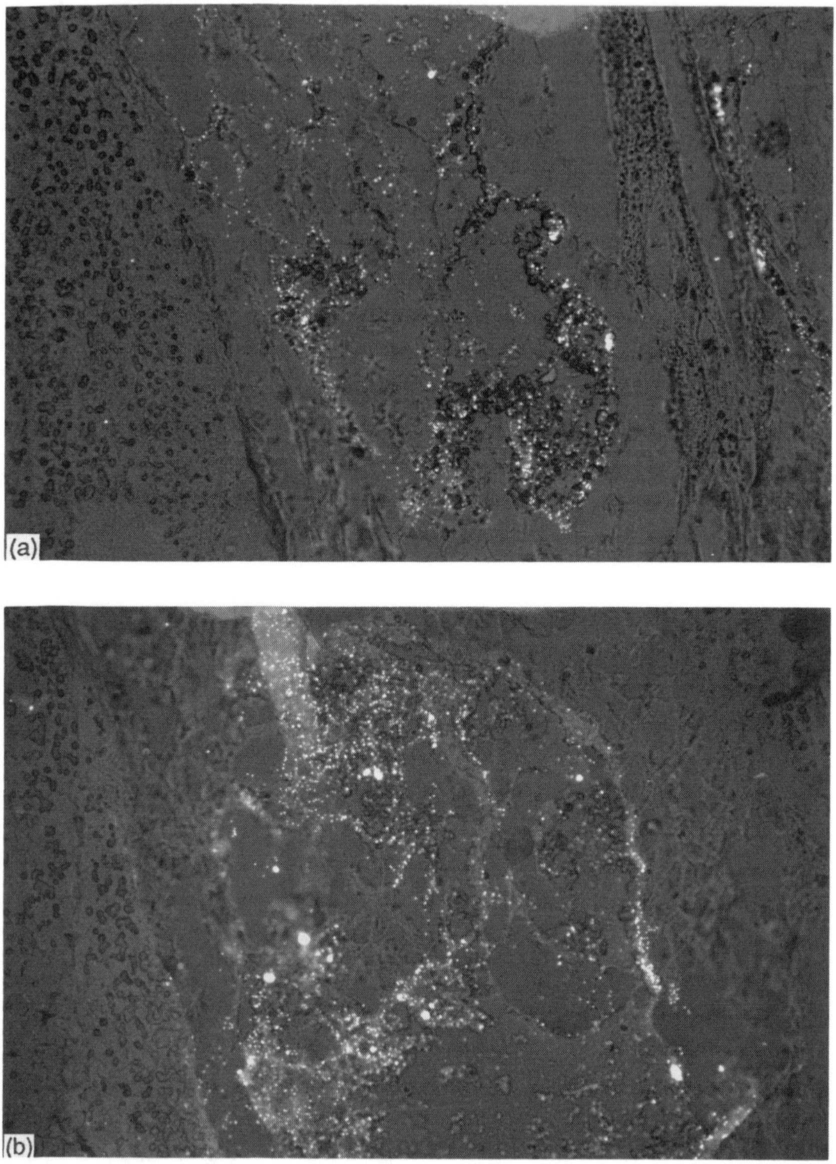

Figure 17 Results of single-dose experiments in female Sprague–Dawley rats. Concentration of 500-nm beads in the mesentery network (a) at 6 hr and (b) after 12 hr, showing clearly that the uptake is rapid and that in the first 24 hr the concentration of particles increases as the time progresses.

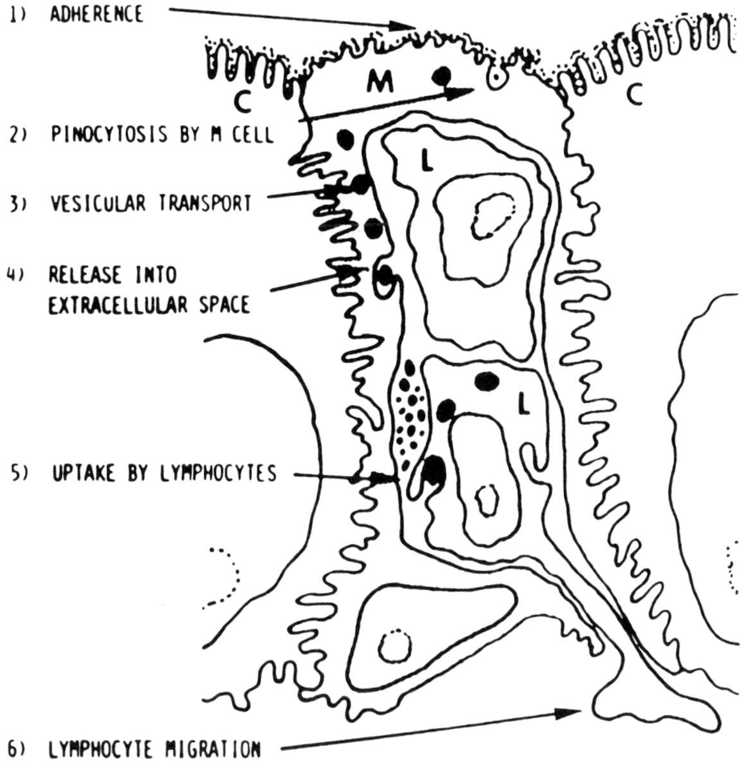

Figure 18 Diagram summarizing the stages observed in the transport of horseradish peroxidase (HRP) by the M cell from the intestinal lumen to the intraepithelial lymphocyte. C, columnar epithelial cells; L, lymphocytes. (From Ref. 13.)

Peptide and Protein Delivery

Peptides and proteins represent an increasingly important group of drugs. Although such products have been administered intravenously, intraperitoneally, intranasally, and transdermally, successful oral therapy has largely eluded solution. The potential for the delivery of labile drugs in a colloidal carrier system via the oral route was the primary motivation of our own work discussed above. To achieve absorption of colloidal carriers *intact* into the systemic circulation via the oral route means that the carriers not only have to survive disruption in a wide range of pH conditions but in terms of the effect of digestive enzymes and other gastric contents, must avoid entrapment by mucus and bacterial interference, and have a significant residence time within the GI tract at the site of uptake before translocation across the

Table 3 Targeted Absorption of 1- to 10-μm Microspheres of Various Polymers by the Peyer's Patches of the Gut-Associated Lymphoid Tissues Following Oral Administration

Microsphere polymer	Biodegradable	Absorption by the Peyer's patch[a]
Poly(styrene)	No	Very good[b]
Poly(methyl methacrylate)	No	Very good
Poly(hydroxybutyrate)	Yes	Very good
Poly(DL-lactide)	Yes	Good
Poly(L-lactide)	Yes	Good
85:15 Poly(DL-lactide-*co*-glycolide)	Yes	Good
50:50 Poly(DL-lactide-*co*-glycolide)	Yes	Good
Cellulose acetate hydrogen phthalate	No	None
Cellulose triacetate	No	None
Ethyl cellulose	No	None

[a]Mice were administered 0.5 ml of a suspension containing 20 mg of coumarin-containing microspheres into the stomach with the aid of an animal-feeding needle. Forty-eight hours after administering the microspheres, three representative Peyer's patches were removed and serial sectioned at 5-μm intervals.
[b]Results denote the efficiency of absorption by the Peyer's patches of the microspheres composed of various polymers where *very good* = 1000–1500, *good* = 200–1000, and *none* = <10 microspheres observed when all the sections from the three Peyer's patches were viewed with a fluorescence microscope.
Source: From Ref. 89.

epithelial barrier. The small target area of the GALT collectively suggests that the statistical chances of success are reduced. The total available target area can be calculated to be less than 0.01% of the total GI area in crude calculations. This low-percentage figure would be offset by any increased affinity of the carrier particles or increased affinity of the lymphoid tissue for the carriers. The long residence time of latex particles in the lumen of the small intestine of the mouse (90) might be of therapeutic interest. Intestinal absorption of a peptide has been shown to be increased from bioadhesive microspheres of Carbopol-coated PHEMA microspheres (90) in isolated gut experiments.

A wide range of techniques to administer proteins and enzymes by the oral route has been attempted. Emulsions and lipid vesicles have been used (100–104); the adsorption and immobilization of the proteins, labile drugs, and hormones on biodegradable polymers and silica surfaces has been used as a technique for both oral and systemic administration, but largely by the intravenous route (105–112). Changes in the membrane surface charge of the GI tract could achieve increased uptake of vesicles and microspheres loaded with drugs (34).

A number of attempts have been made to enhance the GI absorption of peptides by simultaneous administration of the agents with proteolytic inhibitors or by binding the peptide to surfactant micelles (113,114), but with mixed outcomes. Liposomes and other polymeric carriers have been utilized to administer insulin by the oral route (97,109,115–120) but this approach has had variable success. Positive results have sometimes been subjected to skeptical appraisal because of the belief that particulate (including liposomal) uptake was unproven. Oral treatment of hemophilia A by liposomal factor VIII has been attempted by Hemker (121) and Morgenthaler (122). It was claimed by Chiang (123) that oral delivery of factor VIII and insulin by carriers such as liposomes may not have resulted from enhanced absorption via uptake of intact liposomes but from the greater drug availability at the mucosal surfaces; the dose of insulin and factor VIII utilized for such oral delivery is substantially higher than that with any parenteral dose. However, recent evidence showing liposomes associated with Peyer's patches (170), along with the results already achieved, suggest a reinvestigation of liposomal oral delivery to be necessary. Liposomes stable in the GI tract must be employed in future experimental schemes.

Andrieu et al. (124) administered indomethacin containing polyisobutylcyanoacrylate nanocapsules in the size range 200–300 nm. Although indomethacin is normally absorbed from the gut, nanoencapsulation was claimed to accelerate the extravascular distribution of the drug due partly to enhanced uptake of the colloidal carrier by the RES. The results were attributed to an increase in the extent or duration of contact of the encapsulated drug with the gut wall or to absorption via paracellular pathways.

More convincing is the delivery of normally nonabsorbed agents, such as insulin (118) fed orally to diabetic rats. Insulin encapsulated in 200-nm polycyanoacrylate nanocapsules decreased fasting glycemia by 50–60% by the second day, while insulin absorbed at the surface of the nanocapsules did not produce responses (120) (Figure 19). Prolongation of the biological effect similar to that produced by insulin delivered in nanoparticles has been observed. About 4% of the dose of ^{14}C was absorbed after a single oral dose and less than 0.15% was found in lung, spleen, and liver 7 days after administration. Aprahamian and colleagues (97) used 100- to 200-nm polyalkylcyanoacrylate nanocapsules loaded with iodized oil, as a tracer for x-ray probe analysis. Administered to the jejunum the absorption of iodide was improved several fold over levels achieved with a Lipiodol emulsion (Figure 20). The pharmaceutical consequences of the uptake of the colloidal particulate matter and its subsequent concentration in the lymphatic tissues and the tissues of the RES after oral delivery has been stressed above. Biodegradable polymeric

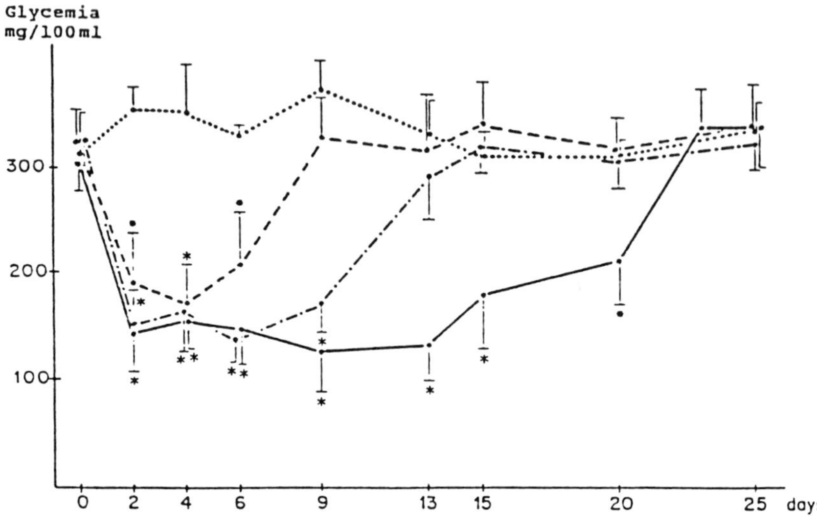

Figure 19 Effect of a single intragastric administration of insulin nanocapsules on glycemia in diabetic rats fasted overnight. Insulin nanocapsules administered with volume of 5, 10, and 20 ml/kg, corresponding to 12.5 (dashed line, $n = 7$), 25 (dashed dotted line, $n = 8$), and 50 (solid line, $n = 7$) U total insulin/kg body weight. Results are mean ± SE. Comparisons calculated at each period against controls (dotted line, $n = 8$): $^*p < .01$, $^*p < .001$. (Ref. 118.)

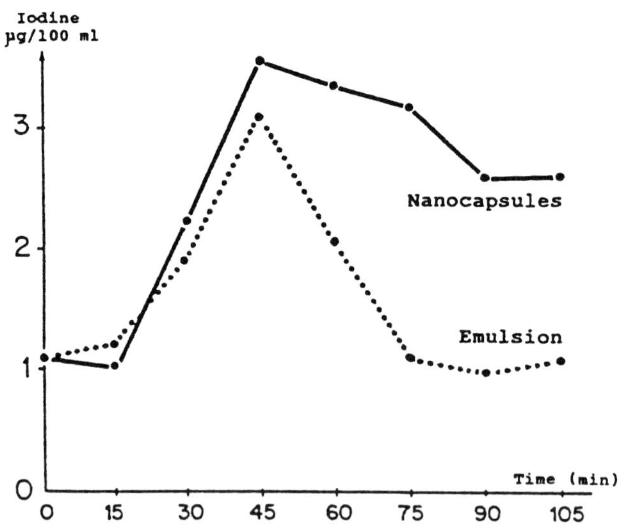

Figure 20 Plasma iodine concentration after intrajejunal administration of a Lipiodol emulsion and nanocapsules, mean size 165 nm (SD 46 nm) containing iodized oil. Mean emulsion droplet size was 258 nm (SD 71 nm). (From Ref. 97.)

drug carriers could be exploited, for instance, in the treatment of experimental models of lymphoproliferative disorders, such as multiple myeloma or Hodgkin's lymphoma, which affect the cells of the reticuloendothelial and lymphatic system (125–127). These are primary lymphomas of the GI tract notably affecting the B- and T-cell population of the lymphoid tissues of the GALT, such as polymorphic small and large B-cell lymphoma, Burkitt-type lymphoma, and pleomorphic small and large T-cell lymphoma (128). As larger particles escaping from Peyer's patches would at least be found in the lymph nodes, there is a higher possibility that these larger-size carriers would become lodged in the mesentery network and not be cleared into the thoracic duct and thence into the venous circulation to the organs of the RES. Thus, microspheres loaded with chemotherapeutic agents such as doxorubicin, bleomycin, and vincristine, which at present are all given by the intravenous route, might be administered orally to treat these lymphomas. A more specific deposition of the microspheres at the site of need with reduced drug concentration elsewhere would be of tremendous advantage. However, there is the problem of drug loading into the microspheres and the key questions of the extent of uptake, and whether the extent of uptake can be augmented from the levels indicated earlier in this chapter.

Tropical diseases of an infectious nature affect almost half a billion people worldwide. There would be considerable benefit from oral immunization against diseases such as malaria, leishmaniasis, schistosomiasis, and lymphatic filariasis (129–135). Oral administration of vaccines in biogradable microspheres have been reported (89,136,137) to produce antibody titers, possibly because of the interaction and presentation of the antigenic substances for immunization directly to the cells of the immune system, thereby making the gut-associated immune tissue, the centers of T and B immunocytes, as the first stop and then further percolating into the lymphatic system as the target (137–140).

It is well known that the antigenicity of antigens is increased in the presence of lipidic as well as metallic and other adjuvants (141–144), so that the administration of a vaccine in a polymeric colloid might offer advantages (129,136) not only because of potential uptake but also because of slow release, since the slow release of antigen from adsorbed products appears to be one reason for their adjuvanticity.

Consequences

It is important to mention the possible toxicological consequences of increasing the extent of the uptake of particulate matter from the gut nonselectively. Wolf et al. (145) showed that the penetration of certain reoviruses through the GI tract occurs *via* the GALT. Although in small numbers such viruses would be dealt with by the formidable immune system of the body, an overwhelming load could give rise to viral as well as bacterial diseases. Dietary ingestion of macromolecules present as food additives and oral ingestion of particulate materials (such as titanium dioxide) or macromolecular additives present in pharmaceutical formulations have been highlighted (93,146,147). The possibilities should be thoughtfully considered by the formulators of oral pharmaceutical products.

Exogenous pigment found in adult Peyer's patches (148) is possibly derived from foodstuffs as suspending, anticaking, and thickening agents. Toothpaste offers another source of aluminum, silicon, and titanium. Barium sulfate crystals have been seen in the mucosa and submucosa of the intestine (149). Questions have been raised about the relationship between exogenous particles and the etiology of Crohn's disease, particularly in a connection to the inadvertent repeated ingestion of toothpaste (150,151). Whatever the final outcome of such speculation, it is clear that in any attempt to exploit the oral route using colloidal carriers the possibility of alteration in permeability of the gut in certain disease states will have to be considered. More detailed studies of the interaction between carrier material and epithelial cells will be required.

Future work requires a greater range of biodegradable microspheres for oral delivery. Preparation techniques and characterization of microspheres of limited size range are well documented (152–159) but balances must always be struck between carrier stability in the gut and an appropriate rate of biodegradation once absorption has been accomplished.

What may be required to modulate uptake is an outer monolayer coating of hydrophobic surfactant or polymer that would first facilitate uptake by Peyer's patches and then, as the first coat is removed, an inner hydrophilic layer that would prevent uptake by the RES. Various distribution and drug elimination profiles must be studied after delivery in polymeric carrier systems (106,160–162) revealing more about the factors that control release and the behavior of the microspheres after uptake.

A plethora of activity is unleased by intestinal bacteria – some, for example, that might increase uptake by Peyer's patches if they could be harnessed. Production of bacterial *N*-formylmethionyloligopeptides induces bacterial adhesion. The oligopeptide also causes the release of lysosomal enzyme especially in the neutrophil leukocytes, producing a pronounced inflammatory response on the mucosal barriers (163) that might lead to increased uptake. Yamamoto et al. (164) examined the adherence of *Vibrio parahaemolyticus* and *Vibrio cholera* in the small intestinal mucosa with mucus coating, villi, and lymphoid follicle epithelium at the mucosal surfaces. The adherence of *V. parahaemolyticus* to human small intestinal mucosa correlated roughly with the cell-associated haemagglutinin (HA) levels of the strain. M cells provided the best platform for adherence and internalization occurred optimally with these cells. *V. cholera 01* can adhere to the formalin-fixed brush border of villus absorptive cells to a similar extent as to freshly isolated brush border, supporting the idea that bacterial pathogens including *V. cholera 01* recognize the sugar residues of receptors with their adhesions. This observation together with the fact that M-cell microfolds lack receptors leads to the conclusion that there are different chemical structures on M cell microfolds and the microvilli of major constituent epithelial cells. Yamamoto's experiments (164) showed that there was adherence but not internalization of *E. coli*. Hence, it might be possible to learn from the tactics of nature how to engineer particles with the appropriate affinity for Peyer's patches. However, it is not a simple task.

CONCLUSIONS

Sternson (165) pointed out that the strategy of delivering peptides and proteins orally utilizing Peyer's patch uptake is compromised by (a) the limited efficiency and capacity of the pathway; (b) the lag before onset of any pharmacological effect, due to the kinetics of processing and slow rate of lymph flow; and (c) the potential loss of drug to local lymphocytes and macrophages. To progress it is vital that the factors affecting uptake of carriers be better understood, that targeting to Peyer's patches be mastered, and that some control over the destination of particles once they are absorbed be achieved. Progress in nanotechnology might also be anticipated so that colloidal products with the appropriate stability, surface character, and drug release profiles, even pulsatile release profiles, are within our grasp.

ACKNOWLEDGMENTS

The collaboration of Dr. Gavin W. Halbert (Department of Pharmacy, University of Strathclyde) in the early stages of our own work was greatly appreciated. The support of Syntex Research Centre, Edinburgh, and in particular of Prof. Calum B. Macfarlane, Dr. John Langridge, and Dr. Alastair Selkirk has been influential in our progress. P.U.J. is currently a Maplethorpe Postdoctoral Fellow of the University of London.

REFERENCES

1. J. E. O'Mullane, P. Artursson, and E. Tomlinson, *Ann. N.Y. Acad. Sci.*, *507*, 120 (1987).
2. A. R. Thompson, J. M. Payne, B. F. Sansom, R. J. Garner, and B. J. Miles, *Nature*, *188*, 586 (1960).
3. E. Sanders and C. T. Ashworth, *Exp. Cell. Res.*, *22*, 137 (1961).
4. R. D. Pontefract and H. M. Cunningham, *Nature*, *243*, 352 (1973).

5. M. E. LeFevre, J. W. Vanderhoff, J. A. Laissue, and D. D. Joel, *Experientia*, *34*, 120 (1978).
6. G. Volkheimer and F. H. Schulz, *Digestion*, *1*, 213 (1968).
7. G. Volkheimer, *Ann. N.Y. Acad. Sci.*, *246*, 164 (1975).
8. J. M. Payne, B. F. Sansom, R. J. Garner, A. R. Thomson, and B. J. Miles, *Nature*, *188*, 586 (1960).
9. M. E. LeFevre and D. D. Joel, *Life Sci.*, *21*, 1403 (1977).
10. M. E. LeFevre, R. Olivo, J. W. Vanderhoff, and D. D. Joel, *Proc. Soc. Exp. Biol. Med.*, *159*, 298 (1978).
11. M. E. LeFevre, D. C. Hancock, and D. D. Joel, *J. Toxicol. Envir. Hlth.*, *6*, 691 (1980).
12. M. E. LeFevre, A. M. Boccio, and D. D. Joel, *Proc. Soc. Exp. Biol. Med.*, *190*, 23 (1989).
13. R. L. Owen, *Gastroenterology*, *72*, 440 (1977).
14. W. A. Walker, R. Cornell, L. M. Davenport, and K. J. Isselbacher, *J. Cell Biol.*, *54*, 195 (1972).
15. W. A. Walker and K. J. Isselbacher, *Gastroenterology*, *67*, 531 (1974).
16. R. Ducroc, M. Heyman, B. Beaufrere, J. L. Morgat, and J. F. Desjeux, *Am. Physiol. Soc.*, *245*, G54 (1983).
17. A. L. Warshaw, W. A. Walker, and K. J. Isselbacher, *Gastroenterology*, *66*, 987 (1974).
18. W. McA. Babour and D. Hopwood, *J. Pathol.*, *139*, 167 (1983).
19. W. A. Hemmings, *Med. Hypoth.*, *6*, 1209 (1980).
20. J. Hall, E. Orlans, J. Reynolds, C. Dean, J. Peppard, L. Gyure, and S. Hobbs, *Int. Arch. Allergy Appl. Immunol.*, *59*, 75 (1979).
21. D. Marcon-Genty, D. Tome, O Kheroua, A. M. Dumontier, M. D. Heyman, and J. F. Desjeux, *Am. J. Physiol.*, *256*, G943 (1989).
22. R. M. Clarke and R. N. Hardy, *J. Physiol.*, *204*, 113 (1969).
23. A. Pusztai, *Adv. Drug Deliv. Rev.*, *3*, 215 (1989).
24. H. Heyman, A. Bonfils, M. Fortier, A. M. Crain-denoyelle, P. Smets, and J. F. Desjeux, *Int. J. Pharm.*, *37*, 33 (1987).
25. V. G. Daniels, *Proc. Physiol. Soc.*, 7, 112P (1972).
26. E. A. Deitch, *Arch. Surg.*, *125*, 403 (1990).
27. R. L. Owen, N. F. Pierce, R. T. Apple, and W. C. Cray, *J. Infect. Dis.*, *153*, 1108 (1986).
28. R. J. Walker and R. L. Owen, *Ann. Rev. Med.*, *41*, 393 (1990).
29. A. Grutzkau, C. Hanski, H. Hahn, and E. O. Riecken, *Gut*, *31*, 1011 (1990).
30. S. M. C. Newton, C. O. Jacob, and B. A. Stocker, *Science*, *70*, 244 (1989).
31. M. Cereijido, L. G. Mariscal, and R. G. Contreras, *Am. Rev. Respir. Dis.*, *138*, S17 (1988).
32. P. M. Quinton and C. W. Philpott, *J. Cell Biol.*, *56*, 787 (1973).
33. G. W. Gullikson, W. S. Cline, V. Lorenzsonn, L. Benz, W. A. Olsen, and P. Bass, *Gastroenterology*, *73*, 501 (1977).
34. T. Ohyashiki, M. Taka, and T. Mohri, *J. Biochem.*, *106*, 584 (1989).
35. S. Phillips, *Gut*, *28*, 650 (1986).

36. R. T. Spychal, J. M. Marrero, S. H. Saverymuttu, and T. C. Northfield, *Gastroenterology*, *97*, 104 (1989).
37. S. G. McClugage, F. N. Low, and M. L. Zimny, *Gastroenterology*, *91*, 1128 (1986).
38. R. L. Owen and A. L. Jones, *Gastroenterology*, *66*, 189 (1974).
39. R. L. Owen and P. Nemanic, *Scann. Electron Microsc.*, *11*, 367 (1978).
40. W. P. Faulk, J. N. McCormick, J. R. Goodman, J. M. Yoffey, and H. M. Fudenberg, *Cell. Immunol.*, *1*, 500 (1971).
41. K. N. Chin and G. Hudson, *Acta Anat.*, *78*, 306 (1971).
42. J. Spencer, T. Finn, and P. G. Isaacson, *Cell Pathol.*, *51*, 509 (1986).
43. P. Sobhon, *J. Morphol.*, *135*, 457 (1971).
44. S. Nash, J. Stafford, and J. L. Madara, *J. Clin. Invest.*, *80*, 1104 (1987).
45. A. M. Mowat, *Spring Sem. Immunopathol.*, *12*, 165 (1990).
46. J. S. Wassef, D. F. Keren, and J. L. Mailloux, *Infect. Immun.*, *57*, 858 (1989).
47. A. Grutzkau, C. Hanski, H. Hahn, and E. O. Riecken, *Gut*, *31*, 1011 (1990).
48. M. W. Smith, P. S. James, and D. R. Tivey, *Am. J. Pathol.*, *128*, 385 (1987).
49. P. Sicinski, J. Rowinski, J. B. Warchol, Z. Jarzabek, W. Gut, B. Szczygiel, K. Bielecki, and G. Koch, *Gastroenterology*, *98*, 56 (1990).
50. W. A. Bye, C. H. Allan, and J. S. Trier, *Gastroenterology*, *86*, 789 (1984).
51. Y. Cho and P. P. H. Debruyn, *Am. J. Anat.*, *177*, 481 (1986).
52. P. Brandtzaeg and K. Bjerke, *Digestion*, *46*, 262 (1990).
53. T. Landsverk, *Immunol. Cell. Biol.*, *65*, 251 (1987).
54. T. Landsverk, W. Trevella, and L. Nicander, *Cell Tissue Res.*, *261*, 239 (1990).
55. H. G. Durkin, S. M. Chice, E. Gaetjens, H. Bazin, L. Tarcsay, and P. Dukor, *J. Immunol.*, *143*, 1777 (1989).
56. G. Bevillacqua, A. Marchetti, and R. Biondi, *Gastroenterology*, *96*, 139 (1989).
57. R. Weltzin, P. Lucia-Jandris, P. Michetti, B. N. Fields, J. P. Kraehenbuhl, and M. R. Neutra, *J. Cell Biol.*, *108*, 1673 (1989).
58. E. A. Elsinghorst, L. S. Baron, and J. D. Kopeko, *Proc. Natl. Acad. Sci. USA*, *86*, 5173 (1989).
59. D. R. Appleton, J. P. Sunter, and A. J. Watson (eds.), *Cell Proliferation in the Gastrointestinal Tract*, Pitman Press, London, 1980.
60. E. Lebenthal (ed.), *Human Gastrointestinal Development*, Raven Press, New York, 1989.
61. L. Diamond, J. T. Doluisio, and W. Crouthamel, *Eur. J. Pharmacol.*, *11*, 109 (1970).
62. H. J. Lenz, *Gastroenterology*, *97*, 216 (1989).
63. S. Kojima, R. B. Smith, and J. T. Doluisio, *J. Pharm. Sci.*, *60*, 1639 (1971).
64. M. Mantle, L. Basaraba, S. C. Peacock, and D. G. Gall, *Infect. Immun.*, *57*, 3292 (1989).
65. K. Chadee, W. A. Petri, D. J. Innes, and J. I. Ravdin, *J. Clin. Invest.*, *80*, 1245 (1987).
66. O. Hanninen, P. L. Seppa, and K. Pelkonen, *Arch. Toxicol.*, *60*, 34 (1987).
67. T. Y. Ma, D. Hollander, P. Krugliak, and K. Katz, *Gastroenterology*, *98*, 39 (1990).

68. D. Nutting, J. Hall, J. A. Barrowman, and P. Tso, *Biochim. Biophys. Acta.*, *1004*, 357 (1989).
69. T. Bolin, L. Franzen, R. Sjodahl, and C. Tagesson, *Scand. J. Gastroenterol.*, *21*, 441 (1986).
70. H. O. Alpar, W. N. Field, R. Hyde, and D. A. Lewis, *J. Pharm. Pharmacol.*, *41*, 194 (1989).
71. M. N. Marsh (ed.), *Immunopathology of the Small Intestine*, John Wiley and Sons, Chichester, 1987.
72. R. L. Owen and T. H. Ermak, *Springer Seminars in Immunopathology*, *12*, 139 (1990).
73. W. E. LeGros Clark, *The Tissues of the Body: An Introduction to the Study of Anatomy*, Oxford University Press, Oxford, 1939.
74. L. Weiss (ed.), *Cell and Tissue Biology: A textbook of Histology*, 6th Ed., Urban & Schwarzenberg, Baltimore, 1988.
75. J. M. Yoffey and F. C. Courtice (eds.), *Lymphatics, Lymph and Lymphomyeloid Complexes*, Academic Press, London, 1970.
76. D. R. Shaw and F. M. Griffin, *Nature*, *289*, 409 (1981).
77. L. Illum and S. S. Davis, *J. Pharm. Sci.*, *72*, 1086 (1983).
78. L. Illum, N. W. Thomas, and S. S. Davis, *J. Pharm. Sci.*, *75*, 16 (1986).
79. E. Egensperger and H. Hollander, *J. Neurosci. Meth.*, *23*, 181 (1988).
80. M. S. Roy, K. S. Harrison, E. Harvey, and T. Mitchell, *Nucl. Med. Biol.*, *16*, 81 (1989).
81. J. Pappo and T. H. Ermak, *Clin. Exp. Immunol.*, *76*, 144 (1989).
82. S. L. Hale, M. T. Vivaldi, and R. A. Kloner, *Am. J. Physiol.*, *215*, H863 (1986).
83. T. Oonishi, and N. Uyesaka, *J. Immunol. Meth.*, *84*, 143 (1985).
84. F. Schroeder, and A. B. Kier, *J. Immunol. Meth.*, *57*, 363 (1983).
85. M. G. Wetzel and E. D. Korn, *J. Cell Biol.*, *43*, 90 (1969).
86. P. U. Jani, G. W. Halbert, J. Langridge, and A. T. Florence, *J. Pharm. Pharmacol.*, *41*, 809 (1989).
87. P. U. Jani, G. W. Halbert, J. Langridge, and A. T. Florence, *J. Pharm. Pharmacol.*, *42*, 821 (1990).
88. P. U. Jani, A. T. Florence, and D. E. McCarthy, *Int. J. Pharm.*, *84*, 245 (1992).
89. J. H. Eldridge, C. J. Hammond, J. A. Meulbroek, J. K. Staas, R. M. Gilley, and T. R. Tice, *J. Controlled Release*, *11*, 205 (1990).
90. J. P. Ebel, *Pharm. Res.*, *7*, 848 (1990).
91. J. Kreuter, U. Muller, and K. Munz, *Int. J. Pharm.*, *55*, 39 (1989).
92. K. Matsuno, T. Schaffner, H. A. Gerber, C. Ruchti, M. W. Hess, and H. Cottier, *J. Reticuloendothel. Soc.*, *33*, 263 (1983).
93. A. T. Florence, P. U. Jani, and D. McCarthy, *Lancet*, *336*, 1580 (1990).
94. S. J. Urbanski, L. Arsenault, F. H. Green, and G. Haber, *Modern Pathol.*, *2*, 222 (1989).
95. M. E. LeFevre, F. H. Green, D. D. Joel, and W. Laqueur, *Hum. Pathol.*, *13*, 1121 (1982).
96. A. D. N. Scott, C. Uff, and R. K. S. Phillips, *Lancet*, 1581 (1990).

97. M. Aprahamian, C. Michel, W. Humbert, J. Ph. Devissaguet, and C. Damgé, *Biol. Cell*, *61*, 69 (1987).
98. M. Kuklan, V. Koprda, and S. Bezek, *Die Pharmazie*, *46*, 37 (1991).
99. C-M. Lehr, J. A. Bouwstra, J. J. Tukker, J. C. Verhoof, and H. E. Junginger, *Bioact. Mater.*, *16*, 418 (1989).
100. J. M. Chobert, C. Bertrand-Harb, M. G. Nicolas, H. F. Gaertner, and A. J. Puigserver, *J. Agric. Food Chem.*, *35*, 638 (1987).
101. M. Cheema, K. J. Palin, and S. S. Davis, *J. Pharm. Pharmacol.*, *39*, 55 (1987).
102. T. Kimura, K. Takeda, A. Kageyu, M. Toda, Y. Kurosaki, and T. Nakayama, *Chem. Pharm. Bull.*, *37*, 463 (1989).
103. Y. Mizushima, T. Hamano, and K. Yokoyama, *Ann. Rheum. Dis.*, *41*, 263 (1982).
104. Y. Mizushima, T. Hamano, and K. Yokoyama, *J. Pharm. Pharmacol.*, *34*, 49 (1982).
105. J. C. Venter, *Pharmacol. Rev.*, *34*, 153 (1982).
106. B. Ekman, C. Lofter, and I. Sjoholm, *Biochemistry*, *15*, 5115 (1976).
107. B. He, P. Jiang, and Y. Qiu, *Biomat. Art. Cells. Art. Org.*, *15*, 709 (1987).
108. C. Drake and W. M. Allen, *Biochem. Soc. Trans.*, *516* (1985).
109. M. Saffran, G. S. Kumar, C. Savariar, J. C. Burnham, F. Williams, and D. C. Neckers, *Science*, 1081 (1986).
110. P. Edman, B. Ekman, and I. Sjöholm, *J. Pharm. Sci.*, *69*, 838 (1980).
111. P. Edman and I. Sjöholm, *J. Pharm. Sci.*, *70*, 684 (1981).
112. P. Edman and I. Sjöholm, *J. Pharm. Sci.*, *71*, 576 (1982).
113. Y. Masuda, H. Yoshikawa, K. Takada, and S. Muranishi, *J. Pharmacobio. Dyn.*, *9*, 793 (1986).
114. M. M. P. Paulssen and B. C. Van Pelt, *Lancet*, *1*, 1310 (1981).
115. H. M. Patel and B. E. Ryman, *FEBS. Lett.*, *62*, 60 (1976).
116. R. C. Oppenheim, N. F. Stewart, L. Gordon, and H. M. Patel, *Drug Dev. Ind. Pharm.*, *8*, 531 (1982).
117. C. Damgé, M. Aprahamian, G. Balboni, A. Hoeltzel, V. Andrieu, and J. P. Devissaguet, *Int. J. Pharm.*, *36*, 121 (1987).
118. C. Damgé, C. Michel, M. Aprahamian, and P. Couvreur, *Diabetes*, *37*, 246 (1988).
119. C. Damgé, C. Michel, M. Aprahamian, P. Couvreur, and J. P. Devissaguet, *J. Controlled Release*, *13*, 233 (1990).
120. P. Couvreur, V. Lenaerts, B. Kante, M. Roland, and P. P. Speiser, *Acta Pharm. Technol.*, *26*, 220 (1980).
121. H. C. Hemker, A. D. Muller, W. Th. Hermens, and R. F. A. Zwaal, *Lancet*, *1*, 70 (1980).
122. J. J. Morgenthaler, *Lancet*, *1*, 546 (1980).
123. C. Chiang and N. Weiner, *Int. J. Pharm.*, *37*, 75 (1987).
124. V. Andrieu, H. Fessi, M. Dubrasquet, J. Ph. Devissaguet, F. Puisieux, and S. Benita, *Drug Design Deliv.*, *4*, 295 (1989).
125. T. Chisesi, M. Vespignani, and G. Capnist, *Int. J. Biol. Marker*, *2*, 173 (1987).
126. A. Hagiwara, T. Takahashi, T. Ueda, A. Iwamoto, and T. Torii, *Anti-Cancer Drug Design*, *1*, 313 (1987).

127. J. Roberts and J. H. Quastel, *Biochem. J.*, *89*, 151 (1963).
128. P. A. Hall, J. R. Jass, D. A. Levison, B. C. Morson, N. A. Shepherd, L. H. Sobin, and A. G. Stansfeld, *Lancet*, *2*, 958 (1988).
129. D. T. O'Hagan, K. Palin, S. S. Davis, P. Artursson, and I. Sjoholm, *Vaccine*, *7*, 421 (1989).
130. D. T. O'Hagan, *Adv. Drug Deliv. Rev.*, *5*, 265 (1990).
131. T. Lehner, J. Haron, L. A., Bergmeier, A. Mehlert, R. Beard, M. Dodd, B. Mielnik, and S. Moore, *Immunology*, *67*, 419 (1987).
132. F. A. Klipstein, R. F. Engert, and W. T. Sherman, *Infect. Immun.*, *39*, 1000.
133. R. Kirsh, P. J. Bugelski, and G. Poste, *Ann. N.Y. Acad. Sci.*, 141 (1988).
134. D. R. Hill, *Infect. Immun.*, *58*, 2683 (1990).
135. D. R. Hill and R. Pohl, *Infect. Immun.*, *58*, 3202 (1990).
136. J. H. Eldridge, J. A. Meulbroek, J. K. Staas, T. R. Tice, and R. M. Gilley, *Adv. Exp. Med. Biol.*, *251*, 191 (1989).
137. J. Mestecky and J. R. McGhee (eds.), *Current Topics in Microbiology and Immunology: New Strategies for Oral Immunisation*, *146*, Springer-Verlag, Berlin, 1989.
138. J. H. Dean, M. I. Luster, A. E. Munson, and H. Amos (eds.), *Immunotoxicology and Immunopharmacology*, Raven Press, New York, 1985.
139. J. Bienenstock, P. B. Ernst, and B. J. Underdowm, *Ann. Allergy*, *59*, 17 (1987).
140. L. Congleton, W. Potts, and S. Mathur, *Fertil. Steril.*, *52*, 106 (1989).
141. A. C. Allison, *Biotechnology*, *5*, 1043 (1987).
142. G. Gregoriadis, *Immunol. Today*, *11*, 89 (1990).
143. B. Morein, B. Sundquist, S. Hoglund, K. Dalsgaard, and A. Osterhaus, *Nature*, *308*, 457 (1984).
144. U. E. Diner, D. Kunimoto, and E. Diener, *J. Immunol.*, *22*, 1886 (1979).
145. J. L. Wolf, R. S. Kauffman, R. Finbert, R. Dambrauskas, B. Fields, and J. S. Trier, *Gastroenterology*, *85*, 291 (1983).
146. H. Ohta, M. V. DeAngelis, and M. G. Cherian, *Toxicol. Appl. Pharmacol.*, *101*, 62 (1989).
147. M. L. Weiner, *Fd. Chem. Toxicol.*, *26*, 867 (1988).
148. N. A. Shepherd, P. R. Crocker, A. P. Smith, and D. A. Levison, *Hum. Pathol.*, *18*, 50 (1987).
149. D. A. Levison, P. R. Crocker, G. Lee, N. A. Shepherd, and A. P. Smith, *J. Pathol.*, *144*, 119 (1984).
150. J. B. Kirsner, *JAMA*, *251*, 80 (1984).
151. S. N. Sullivan, *Lancet*, *336*, 1096 (1990).
152. N. Al Khouri Fallouh, L. Roblot-Treupel, H. Fessi, J. Ph. Devissaguet, and F. Puisieux, *Int. J. Pharm.*, *28*, 125 (1986).
153. A. J. Baillie, A. T. Florence, L. R. Hume, T. G. Muirhead, and A. Rogerson, *J. Pharm. Pharmacol.*, *37*, 863 (1985).
154. R. Bodmeier and H. Chen, *J. Pharm. Pharmacol.*, *40*, 754 (1988).
155. T. M. S. Chang, F. C. MacIntosh, and S. G. Mason, *Can. J. Physiol. Pharmacol.*, *44*, 115 (1966).

156. S. Hosaka, Y. Murao, S. Masuko, and K. Miura, *Immunol. Commun.*, *12*, 509 (1983).
157. P. Stjarnkvist, T. Laakso, and I. Sjoholm, *J. Pharm. Sci.*, *78*, 52 (1989).
158. J. Ugelstad and P. C. Mork, *Adv. Colloid Interf. Sci.*, *13*, 101 (1980).
159. N. Wakiyama, K. Juni, and M. Nakano, *Chem. Pharm. Bull.*, *29*, 3363 (1981).
160. B. Ekman and I. Sjoholm, *J. Pharm. Sci.*, *67*, 693 (1978).
161. T. Ishizaka and M. Koishi, *J. Pharm. Sci.*, *72*, 1057 (1983).
162. M. Nefzger, J. Kreuter, R. Voges, E. Liehl, and R. Czok, *J. Pharm. Sci.*, *73*, 1309 (1984).
163. D. M. Ferry, T. J. Butt, M. F. Broom, J. Hunter, and V. S. Chadwick, *Gastroenterology*, *97*, 61 (1989).
164. T. Yamamoto and T. Yokota, *Infect. Immun.*, *57*, 2410 (1989).
165. L. A. Sternson, *Ann. N.Y. Acad. Sci.*, *507*, 19 (1987).
166. E. H. Gerhart, R. J. Liukkonen, R. M. Carlson, G. N. Stokes, M. Lukasewycz, and A. R. Oyler, *Envir. Pollution*, *25*, 165 (1981).
167. S. K. Bakar and S. Niazi, *J. Pharm. Sci.*, *72*, 1027 (1983).
168. F. G. J. Poelma and J. J. Tukker, *J. Pharm. Sci.*, *76*, 433 (1987).
169. S. Brenner, *Lancet*, *1*, 781 (1989).
170. N. K. Childers, F. R. Denys, N. F. McGee, and S. M. Michalek, *Regional Immunol.*, *3*, 8 (1990).

5

Drug Carriers for the Oral Administration and Transport of Peptide Drugs Across the Gastrointestinal Epithelium

Philip L. Smith, Doris A. Wall, and Glynn Wilson

SmithKline Beecham Pharmaceuticals
King of Prussia, Pennsylvania

INTRODUCTION

The oral route of administration presents formidable challenges to the development of effective, safe, and commercializable delivery systems for peptides and proteins. However, for chronic therapies in all therapeutic areas, the enormous market potential and patient compliance offered by developing systems for oral delivery continues to generate much interest and activity. Advances in our understanding of the specialized transport properties of the cellular barriers of the intestinal epithelium have provided the impetus for these studies. Few examples exist for oral peptide delivery systems in routine clinical practice, and there are no systems that can be cited for systemic delivery of proteins. For a few structural mimics of di- and tripeptides, i.e., β-lactam antibiotics and angiotensin-converting enzyme (ACE) inhibitors, high oral bioavailability appears to result from serendipitous use of dietary peptide transporters in the small intestine. Oral absorption of cyclosporine, a cyclic undecapeptide, is achieved through administration as an olive oil/glyceride formulation, although the reasons for the high (albeit highly variable) bioavailability of this molecule in this specific formulation remain an enigma. Oral delivery systems that have been rationally designed to overcome physiological, metabolic, and physical barriers and to meet pharmacological requirements of specific drug molecules for either peptides or proteins are not on the market.

To successfully deliver peptides and proteins by the oral route, pharmacological requirements of specific molecules, i.e., the dose, dosing regime, duration of therapy, and total dose required, need to be considered. For many of the 100 or more peptides and proteins currently under clinical investigation, only a small number of endocrine hormones, hormone-releasing peptides or receptor antagonists, and enzyme inhibitors that are active at low (μg to low mg) concentrations will be appropriate candidates to exploit intrinsically low oral bioavailabilities that may be currently achievable. While absorption across the gastrointestinal (GI) tract is a central problem in oral peptide delivery, achievement of good absorption may not ultimately translate into adequate bioavailability due to first-pass metabolism, rapid clearance, peptide and protein instability, or susceptibility to proteolysis. Furthermore, the relevance of absorption data in some animal models to absorption in humans is questionable.

Many of the problems associated with oral delivery of peptide and protein drugs and strategies to overcome physiological, metabolic, and physical barriers in the GI tract have been discussed previously (1–6). The aim of this chapter is to review evidence for passage of particulates across the GI epithelium, including a limited number of preliminary studies on use of particulate delivery systems for peroral administration of peptide drugs. Although the phenomenon of particle transport across the GI epithelium has been known for many years, the mechanism and quantitative aspects of this process have only recently begun to be characterized in greater detail with the aim of drug delivery system design. Liposomes, lipid-based microemulsion and micellar systems, microspheres of various compositions, bacteria and viruses, and carriers used to target drugs to the colon or prolong transit time via mucoadhesion will be considered. The status of these approaches will be examined against our current understanding of specific routes and mechanisms of transport across the GI epithelial barrier, and the feasibility of exploiting such pathways for achieving effective, safe, and commercializable delivery systems for peptide absorption and oral immunization will be evaluated.

APPROACHES

Work on oral peptide delivery systems using particulate carriers has had two distinct goals: improvement of absorption efficiency of drugs and/or targeting of antigens to gut-associated lymphoid tissue (GALT) for stimulation of mucosal immunity. Strategies for use of particulate systems are based on the following: (a) Most systems are postulated to provide some level of protection against degradation within the GI tract, although direct evidence to support this is usually lacking. (b) Prolongation of transit time will increase duration of exposure to drug molecules to the epithelial surface and may improve absorption of peptidergic drugs whose intrinsic rates of transepithelial transport are generally very low. Whether particulate carriers can be designed to improve efficiency of translocation across epithelial barriers is controversial since current data on mechanisms and efficiency of particle translocation are limited and contradictory. (c) Use of carriers can target specific absorption pathways (e.g., lipid-based carrier systems) or sites (e.g., azopolymers for colonic delivery). Synthetic microspheres and attenuated pathogens have shown promise in allowing targeting of antigens to GALT for induction of mucosal immunity.

Current research on particulate delivery systems is focused on gaining a better understanding of mechanisms, efficiency, and reproducibility of translocation of both the carrier and/or the drugs across the GI tract, and analysis of the effect of carrier composition on these parameters. Use of such systems to improve absorption of peptide or protein drugs is in very preliminary stages of evaluation, and investigations of their potential toxicity are lacking.

TRANSPORT ACROSS THE GASTROINTESTINAL EPITHELIUM

Transport Mechanisms

Mechanisms by which macromolecules traverse intestinal epithelial cells include carrier-mediated, passive, and endocytotic transport processes (6) (Figure 1). Passive transport across intestinal epithelium can occur either via a transcellular or paracellular route. Passive transcellular transport is dependent on the ability of a molecule to partition into a cell membrane and thus will be influenced by partition coefficient, molecular weight, and net charge. New techniques (e.g., Caco-2 cells/tissues) are enabling early predictions of permeability of drugs and delivery systems (7).

Paracellular transport occurs via aqueous channels with an equivalent pore diameter in humans of 4–8 Å and a somewhat larger pore diameter in animals (~10–15 Å). Morphological studies of the extrusion process for mature enterocytes at the villus tip indicate that the voids produced during this process are too transient to account for the magnitude of paracellular flux observed in intact tissue (8). Thus, the majority of transport via the paracellular pathway appears to result from passage through apical junctional complexes. Recent evidence showed that these complexes can be altered by a variety of cellular regulatory mechanisms (9) that may be responsible for the increase in pore diameter (10–30 Å) associated with an increase in water absorption in canine jejunum (10–12). Thus, molecules with molecular weight greater than 5 kDa or particles with diameters greater than 15–25 Å will

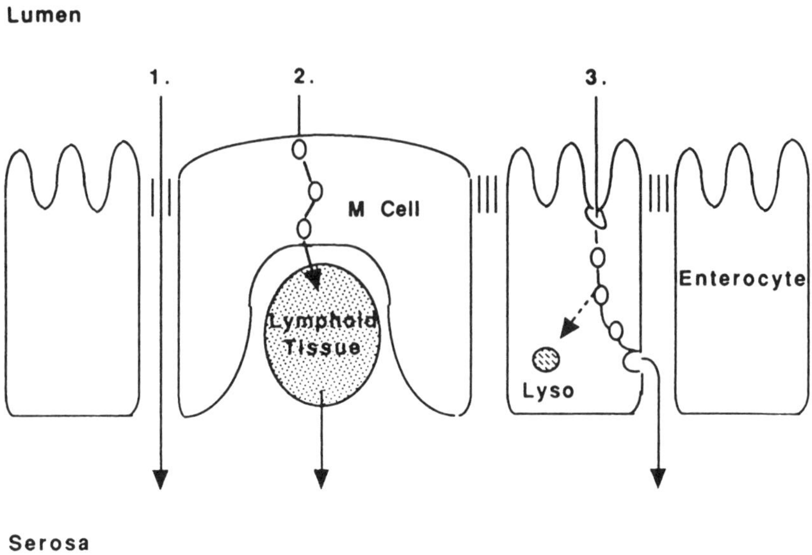

Figure 1 Postulated routes for transport of particles across the gastrointestinal epithelium. 1, Paracellular transport between epithelial cells is the most likely route for "persorption" of particles >10–12 μm, which are probably too large to be included in transcytotic vesicles and may contribute to passive transport of small particles as well. 2, Transcytosis across the M cells overlying Peyer's patches and isolated lymphoid follicles is likely to be the route taken by a wide variety of particles found to be selectively associated with GALT tissue. Particles >10–12 μm do not appear to utilize this pathway. 3, Transcytosis across enterocytes within membrane-bound vesicles has been documented in relatively few cases for small particles of <220 nm diameter: delivery to intracellular lysosomes (Lyso) is a common fate of endocytosed material in enterocytes.

not be able to traverse the epithelial layer via this pathway. Thus, physiological regulation of the paracellular pathway does not enable passage of particulates.

Active transport can involve entry of molecules into the cytoplasm via specific membrane carries (e.g., bile salts, amino acids, small peptides), fluid phase or receptor-mediated endocytic transport via membrane-bound vesicles (vitamin B_{12}, β-lactoglobulin) or the complex sequence of hydrolysis and reassembly involved in lipid absorption (5–6,13–15). We will focus in this chapter on those routes postulated to be involved in transport of natural and synthetic particles through the intestinal mucosa. These include paracellular passage between adjacent epithelial cells, and endocytic transport across enterocytes and across M cells overlying lymphoid tissue. Penetration of the mucosa by enteric pathogens is considered in a separate section of this chapter.

Particle Transport

A wide variety of inert particles can be transported to some extent across the GI epithelium, most likely by transcytotic or paracellular pathway. (16–19) (Figure 1). With few exceptions, efficiency of transport reported is low, although selective targeting of some particles to Peyer's patches appears possible. Evidence for particulate uptake via these routes is reviewed below.

Paracellular Transport: Persorption

The term *persorption* was introduced by Volkheimer to describe absorption of solid particles whose size range (up to 150 μm diameter) and nondeformability make inclusion within transcytotic vesicles very unlikely (20). A wide variety of particles of diverse shapes, sizes, and compositions have been claimed to be transported by this pathway in many different species: those of 7–70 μm diameter and high rigidity exhibit the highest rates of absorption (16,20). The time course of absorption is complex. For example, cornstarch granules of 3–25 μm diameter appeared in the circulation as early as 5 min after oral ingestion by patients, followed by one or two additional "waves" of absorption occurring over the next 5 hr (20). Few starch granules were in the circulation at 12 hr, and none at 24 hr after administration. The extent of persorption correlates with GI motility, increasing with pharmacological agents that stimulate peristalsis (e.g., caffeine), and decreasing with treatments that inhibit motility (e.g., atropine). Light microscopy reveals starch granules in spaces between epithelial cells, the lamina propria, and the lumen of lymphatic and blood vessels, although the possibility of artifact due to tissue damage is difficult to eliminate (21). The means by which such large particles might cross epithelial barriers as well as the endothelium of villous capillaries to reach the portal circulation remains speculative: passage between cells in the "extrusion zone" at tips of villi where enterocytes desquamate or mechanical lesions between adjacent cells have both been postulated as entry sites. There is little direct evidence to support these claims (8). Furthermore, observations of appearance in the portal circulation are difficult to reconcile with current knowledge of size restrictions imposed by capillary endothelia that contain fenestrae of 80–100 nm diameter. Removal of particles from portal blood by hepatic Kupffer cells would be expected to eliminate most particles able to cross the epithelial barrier. It is therefore not surprising that the efficiency of starch granule transport by persorption has been estimated at 0.002% (20), much too low to be considered as a mechanism for systemic drug delivery due to potential problems with reproducibility. Other workers have failed to confirm the rapid appearance in blood of particles greater than 10 μm diameter (22).

Paracellular passage has been proposed as an explanation for penetration of GI mucosa by polycyanoacrylate nanoparticles, since particles were seen in intercellular spaces and not inside enterocyte cytoplasm (23). However, efficiency of transport appears low, based on the low number of particles seen, and factors such as use of a surfactant (Pluronic F68) in preparation of nanocapsules complicate interpretation of data. No published evidence supports the claim of "proteinoid" particle absorption by a paracellular route (24). Therefore, although occasional particles may be transported across GI epithelia by a paracellular route, efficiency of this process is either unknown or low. Absorption enhancement seen with such systems could result from paracellular passage of drug molecules after release from the carrier particle, whose function may be more protection of the drug than increase in transport efficiency.

Transcytotic Transport Across Enterocytes

In neonates, transcytosis across enterocytes contributes to absorption of molecules such as milk proteins and immunoglobulins, and small inert particles such as Percoll can also be internalized via these routes (25,26). Some pathogenic organisms have developed mechanisms to penetrate the mature epithelial barrier by this route (see below). However, evidence for transcytosis of inert particles across normal absorptive epithelium in adults is limited to infrequent reports of small-particle uptake. Ferritin and killed adenovirus particles (27), cationized ferritin (28), and small latex particles of 220 nm diameter (29) have all been reported within vesicular compartments of enterocytes. Although small numbers of particles may appear in the lamina propria, transport to lysosomes is the most common intracellular fate for absorbed material (e.g., 28), and material reaching the circulation is likely to be removed by cells of the reticuloendothelial system (RES). Transport of particles across enterocytes in the adult does not appear to be a route for enhancing drug absorption.

Transcytotic Transport Across M Cells Overlying Peyer's Patches

In contrast to the relatively few reports of particles within intestinal enterocytes, numerous studies of particle transport across the intestine have demonstrated Peyer's patches to be a major site for their uptake. These observations suggest that particles may be transported across M cells in the epithelium overlying intestinal lymphoid tissue. M cells appear to play a role in transepithelial transport of limited quantities of luminal contents for delivery to cells of the immune system. These cells possess a relatively sparse glycocalyx and few lysosomes, and are specialized for endocytosis and transcellular transport of intact macromolecules or particles, with little degradation during transit through the cells (30,31). Morphological studies demonstrated transcytosis of fluid phase tracers (32,33) and of membrane-bound molecules [cationized ferritin (32), IgA (34), poly-D-lysine (35)] and a variety of lectins, including wheat germ agglutinin (WGA), *Ricinus communis* agglutinin (RCA) I and II (32), and *Ulex europeaus* and agglutinin (UEA) (36) across these cells. Except for some enteroinvasive pathogens, direct observation of particles larger than ferritin (10 nm) within M-cell vesicles has not been reported at the ultrastructural level. Since not all epithelial cells over GALT are M cells, passage of particles across other types of cells remains a possibility.

Joel and colleagues have demonstrated cumulative association of 20- to 50-nm carbon particles with Peyer's patches and with isolated lymphoid follicles of mice given carbon suspensions in drinking water over a 1-year period (37). Particles were initially associated primarily with macrophages in subepithelial tissue of patches, and at longer times (2 months or more) were detected in macrophages of mesenteric lymph nodes. These and other observations suggest that macrophages are involved in transport of phagocytosed particles out of GALT (38,39). Particles were still observed for 4 months after administration was stopped. Additional studies with 1.8- to 2-μm latex particles demonstrated a similar accumulation in Peyer's patch regions in both conventional and germ-free mice (40–43). The mechanism of penetration of the mucosa was not determined but presumably involved transcytosis across M cells overlying GALT.

Not all particulates are sequestered within patch tissue; chrysotile asbestos fibers, quartz and carmine particles, and 15.8-μm-diameter carbonized styrenedivinylbenzene particles were not found to accumulate appreciably within GALT, while iron oxide particles and 5.7-μm-diameter styrenedivinylbenzene particles were (44). A correlation between particle hydrophobicity and accumulation was suggested from these data. This conclusion is further supported by the work of Eldridge et al. (45), who compared the uptake of fluorescently labeled microspheres of different composition into Peyer's patches. Polymers with the greatest relative hydrophobicity [poly(styrene), poly(methyl methacrylate), poly(hydroxybutyrate)] were taken up most readily, polymers of poly(DL-lactide), poly(L-lactide), and poly(DL-lactide-*co*-glycolide) were taken up in lower numbers, and microspheres composed of cellulose polymers were rarely found in Peyer's patch tissue. Jani and coworkers (46), also found that fluorescent carboxylated polystyrene microspheres are taken up less efficiently by GALT than more hydrophobic nonionized polystyrene beads. The difference between 5.7- and 15.8-μm-diameter synthetic particles described above suggests a size restriction for accumulation of particles within GALT. A similar size dependence was seen for translocation of polystyrene latex particles ranging from 50 nm to 3 μm diameter into extraintestinal tissues, but not for their initial uptake into Peyer's patch regions (47). Extraintestinal particles were largely associated with cells of the reticuloendothelial system in mesenteric lymph nodes, spleen, and liver. These results contrast with the work of Eldridge et al., who studied effect of particle size on uptake and retention in Peyer's patch regions, mesenteric lymph nodes, and spleen using fluorescently labeled microspheres (48). Microspheres were not observed to penetrate the mucosa at any sites other than Peyer's patches. No selectivity in size of particles taken up into Peyer's patches was seen within a range from 1 to 10 μm. However, poly(DL-lactide-*co*-glycolide) microspheres larger than 5 μm in diameter were selectively retained, within subepithelial lymphoid tissue of the patch, while smaller particles were removed into the circulation to appear sequentially in mesenteric lymph nodes and spleen. Particle number peaked at 4 days in GALT, 7 days in mesenteric lymph nodes, and 14 days in spleen. Loss from Peyer's patches was attributed to inclusion of particles within extravasating macrophages, although no data were shown to support this.

Several attempts to determine efficiency of particle transport have been made using either fluorescently labeled or radiolabeled particles. A recent study of particulate transport across the intestine has used fluorescently labeled carboxylate latex microspheres (1.1 μm diameter) to demonstrate transport into the circulation after oral administration to rats and rabbits. Quantitation of the data gave values as high as 39% of the total administered particles in the blood 45 min after administration (49), much higher transport efficiencies than that seen in other studies. Particles could be detected as early as 10 min after dosing. However, in contrast to earlier studies on persorption (20,21), the number of microspheres in the circulation decreased sharply as particle size increased between 4 and 8 μm diameter. The route of absorption (paracellular between enterocytes or transcytosis across M cells) was not investigated in this work. In contrast to this report, Ebel (50) estimated very low levels (<0.01% of the total administered dose) of fluorescent polystyrene latex microsphere transport across intestinal tissue in mice. Particles of diameters 2.65 and 9.13 μm were quantitated in a fluorescence-activated cell sorter after tissue dissolution. Particles, especially smaller diameter (2.65 μm) ones, were concentrated in Peyer's patches. However, the earliest time examined was 24 hr after microsphere administration, a time when previous studies of persorption revealed few particles remaining in the blood. Another estimate of extent of particle absorption has been made by Jani et al. using gel permeation chromatography to quantitate the amount of polystyrene latex associated with tissues after rats had been dosed with 50-nm to 3-μm microspheres for 10 consecutive days (46,47). These workers found the total accumulated polystyrene in extraintestinal tissue to range from a maximum of 6% for the smallest particles to 1% for 1-μm and 0%

for 3-μm microspheres. Although 10–15% of administered radioactivity was absorbed after peroral dosing of rats with polymethyl-[1-^{14}C]methacrylate nanoparticles, absorbed material was probably low molecular weight components in the polymer (51). The same concern applies to later electron microscope (EM) autoradiographic studies of radiolabeled polyhexyl-[3-^{14}C]cyanoacrylate distribution within mouse intestine (52–53).

Neither transepithelial transport via persorption nor transcytosis across M cells has been conclusively demonstrated to be the mechanism of transport for particulate drug carrier systems described above. Pappo and Ermak (54) investigated the site and time course of particle penetration of the gastrointestinal epithelium by colocalization of fluorescent latex microspheres with M-cell–specific antibodies. Rabbit intestinal loops were inoculated with 600- to 750-nm latex spheres, and incubated for 10, 30, or 90 min of continuous exposure to the particles. Latex microspheres were found in association with M cells at 30 min, when the vast majority of particles were associated with the surface epithelium of Peyer's patches. The time course of transepithelial transport was similar to that reported for transcytosis of EM tracers across the M cell, with particles appearing the subepithelial tissues by 90 min (32). Approximately 5% of the total particles administered were estimated to be taken up. This figure is likely to be high compared to studies in which particles are being translocated down the intestinal tract by peristalsis, and rabbit Peyer's patches appear to be more efficient at particle transport than mouse (54).

Similar results have been reported by Jepson et al. (55), who found that 460- to 670-nm fluorescent latex spheres were predominantly associated with rabbit M cells. These authors used different concentrations of latex particles to demonstrate an apparent saturation of M-cell–binding capacity, suggesting that the overall rate of particle uptake was limited not by binding but by endocytic rate.

Adsorption of anti-M-cell monoclonal antibodies to polystyrene microspheres increased both adherence to and transcytosis across the Peyer's patch epithelium of rabbits by about three-fold compared to control microspheres bound to unrelated antibody (56).

Although numerous studies demonstrate particle uptake into Peyer's patches, determination of the efficiency of this process has been technically difficult and generally appears extremely low, requiring prolonged dosing protocols to demonstrate uptake in many cases. Studies conducted to date are contradictory regarding the effect of particle size and overall efficiency of the process, and incomplete regarding the site and mechanism of transepithelial transport. The influence of variables such as size and particle composition is only beginning to be understood, and will require further investigation before the utility of particulate delivery systems can be completely evaluated. Delivery of transported particles to subepithelial lymphoid cells has been shown to induce an immune response (see below), a potential disadvantage for oral delivery of peptides and proteins requiring chronic administration, which will also need to be addressed.

Enteric Pathogens

A number of pathogenic organisms, including some strains of *Enterobacteriaceae* (57,58), *Vibrio cholerae* (59), reovirus (60–61), HIV-1 virus (62), and poliomyelitis virions (63) can penetrate the GI epithelium through M cells or enterocytes. At least some prion-mediated diseases, including kuru in humans, scrapie in sheep, and Creutzfeldt-Jakob disease in cattle can be orally transmitted, but the mode of penetration of these proteinaceous infectious particles across the GI tract epithelium is not known (64).

Enteroinvasive bacteria utilize several distinct strategies for epithelial penetration by endocytosis (58,65,66). Specific recognition of binding sites on epithelial cell surfaces involves either induction of bacterial surface proteins after exposure to the epithelium (*Salmonella*, *Shigella*), or molecules that are expressed constitutively (*Yersinia*). Identification of the components of both bacterial and host cells that mediate cell–cell recognition has progressed quite far, and may allow design of drug carriers for adhesion to and/or penetration via intestinal epithelial cells. Another well-studied pathogen is reovirus, where the minor outer capsid protein σ1 mediates binding to M-cell plasma membranes (67). The isolated capsid protein has been used to direct liposomes to M cells (see below).

The first step in invasion of the host by pathogenic *Enterobacteriaceae* is adhesion of the bacterium to the host cell surface. Many pathogens utilize several distinct strategies simultaneously to accomplish this, making genetic dissection of the process complex. Several fimbrial proteins are involved in recognition of host cell carbohydrate residues, and a single strain of bacteria may have the ability to express more than one. However, other mechanisms for cell attachment must exist, since nonfimbriated strains of *S. typhimurium* are as virulent as fimbriated strains after oral administration. Synthesis of some bacterial polypeptides required for *Salmonella* adherence is induced by interaction with the epithelial cell surface. This contrasts with *Yersinia*, which can invade epithelial cells in the absence of prokaryotic protein synthesis since adherence mediators are expressed constitutively. At least three gene products are involved in *Yersinia* attachment; best characterized is the invasin protein, which binds to members of the β_1-chain integrin family on the host cell surface. Interaction with these receptors may induce cytoskeletal rearrangements that result in internalization of the bacterium through movement of the host cell plasma membrane to surround the pathogen (66). Incorporation of proteins such as invasins into carrier systems might promote more efficient transcytosis across epithelial cells, but efficiency of translocation of *Yersinia* or other enteric pathogens is not established.

Most *Salmonella* species enter cells of the terminal ileum, where penetration may occur across enterocytes or M cells. These bacteria are facultative intracellular pathogens, which replicate within endocytic vacuoles but also undergo transcytosis with entry into the circulation and reticuloendothelial cells to establish a systemic infection. Intracellular survival is mediated at least in part by bacterial factors that provide protection against host cell cationic peptides (defensins) that lyse unprotected bacteria (68). Since *Salmonella* has the ability to survive within macrophages, migration of macrophages carrying ingested bacteria likely plays an important role in their colonization of the RES and subsequent induction of a secretory immune response. Attenuated strains of *Salmonella* that lack the ability to establish infection but retain immunogenicity are being investigated as carriers for induction of secretory immunity after oral challenge (see below). Pathogenic *Listeria* species have also been observed in macrophages and in polymorphonuclear leukocytes in the lamina propria (65).

A different invasive strategy is employed by *Yersinia* species, which become preferentially associated with macrophages underlying Peyer's patches. These bacteria replicate slowly, if at all, within epithelial cells, and the exact site of penetration (M-cell versus absorptive cell) is unclear. Once across the epithelial barrier, *Yersinia* appear to survive and even multiply within phagocytic cells, which is the major site for their replication. This species therefore has promise as a vehicle for delivery of antigens in a manner similar to *Salmonella*, if attenuated strains can be developed. Although use of bacterial strains as drug carriers is theoretically possible, the present lack of knowledge of their translocation efficiency and the possibility of inducing an immune response through repeated exposure and cost and complexity of manufacture will need to be investigated before oral delivery systems based on this strategy can be feasible.

LIPOSOMES

Liposomes may transiently protect some peptides and proteins from degradation in the GI tract. However, although some authors have suggested that intact liposomes can be taken up by epithelial cells, evidence to support this is weak, and the majority of data are consistent with the view (69,70) that intact liposomes are not taken up by epithelial cells and do not facilitate uptake of entrapped contents.

SYNTHETIC NANOPARTICLES

Polycyanoacrylates, Polyacrylic Polymers

Several laboratories have investigated the use of particulate delivery systems for oral drug delivery. Intragastric administration of polyisobutylcyanoacrylate nanocapsules, containing insulin, to fasted diabetic rats resulted in prolonged hypoglycemic effects (71–73). The pharmacokinetics were highly unusual in that hypoglycemia was detected only after a 2-day delay but persisted for up to 20 days at a dose of 50 U/kg. These effects were less pronounced in fed or nondiabetic animals. Subcutaneous administration of equivalent doses of encapsulated insulin resulted in the rapid onset of a two- to three-fold more pronounced glycemia, but this effect lasted for only a day. Absolute bioavailability of insulin administered by intragastric versus subcutaneous routes cannot be determined from these experiments. Oral administration of nanocapsules, which had insulin adsorbed to their surfaces, produced no hypoglycemic effect in rats (74), suggesting that incorporation within nanocapsules was required to protect the peptide from degradation within the gut. Similar experiments using 0.1- to 0.3-μm-diameter nanocapsules loaded with Lipiodol have also demonstrated a more prolonged elevation of plasma iodine levels in dogs after intrajejunal administration. Although x-ray emission analysis of the distribution of iodine at different levels within the intestinal mucosa suggested transcellular transport, nanocapsules were seen in paracellular spaces, but not in intracellular compartments by electron microscopy (72,75). In a more recent report, insulin-loaded isobutylcyanoacrylate nanoparticles did not protect the peptide from degradation or display any biological effect after oral administration in an aqueous medium (76). Dispersion of the same nanoparticles in a medium containing miglyol 812, deoxycholic acid, and Poloxamer 188 resulted in a 50% decrease in fasted glycemia from 2 hr to 8 days following oral administration of 100 U/kg insulin to diabetic rats (76). Substantial degradation of polybutylcyanoacrylate nanoparticles was found after incubation with esterase *in vitro*, raising doubt about their stability within the gastrointestinal tract (77).

Oral administration of a combination of insulin and protease inhibitors incorporated into polyacrylic polymer (Eudragit L100) microspheres resulted in a two- to three-fold increase in the efficacy of induction of hypoglycemia in normal and diabetic rats (78). This enhancement was not seen when protease inhibitors were omitted.

Reproducibility, oral bioavailability, mechanism of transmucosal transport, and efficiency of these delivery systems remain to be determined. If typical, the extremely long time course of hypoglycemia caused by insulin within nanocapsules will preclude use of this system for nonchronic administration or for drugs with a narrow therapeutic index.

Proteinoids

Encapsulation of peptides or proteins in amino acid microspheres of approximately 10 μm in diameter (proteinoids) has been described as a method for protecting and achieving the oral absorption of macromolecules such as insulin or heparin. In an initial study (24), orally administered insulin produced a significant hypoglycemic effect in diabetic rats and had a longer duration of action than when administered subcutaneously. The absolute bioavailability of insulin administered in proteinoids was not determined (24). Studies to support the claim that these particles are efficiently taken through the GI mucosa have not yet been critically interrogated by the scientific community and are viewed as speculative.

Mucoadhesive Microspheres

In another approach, polyhydroxylethyl methacrylate microspheres coated with poly(acrylic acid) copolymers have been examined for their ability to alter GI transit through their mucoadhesive properties using an *in situ* rat model (79,80). The GI transit time of particles was substantially prolonged by a polymeric coating consisting of a blend of Carbopol 934P and Eudragit RL100. Increased retention time correlated well with relative strength of mucoadhesion measured by a tensiometer technique *in vitro*. In one series of experiments, the rate of transport of the octapeptide 9-desylycinamide, 8-arginine vasopressin (DGAVP) from the mucosal to the serosal side of rat ileum *in vitro* was improved approximately fourfold by administration as mucoadhesive versus nonadhesive microspheres, but data from *in vivo* experiments are currently lacking (79). This delivery system may improve drug absorption due to longer contact time with the GI epithelium, higher concentration of peptide near the apical membrane, and protection of encapsulated drug from inactivation within the gut lumen. However, the utility of this approach may be limited by the rate of turnover of the mucous layer, a complex process that remains poorly understood (81,82).

MICROEMULSIONS/MICELLES

Many lipid carrier systems have been studied for their ability to promote oral absorption of conventional drugs, peptides, and proteins. These systems include oil-in-water (83), water-in-oil (84,85), and water-in-oil-in-water (86,87) emulsions and lipid-surfactant mixed micelles (88). In addition, covalent attachment of peptides to lipids has been proposed as a means for absorption enhancement. For example, covalent complexes between insulin and cationic triglyceryl carriers show a hypoglycemic effect when given orally (89) although few details on reproducibility or bioavailability are available. Suggestions that lipid transport pathways can be used to "smuggle" peptides and proteins across membranes and cellular barriers "disguised" as lipid are not clearly supported by the available data on lipid absorption. Dietary lipid appears to be hydrolyzed to monomeric constituents (fatty acids, monoglycerides, lysophospholipids) that are reassembled within the enterocyte for exocytotic release as chylomicrons and lipoprotein particles (13–15). The means by which micelles or microemulsion "particles" might be absorbed intact, if at all, are therefore unrelated to our present understanding of lipid transport pathways (13–15).

Lipid carrier systems may increase oral bioavailability of proteins by protecting them from proteolysis, by perturbing cellular membranes or intracellular junctions, or by utilizing lipid transport systems. Most experimental approaches have, however, failed to distinguish between these possibilities. In addition, in many cases it is not clear whether absorption enhancement is achieved via physiological pathways or by transient or permanent disruption of the epithelial layer with the potential for toxicity.

Evidence from diverse sources shows that incorporation of drug molecules within micelles or emulsions may improve absorption of certain peptides as well as other drugs. Use of dispersed lipid-based carriers has the benefit of not necessarily requiring covalent modification of the drug. Dispersed systems for which some peptide absorption data are currently available include single surfactant micellar systems (acylcarnitines), mixed micelles, and microemulsions.

Micellar Systems

Absorption enhancement of a tetradecapeptide somatostatin analog after both duodenal and rectal coadministration with long-chain acylcarnitines capable of forming micelles was reported by Fix and coworkers (90,91). In this work, palmitoylcarnitine increased bioavailability from 1% to 11% (relative to intravenous administration) after direct duodenal infusion in rats. Preliminary evidence from Caco-2 cell studies suggest that at least one effect of palmitoylcarnitine may be to increase tight junctional permeability (91). Tissue morphology appeared normal after a single exposure of tissue to this surfactant, although the effect of chronic exposure remains to be determined (90).

Mixed micelles consisting of linoleic acid and polyoxyethylene-hardened castor oil (HCO-60) have been reported to aid absorption of the 23-kDa polypeptide α-interferon from rat large intestine (92–94). The drug appeared to be delivered selectively to lymphatics, since concentration in thoracic lymph was considerably higher after administration with fatty acid and surfactant than with either component alone. Increases in lymphatic absorption of the 1.2-kDa hydrophobic peptide cyclosporine in the same lipid–surfactant mixture have also been claimed using intragastric, intraduodenal, or rectal dosing (95). However, bioavailability, tissue damage, and the mechanism of absorption promotion were not addressed. Further work from this group has used HCO-60 by itself or in combination with enteric coating for cyclosporine delivery, claiming a 1.5-fold improvement in oral bioavailability in rabbits compared to a conventional olive oil formulation (96). Interpretation of these data is complicated by the extremely variable bioavailability of cyclosporine after peroral administration (84).

Emulsions/Microemulsions

Evidence for the utility of stable microemulsions as delivery vehicles for improvement of peptidergic drug absorption is even more limited. Native insulin when orally administered to humans is poorly (approx. 0.5%) bioavailable (97). Reports that absorption of insulin is enhanced when administered as a complex water-in-oil emulsion in which the oil phase contains lecithin, nonesterified fatty acids, and cholesterol together with the protease inhibitor aprotinin require further verification with appropriate controls (98). Furthermore, the transport pathways involved cannot be determined from this study, although the relatively short time to achieve peak plasma levels indicates that lymphatic absorption is not involved. The proposed mechanism, i.e., the absorption of chylomicron-like particles (99), is not consistent with current views on fat absorption as discussed above.

A microemulsion form of cyclosporine consisting of long-chain fatty acid esters, surfactants, and a low molecular weight alcohol as cosurfactant improved oral absorption of this molecule in rats (2.6-fold increase in absolute bioavailability relative to a commercially available preparation in polyethylated castor oil). However, a similar formulation in which branched chain fatty esters were substituted for long-chain fatty acid esters gave no significant improvement over the standard oral solution in olive oil and pegliol-5-oleate (85). The latter result suggests that the specific composition of microemulsions may be critical for success of this approach. The optimal site for absorption of cyclosporine and insulin from a microemulsion was the small intestine (100,101). Absorption of vasopressin in rats from ligated small intestinal segments was increased only about twofold by formulation of the peptide in a microemulsion versus an aqueous solution (102). Potency of insulin within one of the microemulsion formulations was lost between 3 and 4 weeks (103).

Thus, while the use of microemulsions shows some promise, many issues, such as stability of formulations, reproducibility of absorption enhancement, route of absorption (i.e., lymphatic versus portal circulation), and potential for tissue damage, require further investigation.

The mechanism by which lipid formulations can promote the transport of peptides and proteins across membranes remains to be established. Enhancement of the absorption of peptides coadministered with lipid-based carriers such as mixed micelles or microemulsions may involve transport via lipid absorption pathways, or protection of the molecule from proteolysis and absorption via one of the mechanisms previously discussed (e.g., transcytosis or paracellular transport). The concept that formulation of peptides within micelle- or emulsion-based lipid systems can promote their absorption via lipid transport pathways is difficult to reconcile with current views on lipid transport, which indicate permeation of the epithelial layer by monomeric lipid molecules.

COLONIC ABSORPTION/AZOPOLYMERS

Interest in targeting drugs to the colon results from the need to treat colonic diseases (e.g., irritable bowel syndrome, ulcerative colitis, inflammatory bowel disease) or to test the proposal that bioavailability may be increased by avoidance of degradation due to enzymatic hydrolysis or pH instability in the upper GI tract (104). The use of impermeable coatings designed to achieve colonic release after transit of drug-containing capsules through the upper GI tract has been applied to the oral delivery of insulin and vasopressin (105–107).

Another approach to site-specific colonic delivery has been taken by Kopecek (108) using *N*-(2-hydroxypropyl)methacrylamide (HPMA) copolymers containing fucosylamine residues. The strategy employed in this work is analogous to that used by a number of pathogenic bacteria, such as *Shigella flexneri* and *E. coli* 0/24, which bind to the colonic mucosa via sugar residues at their surfaces (109). Fucosylamine residues mediate binding to lectins in the mucus layer of the colon, the HPMA copolymers derivatized in this manner appeared to be selectively retained in the colon after oral administration to guinea pigs (108). Linkage of drug molecules to fucosylamine-derivatized HPMA copolymers via azo bonds might therefore provide an efficient, site-specific release in the colon.

Although potential colonic absorption of peptides and proteins has been suggested by many workers, to our knowledge there are no reported bioavailability studies that demonstrate efficient absorption of these molecules from the colon in humans. Experiments with a number of peptidergic drugs demonstrated very poor absorption across colonic tissue *in vitro* (Smith et al., unpublished observations). Thus, although particulate carriers may allow delivery of molecules to the colon, systemic bioavailability may not be enhanced.

CONJUGATION TO SOLUBLE CARRIERS

Vitamin B_{12}

In the adult, a receptor-mediated transcytotic pathway is involved in systemic absorption of vitamin B_{12} (cyanocobalamin, Cbl) from the lumen of the GI tract. Absorption of Cbl is a complex process that involves transfer of the ligand between a number of transport proteins (110).

The relative lack of steric specificity of the intrinsic factor cobalamin receptor for ligands attached at the Co-β position of the corrin ring suggests that attachment of drug molecules at this site may allow them to be directed into this endocytic pathway. Russell-Jones et al. (111) using monocarboxylic acid derivatives of Cbl demonstrated pharmacological activity of Lys-6-LHRH-B_{12} conjugates, and systemic immunity to BSA-B_{12} conjugates when given orally to mice and rats. In all cases there was no oral activity of the free molecules or molecules administered together with free Cbl. No details on the extent of bioavailability of the conjugates or data to support transport via the cobalamin pathway are yet available. However, the probability of successful exploitation of this pathway for enhancing absorption of peptide drugs would appear to be low due to extremely low capacity (<2 μg Cbl/day in humans), the possibility of degradation during intracellular passage (similar to intrinsic factor), and the requirement for either a pharmacologically active conjugate or release of the drug molecule from the conjugate.

β-Lactoglobulin

β-Lactoglobulin has been proposed as a carrier molecule for the transepithelial transport to bound hydrophobic drug molecules (112). Although plasma membrane receptors for this protein have been demonstrated in neonates, e.g., (113), their presence in adults is doubtful (113,114), and transport of intact β-lactoglobulin across the gastrointestinal epithelium is inefficient [less than 0.01% of load/hr: (115,116)]. Therefore, the probability of exploiting this route in the adult would also appear to be very low.

STIMULATION OF MUCOSAL IMMUNITY WITH PARTICULATE ANTIGENS

Since mucosal surfaces are the site of entry of organisms causing most infectious diseases, the achievement of high levels of mucosal immunity is highly desirable. GALT of the GI mucosa is the largest lymphoid organ of the body, providing protection against microbial and food antigens. Oral immunization results in a disseminated immune response due to migration of antigen-stimulated lymphoid cells from GALT to other mucosal sites, where maturation to IgA-secreting plasma cells occurs. Attempts to achieve mucosal immunity are limited by poor immunogenicity of soluble antigens and the failure of parenteral immunization strategies to induce production of secretory antibodies that are predominantly made by mucosa-associated lymphoid tissues. Oral immunization is further complicated by destruction of the antigen by proteolytic enzymes of the digestive tract. Several strategies have been taken to address these limitations, including use of particulate antigens incorporated into synthetic microsphere carriers, use of attenuated avirulent strains of bacteria or viruses that have been genetically engineered to express foreign antigens, and coadministration of adjuvants such as cholera toxin B subunit (48,117–123).

Microspheres

Microspheres of varying compositions have been used to improve the ability of antigens to provoke a mucosal immune response. Poly(DL-lactide-*co*-glycolide) microspheres were chosen as antigen carriers (45,48,122) because of their biocompatibility, biodegradability, and established precedent for safe use in humans (45). Poly(DL-lactide-*co*-glycolide) microspheres of 1–10 μm diameter carrying staphylococcal enterotoxoid were also used for oral immunization of mice, and were superior to soluble antigen in eliciting both systemic and mucosal immunity (monitored by increased IgA levels in saliva and in fluid obtained by intestinal lavage) during an oral immunization protocol. Microspheres of >10 μm diameter were less effective, consistent with the finding that polystyrene microspheres of >10 μm and latex microspheres of 15 μm were not taken up within Peyer's patches (22,45). In similar work from the same laboratory, microencapsulated influenza type A virus resulted in higher and more prolonged levels of specific antibodies in serum and saliva of mice than oral immunization with free viral suspensions, and these animals had better survival when challenged with the virus (124).

Although no other laboratory has investigated the synthetic microparticle approach in as much detail as these studies, preliminary data supporting the concept of enhanced induction of an IgA response by this strategy has been reported by O'Hagan and coworkers (125,126). In this work, nonbiodegradable polyacrylamide and poly(butyl-2-cyanoacrylate) particles have been used as adjuvants for oral immunization of rats with ovalbumin. Association of the antigen with these particles resulted in a rather small but significant increase in the production of secretory antibodies when compared to soluble antigen.

Microspheres are likely to serve two functions in oral vaccination: (a) protecting of the peptide from degradation during transit through the digestive tract and (b) targeting antigen to GALT, where particle retention allows it to serve as a depot for controlled release of antigen during the process of particle dissolution. After absorption has occurred, association of particles by MHC class II expressing macrophages may play an important role in antigen processing and presentation, while extravasation of cells carrying particles results in dissemination of the processed antigen.

A potential advantage of synthetic biodegradable microspheres as carriers for antigens is the ability to vary the size range of particles, which may allow some selectivity in induction of systemic versus mucosal immunity. This hypothesis has yet to be directly tested.

Bacterial and Viral Pathogens

A related approach to oral vaccination has used live attenuated strains of bacteria that retain the immunogenicity of wild-type strains but cannot establish infection of the host (119,127). *Salmonella* species carrying nonreverting mutations in genes coding for enzymes of galactose (*galE*) or aromatic amino acid (*aroA*) metabolism have been shown to elicit both secretory and systemic immune responses after oral administration. Such strains have limited ability to colonize the GALT and other components of the RES, multiplying at slower rates which result in a controlled infection and eventual clearance by normal defense mechanisms. Genetic manipulation of these auxotrophic strains allows introduction of antigens from other bacteria, viruses, or eukaryotic parasites into the vaccine strain. For example, *S. typhimurium aroA* mutants expressing the *E. coli* K88 fimbrial antigen and heat-labile toxin B subunit induced both serum and GALT-associated immune responses after oral infection in mice (119,127). These results suggest that multivalent vaccines could be developed. However, this use of bacterial carriers may be limited by the immunogenicity of the carrier itself, which restricts frequency of administration to a single treatment or to very infrequent intervals, since induction of an IgA response could inhibit subsequent colonization of the GI lymphoid tissue.

An analagous approach has been taken with construction of a chimera of the Sabin type 1 vaccine strain of poliovirus engineered to express an epitope from the transmembrane glycoprotein gp41 of HIV-1. This strain has been used for subcutaneous immunization of rabbits, where a response specific for gp41 was elicited (128). The evaluation of the utility of such an antigen for oral immunization may provide another strategy for the induction of secretory immunity.

Liposomes

Liposomes have also been used as antigen carriers for oral immunization (129–132). Both local and systemic immune responses were provoked by intragastric immunization of rats with liposome-associated *S. mutans*–soluble cell wall antigen, while animals immunized with free antigen or with liposomes alone gave no immune response (130). Immunocytochemical localization of antigen demonstrated its presence in Peyer's patches and mesenteric lymph nodes at days 8–17 postimmunization. Response to the first antigen lasted for about 30 days; a second dose of antigen caused a more rapid response, but antibody titers declined rapidly after both exposure to antigen (130). A more prolonged secretory immune response lasting for at least 50 days was provoked by oral administration of BSA in liposomes supplemented with muramyl dipeptide as adjuvant (129). Preliminary data on the immunogenicity of *S. mutans* serotype-specific carbohydrate (CHO) antigen incorporated into 100-nm unilamellar liposomes compared to free CHO suggests that the carrier improves the salivary IgA response of rats after gastric intubation of the antigens (131). Preferential targeting of liposomes to M cells has been achieved by incorporation of the reoviral attachment protein σ1 into the membrane bilayer of liposomes during their formation. However, phagocytosis of liposomes by macrophages within the underlying lymphoid tissue of the Peyer's patch will result in considerable loss of antigen through lysosomal degradation (133). Incorporation of *Vibrio cholerae* antigens into osmium-stabilized liposomes has been reported to provoke a higher level of mucosal immunity than free antigen after oral immunization of rats (132).

Given the relatively few studies performed to date, the lack of understanding of mechanisms for stimulation of antigenicity, and the instability of liposomes within the environment of the gastrointestinal tract, it is difficult to evaluate the potential use of liposomal carriers for oral vaccination without further work addressing these issues.

CONCLUSIONS

Experimental evidence reviewed above clearly indicates that a variety of particles can be transported across the gastrointestinal epithelium. However, most studies indicate that the extent of absorption of particulate materials is extremely low. The few exceptional reports that suggest otherwise still require experimental verification. Thus, unless particulates with extremely high drug capacity can be formulated this approach would appear to be restricted to the most potent peptides. For progression to clinical trials the ability to reproductively control the time course and extent of absorption and the administration of such systems in a safe manner will be required. In addition, the ability to scale up and manufacture such systems is a prerequisite for commercialization.

The information on the efficacy, reproductivity, and safety of oral particulate peptide delivery systems is severely limited. In new approaches, e.g., lipid-based systems where it is postulated that components of the carrier may be absorbed by normal physiological routes, it clearly remains to be established whether absorption of peptidergic molecules can be enhanced through carrier association. However, in contrast the use of particulate antigens to target the GALT for induction of a secretory immune response shows much promise for the development of oral vaccines. Recent advances in our understanding of the molecular basis of adherence and endocytosis of enteric pathogens may allow design of particulate carriers which are more efficiently targeted for M-cell transcytosis. The risk of inducing an immune response from the preferential sequestration of particulates within the GALT would appear to rule out this route for the systemic absorption of peptidergic molecules for chronic therapies.

REFERENCES

1. V. H. L. Lee, S. Dodda-Kashi, G. M. Grass, and W. Rubas, in *Peptide and Protein Drug Delivery* (V. H. L. Lee, ed.), Marcel Dekker, New York, 1991, p. 691.
2. V. H. L. Lee and A. Yamamoto, *Adv. Drug Deliv. Rev.*, *4*, 171 (1990).
3. M. J. Humphrey, in *Delivery Systems for Peptide Drugs* (S. S. Davis, L. Illum, and E. Tomlinson, eds.), Plenum Press, New York, 1986, p. 139.
4. U. B. Kompella and V. H. Lee, *Adv. Drug Deliv. Rev.*, *8*, 115 (1992).
5. W. C. Shen, J. Wan, and H. Ekrami, *Adv. Drug Deliv. Rev.*, *8*, 93 (1992).
6. P. L. Smith, D. A. Wall, C. H. Gochoco, and G. Wilson, *Adv. Drug Deliv. Rev.*, *8*, 253 (1992).
7. P. Artursson and J. Karlsson, *Biochem. Biophys. Res. Commun.*, *175*, 880 (1991).
8. J. L. Madara, *J. Membrane Biol.*, *116*, 177 (1990).
9. J. L.Madara, *Am. J. Pathol.*, *137*, 1273 (1990).
10. D. G. Levitt, A. A. Hakim, and N. Lifson, *Am. J. Physiol.*, *217*, 777 (1969).
11. H. N. Nellans, *Adv. Drug Deliv. Rev.*, 7, 339 (1991).
12. A. A. Hakim and N. Lifson, *Am. J. Physiol.*, *216*, 276 (1969).
13. Y.-F. Shiau, in *Physiology of the Gastrointestinal Tract* (L. R. Johnson, ed.), Raven Press, New York, 1987, p. 1527.
14. H. Westergaard and J. M. Dietschy, in *Membrane Transport Processes In Organized Systems* (T. E. Andreoli, J. F. Hoffman, D. D. Fanestil, and S. G. Schultz, eds.), Plenum, New York, 1987, p. 213.
15. J. S. Patton, in *Physiology of the Gastrointestinal Tract* (L. R. Johnson, ed.), Raven Press, New York, 1981, p. 1123.
16. K. Baintner, in *Intestinal Absorption of Macromolecules and Immune Transmission from Mother to Young*, CRC Press, Boca Raton, 1986.
17. D. T. O'Hagan, *Adv. Drug Deliv. Rev.*, *5*, 265 (1990).
18. J. Kreuter, *Adv. Drug Deliv. Rev.*, 7, 71 (1991).
19. W. Rubas and G. M. Grass, *Adv. Drug Deliv. Rev.*, 7, 15 (1991).
20. G. Volkheimer, *Adv. Pharmacol. Chemother.*, *14*, 163 (1977).
21. G. Volkheimer, *Gut*, *10*, 32 (1969).

22. M. E. LeFevre, D. C. Hancock, and D. D. Joel, *J. Toxicol. Envir. Health*, *6*, 691 (1980).
23. M. Aprahamian, C. Michel, W. Humbert, J.-P. Devissaguet, and C. Damge, *Biol. Cell*, *61*, 69 (1987).
24. S. Steiner and R. Rosen, Delivery systems for pharmacological agents encapsulated with proteinoids, U.S. Patent 4,925,673.
25. P. A. Gonnella and W. A. Walker, *Adv. Drug Rev.*, *1*, 235 (1987).
26. K. Matsuno, T. Schaffner, H. A. Gerber, C. Ruchti, M. W. Hess, and H. Cottier, *J. Reticuloendothel. Soc.*, *33*, 263 (1983).
27. B. S. Worthington and J. Syrotuck, *J. Nutr.*, *106*, 20 (1976).
28. W. McA. Barbour and D. Hopwood, *Pathology*, *139*, 167 (1983).
29. E. Sanders and C. T. Ashworth, *Exp. Cell Res.*, *22*, 137 (1961).
30. R. L. Owen, R. T. Apple, and D. K. Bhalla, *Anat. Record*, *216*, 521 (1986).
31. M. E. LeFevre and D. D. Joel, in *Intestinal Toxicology* (C. M. Schiller, ed.), Raven Press, New York, 1984, p. 45.
32. M. R. Neutra, T. L. Phillips, E. L. Mayer, and D. J. Fishkind, *Cell Tissue Res.*, *247*, 537 (1987).
33. R. L. Owen, *Gastroenterology*, 72, 440 (1977).
34. R. Weltzin, P. Lucia-Jandris, P. Michetti, B. N. Fields, J. P. Kraehenbuhl, and M. R. Neutra, *J. Cell Biol.*, *108*, 1673 (1989).
35. N. F. H. Ho, J. S. Day, C. L. Barsuhn, P. S. Burton, and T. J. Raub, *J. Controlled Release*, *11*, 3 (1990).
36. M. J. Roy, *Cell Tissue Res.*, *248*, 483 (1987).
37. D. D. Joel, J. A. Laissue, and M. E. LeFevre, *J. Reticuloendothel. Soc.*, *24*, 477 (1978).
38. C. L. Wells, M. A. Maddaus, S. L. Erlandsen, and R. L. Simmons, *Infect. Immun.*, *56*, 278, (1988).
39. S. H. M. Jeurissen, G. Kraal, and T. Sminia, *Adv. Exp. Biol. Med.*, *216A*, 257 (1987).
40. M. E. LeFevre, J. W. Vanderhoff, J. A. Laissue, and D. D. Joel, *Experientia*, *34*, 120 (1978).
41. M. E. LeFevre, R. Olivo, J. W. Vanderhoff, and D. D. Joel, *Proc. Soc. Exp. Biol. Med.*, *159*, 298 (1978).
42. M. E. LeFevre, J. J. Joel, and G. Schidlovsky, *Proc. Soc. Exp. Biol. Med.*, *179*, 522 (1985).
43. M. E. LeFevre, A. M. Boccio, and D. D. Joel, *Proc. Soc. Exp. Biol. Med.*, *190*, 23 (1989).
44. M. E. LeFevre, J. B. Warren, and D. D. Joel, *Exp. Cell Biol.*, *53*, 121 (1985).
45. J. H. Eldridge, C. J. Hammond, J. A. Muelbroek, J. K. Staas, R. M. Gilley, and T. R. Tice, *J. Controlled Release*, *11*, 205 (1990).
46. P. Jani, G. W. Halbert, J. Langridge, and A. T. Florence, *J. Pharm. Pharmacol.*, *41*, 809 (1989).
47. P. Jani, G. W. Halbert, J. Langridge, and A. T. Florence, *J. Pharm. Pharmacol.*, *42*, 821 (1990).

48. J. H. Eldridge, J. A. Meulbroek, J. K. Staas, T. R. Tice, and R. M. Gilley, *Adv. Exp. Biol. Med.*, *251*, 191 (1989).
49. H. O. Alpar, W. N. Field, R. Hyde, and D. A. Lewis, *J. Pharm. Pharmacol.*, *41*, 194 (1988).
50. J. P. Ebel, *Pharm. Res*, 7, 848 (1990).
51. M. Nefzger, J. Kreuter, R. Voges, E. Liehl, and R. Czok, *J. Pharm. Sci.*, *73*, 1309 (1984).
52. J. Kreuter, U. Miller, and K. Munz, *Int. J. Pharm.*, *55*, 39 (1989).
53. M. Kukan, S. Bezek, V. Koprda, J. Labsky, J. Kalal, K. Bauerova, and T. Trnovec, *Pharmazie*, *44*, 339 (1989).
54. J. Pappo and T. H. Ermak, *Clin. Exp. Immunol.*, *76*, 144 (1989).
55. M. A. Jepson, N. L. Simmons, T. C. Savidge, P. S. James, and B. H. Hirst, *J. Physiol.*, *452*, 258P (1992).
56. J. Pappo, T. H. Ermak, and H. J. Steger, *Immunology*, *73*, 277 (1991).
57. L. R. Inman and J. R. Cantey, *J. Clin Invest.*, *71*, 1 (1983).
58. B. B. Finlay and S. Falkow, *Microbiol. Rev.*, *53*, 210 (1989).
59. R. L. Owen, N. F. Pierce, R. T. Apple, and W. C. Cray, Jr., *J. Infect Dis.*, *153*, 1108 (1986).
60. J. L. Wolf, D. H. Rubin, R. Finberg, R. S. Kauffman, A. H. Sharpe, J. S. Trier, and B. N. Fields, *Science*, *212*, 471 (1981).
61. J. L. Wolf, R. S. Kauffman, R. Finberg, R. Dambrauskas, B. N. Fields, and J. S. Trier, *Gastroenterology*, *85*, 291 (1983).
62. H. M. Amerongen, R. Weltzin, C. M. Farnet, P. Michetti, W. A. Haseltine, and M. R. Neutra, *J. Acquir. Immune Def. Syndromes*, *4*, 760 (1991).
63. P. Sicinski, J. Rowinski, J. B. Warchol, Z. Jarzabek, W. Gut, B. Szczygiel, K. Bielecki, and G. Koch, *Gastroenterology*, *98*, 56 (1990).
64. S. B. Prusiner, *Science*, *252*, 1515 (1991).
65. C. L. Wells, M. A. Maddaus, and R. L. Simmons, *Rev. Infect. Dis.*, *10*, 958 (1988).
66. R. R. Isberg, *Science*, *252*, 934 (1991).
67. P. W. K. Lee, E. C. Hayes, and W. K. Joklik, *Virology*, *108*, 156 (1981).
68. E. A. Groisman and M. H. Saier, Jr., *Trends In Biological Sciences*, *15*, 30 (1990).
69. C.-M. Chiang and N. Weiner, *Int. J. Pharm.*, *37*, 75 (1987).
70. J. F. Woodley, *CRC Crit. Rev. Ther. Drug Carrier Sys.*, *2*, 1 (1985).
71. P. Couvreur, *CRC Crit. Rev. Ther. Drug Carrier Sys.*, *5*,1 (1988).
72. C. Damge, C. Michel, M. Aprahamian, P. Couvreur, and J. P. Devissaguet, *J. Controlled Release*, *13*, 233 (1990).
73. C. Damge, C. Michel, M. Aprahamian, and P. Couvreur, *Diabetes*, *37*, 246 (1988).
74. P. Couvreur, V. Lenaerts, B. Kante, M. Roland, and P. Speiser, *Acta Pharm. Technol.*, *26*, 220 (1980).
75. C. Damge, M. Aprahamian, G. Balboni, A. Hoeltzel, V. Andrieu, and J. P. Devissaguet, *Int. J. Pharm.*, *36*, 121 (1987).
76. C. Michel, M. Roques, P. Couvreur, H. Vranckx, P. Balschmidt, and C. Damgé, *Proc. Int. Symp. Controlled Release Bioact. Mater.*, *18*, 97 (1991).

77. D. Scherer, J. Kreuter, and J. R. Robinson, *Proc. Int. Symp. Controlled Release Bioact. Mater.*, *18*, 159 (1991).
78. I. Morishita, M. Morishita, K. Takayama, Y. Machida, and T. Nagai, *Int. J. Pharm.*, *78*, (1992).
79. C.-M. Lehr, J. A. Bouwstra, J. J. Tukker, J. C. Verhoef, and H. E. Junginger, *Proc. Int. Symp. Controlled Release Bioact. Mater.*, *16*, 418 (1989).
80. C.-M. Lehr, J. A. Bouwstra, J. J. Tukker, and H. E. Junginger, *J. Controlled Release*, *13*, 51 (1990).
81. C.-M. Lehr, J. A. Bouwstra, J. J. Tukker, J. C. Verhoef, A. G. deBoer, H. E. Junginger, and D. D. Breimer, *Proc. Int. Symp. Controlled Release Bioact. Mater.*, *17*, 67 (1990).
82. M. R. Neutra and J. F. Forstner, in Physiology of the Gastrointestinal Tract, 2nd Ed. (L. R. Johnson, ed.), Raven Press, New York, 1987, p. 975.
83. R. H. Engel and S. J. Riggi, *J. Pharm. Sci.*, *58*, 1372 (1969).
84. W. A. Ritschel, G. B. Ritschel, A. Sabouni, D. Wolochuk, and T. Schroeder, *Meth. Find. Exp. Clin. Pharmacol.*, *11*, 281 (1989).
85. W. A. Ritschel, S. Adolph, G. B. Ritschel and T. Schroeder, *Meth. Find. Exp. Clin. Pharmacol.*, *12*, 127 (1990).
86. M. Schichiri, R. Kawamori, M. Yoshida, W. Etani, M. Hoshi, K. Izumi, Y. Shigeta, and H. Able, *Diabetes*, *24*, 971 (1975).
87. M. Shichiri, Y. Shimizu, Y. Yoshida, R. Kawamori, M. Fukuchi, Y. Shigeta, and H. Abe, *Diabetologia*, *10*, 317 (1974).
88. S. Muranishi, *CRC Crit. Rev. Therapeut. Drug Carrier Syst.*, 7, 1 (1990).
89. L. L. Teng, Enteric compounds and complexes, U. S. Patent 4,604,376.
90. J. A. Fix, K. Engle, P. A. Porter, P. S. Leppert, S. J. Selk, C. R. Gardner, and J. Alexander, *Am. J. Physiol.*, *251*, G332 (1986).
91. J. A. Fix, J. Hochman, and E. LeCluyse, *FASEB J.*, *4*, A984 (1990).
92. H. Yoshikawa, K. Takada, S. Muranishi, Y. Satoh, and N. Naruse, *J. Pharmacobio. Dyn.*, 7, 59 (1984).
93. H. Yoshikawa, Y. Satoh, N. Naruse, K. Takada, and S. Muranishi, *J. Pharmacobio. Dyn.*, *8*, 206 (1985).
94. H. Yoshikawa, K. Takada, Y. Satoh, N. Naruse, and S. Muranishi, *Pharm. Res.*, *2*, 249 (1985).
95. K. Takada, H. Yoshimura, N. Shibata, Y. Masuda, H. Yoshikawa, S. Muranishi, T. Yasumura, and T. Oka, *J. Pharmacobio. Dyn.*, *9*, 156 (1986).
96. K. Takada, M. Oh-hashi, Y. Furuya, H. Yoshikawa, and S. Muranishi, *Chem. Pharm. Bull.*, *37*, 2542 (1989).
97. C. W. Crane, M. C. Path, and G. R. W. N. Luntz, *Diabetes*, *17*, 625 (1968).
98. Y. W. Cho and M. Flynn, *Lancet*, *2*, 1518 (1989).
99. Y. W. Cho, Pharmaceutical compositions, U.S. Patent 4,849,227.
100. W. A. Ritschel, D. G. Patel, P. Chalasani, and T. Schroeder, *Pharm. Res.*, 7, S-119 (1990).
101. W. A. Ritschel, D. G. Patel, P. Chalasani, and T. Schroeder, *Pharm. Res.*, 7, S-120 (1990).
102. W. A. Ritschel, *Meth. Find. Exp. Clin. Pharmacol.*, *13*, 205 (1991).

103. D. G. Patel, W. A. Ritschel, P. Chalasani, and S. Rao, *J. Pharm. Sci.*, *80*, 613 (1991).
104. D. R. Friend, *Adv. Drug Deliv. Rev.*, 7, 149 (1991).
105. M. Saffran, G. S. Kumar, C. Savariar, J. C. Burnham, F. Williams, and D. C. Neckers, *Science*, *233*, 1081 (1986).
106. E. Touitou and A. Rubenstein, *Int. J. Pharm.*, *30*, 95 (1986).
107. M. E. McNeill, A. Rashid, and H. N. E. Stevens, *Dispensing Device*, WO 90/09168.
108. J. Kopecek, *J. Controlled Release*, *11*, 279 (1990).
109. D. Mirelman, M. Izhar, and Y. Eshdat, *Tokai J. Exp. Clin. Med.*, 7 *(Suppl.)*, 177 (1982).
110. R. M. Donaldson, in Physiology of the Gastrointestinal Tract, 2nd Ed. (L. R. Johnson, ed.), Raven Press, New York, 1987, p. 959.
111. G. J. Russell-Jones and H. J. Aizpurna, *Proc. Int. Symp. Controlled Release Bioact. Mater.*, *15*, 142 (1988).
112. J. Godovac-Zimmerman, *Trends Biochem. Sci.*, *13*, 64 (1988).
113. M. Z. Papiz, L. Sawyer, E. E.Eliopoulos, A. C. T. North, J. B. C. Findlay, R. Sivaprasadarao, T. A. Jones, M. E. Newcomer, and P. J. Kraulis, *Nature*, *324*, 383 (1986).
114. M. Stern and W. A. Walker, *Am. J. Physiol.*, *246*, G556 (1984).
115. J. F. Huneau and D. Tome, *Reprod. Nutr. Dev.*, *30*, 237 (1990).
116. D. Marcon-Genty, D. Tome, O. Kheroua, A. M. Dunomtier, M. Heyman, and J. F. Desjeux, *Am. J. Physiol.*, *256*, G943 (1989).
117. J. Mestecky, *J. Clin. Immunol.*, 7, 265 (1987).
118. G. Dougan, C. E. Hormaeche, and D. J. Maskell, *Parasite Immunol.*, *9*, 151 (1987).
119. S. N. Chatfield, R. A. Strugnell, and G. Dougan, *Vaccine*, 7, 495 (1989).
120. J. R. McGhee, J. Mestecky, M. T. Dertzbaugh, J. H. Eldridge, M. Hirasawa, and H. Kiyono, *Vaccine*, *10*, 75 (1992).
121. C. A. Gilligan and A. Li Wan Po, *Int. J. Pharm.*, *75*, 1 (1991).
122. J. H. Eldridge, J. K. Staas, J. A. Meulbroek, J. R. McGhee, T. R. Tice, and R. M. Gilley, *Mol. Immunol.*, *28*, 287 (1991).
123. D. T. O'Hagan, *Clin. Pharmacokinet.*, *22*, 1 (1992).
124. Z. Moldoveanu, J. K. Staas, R. M. Gilley, R. Ray, R. W. Compans, J. H. Eldridge, T. R. Tice, and J. Mestecky, *Curr. Topics Microl. Immunol.*, *146*, 91 (1989).
125. D. T. O'Hagan, K. J. Palin, and S. S. Davis, *Vaccine*, 7, 213 (1989).
126. D. T. O'Hagan, K. Palin, S. S. Davis, P. Artursson, and I. Sjöholm, *Vaccine*, 7, 421 (1989).
127. G. Dougan, S. Chatfield, M. Roberts, I. Charles, S. Comerford, L. J. Li, and N. Fairweather, *Biochem. Soc. Trans.*, *18*, 746 (1990).
128. D. J. Evans, J. McKeating, J. M. Meredith, K. L. Burke, K. Katrak, A. John, M. Ferguson, P. D. Minor, R. A. Weiss, and J. W. Almond, *Nature*, *339*, 385 (1989).
129. R. J. Genco, R. Linzer, and R. T. Evans, *Ann. N.Y. Acad. Sci.*, *409*, 650 (1983).

130. D. Wachsmann, J. P. Klein, M. Schöller, and R. M. Frank, *Immunology*, *54*, 189 (1985).
131. S. M. Michalek, N. K. Childers, J. Katz, F. R. Denys, A. K. Berry, J. H. Eldridge, J. R. McGhee, and R. Curtiss III, *Curr. Topics Microbol. Immunol.*, *146*, 51 (1989).
132. W. Chaicumpa, J. R. Parairo, R. C. New, E. Pongponratn, Y. Ruangkunaporn, P. Tapchaisri, and M. Chongsa-nguan, *Asian Pac. J. Allergy Immunol.*, *8*, 87 (1990).
133. W. Rubas, A. C. Banerjjea, H. Gallati, P. P. Speiser, and W. K. Joklik, *J. Microencapsulation*, 7, 385 (1990).

6

Applications of Particulate Carriers in Intratumoral Drug Delivery

P. K. Gupta

Abbott Laboratories
North Chicago, Illinois

C. T. Hung

Zenith Technology Corporation, Ltd.
Dunedin, New Zealand

F. C. Lam

University of Otago
Dunedin, New Zealand

INTRODUCTION

Cancer is a disease of cells characterized by a reduction or loss of normal cellular control and maturation mechanisms. Its main features include excessive cell growth, undifferentiated cells and tissues, and the ability to grow into adjacent tissues and metastasize. The malignant tumor typically infiltrates adjacent nervous tissue and spreads beyond the primary site. For example, malignant brain tumors spread through parenchyma, corpus callosum, or subarachnoid space (1). Cancer treatment is largely determined by the stage of diagnosis, treatment(s) available, and patients' compliance toward these treatments (2). The oncological characteristics of most tumors require one of the following treatments: (a) total excision of tumor tissue and possibly part of the adjacent tissue(s), (b) combination chemotherapy, (c) immunotherapy, (d) radiation therapy, and (e) a combination of surgery, chemotherapy, immunotherapy, and/or radiation therapy. Wherever possible, total excision is the treatment of choice as it ensures complete removal of tumor cells. However, depending on the location of tumor and its involvement with surrounding tissues, surgery may not always be advisable due to possible postoperative disturbances. Subtotal or curative excision of tumor will lead to relapse and recurrence of oncological symptoms. Hence under such circumstances, radiotherapy and/or chemotherapy becomes necessary (1).

General complications as a result of radiotherapy have been reported (3). For example, external irradiation for the treatment of brain tumors has been known to cause necrosis of associated tissue; in other instances, even megavoltage irradiation treatments have been ineffective in arresting the growth of glioblastoma (1). Techniques involving implantation of radioactive substances, i.e., brachytherapy, which allows concentrated irradiation of tumor tissue, have also been attempted; however, they involve relatively large surgical manipulation (1). Chemotherapy has been advised for rapidly progressing malignant brain tumors like grade III and IV astrocytomas and oligodendrogliomas (1); however the blood–brain barrier restricts effective transport of drug to the target tissues.

From the standpoint of chemotherapy, tumor cell destruction is generally believed to follow first-order kinetics, i.e., a specified dose kills a specific fraction of cells, not a specific number of cells. Hence, an increase in the fraction of cells killed requires an increase in drug dose (4). Unfortunately, due to their toxic effects on the normal cells, only limited doses of chemotherapeutic agents can be administered. Despite the systemic administration of maximum acceptable dose of drug as a solution, the fraction of dose or the drug concentration at the target site is generally low. In addition, these levels are rarely maintained at the target site for sufficient time to be of real benefit without concurrent increase in side effects and toxicities. Hence oncologists to date face the challenge of developing treatment schedules that may allow selective eradication of tumor cells without causing comparable damage to the nontarget cells of the body (5,6).

Pharmacological rationales and implications of different routes of administration of chemotherapeutic agents have been proposed and reviewed by several workers (7,8). Based on the promise of regional drug delivery, efforts have been made to link chemotherapeutic agents, radionuclides, toxins, and enzymes to monoclonal antibodies and other macromolecules with the hope that it will enhance tumor localization with reduced systemic exposure (6,9,10). However, in most instances the accumulation of drug or drug-carrier in tumors has been nonuniform and inadequate. Several explanations have been offered in response to this process (5,6,9–11):

1. Lack of sufficient drug/antibody concentration ratio
2. Loss of drug-carrier linkage upon administration
3. Heterogeneity in the tumor cell–associated antigens
4. Heterogeneous growth characteristics of tumor cells and their response to chemotherapeutic agents

5. Heterogeneous blood supply to tumor tissue
6. Elevated interstitial pressure in tumor tissue, and
7. Large transport distance in the interstitium

These barriers vary from tumor to tumor, from one location to another location within the same tumor, and change from one day to the next in the same tumor (12).

More recently, some studies have demonstrated a moderate to high degree of success following the administration of drug directly into or near the tumor (13–17). The intention is to ensure delivery of drug in maximum acceptable concentrations directly into the target tissue with minimal or no exposure to the healthy tissues. In order to prolong drug levels in tumor tissue, different drug delivery systems, including implants and particulate carriers, have been studied for local drug administration. This chapter will review the information on different strategies which have been explored for local or intratumoral (i.t.) drug delivery. This interesting area of research will be discussed in light of limited experimental and clinical data that have been collected to date using various drug delivery systems. The scope and limitations of this route of drug delivery, possible means of its optimization, and its future prospects in therapeutics will also be outlined.

INDICATIONS FOR INTRATUMORAL DRUG DELIVERY

Perhaps the most attractive feature of i.t. drug delivery is its overall simplicity. This mode of drug administration guarantees, at least initially, local drug delivery and hence minimizes the extent of drug toxicity to nontarget tissues. In addition, the application of an optimal formulation can sustain drug levels in tumor tissues and thus improve therapeutic outcome. In fact, existing data suggest that i.t. drug delivery prior to surgery can reduce the extent of its manipulation, i.e., cytoreductive surgery, and hence improve survival (16,18). Alternatively, the i.t. drug delivery following surgery may increase the probability of complete regression of residual tumor cells (17,18). However, it must be appreciated that i.t. drug delivery is not applicable to all kinds of tumor therapy. In fact, this mode of treatment is useful in limited cases that meet one or more of the following criteria:

1. The total excision of tumor is desirable to prevent relapse. However, its centralized location prohibits surgical intervention due to high risk of postoperative complications and mortality, e.g., neurological damage and/or hormone deficiency following surgical treatment of craniopharyngioma (1,17).

2. Systemic chemotherapy or radiotherapy after partial excision is unacceptable either due to the associated side effects or because it does not guarantee complete regression of tumor (17).
3. Systemic treatments require circumvention of natural obstructions for the transport of drug to target cells, e.g., overcoming the blood–brain barrier for the transport of hydrophilic chemotherapeutic agents in the treatment of malignant brain tumors.
4. The tumor geometry is reasonably well defined, tumor deposits are easily accessible, and the target area can be injected without compromising the normal functions of the adjacent normal tissues.
5. The tumor is highly sensitive to chemotherapy, immunotherapy, or the other locally administered agent.
6. The tumor growth cannot be controlled by existing systemic drug delivery techniques.

Generally, this mode of drug delivery is desirable for patients who are poor candidates for surgery and high-dose systemic chemotherapy or immunotherapy, e.g., those with small hepatocellular carcinoma and hence possible liver cirrhosis. Accordingly, there are several inherent limitations to this approach of drug administration, e.g.:

1. This mode of treatment is not primary; it can rather be applied only to ensure better palliation of tumor (17).
2. Its probability of success is dependent on the nature of tumor, the tumors with cyst being more easily treatable than the solid tumors (17).
3. If the drug is delivered as a solution, this technique may require continuous microinfusion over an extended period of time (15).
4. The resistance offered by tumor tissue upon injection may limit the total volume that could be administered, and it may warrant use of a low-pressure delivery device to prevent drug leakage (19).
5. This technique may cause fibrosis of the underlying skin (20).

INTRATUMORAL DELIVERY OF DRUG AS A SOLUTION

If tumor tissue is accessible for catheterization and its geometry is reasonably well defined, then part of the physiological barriers can be overcome by continuous i.t. infusion of chemotherapeutic agents. However, the continued exposure of tumor cells to chemotherapeutic agents locally administered as a solution almost always requires their chronic infusion. Often the drug is injected at multiple sites to ensure greater coverage of tumor mass. Over the last several years special methods have been developed to meet these drug delivery challenges. In 1982, Kroin and Penn (13) demonstrated the use of an osmotic minipump for the i.t. delivery of cisplatin into the cerebellum and frontal cortex of rats. This pump allowed i.t. delivery of cisplatin and 5-fluorouracil at rates of 0.9 μl/hr for 7 days (21). More recently techniques have been developed to allow drug delivery via multiple stereotaxically placed intratumoral catheters (14). Each catheter bundle has several microcatheters for dose administration at multiple sites. In some cases, up to 25 catheter bundles have been placed in tumor allowing 100 sites of injection. The catheters can be left attached for several days and can be controlled to provide a desirable rate of drug infusion (14,19). In one study, the i.t. delivery of 0.82 mg/day cisplatin for 10 days via this technique was found to be comparable to the systemic delivery of 36 mg/day drug and it therefore suggested possible dose reduction by a factor of 40 (14).

An interesting study by Takahashi et al. (17) evaluated postoperative i.t. chemotherapy of craniopharyngioma in pediatric patients. Total excision and/or radiotherapy were not attempted due to post-therapy complications. Instead, 2 weeks after partial excision, intensive local chemotherapy was tried by i.t. administration of 1–5 mg bleomycin every other day. The results of this study are summarized in Table 1. Of the seven patients with cystic type tumors, with little or no solid tissue, four demonstrated excellent clinical outcome. Over the 7 years of follow-up period they revealed no sign of tumor recurrence. In the remaining three patients, the tumor regressed initially; however, it recurred in two cases. In fact, these three patients had huge tumors involving the third ventricle and hence they died 0.12–2 years after local chemotherapy (17).

Several other studies involving i.t. chemotherapy, photodynamic therapy, and/or immunotherapy are reported in the literature. For example, the i.t.

Table 1 Clinical Summary of the Intralesional Delivery of Bleomycin for the Treatment of Craniopharyngioma

Case no.	Sex	Age (yr)	Tumor type	Total drug dose (mg)	Follow-up period (yr)	Recurrence
1	F	7	Solid	39	0.12	No
2	M	12	Cystic	20	7	No
3	F	3	Cystic + solid	25	2	Yes
4	M	2	Cystic	13	7	No
5	M	13	Cystic + solid	50	1.5	Yes
6	M	10	Cystic	95	4	No
7	F	10	Cystic	79	2	No

Source: Adapted in part from Ref. 17, with permission.

delivery of vinblastine sulfate solution has been followed as a postoperative adjuvant chemotherapy in patients with angiosarcoma of the breast. An i.t. dose of 4 mg/week in 2 patients for 15 and 12 months maintained them disease-free for 12 and 10 years, respectively (22). In a separate study, the i.t. delivery of photofrin II, a derivative of hematoporphyrin, followed by photoradiation considerably enhanced the efficacy of photodynamic therapy compared to control treatments involving photoradiation alone, i.t. photofrin II without radiation, or i.p. photofrin II with radiation (23). Encouraging results have also been reported in a phase I clinical i.t. photodynamic study (24). D'Amico et al. (25) investigated the usefulness of localized delivery of thrombin. Following transplantation of solid tumor beneath the kidney capsules, the i.t. delivery of thrombin increased the survival of animals by 85% that of animals receiving no treatment or an equivalent volume of saline. The i.t. delivery of recombinant human necrosis factor alpha have been attempted clinically with successful tumor necrosis in 36% (5/14) patients; however the duration of action was reported to be short-lived (26). Ultrasound-controlled techniques have been developed to allow i.t. delivery of recombinant human tumor necrosis factor in patients with liver metastasis, and the phase I results have been very encouraging (27).

Attempts have been made to use this mode of treatment for tumor immunotherapy and immunochemotherapy. For example, i.t. administration of ricin and mafosfamid has been shown to allow 72–100% reduction in tumor volume in 3–4 days after treatment (28). Although a single-dose treatment was associated with tumor relapse within 5–13 days, repeated i.t. chemoimmunotherapy led to persistent tumor regression. Seventy one percent of mice (5/7) were tumor free after 30 days (28). Intratumoral immunotherapy has also been investigated for the treatment of primary melanoma of skin (16,29). In one study, the i.t. delivery of bacillus Calmette-Guerin (BCG) in metastatic melanoma lesions of skin demonstrated regression of 90% of the injected lesions and 17% of the noninjected surrounding lesions (29). In another study, the patients receiving i.t. vaccination with live virus, 2 weeks prior to excision, demonstrated improved survival and reduced incidence of metastatic recurrence compared to the patients treated by local excision alone (16). The stimulation of antitumor cell–mediated immunity was suggested as a possible mechanism for this favorable response (16).

Ethanol has been historically employed as a thromboembolic agent and for treatment of a variety of benign conditions. Promising results from these studies have encouraged some workers to investigate i.t. delivery of ethanol for the treatment of adenocarcinomas (30). Due to nonhomogeneous i.t. delivery of ethanol in some instances, Sheu et al. (31) developed multiple side-hole needles that were shown to allow extended necrosis of tumor tissue.

NOVEL STRATEGIES IN INTRATUMORAL DRUG DELIVERY

As discussed earlier, i.t. drug delivery offers certain distinct advantages over the treatment approaches involving radiation therapy or systemic administration of antitumor agents. However, the local administration of native drug as a solution or suspension rarely allows maintenance of its effective levels in tumor tissue for a sufficiently long time. For example, microinfusion techniques requiring up to 100 sites of injection and continuous infusion of drug solution for 10–20 days at rates of 0.5–1 ml/day have been adopted with the local delivery of cisplatin, 5-fluorouracil, and other compounds (15,19). This is not only expensive but also inconvenient. In addition, several studies have shown that tumor blood vessels are generally more disorganized, tortuous, and dilated than blood vessels in normal tissues (5,12,32–34). Due to these characteristics, the rate of loss of drug from tumor tissue to general circulation is usually high (8,33,35,36). The i.t. delivery of drug as a solution becomes particularly unacceptable in instances where the high rate of tumor cell proliferation leads to its recurrence in the interval between the administration of two sequential doses of drug (1). However, if an encapsulated or a chemically conjugated drug could be directly administered to the tumor in such a manner that the size of the delivery device prevents its shunting through the endothelial cell junctions or gaps to general circulation, and its release rate allows protracted local drug levels over an extended period of time, then the probability of therapeutic cure is appreciably high. This approach can therefore be expected to increase therapeutic benefit and hence allow improved tumor treatments.

A great deal of effort has been made to modify the chemical structure of drugs with a view to enhance target cell recognition and thus improve their disposition in favor of target cells. Ideally, the chemically modified molecule is relatively less active, and less toxic, in nontarget tissues than in the target tissue (9). The route of administration and its molecular size and physicochemical characteristics significantly influence the site of deposition. However, once it reaches the target issue, the interaction between conjugate and tumor tissue environment favor its prolonged local residence and selective pharmacological response (9). Due to these exciting possibilities, macromolecular drug conjugates have been investigated for i.t. chemotherapy (37,38).

In one study, mitomycin C was conjugated with dextran (37). The i.t. administration of 5 and 10 mg/kg drug conjugate, 4 days after the subcutaneous (s.c.) implantation of B16 melanoma in mice, resulted in appreciable suppression of tumor growth. At 11 days after dosing, the drug conjugate controlled the tumor growth to about one half and one fifth of that in the groups receiving unconjugated drug and no treatment, respectively. The reduction in tumor growth with drug conjugate was in turn suggested to enhance the life span of animals by about 30% (37). In another study, Morimoto et al. (38) conjugated melphalan to 60 and 14K poly(l-glutamic acid) at a ratio of 1 drug molecule per 23 glutamate residues. Four days after the s.c. implantation of Yoshida sarcoma in rats, about 50 μg of free or conjugated drug in 0.1 ml saline was injected directly into the tumor. The tumor growth rate over the period of 9 days, following the administration of drug conjugate, was restricted to about 0.42 g/day. This growth rate was about one third of that in animals receiving unconjugated drug or drug mixed with poly(glutamic acid), and about one fifth of the animals treated with saline only (see Figure 1). The multiple local injections of drug conjugate on days 4, 5, and 6 further reduced the tumor growth rate to about 0.25 g/day, which was again at least 3 times more effective that the multiple i.t. doses of unconjugated drug (see Figure 1) (38). More recently, Bures et al. (39) investigated the feasibility of conjugation of methotrexate with human serum albumin for improved chemotherapy. About 26 molecules of drug were coupled to a single molecule of albumin and its i.t. delivery was reported to be more effective than intraperitoneal (i.p.) administration.

Limited work with polymeric implants also suggests the possibility of optimizing i.t. drug delivery. In one study the i.p. administration of cisplatin in collagen matrix was found to require 11.1 ± 2.5 days to allow the fibrosarcoma to grow to 3 times its pretreatment size. However, the same drug–

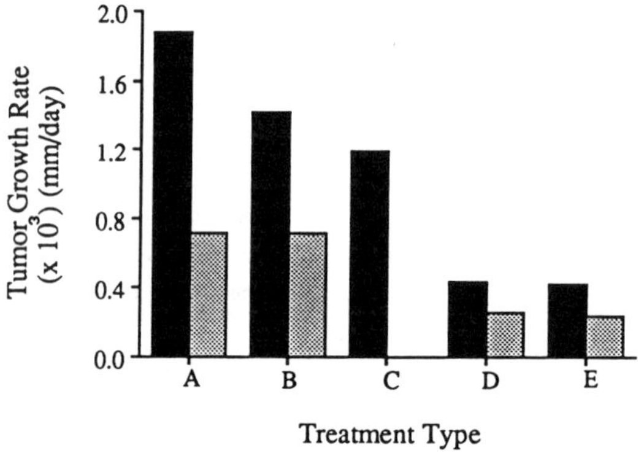

Figure 1 Effect of i.t. single shot (■) and multiple shot (▩) treatments on the growth rate of s.c. implanted Yoshida sarcoma in rats. All animals, except those in group A, received i.t. 50 μg of drug. Key: (A) saline, (B) drug solution, (C) drug solution plus poly(glutamic acid), (D) drug conjugated with 14K Poly(glutamic acid) and (E) drug conjugated with 60K Poly(glutamic acid). Adapted in part from ref. 38, with permission from the copyright owner, The Pharmaceutical Society of Japan, Tokyo, Japan.

polymer matrix required 17.2 ± 1.7 days for threefold growth of sarcoma following its i.t. implantation (40). In addition, the toxicity studies indicated LD_{50}, at 10 days post-treatment, to be 14 mg/kg and ≥30 mg/kg for the i.p. and i.t. treatments, respectively (40). Sutton et al. (41) demonstrated a threefold increase in tumor exposure to vinblastine following its delivery as a collagen implant as opposed to the administration of drug without the polymer. Langer and coworkers demonstrated that ethylene–vinyl acetate or polyanhydrides can be used for the controlled release of macromolecules, e.g., tumor angiogenesis factor (42). Following the implantation of $1 \times 1 \times 0.5\ mm^3$ polyanhydride sample containing 2% w/w heparin and 16.6% w/w cortisone in rabbit cornea with V2 carcinoma, the polymer was found to dissolve completely over a 3-week period with concurrent inhibition of tumor neovascularization (42). More recently, Brem (43) reported that the intracranial implantation of BCNU (bis-2-chloroethyl-1-nitrosourea) loaded polyanhydride in rats resulted in total regression of tumor with mean survival time of 69 ± 10 days. The i.p. administration of BCNU demonstrated a mean survival of 27 ± 2 days. The animals in all other control groups died in 11 ± 1 days. These results formed the basis of the clinical evaluation of these implants, and so far the results from phase I and II studies involving treatment of patients with recurrent malignant gliomas have been encouraging (43). Stricker and coworkers (44) evaluated the potentials of polylactide implants in the localized delivery of methotrexate. Compared to the untreated controls, a treatment schedule involving administration of 4.5 mg/kg drug as an i.t. implant retarded the growth of glioma in rats by 61%. This therapy was considerably superior to the i.p. administration of four doses of 4 mg/kg drug as a solution, but comparable to the i.t. delivery of three doses of 4 mg/kg drug solution (44).

Intratumoral implantation chemotherapy, however, is not always feasible. Apart from relatively large surgical manipulation required to place implants, a major limitation with these systems is that only the tumor cells that come into direct contact with the drug-containing fluid are optimally exposed to its high concentrations. Hence it offers limited therapeutic advantage to deeper layers of tumor cells. This becomes even more limiting when the interstitial diffusion and transport of drug across cell membrane are inherently low (12). Indeed, lack of therapeutic advantage to deeper cells following the i.p. and intrathecal drug delivery have been demonstrated in some studies (45,46). However, if the mass equivalent to an implanted system could be administered at multiple sites in tumor tissue, in the form of a particulate carrier, it may enhance the exposure of overall tissue to the delivered drug and thus increase the probability of therapeutic cure. In addition, the use of particulate drug carriers can be expected to minimize major surgical intervention. Due to these reasons, several particulate carrier systems have been investigated for i.t. drug delivery, and these are discussed in the next section of this chapter.

APPLICATION OF PARTICULATE CARRIERS IN INTRATUMORAL DRUG DELIVERY

Conceptually, the maintenance of therapeutic levels of drug in tumor tissue without surgical implantation or continuous microinfusion warrants drug encapsulation or its chemical conjugation in such a manner that (a) the final product could be injected via a routinely used needle with minimal shunting or leakage to general circulation and (b) it allows local sustained release over an extended period of time. Depending on the type of macromolecule used for encapsulation or conjugation, and the characteristics of tumor tissue, these systems can enhance target tissue recognition and hence modulate the disposition of drug in favor of target tissue. Numerous microencapsulation techniques have been developed over the last few years that offer the convenience of controlling the product size, its injectability, biodegradation, and sterile processing, and, above all, controlled release at predetermined rates over a period of several hours to several days (47). Attempts have also been made to develop high molecular weight (MW) chemical conjugates of chemo- and immunotherapeutic agents (9). Once the retention of a particulate carrier in tumor tissue is optimized, it acts as a depot allowing sustained release of the delivered tumoricidal agent. The prolonged exposure of tumor cells in high concentration in turn ensures complete tumor regression. Hence, these systems offer drug delivery possibilities similar to that achievable with continuous microinfusion systems. Due to these exciting possibilities, emulsions, polymeric microspheres, and liposomes have been investigated for i.t. drug delivery. The current status of these systems in i.t. drug delivery is described below.

Emulsions

The ease with which hydrophilic and lipophilic agents can be incorporated into water-in-oil (w/o) and oil-in-water (o/w) emulsions, respectively, to temporarily sustain their release is key to early interest in these systems for i.t. chemotherapy. The early reports include work by Takahashi and coworkers (48), who evaluated the comparative efficacy of water-in-oil-in-water (w/o/w), w/o, and o/w emulsions in the intratesticular delivery of [^{3}H]5-fluorouracil for the management of regional lymph node metastasis. The results were compared with local and intravenous (i.v.) delivery of drug as an aqueous solution. The comparison of different delivery systems indicated that the localization of drug in lymph nodes increases in the following rank order: w/o/w/ > w/o > o/w > water (see Figure 2). The systemic delivery of drug as an aqueous solution demonstrated the least drug-targeting potential. Compared to the localized delivery of drug as an aqueous solution, the w/o/w emulsion improved drug localization in lymph nodes by a factor of 8. In addition, the w/o/w emulsion maintained appreciably high drug levels in

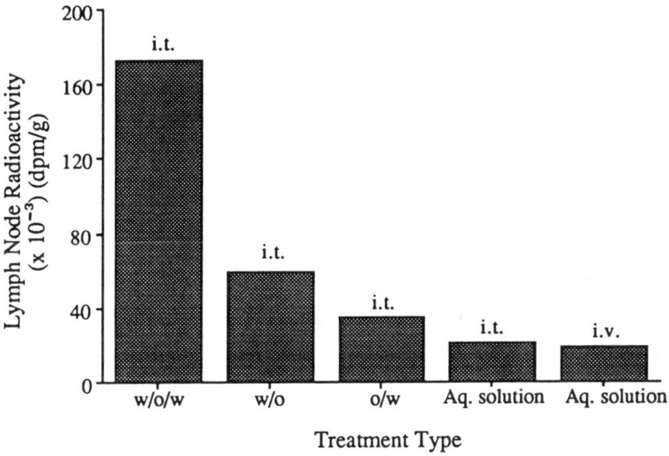

Figure 2 Effect of treatment type on the localization of ^{3}H-5-fluorouracil in the regional lymph node of rats. The animals received intra-testicular injections of drug formulated into w/o/w, w/o or o/w emulsion or aqueous solution, and the data is compared with i.v. administration of drug solution. Adapted in part from ref. 48, with permission.

target tissue for up to 24 hr after local administration which were two- to threefold higher than that with aqueous drug solution at similar time. Most chemotherapeutic agents, being hydrophilic, are readily absorbed into the vascular system, and rarely are absorbed and retained in the lymphatic tissue. However, the formulation of such compounds as an emulsion was suggested to improve drug delivery due to the inherent lipid-absorbing capability of the lymphatic capillaries (48).

Their next study investigated the clinical usefulness of emulsions in i.t. delivery of bleomycin and mitomycin C for treating squamous cell carcinoma of skin or adenocarcinoma of breast (18). Either a w/o/w or a w/o emulsion system was used. Two to 13 doses of formulation were administered at an interval of 3–7 days. The total drug dose varied between 20 and 65 mg. In three of the eight patients treated, the tumor regressed completely (see Figure 3). In other cases, the i.t. drug delivery as an emulsion, before or after surgery, enhanced the extent of regression and minimized tumor spread. In one case, where the patient was diagnosed with postoperative recurrence of the carcinoma of penis, the localized bleomycin emulsion therapy did not respond and the patient died within 3 months of treatment.

In order to achieve additional control on the release of drug from emulsions, Hashida and coworkers developed a w/o emulsion containing gelatin and called it sphere-in-oil (s/o) emulsion (49,50). In one study, these workers

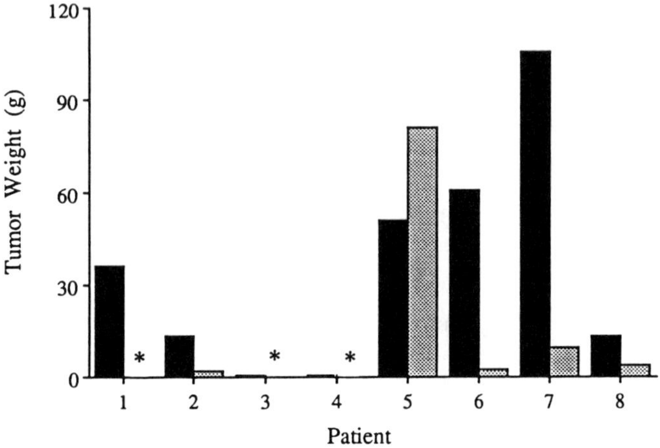

Figure 3 Effect of local administration of drug via w/o or w/o/w emulsion on the tumor size of patients with carcinoma of skin, cheek, or breast. Patients 1 through 5 received 25 to 65 mg bleomycin and patients 6 through 8 received 20 to 26 mg mitomycin C. Patient 5 was a 76 yr old male, and all others 47 to 97 yr old females. Key: (■) tumor size before treatment, (▩) tumor size after treatment, (*) tumor free patients after treatment. Adapted in part from ref. 18, with permission from the copyright owner, J. B. Lippincott Co., Philadelphia, PA.

injected w/o and s/o emulsions into the stomach wall of rats to investigate their comparative efficacy in drug delivery to lymphatics. Radiolabeled water- and oil-soluble tracers were used to monitor the distribution of emulsions in blood, regional lymph nodes, thoracic lymph, and stomach (49). Compared to the local delivery of an aqueous solution of marker, the w/o and the s/o emulsions increased its area under curve (AUC) in the regional lymph nodes by a factor of 1.7 and 5.5, respectively (49). These workers later evaluated the postoperative usefulness of emulsion delivery systems in the treatment of VX2 tumors transplanted in the vermiform appendix of rabbits (51). Ten days after transplantation, the primary tumor was resected and animals treated with 3 mg/kg i.v. or i.t. bleomycin via s/o emulsion, w/o emulsion, or as an aqueous solution. The i.t. treatment via s/o emulsion enhanced drug delivery to lymph nodes and sustained its local release, and thus appreciably improved the survival of the animals (see Table 2).

S/o emulsions have also been evaluated for treating cystic hygroma and lymphangioma of pediatric patients (52–54). Complete removal of this tumor by surgery is difficult due to infiltration to surrounding tissues, and hence a high recurrence rate. However, the i.t. delivery of bleomycin via s/o emulsion was found to be very effective. In one study, the effectiveness of localized

Table 2 Drug Concentrations in Model Metastatic Lymph Node at 24 hr After Delivery of 3 mg/kg Bleomycin in Various Forms

Delivery system (route)	(n)[a]	Survival time (weeks)	Drug conc. (μg/g)
Sphere-in-oil emulsion (i.t.)	7	>28	46.8
w/o emulsion (i.t.)	7	>28	17.2
Aqueous solution (i.t.)	3	13	—[b]
Aqueous solution (i.v.)	3	12	—[c]
Untreated control	3	10	—

[a]Number of rabbits receiving treatment.
[b]Not detectable; <0.7 μg/g at 1 hr after dosing.
[c]Not detectable; trace levels at 0.5 hr after dosing.
Source: Adapted in part from Ref. 51, with permission.

bleomycin s/o emulsion therapy, with or without subsequent surgery, was compared with surgery followed by localized emulsion chemotherapy (53). In the first group, 33 patients received 0.3 to 0.5 ml of 0.89% w/v bleomycin s/o emulsion every 4–6 weeks. Of the total patients, 40% (13/33) demonstrated an excellent response and another 42% (14/33) demonstrated a fairly good response. These patients had tumors on their neck and cheek. Six patients with tumors on their chest and leg, however, did not respond. Four of these patients later received surgery, with complete resection in only one case. In the second group, nine patients underwent surgery with complete resection in only three cases. The four out of the remaining six patients later received localized bleomycin s/o emulsion treatment with good to excellent response in all cases (53). The radiological examination of the tumors indicated the less responsive tumors to be the noncystic type. Hence this study suggested that localized chemotherapy is more effective in the treatment of cystic lymphangiomas than capillary or cavernous lymphangiomas (53). The major side effect of the localized emulsion therapy was fever (≥37.5°C), with diarrhea and local infection in a few instances (53). More recently, another study has confirmed similar clinical efficacy of bleomycin s/o emulsion (54). Forty-seven children with lymphangioma received 0.3 to 0.6 mg/kg bleomycin via the s/o emulsion with excellent response in 43% cases and good response in 45% cases. The patients which demonstrated excellent response required only 0.6–4 mg/kg total dose. However, this therapy proved dangerous for the treatment of large cervicomediastinal lymphangiomas due to swelling and airway compression (54).

Emulsion systems have also been used to improve tumor immunotherapy. The immunotherapeutic treatments with live BCG are often complicated by the occurrence of allergic immune response. Hence efforts have been made to isolate and inject the component(s) that are responsible for its tumoricidal activity and thus avoid administration of live BCG. The cell wall skeleton (CWS) of BCG has been identified as its key tumoricidal component. Interestingly, the i.t. delivery of CWS as an aqueous suspension, without oil, lacks the desired tumoricidal effect (55,56). However, when delivered as an o/w emulsion, CWS preparations have demonstrated high efficacy in tumor regression. In fact, a study aimed at optimization of i.t. immunotherapy with CWS of BCG showed the oil concentration in the formulation to be more important than the dose of active component (56). It was suggested that the presence of oil enhances the tumoricidal activity of CWS by facilitating interaction between the bacterial component and the mycobacterial adjuvant. Similarly, formation of a tertiary complex between oil, muramyl dipeptide, and trehalose dimycolate (mycobacterial adjuvant) has been suggested as an important requirement for the local immunotherapeutic efficacy of this preparation (57). Another study comparing the surgical treatment of patients having squamous cell carcinoma of head and neck with preoperative i.t. o/w CWS therapy demonstrated complete tumor remission in only 39% of cases with the former treatment as opposed to remission in 69% cases with the latter (58). In addition, the cumulative proportion of patient survival was 50% in the group receiving surgery alone as opposed to 73% in the group receiving preoperative i.t. o/w immunotherapy (58).

Albumin Microspheres

Albumin microspheres are biodegradable particles that have been shown to allow incorporation and sustained release of a variety of drugs (59). In addition, these particles offer the advantages of excellent pharmaceutical stability and good biocompatibility. Hence they have been clinically investigated for the intra-arterial delivery of chemotherapeutic agents (60). Typically, the preparation of albumin microspheres involves emulsion and suspension technology, the particle solidification being achieved by heat or chemical cross-linking (59). Figure 4 displays a representative scheme of the preparation of albumin microspheres as developed and tested in our own as well as several other laboratories. Following hydration, these particles swell 50–100% of their size with concurrent slow matrix erosion (see Figures 4A and 4B) (61,62). The release of drug from the microspheres usually follows a biexponential pattern, with the initial fast-release phase followed by the terminal slow-release phase, and these can be controlled by appropriate selection of the processing conditions (see Figure 4C) (63).

An early work compared the efficacy of 3.3% w/w 5-fluorouracil-loaded albumin microspheres in i.t. drug delivery against the delivery of an equivalent dose of drug solution and saline (64). The microspheres were prepared by

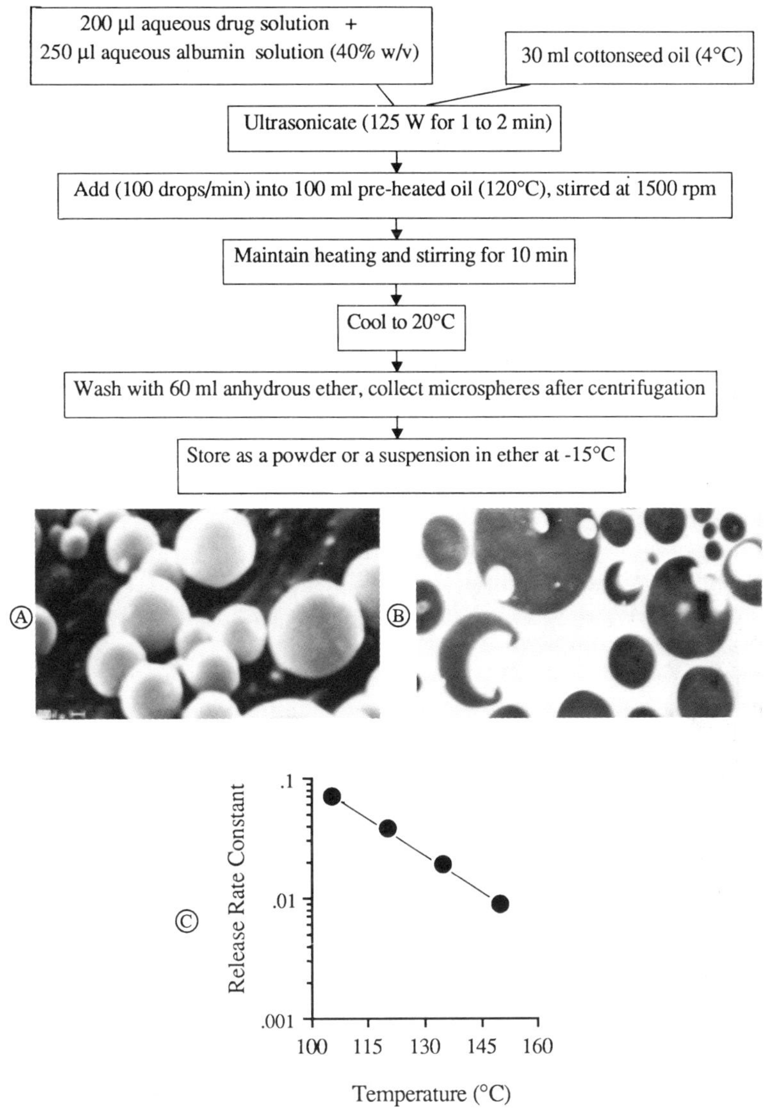

Figure 4 A schematic representation of the steps involved in the preparation of albumin microspheres by heat-denaturation. Fig. 4A shows a scanning electron photomicrograph (SEM) of adriamycin-loaded albumin microspheres (Magnification: 7,000x; Note 0.1 µm line). Fig. 4B shows a SEM of adriamycin-loaded albumin microspheres after hydration in an aqueous medium for 2 hr (Magnification: 7,000x), and Fig. 4C shows a log-linear relationship between the drug release rate constant and stabilization temperature. Adapted in part from ref. 61.

the heat denaturation technique. The i.t. drug delivery experiments were initiated 1 day after the s.c. transplantation of ascite cells on the scapular region of mice. The microsphere-delivered drug reduced the tumor size to about one fifth after 10 days, and to about one tenth after 20 days of its inoculation, as compared to saline or drug solution treatments. The tumor growth rate* over 20 days after the i.t. microspheric drug delivery was at least one third of that with drug delivered as a solution and one fifth of that in animals receiving only saline (see Figure 5) (64). In fact, all mice in the latter two groups demonstrated gross tumor involvement, indicating little benefit with the local delivery of drug as a solution. The multiple-shot i.t. delivery of microsphere-loaded drug 1, 5, and 9 days after tumor cell inoculation demonstrated even greater suppression of tumor growth. Once again, the multiple i.t. doses of drug as a solution failed to demonstrate significant effect on the tumor growth (64).

Other workers demonstrated similar results using adriamycin-loaded albumin microspheres. In one study, the microspheres were prepared by for-

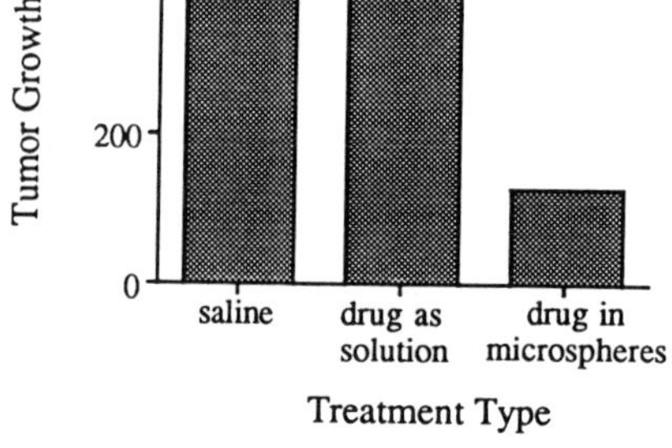

Figure 5 Effect of treatment type on the growth rate of s.c. implanted solid tumor in mice. The animals received i.t. injections of 0.15 ml saline, 5-fluorouracil solution or 5-fluorouracil via microspheres. Adapted in part from ref. 64, with permission from the copyright owner, The Pharmaceutical Society of Japan, Tokyo, Japan.

$$\text{*Tumor growth rate} = \frac{\text{(tumor weight or volume at the end of treatment)} - \text{(tumor weight or volume at the time of initiation of treatment)}}{\text{duration of treatment}}$$

mation of a w/o emulsion followed by droplet crosslinking with glutaraldehyde (65). The microspheres were 14–17 μm in diameter and contained about 1% w/w of drug. Following the i.t. delivery of 85 μg drug via the microspheres, the average tumor growth rate over a period of 30 days approximated 0.44 g/day. This rate was about threefold slower than the tumor growth rate observed following the i.t. delivery of 100 μg drug as a solution and fourfold slower than the rate in the untreated animals or those receiving placebo treatment (see Figure 6). The determination of drug concentration in tumor tissue revealed comparable drug levels during the first 24 hr after the i.t. delivery of drug as a solution or via the microspheres. However, at 40, 48, and 72 hr after dosing, the drug concentrations were two- to threefold higher with microspheres than those monitored after the administration of drug as a solution (1.6, 1.6, and 1.0 μg/g, respectively, with drug solution and 4.1, 3.6, and 2.6 μg/g with microspheres). The increase in antitumor activity with the microsphere delivery system was attributed to increased duration of exposure of tumor tissue to the drug (65).

Yet another study has evaluated the effect of age of tumor on the efficacy of 10 mg/kg epirubicin delivered locally as a solution or via albumin microspheres (66). The microspheres were about 20 μm in diameter and con-

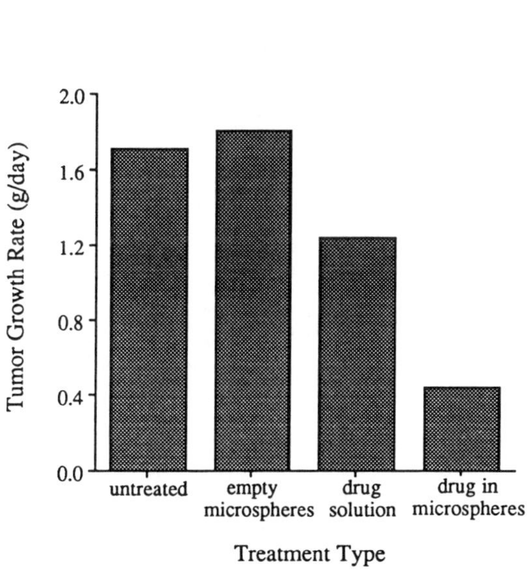

Figure 6 Effect of treatment type on the growth rate of s.c. implanted Sp107 carcinoma in rats. The animals received i.t. injections of 0.5 ml saline, placebo microsphere suspension, 100 μg adriamycin solution or 85 μg adriamycin via albumin microspheres. Adapted in part from ref. 65, with permission from the copyright owner, Pergamon Press, PLC, Oxford, UK.

tained about 12.5% w/w of drug. The mean survival time of animals receiving microsphere-loaded drug on the day of tumor inoculation was found to be significantly longer than the mean survival time of animals receiving the same drug dose via microspheres at 7 days after tumor inoculation. In addition, the survival time of animals receiving i.t. drug as a solution on the day of tumor inoculation was found to be significantly longer than those receiving either drug solution or drug microspheres at 7 days after tumor inoculation. This study therefore highlighted the importance of early i.t. treatment for effective therapeutic outcome (66).

Recently we investigated the efficacy of heat-denatured albumin microspheres in the i.t. delivery of adriamycin (67). The microspheres were about 1 μm in diameter (see Figure 4A) and contained about 4% w/w of drug. A nonmetastasizing sarcoma was chemically induced on the back of male AS inbred rats. When the sarcoma was in its fourth passage and 2 weeks old, with an average doubling time of ~3 days, the rats were randomly divided into seven groups. The first group with five animals did not receive any treatment and served as a control. The second, third, and fourth groups, with five to six animals each, received i.t. injections of 2, 8, and 20 mg/kg adriamycin HCI, respectively, as a solution in 0.2 ml normal saline. The fifth, sixth, and seventh groups of five to six animals received i.t. injections of 2, 8, and 20 mg/kg adriamycin HCI, respectively, via albumin microspheres suspended in 0.2 ml saline. Thereafter the animals were periodically monitored for changes in the tumor dimensions. The tumor dimensions were converted to tumor weights using the relationship (65):

$$\text{Tumor weight (g)} = \frac{w^2\, l}{2} \qquad (1)$$

where w and l are the width and length of tumor at right angles, in cm, respectively.

Figure 7 compares the tumor growth rate in various groups of animals. Neither 2, 8, nor 20 mg/kg drug administered locally as a solution was effective in managing the tumor growth. When the drug was locally administered via albumin microspheres, the 8 and 20 mg/kg doses reduced the tumor growth rate by ~30% (p = .0311) and 50% (p = .0001), respectively, than that observed in the untreated animals. The comparison of the antitumor efficacy of an equivalent dose of drug locally administered as a solution and via the microspheres revealed no significant difference at 2 mg/kg levels; however, significant differences were observed at 8 (p = .0049) as well as 20 mg/kg (p = .0280). At these dose levels, the microsphere-delivered drug was considerably more effective than the drug delivered as a solution in saline.

As anticipated, the increase in the dose of the microsphere-encapsulated drug resulted in a corresponding decrease in the tumor growth rate, and a log-linear relation was established with $R^2 \geq .99$ (see Figure 8). The extra-

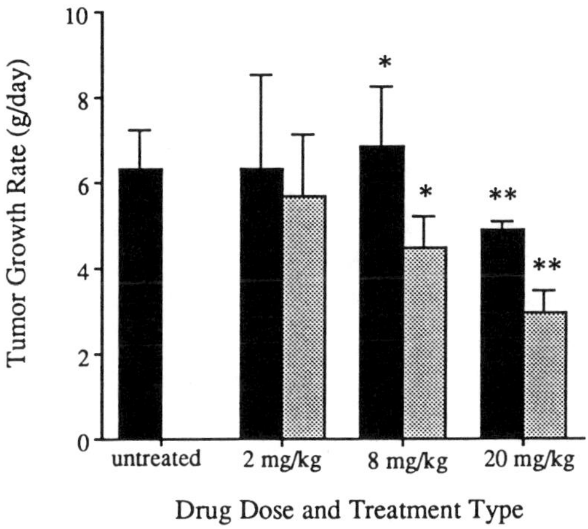

Figure 7 A plot showing the effect of dose of adriamycin HCl, intra-tumorally administered as a solution or via albumin microspheres, on the tumor growth rate over a period of 32 days. Key: (■) drug solution (▒) drug encapsulated in albumin microspheres. Symbol (*) indicates significant difference between 8 mg/kg drug administered as a solution and via the microspheres, and (**) indicates similar difference at dose of 20 mg/kg. Significant difference was not observed for the 2 mg/kg dose given as a solution and via microspheres.

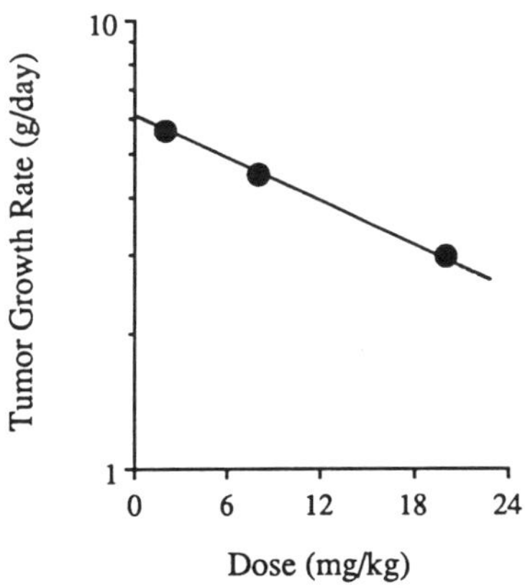

Figure 8 A plot showing log-linear relationship between dose of intra-tumorally administered adriamycin HCl, via albumin microspheres, and tumor growth rate over a period of 32 days. $R^2 \geq 0.99$. P-value for 2 and 8 mg/kg doses = 0.1607; P-value for 8 and 20 mg/kg doses = 0.0572; and P-value for 2 and 20 mg/kg doses = 0.0016.

polation of tumor growth rate for "zero dose" resulted in a value of about 6.1 g/day, which closely resembled the experimentally determined tumor growth rate in the untreated animals (see Figure 7).

Caesin Microspheres

Caesin microspheres have also been investigated for the i.t. delivery of adriamycin. In one reported study, the microspheres were prepared by emulsification followed by polymer crosslinking with glutaraldehyde (68). Thereafter the efficacy of 11 μg drug delivered via microspheres was compared against 25 and 50 μg drug administered as a solution and that with the untreated control animals. The tumor weight was ≤1 g at the time of initiation of the study. In the untreated animals, the tumor weight increased to 12 g in about 19 days. However, following the i.t. administration of drug via caesin microspheres, 40 days were required for the tumor to grow to the same mass. Hence the microspheres were demonstrated to cause a considerable delay in tumor growth (68).

Figure 9 displays the tumor growth rate in untreated animals and those receiving i.t. 25 or 50 μg adriamycin as a solution, or 11 μg drug via the microspheres. The local administration of 25 μg drug as a solution reduced the tumor growth rate by about 35%; however, doubling the dose of drug

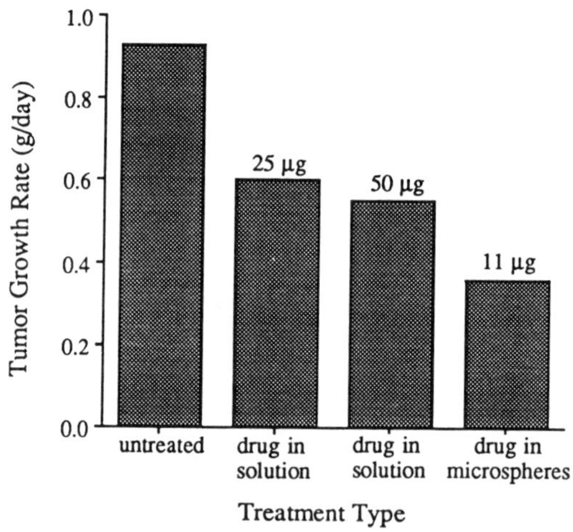

Figure 9 Effect of treatment type on the growth rate of s.c. implanted Sp107 carcinoma in rats. The animals received i.t. injections of 0.5 ml saline, 25 μg adriamycin solution, 50 μg adriamycin solution or 11 μg adriamycin via caesin microspheres. Adapted in part from ref. 68, with permission from the copyright owner, The Pharmaceutical Society of Great Britain, London.

administered as a solution did not improve antitumor response. Interestingly, compared to the animals receiving no treatment and those receiving i.t. drug as a solution, the i.t. administration of 11 μg drug via caesin microspheres reduced the tumor growth rate by 70% and 40%, respectively (68). If linear pharmacological response were possible, these data suggest that 25 μg drug administered via microspheres may potentially reduce the tumor growth rate by >85% and 70%, respectively. Overall, these results are very encouraging and clearly suggest the possibility of therapeutic improvement with i.t. drug delivery via an appropriate particulate carrier.

Gelatin Microspheres

There is increasing evidence to support the assertion that interferons modulate host response to inhibit the in vivo growth of tumor. Hence attempts have been made to evaluate i.t. delivery of encapsulated interferons. In one study, Tabata and coworkers (69,70) developed gelatin microspheres for the localized delivery of recombinant interferon-alfa A/D (INF-αA/D). The microspheres were prepared by w/o emulsification and glutaraldehyde crosslinking. The diameter of the microspheres was ≤2 μm and contained 740 IU of INF-αA/D per mg. BALB/c mice were inoculated with 1×10^6 Meth-A cells into hind foot pads. Three days after tumor inoculation, the animals received i.t. injection of (a) 177 μg placebo microspheres, (b) a physical mixture of 177 μg placebo microspheres plus 260 IU free INF-αA/D, (c) 177 μg gelatin microspheres containing 260 IU INF-αA/D, or (d) saline. The dosings were repeated every 3 days for up to 3 weeks. All groups of animals, except those receiving INF-αA/D-loaded microspheres, demonstrated progressive growth in their tumor (see Figure 10) (69). However, the i.t. drug delivery via gelatin microspheres was effective in inhibiting the in vivo growth of tumor (Figure 11) (70). It was suggested that the local delivery of INF-αA/D allows extended activation of macrophages in and around the neoplastic tissue, which in turn prevents the tumor growth (69).

Liposomes

Liposomes represent one of the most enthusiastically and exhaustively pursued classes of particulate drug delivery system. These consist of one or more concentric bilayer structures of lipid and aqueous compartment alternate to one another. Perhaps the most interesting feature of these particles is their ability to incorporate water-soluble and oil-soluble compounds in the aqueous and the lipid layers, respectively (5,47). Liposomes have been evaluated for the i.t. delivery of an x-ray contrast material, metrizamide (71). The formulations were composed of 1:1 phospholipid and cholesterol. Their processing involved sonication and extrusion to produce small unilamellar (100

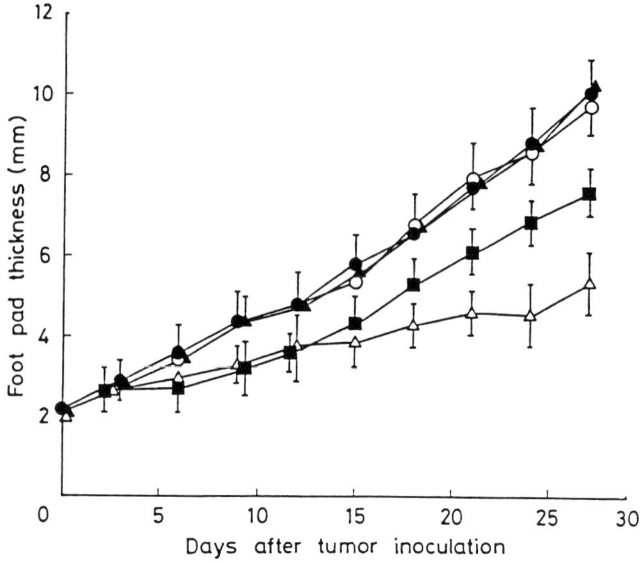

Figure 10 Effect of treatment type on the growth of solid R1 tumors in foot pad of mice. The animals received i.t. injections of (○) saline, (▲) 177 μg of drug free microspheres, (●) 177 μg of drug free microspheres plus 260 IU of free INF-αA/D, (■) 5×10^4 IU of free INF-αA/D, or (△) 177 μg of gelatin microspheres containing 260 IU of INF-αA/D. Reproduced from ref. 69, with permission from the copyright owner, Elsevier Science Publishers BV, Amsterdam.

nm) and large multilamellar (480 nm) liposomes, respectively. The in vitro release rates for the two kind of liposomes were found to be 20 and 10%/48 hr, respectively. The efficacy of i.t. drug delivery via liposomes was evaluated in s.c. gliomas of rats that were 10–30 cm^3 at the time of the initiation of the study. Compared to the tumor tissue residence time (t_{50}%) of metrizamide solution, the small and the large liposomes demonstrated 8- and 300-fold longer tumor residence time, respectively (see Figure 12). The larger liposomes were suggested to act as a depot in the tumor tissue and thus offer superior antitumor response (71). This study therefore demonstrated the drug release rate and the size of particulate carrier as two key determinants of the efficacy of i.t. drug delivery.

Recently these particles have also been tested for the i.t. delivery of methotrexate (72). A single i.t. dose of 4 mg/kg drug via liposomes, in rat gliomas, was shown to be as effective as three repeated i.t. doses of 4 mg/kg drug administered as a solution. Hence this study suggested a possibility of a threefold dose reduction by its localized delivery via liposomes (72).

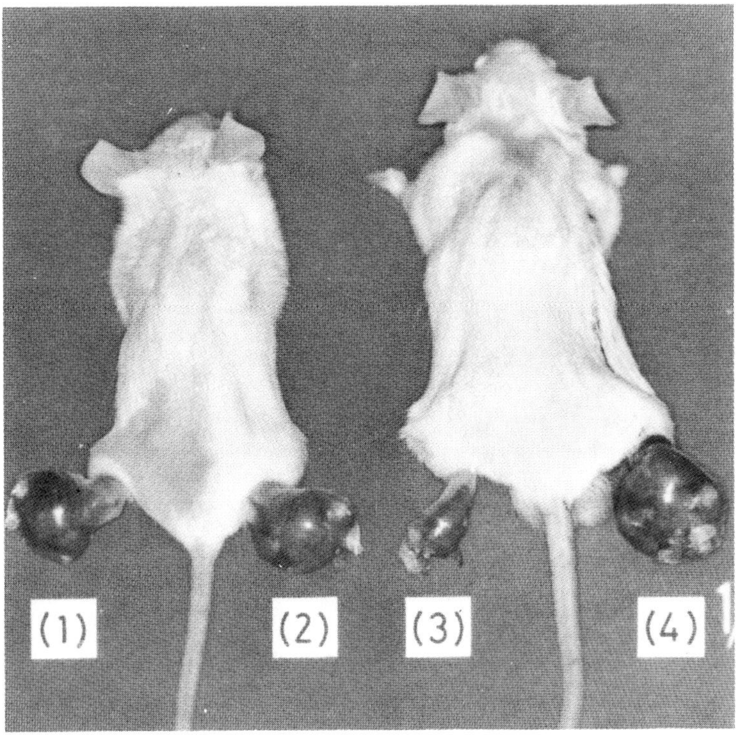

Figure 11 A photograph of mice bearing solid tumor at 24 hr after various treatments: (1) 177 μg of drug free microspheres, (2) 177 μg of drug free microspheres plus 260 IU of free INF-αA/D, (3) 177 μg of gelatin microspheres containing 260 IU of INF-αA/D, and (4) saline. All treatments were repeated every third day for 3 weeks. Reproduced from ref. 70, with permission from the copyright owner, The CRC Press, Inc. Boca Raton, Florida.

CURRENT AND FUTURE CONSIDERATIONS

With regard to the overall success of i.t. drug delivery, the studies to date suggest that its efficacy is a function of several factors, notably those related to the characteristics of drug and drug carrier, the tumor tissue, the dosing protocol adopted in the treatment, and the immune response of the patient (see Table 3).

The physicochemical properties of the drug, e.g., its partition coefficient, molecular weight, protein binding, and affinity to the tumor tissue, govern its residence time in tumor tissue after local administration. Generally, a drug with low molecular weight and partition coefficient cannot be expected to

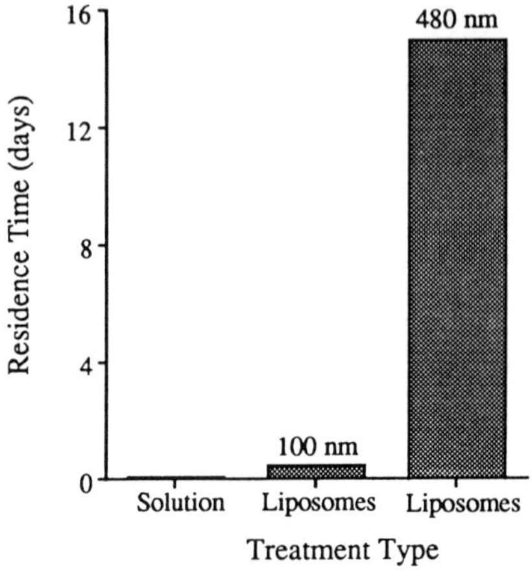

Figure 12 Effect of solution and liposomal delivery of metrizamide on its tumor tissue residence time in rats. Adapted in part from ref. 71, with permission from author and copyright owner, Editio Cantor GmbH.

offer prolonged tumor tissue localization. Its vascular permeability, interstitial diffusion coefficient, and transport across the cell membrane are important for the effective therapeutic response. If the interaction between drug and vascular and/or interstitial components is high, the flux of drug to tumor cells would be low (12). For example, the cell membrane transport with bleomycin has been reported to be low (73). Also, the drugs with MW > 2000 generally exhibit poor membrane permeability; however, some drugs can elicit their response without entering the cells, e.g., adriamycin (74). This situation, in general, becomes even more complicated when the antitumor activity of the delivered compound is relatively low, i.e., large doses are required for the desired response. Under such circumstances, the immunotherapeutic agents, which typically represent compounds with high molecular weight and are also more potent than classical chemotherapeutic agents, may allow superior therapeutic response with relatively low side effects. As mentioned earlier, these compounds have demonstrated excellent antitumor response following localized delivery.

New information relevant to the optimization of chemotherapeutic treatments is constantly emerging. For example, i.t. delivery of a calcium channel blocker, nicardipine HCl, has been shown to enhance acquisition and retention of adriamycin in the tumor tissue (75). The i.t. administration of thymidine has been found to enhance the incorporation of 5-fluorouracil into tumor RNA and hence improve its cytotoxic activity (76). Localized hy-

Table 3 Factors Influencing the Efficacy of Intratumoral Drug Delivery

A. Characteristics of drug
   1. Solubility and partition coefficient
   2. Molecular weight
   3. Protein binding
   4. Vascular permeability, interstitial diffusion coefficient, and cell membrane transport
   5. Overall antitumor activity
   6. Synergism or potentiation
B. Characteristics of encapsulated drug carrier
   1. Particle size
   2. Drug loading
   3. Drug release rate
   4. Sterility
   5. Biocompatibility
C. Characteristics of tumor tissue
   1. Severity of tumor spread and tumor type
   2. Permeability of tumor tissue
   3. Rate of blood flow
   4. Response to drug, carrier, and drug–carrier complex
   5. Microenvironment characteristics
   6. Rate of emergence of cellular variants
D. Treatment schedule
   1. Time of initiation of treatment relative to tumor spread
   2. Frequency or length of treatment
E. Patient condition
   1. Immune response

perthermia lowers pH, at temperatures >41°C it inhibits cell repair, and at >42.5°C it has a cytotoxic effect (1). Application of these approaches, along with i.t. delivery of relevant chemo- or immunotherapeutic agent as a solution or via a particulate carrier, may therefore allow synergism or potentiation of the antitumor activity.

If the drug is administered in encapsulated, conjugated, or similar other forms, then the characteristics of the delivery system also become a key consideration. The physicochemical properties of the encapsulating polymer(s) or conjugating carrier, and the processing technique used in fabrication of delivery system, govern drug loading, particle size, and other surface properties which in turn govern its injectability and perhaps biocompatibility. The solubility of drug and its partition coefficient or diffusion into the encapsulating polymer, and the rate of hydration, swelling, and/or erosion of the encapsulating polymer layer, collectively govern the rate of release of drug and hence control its rate of appearance and duration of activity in the tumor tissue. In case of drug conjugates, factors like intrinsic properties of drug, the molecular weight of the conjugate, and the microenvironment of the tumor tissue, particularly its enzyme activity and pH, govern the therapeutic response. If the formulation process inherently does not allow development of a sterile product, this may need to be achieved by including an appropriate sterilization step. In that case it would also be important to ascertain that the sterilization step does not alter the drug and/or carrier stability, drug release, and other functional parameters of the delivery system. The relevance of particle size and release rate on the tumor tissue residence time of the delivered agent was outlined in an earlier section of this chapter. The possibility of i.t. delivery of particulate carriers capable of emitting radiation before, during, and/or after localized chemo- or immunotherapy remains to be evaluated.

As mentioned earlier, cystic tumors allow better diffusion of drug to the tumor fluid and hence allow superior antitumor response following localized delivery (17). The extent of tumor spread or severity of tumor growth, the permeability of tumor tissue, local blood flow, and response of tumor cells toward the drug, the carrier, and the drug–carrier complex determine the extent and duration of the tumoricidal response. Extensive tumor burden often compromises the probability of therapeutic success (77). Higher permeability of tumor tissue will warrant comparatively larger particle size of drug carrier in order to maintain i.t. localization. Similarly, high blood flow rates will tend to increase the distribution of drug from tumor tissue to the systemic tissue. Strategies involving control in blood flow of tumor tissue have been suggested (78,79).

The tumor tissue microenvironment can also influence the success of i.t therapy. For example, the change in pH with age and the extent of accumulation of toxic catabolites from dying cells may hamper response to chemotherapy (80). On the other hand, information on the pH and/or the enzyme activity of tissue may allow development of drug conjugates with appropriate selectivity and sensitivity (9,10). The rate of emergence of cellular variants that resist ongoing or new treatment schedules may severely limit successful i.t. therapy. However, if this is due to the lack of maintenance of high drug levels in the tumor tissue, high-dose localized treatments via an appropriate particulate carrier may considerably improve the probability of success. For example, the i.t. administration of 11 μg adriamycin via caesin microspheres into Sp 107 mammary carcinoma was shown to control the tumor growth rate to about 0.36 g/day, which was about 40% slower than the local delivery of 25 or 50 μg drug solution (see Figure 9) (68). Interestingly, however, the i.t. administration of 85 μg adriamycin via albumin microspheres into the same carcinoma could not control its growth rate to less than 0.45 g/day (see

Figure 6) (65). These differences in the antitumor activity of the two particulate drug carriers may have been simply due to the differences in their particle size (~15 μm for albumin versus 30 μm for caesin microspheres), drug content (1% in albumin versus 0.7% w/w in caesin microspheres), drug release rate, and/or interaction of the carrier material with the tumor tissue microenvironment. Nonetheless, these results suggest the importance of appropriate selection of polymer for the synthesis of particulate carrier.

Apart from the above characteristics, factors such as the length or the frequency of i.t. dosing of drug solution, or single- or multiple-shot administration of drug via particulate carrier, also influence the efficacy of treatment. However, like any other chemotherapeutic treatment scheduling, these characteristics need to be optimized for each tumor treatment. Previous work suggests that the multiple-shot i.t. drug delivery is more efficacious than the single-shot treatment (38,64,69).

Based on the above discussion, it can be concluded that i.t. drug delivery, especially via particulate carriers, offers unique challenges and possibilities in treating accessible carcinomas. Noninvasive techniques based on magnetic resonance spectroscopy have been developed to assess the extent of cyst in tumor and the kinetics of drug in tumor tissue (81). Hence, if practiced with sound knowledge of the characteristics of the drug, the drug carrier, the nature of the tumor, and patients' immunoresponse, this approach may offer a relatively high degree of success in routine tumor therapy.

ACKNOWLEDGMENTS

The permission granted by several authors and copyright owners for the use of their data and information in this chapter is gratefully acknowledged. The experimental work pursued by the authors and reported in this chapter was conducted at the Department of Pharmacy, University of Otago, Dunedin, New Zealand, and was supported by the Medical Research Council of NZ. The gift of adriamycin HCI from Farmitalia Carlo Erba, Milan (Italy), the technical assistance of Mrs. C. Morris, and the facilities provided by the Department of Pathology, University of Otago, for that work are also acknowledged.

REFERENCES

1. R. Sauer, in *Therapy of Malignant Brain Tumors* (K. Jullinger, ed.), Springer-Verlag, New York, 1987, p. 195.
2. S. E. Salmon and M. Apple, in *Review of Medical Pharmacology* (F. H. Meres, E. Jawetz, and A. Gold, eds.), Lange, 1972, p. 446
3. S. Waga and H. Handa, *Surg. Neurol.*, *5*, 215 (1976).

4. D. Olive and M. Peeters, in *Cancer in Children* (P. A. Vaute, A. Barrett, H. T. G. Bloom, J. Lemerle, and M. K. Neidhardt, eds.), Springer-Verlag, New York, 1986, p. 21.
5. G. Poste and R. Kirsh, *Biotechnology, 10*, 869 (1983).
6. P. K. Gupta, *J. Pharm. Sci., 79*, 949 (1990).
7. J. M. Collins, *J. Clin. Oncol., 2*, 498 (1984).
8. V. A. Levin, *Cancer Treat. Rev., 13*, 61 (1986).
9. H. Sezaki and M. Hashida, *CRC Crit. Rev. Ther. Drug Carrier Syst., 1*, 1 (1984).
10. M. V. Pimm, *CRC Crit. Rev. Ther. Drug Carrier Syst., 3*, 189 (1988).
11. P. Couvreur and J. Aubry, in *Topics in Pharmaceutical Sciences* (D. D. Breimer and P. Speiser, eds.), Elsevier, Amsterdam, 1983, pp. 305–315.
12. R. K. Jain, *Cancer Metas. Rev., 8*, 133 (1989).
13. J. S. Kroin and R. D. Penn, *Neurosurgery, 10*, 349 (1982).
14. G. Bouvier, R. D. Penn, J. S. Kroin, R. A. Beique, and M. J. Guerard, *Neurosurgery, 20*, 286 (1987).
15. W. F. Chandler, H. S. Greenberg, W. D. Ensminger, R. F. Diaz, L. R. Junck, T. W. Hood, S. G. Gebarski, and M. A. Page, *Ann. N.Y. Acad. Sci., 531*, 206 (1988).
16. J. D. Everall, C. J. O'Doherty, J. Wand, and P. M. Dowd, *Lancet, 2*, 583 (1975).
17. H. Takahashi, S. Nakazawa, and T. Shimura, *J. Neurosurg., 62*, 120 (1985).
18. T. Takahashi, S. Ueda, K. Kono, and S. Majima, *Cancer, 38*, 1507 (1976).
19. G. Bouvier, R. D. Penn, J. S. Kroin, R. A. Beique, M. J. Guerard, and J. Lesage, *Ann. N.Y. Acad. Sci., 531*, 213 (1988).
20. P. M. Andrews and C. L. Johnson, Jr., *Cancer Chemother. Pharmacol., 23*, 31 (1989).
21. R. D. Penn, J. S. Kroin, J. E. Harris, K. M. Chu, and D. P. Braun, *Appl. Neurophysiol., 46*, 240 (1983).
22. D. Rosner, *J. Surg. Oncol., 39*, 90 (1988).
23. S. L. Gibson, K. R. vanderMeid, R. S. Murrant, and R. Hilf, *Br. J. Cancer, 61*, 553 (1990).
24. R. A. Gatenby, W. H. Hartz, P. F. Engstrom, J. S. Rosenblum, N. D. Hammond, H. B. Kessker, P. J. Moldofsky, M. R. Clair, E. Unger, and G. J. Broder, *Radiology, 163*, 172 (1987).
25. A. V. D'Amico, A. C. Nelson, and R. K. Babayan, *J. Cancer Res. Clin. Oncol. 114*, 129 (1988).
26. H. H. Bartschl, K. Pfizenmaier, M. Schoreder, and G. A. Nagel, *Eur. J. Cancer Clin. Oncol., 25*, 287 (1989).
27. J. N. Ijzermans, T. A. Splinter, M. Scheringa, R. L. Marquet, and J. Jeekel, *Proc. Ann. Meet. Am. Assoc. Cancer Res., 31*, A1104 (1990).
28. G. Weil-Hillman, F. M. Uckun, J. M. Manske, and D. A. Vallera, *Cancer Res., 47*, 579 (1987).
29. D. L. Morton, F. R. Eilber, E. C. Holmes, J. S. Hunt, A. S. Ketcham, M. J. Silverstein, and F. C. Sparks, *Ann. Surg., 180*, 635 (1974).
30. F. A. Burgener and S. D. Steinmetz, *Invest. Radiol., 22*, 472 (1987).

31. J.-C. Sheu, G.-T. Huang, D.-S. Chen, J.-L. Sung, P.-M. Yang, T.-C. Wei, M.-Y. Lai, C.-T. Su, Y.-M. Tsang, H.-C. Hsu, I.-J. Su, T.-T. Wu, J.-T. Lin, and C.-N. Chuang, *Radiology*, *163*, 43 (1987).
32. K. Kuroda, T. Olden and N. Lassen, *Acta. Neurol. Scand.*, *66*, 160 (1982).
33. J. B. Blacklock, D. C. Wright, R. L. Dedrick, R. G. Blasberg, R. J. Lutz, J. L. Doppman and E. H. Oldfield, *J. Neurosurg.*, *64*, 284 (1986).
34. H. I. Peterson, in *Tumor Blood Circulation: Angiogenesis, Vascular Morphology and Blood Flow of Experimental and Huma Tumor (H. I. Peterson, ed.), CRC Press, Boca Raton, 1979.*
35. N. H. Greig and D. L. Trainer, *Cancer Metas. Rev.*, *5*, 3 (1986).
36. R. K.-Y. Cheng and C. C. Cheng, *Meth. Find.*, *11*, 439 (1989).
37. M. Hashida, A. Kato, T. Kojima, S. Muranishi, H. Sezaki, N. Tanigawa, K. Satomura, and Y. Hikasa, *Gann*, *72*, 226 (1981).
38. Y. Morimoto, K. Sugibayashi, S. Sugihara, K. I. Hosoya, S. Nozaki, and Y. Ogawa, *J. Pharmacobio. Dyn.*, 7, 688 (1984).
39. L. Bures, J. Bostik, K. Motycka, M. Spundova, and L. Rehak, *Neoplasma*, *35*, 329 (1988).
40. N. Y. Yu, F. K. Conley, E. E. Luck, and D. M. Brown, NCI Monogr., #6, 137 (1988).
41. R. Sutton, N. Yu, E. Luck, D. Brown, and F. Conley, *Sel. Cancer Ther.*, *6*, 35 (1990).
42. R. Langer, D. Lund, K. Leong, and J. Folkman, *J. Controlled Release*, *2*, 331 (1985).
43. H. Brem, 5th Int. Symp. Recent Adv. Drug Deliv. Syst., Feb. 25–28, 1991, Salt Lake City, UT.
44. B. Wowra, W. J. Zeller, T. Remmele, V. Strum, and H. Stricker, *Reg. Cancer Treat.*, *2*, 26 (1989).
45. R. Blasberg, C. S. Patlak, and J. D. Fenstermacher, *J. Pharmacol. Exp. Ther.*, *195*, 73 (1975).
46. R. F. Ozols, G. Y. Locker, J. W. Doroshow, K. R. Grotzinger, C. E. Myers, and R. C. Young, *Cancer Res.*, *39*, 3209 (1979).
47. M. J. Poznansky and R. L. Juliano, *Pharmacol. Rev.*, *36*, 277 (1984).
48. T. Takahashi, M. Mizuno, Y. Fujita, S. Ueda, B. Nishioka, and S. Majima, *Gann*, *64*, 345 (1973).
49. M. Hashida, Y. Takahashi, S. Muranishi, and H. Sezaki, *J. Biopharm. Pharmacokinet.*, *2*, 241 (1977).
50. M. Hashida, T. Yoshioka, S. Muranishi, and H. Sezaki, *Chem. Pharm. Bull.*, *28*, 1009 (1980).
51. N. Tanigawa, K. Satomura, Y. Hikasa, M. Hashida, S. Muranishi, and H. Sezaki, *Surgery*, *148*, 147 (1980).
52. H. Sezaki, M. Hashida, and S. Muranishi, in *Optimization of Drug Delivery* (H. Bundgaard, A. B. Hansen, and H. Kofod, eds.), Munksgaard, Copenhagan, 1982, pp. 316–332.
53. N. Tanigawa, T. Shimomatsuya, K. Takahashi, Y. Inomata, K. Tanaka, K. Satomura, Y. Hikasa, M. Hashida, S. Muranishi, and H. Sezaki, *Cancer*, *60*, 741 (1987).

54. K. Tanaka, Y. Inomata, H. Utsunomiya, S. Uemoto, K. Asonuma, T. Katayama, K. Ozawa, and M. Hashida, *Pediatr. Surg. Int.*, *5*, 270 (1990).
55. B. Zbar, H. J. Rapp, and E. E. Ribi, *J. Natl. Cancer Inst.*, *48*, 831 (1972).
56. E. Yarkoni and H. J. Rapp, *Cancer Res.*, *39*, 535, (1979).
57. C. A. McLaughlin, S. M. Schwartzman, B. L. Horner, G. H. Jones, J. G. Moffatt, J. J. Nestor, Jr., and D. Tegg, *Science*, *208*, 415 (1980).
58. J. Bier, H. J. Rapp, T. Borsos, B. Zbar, S. Kleinschuster, H. Wagner, and M. Rollinghoff, *Cancer Immunol. Immunother.*, *12*, 71 (1981).
59. P. K. Gupta and C. T. Hung, *J. Microencapsulation*, *6*, 427 (1989).
60. S. Fujimoto, M. Miyazaki, F. Endoh, O. Takahashi, K. Okui, and Y. Morimoto, *Cancer*, *56*, 2404 (1985).
61. P. K. Gupta, C. T. Hung, F. C. Lam, and D. G. Perrier, *Int. J. Pharm.*, *43*, 167 (1988).
62. P. K. Gupta, C. T. Hung, and D. G. Perrier, *Int. J. Pharm.*, *33*, 137 (1986).
63. P. K. Gupta, C. T. Hung, and D. G. Perrier, *Int. J. Pharm.*, *33*, 147 (1986).
64. Y. Morimoto, M. Akimoto, K. Sugibayashi, T. Nadai, and Y. Kato, *Chem. Pharm. Bull.*, *28*, 3087 (1980).
65. N. Willmott and J. Cummings, *Biochem. Pharmacol.*, *36*, 521 (1987).
66. S. E. Leucuta, R. Risa, D. Daicovicin, and D. Porutiu, *Int. J. Pharm*, *41*, 213 (1988).
67. P. K. Gupta, C. T. Hung, and F. C. Lam, *J. Pharm. Sci.*, *79*, 634 (1990).
68. Y. Chen, N. Willmott, J. Anderson, and A. T. Florence, *J. Pharm. Pharmacol.*, *39*, 978 (1987).
69. Y. Tabata, K. Uno, S. Muramatsu, and Y. Ikada, *Jpn. J. Cancer Res.*, *80*, 387 (1989).
70. Y. Tabata and Y. Ikada, *CRC Crit. Rev. Ther. Drug Carrier Syst.*, 7, 121 (1990).
71. E. Mentrup, B. Wowra, W. J. Zeller, V. Sturm, and H. Stricker, *Arzneim. Forsch./ Drug Res.*, *39*, 421 (1989).
72. H. Stricker, E. Mentrup, R. Krotz, W. J. Zeller, V. Sturm, and B. Wowra, *Eur. J. Pharm. Biopharm.* (in press).
73. T. Hayakaw, Y. Ushio, and K. Morimoto, *J. Neurol. Neurosurg. Psychiatry*, *39*, 341 (1976).
74. T. R. Tritton and G. Yee, *Science*, *217*, 248 (1982).
75. H. Matsuoka, K. Sugimachi, H. Kuwano, and K. Yano, *Eur. J. Surg. Oncol.*, *15*, 224 (1989).
76. H. Matusoka, H. Masuda, Y. Maehara, K. Sugimachi, and K. Inokuchi, *Cancer Treat. Rep.*, *70*, 851 (1986).
77. S. A. Risenberg and H. J. Rapp, *Med. Clin. North Am.*, *60*, 419 (1976).
78. W. D. Ensminger and J. W. Gyves, *Pharmacol. Ther.*, *21*, 277 (1983).
79. H. J. Keizer and H. M. Pinedo, *Cancer Drug Deliv.*, *2*, 147 (1985).
80. I. F. Tannock, *Cancer Metas. Rev.*, *8*, 146 (1989).
81. C. A. Present, W. Woil, C. Wisemmann, D. Blayney, and M. King, *Proc. Ann. Meet. Am. Soc. Clin. Oncol.*, *9*, A289 (1990).

7

Liposomes as Intramuscular and Subcutaneous Injection Drug Delivery Systems

Farida Kadir, Jan Zuidema, and Daan J. A. Crommelin

University of Utrecht, Utrecht, The Netherlands

INTRODUCTION

After intramuscular (i.m.) or subcutaneous (s.c.) injection, therapeutically effective drug levels can be reached fairly rapidly if the drugs are administered dissolved in an aqueous solution. In cases of drugs that are quickly eliminated the initial therapeutic drug concentration may be followed by a rapid decline in blood drug level. As a result of this rapid elimination the therapeutic response is of short duration. Another disadvantage associated with the intravenous (i.v.) route is that initial peak levels might provoke severe side effects.

In many cases it is desirable to maintain the concentration of drug in blood within the therapeutic range for a long time. Slow intravenous infusion employing an initial loading dose whereafter the drug is administered to the patient at a rate about equal to its rate of elimination was introduced to maintain therapeutic steady-state drug concentrations in blood. This method is particularly useful for drugs with a narrow therapeutic index. However, it is not always possible to administer a drug intravenously. This is the case, for example, when drugs have a low aqueous solubility and high doses are required. There are other disadvantages of the i.v. route, e.g., it requires skilled medical staff and its administration implies inconvenience for patients. These considerations do not stimulate the use of the i.v. route of administration in chronic treatment. They rather call for the development of other, nonintravenously administered drug formulations that release active drug slowly. Such systems, which can be administered by several routes, including the parenteral ones, e.g., i.m. and s.c., should allow for a low dosing frequency and therefore may improve patient compliance.

The term *sustained-release system* generally refers to drug delivery systems that are able to provide a slow release of a drug over an extended period. In summary, pharmacotherapeutic reasons for developing dosage forms with sustained or controlled drug release may be (a) improvement of patient compliance, (b) avoidance/reduction of adverse effects, (c) maintenance of an optimal therapeutic drug concentration profile (with only small fluctuations) for a long time (1-3). There are also arguments against the development of dosage forms with sustained or controlled drug release. These include the risk for dose dumping due to poor drug release control, the inability to withdraw the drug once administered, increased irritation at the injection site as a result of continuous exposure of the tissue to the drug, and the development of tolerance.

A large number of methods for prolonging drug action after parenteral administration have been developed, tested, and in some cases marketed. Three main approaches that can be used to prolong a drug's tissue level or its biological action can be distinguished. These include (a) chemical modification (e.g., altering the solubility of a drug through salt formation or derivatization), (b) physicochemical manipulation (e.g., manipulation of particle size, use of additives such as viscosity enhancers, encapsulation of the drug, use of implants), and (c) biological manipulation (coadministration of vasoconstrictors or absorption modifiers, decrease of elimination, e.g., by simultaneous administration of probenecid in the case of mainly renally excreted drugs). Application of these approaches has produced a variety of preparations with different complex "overall" mechanisms of drug release. Preparation techniques as well as the release properties of most of these formulations have been reviewed extensively (2,4-9) and will therefore not be subject of this chapter.

A relatively new approach for prolonging drug release after i.m. or s.c. injection is the use of liposomes. At present sufficient experimental data concerning this subject have become available to warrant a discussion about the potentials of the use of i.m. or s.c. injected drug-laden liposomes. This subject has not been covered in detail before. The main part of this chapter will therefore focus on the applicability of liposomes as carrier systems for the i.m. or s.c. administration of drugs. A brief overview of biodegradable microparticulate drug delivery systems for i.m. or s.c. administration including microspheres (μm range) and nanoparticles (nm range) describes the position of liposomes among alternative carrier systems for parenteral prolonged release. This will be followed by a discussion of some anatomical and physiological aspects of the i.m. and s.c. injection. A section describing the influence of liposomes on drug disposition, mainly focused on prolonged release of the encapsulated drug and lymphatic delivery after i.m. or s.c. administration, follows. Finally, conclusions concerning the potential usefulness of liposome-encapsulated drugs are drawn and prospects for optimizing our therapeutic arsenal through i.m. or s.c. liposomal delivery of drugs are dealt with.

MICROSPHERES AS DRUG CARRIERS

Particulate carriers are used in the formulation of depot preparations for prolonged drug release. Many of these carriers are based on polymers. Basically, drug–polymer combinations can be divided into two types with different release mechanisms. The polymer can either surround a core of drug (encapsulation type) or the drug can be dissolved or homogeneously dispersed in the polymer (matrix device). In addition to coating solid drug particles with a polymeric membrane, aqueous solutions of drugs can be encapsulated in vesicles consisting of lipid bilayers as well. The term *microspheres* is used to describe small particles designed as carriers for drugs or other therapeutic agents. Their size ranges from 10 nm up to 100 μm or more. The smaller particles (below about 500 nm) are sometimes termed *nanospheres* or *nanoparticles*.

Microcapsules are similar in many respects to microspheres but they consist of small spheres that have an outer layer or membrane enclosing a core material that can be the drug itself. Microcapsules are generally large compared to the microspheres (5-7,10-12).

The size of the particulates is mainly governed by the manufacturing procedure and the materials used. Microspheres designed for parenteral drug delivery can be composed of a variety of materials with different physical characteristics. A selection of materials for microsphere delivery systems being used or under investigation for parenteral application is presented in Table 1. The choice of microsphere material is determined by a variety of factors such as drug properties, duration of action, intended destination, biocompatibility properties, and toxicity considerations. Many methods, ranging from laboratory scale to large production scale, are available for the preparation of microspheres. The release kinetics of the drug from a microsphere are mainly determined by the manufacturing procedure, the type of matrix used, or the encapsulated material. Excellent reviews about the possible release mechanisms of drugs from microcapsules and microspheres are available (2,5,10,11,13-16).

The biodegradability of the matrix is an important selection criterion if parenteral use is considered. All matrix materials are, or are assumed to be,

Table 1 Materials Used or Under Investigation for Parenteral Administration of Drugs

Chylomicrons (18,19)	Collagen (20)
High-density lipoproteins (21,22)	DNA (23)
Low-density lipoproteins (24,25)	(DL)-polylactice acid (26-28)
Albumin (29,30-32)	Polyglycolic acid (33-35)
Fibrinogen (36)	Polycaprolactone (37,38)
(Phospho)lipids (6,39,40,60)	(Cyclo)dextrins (41)
Gelatin (42-44)	Polyaminoacids (45-48)
Starch (derivatives) (49)	Polyalkyl(cyano)acrylates (50-53)
Alginates (54-57)	Poly(acryl)amides (58-61)
Chitosan (62,63)	

nontoxic and the majority are biodegradable (17). Polylactic/polyglycolic acids are the most frequently tested and used materials nowadays; they are approved by the FDA for parenteral formulations.

As indicated in Table 1, lipids are used as constituents of vehicles for drug delivery systems for parenteral application. They differ from the other materials listed as being nonpolymeric in nature. Particulate systems based on phospholipids and/or cholesterol are kept together by physicochemical interactions between the low molecular weight building stones.

Phospholipids and cholesterol are the major structural components of biological membranes. They are used in a variety of drug delivery devices such as lipid microspheres (64,65), emulsions (66,67), mixed micelles (68,69), and liposomes. This chapter will exclusively deal with the last category. Only lately have studies started to appear in which liposomes were used for prolonged delivery via the i.m. or s.c. route of administration and the results of these studies have not yet been critically discussed. A number of reviews on liposome chemistry and formulation have been published (70-75). For more detailed information the reader is referred to these publications, which include detailed information on the behavior of liposomes after intravenous administration as well. For this reason the intravenous administration of liposomes is not discussed in detail in this chapter. In the following paragraph a brief introduction to liposomal vesicles and some general properties of these vesicles will be given.

When phospholipids are dispersed in water they can form a variety of structures depending on the lipid-to-water ratio. At relatively low ratios, the liposomal structure is the predominant one. Liposomes are closed structures composed of at least one phospholipid bilayer surrounding water-filled compartments. The common lipids used are phosphatidylcholines (PC) obtained from natural sources that contain mainly unsaturated fatty acyl chains. Dimyristoylphosphatidylcholine (DMPC), dipalmitoylphosphatidylcholine (DPPC), and distearoylphosphatidylcholine (DSPC) are also used and these are synthetic saturated fatty acyl chain–carrying phospholipids. Phosphatidic acid (PA), dipalmitoylphosphatidic acid (DPPA), phosphatidylglycerol (PG), and dipalmitoylphosphatidylglycerol (DPPG) can be included in the bilayer when negatively charged liposomes are desired. These lipids were found to be nontoxic in animal studies (76). Stearylamine (SA) can be used when a positive charge is desired, although adverse effects have been reported to occur with stearylamine (60,76,77).

In addition to phospolipids, the membrane structure may also include other compounds such as sterols (e.g., cholesterol), fatty acids, glycolipids, or even proteins. Depending on the phospholipids used and the ionic strength of the medium, liposomes of various size ranges can be obtained. Several techniques have been described to prepare liposomes with well-defined physical properties such as size, number of lamellae, charge, and fluidity. Liposomes can be classified according to number of bilayers, diameter of the vesicles, and preparation technique. Major types are multilamellar (MLV), large unilamellar (LUV), small unilamellar (SUV), reverse phase evaporation (REV), or stable plurilamellar vesicles (SPLV)(11,78,79).

Basically, drugs can be incorporated into liposomal carriers in two different ways. Water-soluble drugs can be entrapped in the internal aqueous compartment(s) of the liposome while lipophilic drugs can intercalate into or adsorb to the liposome membrane. Thus for carrying water-soluble drugs in lipid vesicles, it is important that the vesicles have a large internal volume so as to incorporate the drug solution as efficiently as possible. This favors liposomes with a large volume/surface ratio, i.e., large unilamellar vesicles. For lipophilic drugs the total amount of drug incorporated will mainly depend on the amount of lipid available.

INTRAMUSCULAR AND SUBCUTANEOUS INJECTION

Intramuscular and Subcutaneous Injection Sites

Prolonged action delivery systems can be administered, applied, or affixed to several regions of the human body such as mouth, nose, eye, skin, muscle, blood vessels, joint, vagina, and rectum. Each of these regions is unique in terms of its anatomy, physiology, and physiological properties and not all of them are suitable as routes for controlled delivery. The fate of drugs and carriers is partially determined by the anatomy of the area of the body to which drug is applied or administered (80-83). Thus far, efforts in developing controlled-release parenteral dosage forms have been concentrated on the s.c. and i.m. routes. These are the two major routes of administration for injectable depot formulations.

Subcutaneous injections are given into connective tissue. Compared to muscular tissue, connective tissue is loosely organized. It is constructed of a network of woven fibrils. In its interstices a finer network is formed by cells and interstitial substance (proteins and hyaluronic acid). The state of this tissue determines to what extent injected solution is exposed to the local capillary membrane for absorption. Connective tissue may be altered by systemic factors or by locally acting stimuli.

Body sites suitable for s.c. administration include most portions of the arms, legs, and abdomen. Subcutaneous injections are easy to administer, which makes them applicable for chronic self-administration of drugs such as insulin.

Subcutaneous administration is generally limited to nonirritating, water-soluble drugs since s.c. tissue is more innervated than muscle tissue and therefore more sensitive to irritating agents. The volume of a single s.c. injection is usually small, ranging from 0.5 to 2.0 ml (2,15,84,85). In cases of s.c. injections into the s.c. fat layer of the buttock volumes up to 5–10 ml can be administered.

Intramuscular injections are made deep into the skeletal muscle fibers that lie beneath the s.c. layer. The principal sites of injections are the gluteal (buttock), deltoid (upper arm), and vastus lateralis (lateral thigh) muscles. The gluteal muscle has become the most common site for i.m. injection, since it has greater muscle mass than arm and thigh, permitting the injection of larger volumes of fluid. The usual volume injected ranges from 1 to 3 ml with occasional volumes up to 10 ml. The most suitable area to inject is the upper and outer quadrant of the gluteus maximus, since danger of the needle piercing the sciatic nerve or the superior gluteal artery is minimal in this area. For children, the deltoid muscle of the upper arm or the midlateral muscles of the thigh are usually preferred (2,86). The deltoid is thick and also has a good blood supply, which provides the fastest absorption of all the i.m. sites. Since only a small area in the center provides a satisfactory site for drug administration and the maximal volume of a single injection is smaller (about 2 ml) than for the gluteal muscle, the deltoid muscle is not the most preferable injection site.

Factors Affecting Absorption

Absorption of drugs from i.m. and s.c. injection sites in the general circulation is a critical step in the sequence of events following injection of a drug and therefore it deserves close attention. The overall absorption from the i.m. and s.c. injection site depends heavily on the combined effect of (a) drug-related factors, (b) formulation-related factors, and (c) physiological factors. Most of these factors have been reviewed before (1,2,15,81,87). In the following section these factors together with some examples from our own studies will be discussed.

Drug-related factors include, e.g., aqueous solubility of the drug, molecular size, drug pK_a value in combination with the pH value of the formulation, and drug lipophilicity.

The role of drug lipophilicity is shown in Figure 1 (88). The test solutions were administered intra-adiposely, i.e., into the adipose fat layer. Intra-adipose injections are often administered, e.g., by accident as a result of too shallow injection into muscular tissue. In this study aqueous solutions of the β-blocking agents propranolol, alprenolol, carazolol, metoprolol, and atenolol were used as test compounds. These drugs are good model compounds since they differ markedly in lipophilicity but only slightly in pK_a value. Initial absorption rate constants were calculated from the fraction not absorbed versus time profiles. It was demonstrated that the fat/buffer partition coefficient of the drug correlated negatively with the release rate after intra-adipose injection. The results could be explained by assuming that adipose fat tissue exerts strong retarding effects on the availability of more lipophilic drugs. It was suggested that moderately hydrophilic substances are probably transported through the intercellular aqueous medium, whereas the transcellular route is more important for lipophilic substances.

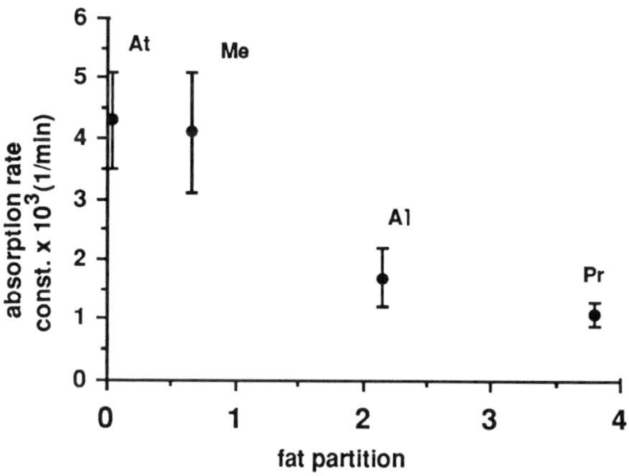

Figure 1 Relation between the fat/phosphate buffer partition constants and initial absorption rate constants after intra-adipose injection (0.2 mg/kg, injection volume 1 ml) of propranolol (Pr), alprenolol (Al), metoprolol (Me), and atenolol (At) into pigs. (From Ref. 88.)

Formulation-related factors include, e.g., dissolution rate of a solid drug in the vehicle used or, in the case of suspensions, particle size of the solid drug, the presence of solubilizers, viscosity of the formulation, properties of the vehicle (oily or aqueous), initial drug concentration, and injection volume. The role of the injection volume for the lipophilic drug propranolol was investigated in a study in rats (89). The aim of that study was to investigate whether drug release could be manipulated by the injection volume. Rats were given the same amount (3 mg) of propranolol in different injection volumes (50, 100, and 200 μl) intramuscularly in the hind leg. Initial absorption rate constants were calculated from the fraction not absorbed versus time curve. It was shown that absorption rate constants as well as drug availability during the first 8 hr after injection increased with the injection volume. Results are presented in Figure 2 a and 2 b.

Drug- and formulation-related parameters are of great importance in determining the onset time, the magnitude, and the duration of a therapeutic response. In addition, physiological conditions are also important factors in affecting the extent of drug availability and drug activity. They relate to the site of injection, muscle activity, injection depth, blood flow at or around the site of injection, age, sex, condition of the tissue at the injection site (e.g., necrosis and inflammation), and condition of the patient (e.g., the presence of fever).

The importance of injection depth was recognized by Dundee et al. (90). They described the differences in plasma diazepam levels after injections given by nurses (using a 3-cm needle) and physicians (using a 4-cm needle) to female subjects. The resulting concentrations were much lower in the 3-cm-needle group than in the 4-cm-needle group (Figure 3). The results were explained by assuming that with the 3-cm needle injections were given too shallowly: the drug was deposited in the adipose layer instead of muscle tissue. As adipose tissue has a retarding effect on drug release, diazepam was absorbed more slowly from the intra-adipose injection site than from the i.m. injection site.

A similar phenomenon was reported by Modderman et al. (91) in a study with subjects of different sexes. They were given i.m. injections (in the gluteal maximus region) of an aqueous suspension of dapsone. Women were found to have more sustained absorption profile than most of the men, resulting into lower initial peak plasma concentrations and higher subsequent concentrations over extended periods (Figure 4). The results were explained by assuming that in most women, having thicker subcutaneous fat layers than men, the injection was actually placed into the fat layer. In a subsequent experiment sex differences in plasma concentration profiles disappeared when the injection depth was related to the thickness of the s.c. fat layer and thus deliberately placed in the adipose layer (1).

INTRAMUSCULAR AND SUBCUTANEOUS INJECTION OF LIPOSOMES

Interactions Between Liposomes and Cells; Liposome Fate at the Injection Site

The stability of liposomes in the body fluid, the clearance of liposomes from the site of injection and their disposition in the body, the release of drug from liposomes, and the subsequent distribution of the released drugs throughout the body determine the outcome of liposomal drug delivery (92–94). A number of variables such as liposome composition (i.e., type of phospolipids, inclusion of cholesterol), liposome size, number of lamellae, bilayer rigidity (fluid or gel crystalline state), osmotic fragility, amount of lipid administered, and liposome surface charge control the stability of liposomes in body fluids.

Insight into the interaction of liposomes with cells is important to understand their behavior and the requirements that must be met before liposomes can act as effective drug delivery vehicles in vivo. Several mechanisms concerning the interaction between liposomes and cells (schematically presented in Figure 5) have been proposed (95,96).

First, liposomes can be adsorbed to cell surfaces through electrostatic and/or hydrophobic interactions or specific receptor binding. While adsorbed they can become leaky and deliver their contents to the cytoplasm (97). Second, exchange of lipids and proteins may occur; proteins and other biomolecules in body fluids may interact with the liposome bilayer and induce lipid exchange between vesicle membrane constituents and lipid components in the fluids or induce the formation of liposomal lipid-protein complexes (73,98-101). These interactions may change liposome fluidity and size and therefore the liposomes might become different from those added initially. Also these interactions may result into leakage of the entrapped drug from the liposomes. Neither the mechanism of lipid transfer nor protein transfer from cell membranes to liposomes are completely clarified (102-104). Third, fusion of liposomal membranes with the cell membrane may take place. Initially fusion was considered to be the predominant mechanism of cell interaction with liposomes. At present it is clear that spontaneous fusion between regular liposomes and the plasma membrane is a rare event (105-107). Finally, liposomes, as any small particles, may be endocytosed by cells (65,108-111). Endocytosis of a lipid vesicle depends on its size. It has been shown in vitro that the optimal size range of LUV (large unilamellar vesicles) liposomes for endocytosis is 0.05–0.10 μm (Heath et al., 112).

Absorption of liposome-encapsulated drug from i.m. and s.c. injection sites can occur in several ways. Both liposome-encapsulated and free drug can reach the blood compartment via lymphatic absorption and by absorption

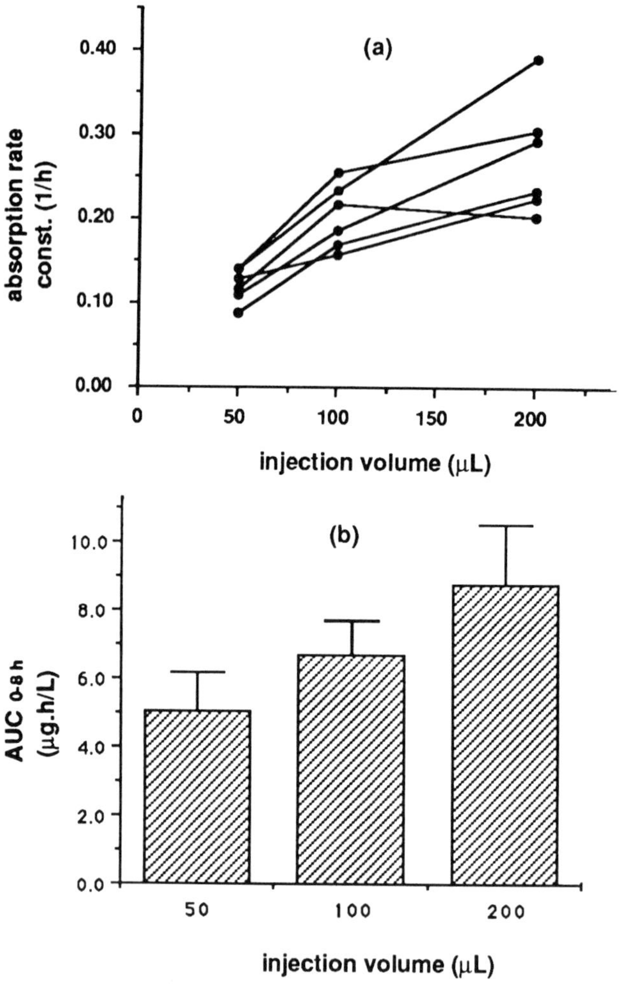

Figure 2 Relation between (a) injection volume and absorption rate constant and (b) injection volume and area under the curve (AUC_{0-8h}), respectively, after i.m. injection of 3 mg propranolol HCl into rats. Lines connect individual values of the same rat. (From Ref. 89.)

through the capillary wall of blood vessels, respectively. The dynamic flow is able to drain the particles away from the site of administration and drainage of the lymph into the bloodstream will carry the particulates into the blood circulation. Whether the liposomes are drained through the lymph or stay at the injection site depends on their size and resistance against destabilization

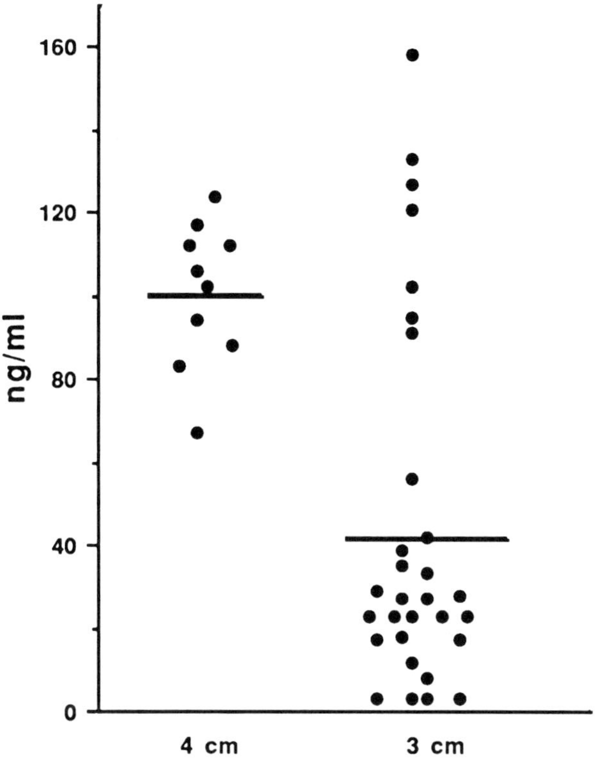

Figure 3 Plasma diazepam concentrations 90 min after intramuscular injection with a 3-cm or a 4-cm needle to humans. (Adapted from Ref. 90.)

(see below). Liposomes remaining at the injection site might be subjected to various destabilizing factors such as degradation resulting from lipid depletion (caused by, e.g., lipid exchange or lipid–protein complexation), attack by lipolytic enzymes, or destruction by neutrophils. All these processes may contribute to the release of the entrapped drug, which is finally absorbed as free drug. As mentioned above, a considerable fraction of liposomally entrapped drug administered i.m. or s.c. is supposed to be released by diffusion through the liposomal bilayer at or close to the site of injection. The exact mechanism of continuous drug liberation from liposomes at the injection site is not totally clear. Interesting information on the fate of egg PC liposomes was provided by Ohsawa et al. (113). They studied the fate of egg PC liposomes labeled with a small amount of [^{3}H]dipalmitoylphosphatidylcholine ([^{3}H]DPPC, marker of the lipid phase) containing [^{14}C]inulin (marker of the

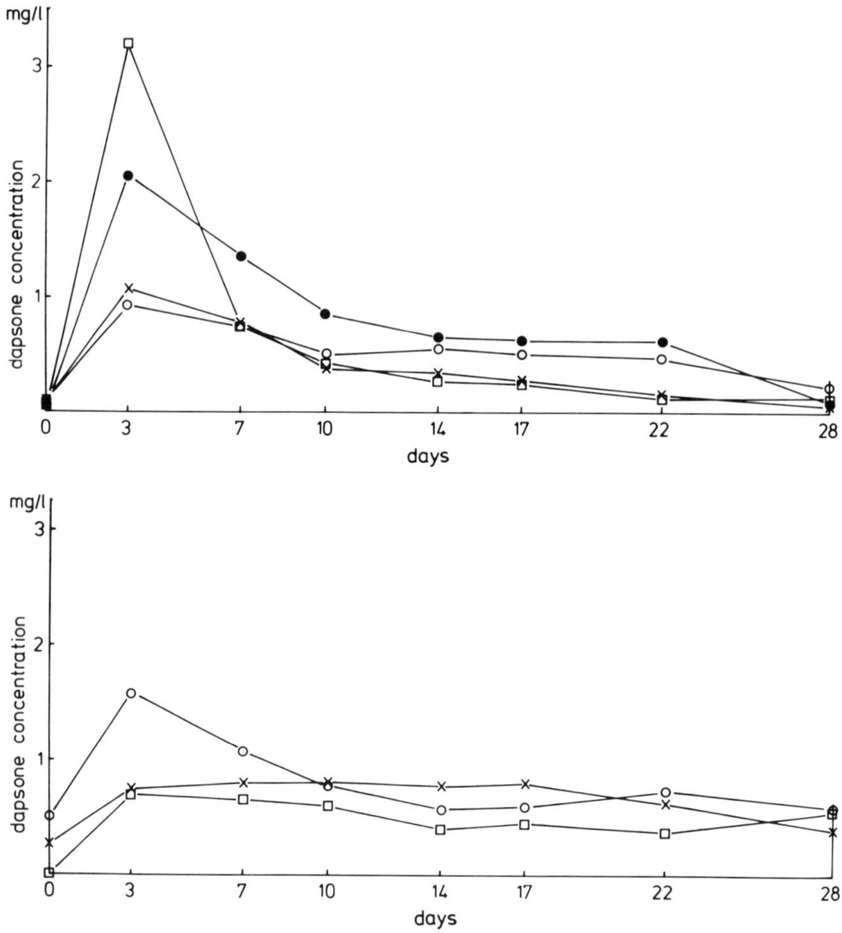

Figure 4 Some representative dapsone concentration–time curves in serum of a group of 47 males and 30 females given i.m. injections of 900 mg dapsone as an aqueous suspension. (From Ref. 91.)

aqueous phase) after i.m. injection in rats. It was shown that about half of the injected amount of [^{3}H]DPPC remained at the injection site. A considerable amount of [^{3}H]DPPC was detected in the regional lymph nodes suggesting a significant role of the lymphatic route in the absorption of liposomes or liposomal components. Accumulation of [^{14}C]inulin in the lymph nodes was also observed. The concentration ratios of [^{14}C]inulin to [^{3}H]DPPC in the muscle and regional lymph nodes increased with time, indicating that

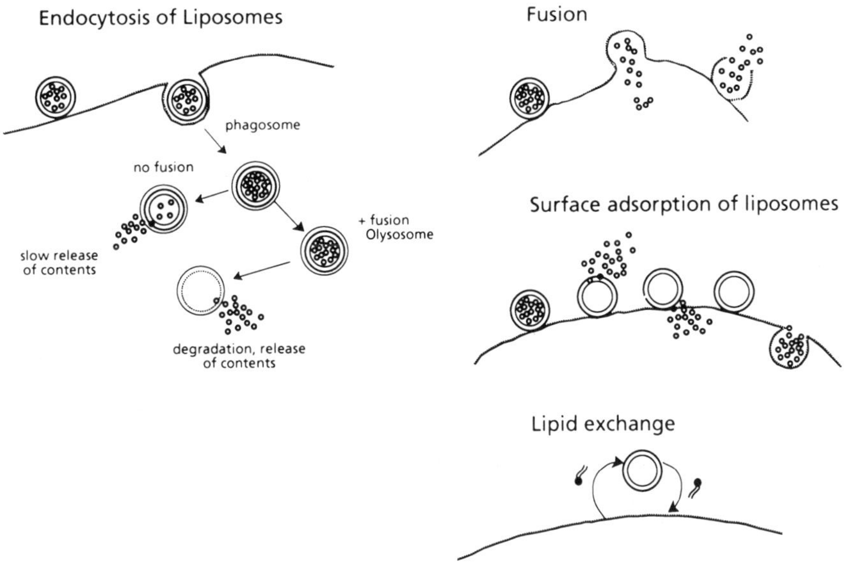

Figure 5 Proposed mechanisms of liposome–cell interaction. (Adapted from Ref. 96.)

the lipid was absorbed faster than [^{14}C]inulin. Based on their findings the authors presented a schematic illustration of drug absorption following i.m. injection of drug laden liposomes (Figure 6). The model shown aims to describe the fate of intramuscularly administered multilamellar egg PC liposomes. These liposomes (0.1–1 μm in size) were prepared by the freeze-thaw method and the film hydration method. Liposomes prepared with other types of lipids or by other methods of preparation will behave differently.

A number of authors attempted to correlate the in vivo (after i.m. and s.c. injection) and in vitro behavior of liposomes. Arrowsmith et al. (114) determined the release rates of cortisone hexadecanoate from liposomes in the presence of a rat skeletal homogenate. They showed that the presence of such biological material had very little effect on the stability of the haxadecanoate–DPPC liposome complex, which suggests that enzymes encountered at the injection site are unlikely to influence the release of entrapped drugs.

Earlier Arakawa et al. (115) performed a similar study. They found that the in vitro stability of cefazolin-sodium liposomes was unaffected by the presence of a rat muscle homogenate. Intramuscular injection of these cefazolin-sodium liposomes into rats led to a five-fold increase in the rate of efflux

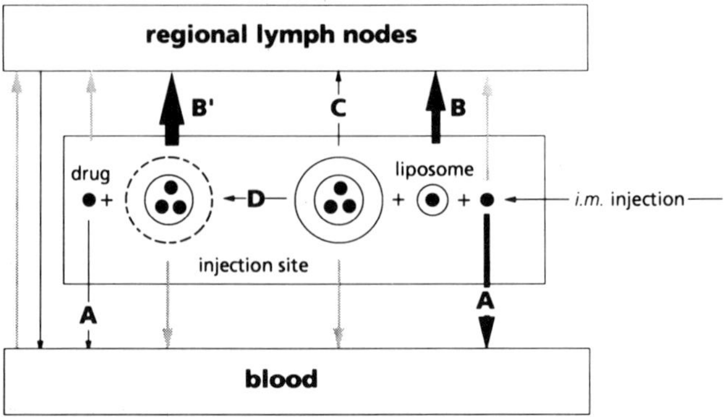

Figure 6 Schematic representation of the drug transfer following i.m. injection of liposomes. A, transfer of free drug; B, transfer of small liposomes included in the liposomal preparation from the beginning and those generated from larger liposomes by erosion at the injection site, respectively; C, transfer of larger liposomes; D, erosion of liposomes. (Adapted from Ref. 113.)

compared with in vitro rates. This deviation from the in vitro results illustrates the poor predictability of in vitro data for in vivo release kinetics. With the aid of a kinetic model, Hunt (116) was able to show that the effectiveness of a liposome carrier cannot be predicted simply based on its plasma stability. It could be concluded that in vitro conditions apparently do not adequately mimick the relevant in vivo conditions.

Liposomes for Sustained Drug Release After Intramuscular and Subcutaneous Administration

Liposomes have a number of characteristics that make them good candidates as drug carriers. For example, they are biodegradable, relatively nontoxic, and nonimmunogenic, depending on the type of lipids used. Both lipophilic as well as hydrophilic drugs can be encapsulated. Last but not least, liposome encapsulation has been shown to increase the therapeutic index of a number of drugs. This is probably accomplished by altered pharmacokinetics and tissue distribution of the applied drugs (117-121).

In experimental designs liposomes have been administered by several routes. These include, e.g., the intravenous, intra-arterial, intra-articular, topical (e.g., dermal and ophthalmic), oral, intrapulmonary, interstitial, intratesticular, intracerebral, intratumoral, intraperitoneal, i.m., and s.c. routes (96,100,122-127).

Liposomes that are introduced directly into the circulation are preferentially cleared by the cells of the mononuclear phagocyte system (MPS) located in the liver, spleen, bone marrow, lymph nodes, and circulating in the blood (12,16,128,129,133). This interaction with the cells of the MPS depends on factors such as size, charge, lipid composition, lipid fluidity, and dose. Because of the rapid clearance from the circulation the capacity of intravenously administered liposomes for sustained drug release while circulating in the blood compartment is very limited. The circulation time of liposomes can be prolonged by including in the bilayer certain lipids (e.g., polyethyleneglycolphosphatidylethanolamine) thereby aiming to change liposome behavior by modification of the liposome surface with the intent of reducing the interaction with MPS cells. These long-circulating liposomes are presently being extensively investigated. They may provide interesting opportunities to deliver "short-living" peptide-based therapeutic agents with a prolonged duration of bioactivity (130,131).

Administration of liposomes at extravascular sites such as i.m. and s.c. can reduce or even eliminate the problem of rapid interception by the MPS system. Information on the passage of liposomes through the capillaries prevalent in skeletal muscles and subcutaneous tissue is not available in the literature. However, because of the dimensions of the open fenestrae (about 40–60 nm in diameter) and the presence of a basal lamina surrounding the capillaries, the chance for liposomes to penetrate the wall of the capillaries in nonpathological situations is small (83,132,133).

The potential of liposomes as drug carrier systems for i.m. or s.c. injection has been reported repeatedly in literature. Most studies demonstrate the ability of liposomes to release the laden drug slowly over a prolonged period of time after i.m. or s.c. injection. Depending on the drug used, the aim of the therapy might be either decreased drug toxicity or sustained systemic therapy in the case of chronic treatment.

The ability of liposomes to release the encapsulated drug slowly after i.m. injection was demonstrated by Arrowsmith et al. (134) who used multilamellar liposomes (MLVs, size range 5.5–7.1 μm) as a vehicle to administer ^{3}H-labeled cortisone esters to rabbits. Liposomes were composed of DPPC and contained less than 1 mol % partially iodinated egg phosphatidylcholine. Plasma tritium level–time profiles were monitored to follow the absorption of the encapsulated drug and ^{131}I activity was measured at the injection site to study the fate of lipid after administration. Iodinated material was cleared from the i.m. injection site according to a zero-order process, with a half-life of 8.5 days. No detectable removal of intact MLVs from the injection site was observed and it was shown that encapsulation in liposomes significantly prolonged the absorption of cortisone hexadecanoate from the injection site.

Prolonged release after subcutaneous administration of liposomes was reported by Kim and Howell (135). These authors used a different type of

MLV (average size of 19.4 μm, SD 6.5 μm): liposomes were composed of dioleoylphosphatidylcholine, DPPG, cholesterol, and triolein in a molar ratio of 9.3:2.1:15:1.8. Cytarabine, a hydrophilic cytostatic drug, was incorporated. The half-life of release from the skin (s.c. injections to mice) was 10 min for nonencapsulated drug and 4 days for liposomal cytarabine.

The efficacy of liposome-entrapped drug can differ from that of free drug (111,136,137). A large number of studies have been performed to demonstrate an increase in therapeutic or prophylactic efficacy or decrease in drug toxicity as a result of liposome encapsulation. Some illustrative examples are discussed here.

Therapeutic Results

The treatment of radiofibrosis with i.m. liposomal superoxide dismutase (SOD) in humans was studied by Baillet et al. (138). Multilamellar liposomes with an average diameter of 300–400 nm were used. It was found that after twice-a-week 5-mg SOD administration for 3 weeks, the average size of the fibrosis was reduced by one third and significant softening occurred in 82% of the cases.

Increased efficacy upon administration of liposomal drug was also reported by Eppstein et al. (139). MLVs composed of DSPC and DPPG (molar ratio 9:1) containing human interferon-β (IFN-β) were administered intramuscularly to monkeys infected with simian varicella virus. Intramuscularly injected liposomal IFN-β did not only result in prolonged release from the injection site but also showed superior antiviral efficacy compared to an identical dosing regimen of aqueous interferon-β.

Another example of increased efficacy was provided by Rutenfranz et al. (140). Human interferon-γ (HuIFN-γ) was entrapped in liposomes and it was investigated whether this preparation prevented the rapid decay of IFN-γ in serum. Titers were determined by two test systems – a bioassay and an ELISA. Liposomes composed of dimyristoylphosphatidylcholine (DMPC), dicetylphosphate (DCP), and cholesterol in a molar ratio of 0.45:0.45:0.1 were given i.m. or i.v. to mice. The average particle size was 333 nm (SD 24 nm). Serum decay curves of liposomal and free IFN-γ were compared. Intramuscular injection of the liposomes resulted in longer lasting serum titers than i.v. injected liposomes. The titers of antiviral activity after i.m. injection of free and liposomal interferon are shown in Figure 7.

Both an increased therapeutic effect and decreased drug toxicity after administration of liposomally entrapped drug was demonstrated by Roy and Kim (141). They tested liposomes containing bleomycin (B) s.c. administration in mice bearing B16 melanomas. MLV (average diameter 19.1 μm) were made with dioleoylphosphatidylcholine, DPPG, cholesterol, and triolein (31:7:30:6 molar ratio). No size distribution was reported. The s.c. half-life

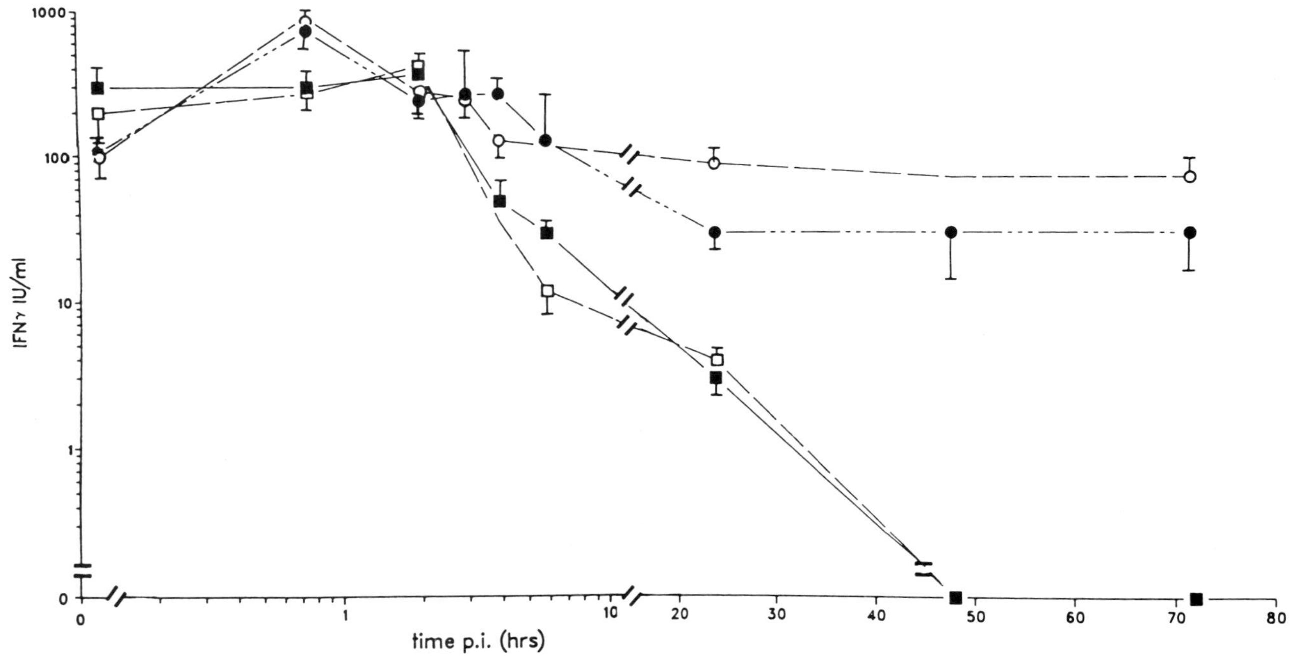

Figure 7 Black filled squares show the antiviral activity in serum after i.m. application of (1×10^5 IU) IFN-γ per mouse. The titers were also determined by ELISA (open squares). Serum titers could be monitored over prolonged periods by the i.m. application of IFN-γ liposomes with a lower dose (4.3×10^3 IU per mouse) presented by bioassay (black filled circles) and ELISA (open circles). (From Ref. 140.)

of liposomal B was 31.8 min whereas that for the nonencapsulated B was 8 min. In mice bearing B16 melanomas greater inhibitions in tumor growth and less toxicity were observed with liposomal B than with free B.

A protective effect of liposomal cephapirin when given prophylactically to mice infected with *Staphylococcus aureus* was reported by Swenson et al. (142). Vesicles were made by hydrating the Tris salt of cholesterol hemisuccinate (CHS) in the presence of sodium cephapirin. Compared with the s.c. injection of free drug this formulation provided prolonged serum levels of cephapirin. Treatment with free cephapirin at 4 hr before infection had no protective effect, whereas treatment with CHS liposomes protected the majority of mice.

Several variables are known to influence the rate and extent of release of drugs from liposomes following i.m. or s.c. injection. Through variation in size, number of bilayers, lipid composition, bilayer rigidity, charge and surface characteristics, injection volume, and amount of liposome administered, the efficacy of the encapsulated agents can be manipulated. These variables have been addressed in various studies.

Size

The influence of liposome size on clearance from the injection site was studied by Zierenberg and Betzing (143). Polyenylphosphatidylcholine (PPC) liposomes were administered i.m. to rats. The liposomal preparation used in these experiments contained 59% unilamellar vesicles with diameter of about 25 nm and 41% multilamellar particles with a diameter above 25 nm. Radiolabels were used to monitor the lipid fraction remaining at the injection site. The clearance profile from this site was biphasic. The initial first phase was attributed to the rapid removal of small vesicular lipids while the second slower removal phase was attributed to the clearance of the larger sized liposomes. It was hypothesized that larger liposomes remain longer at the injection site. Another explanation might have been the formation of liposomes aggregates at the injection site as suggested by Stevenson et al. (148).

Information on the relation between liposome size and lymphatic uptake was collected by Jackson (144). Liposomes were composed of phosphatidylcholine, cholesterol, phosphatidylserine, α-tocopherol, and [^{14}C]dipalmitoylphosphatidylcholine 4:5:1:0.01:0.01 (molar ratio). [^{14}C]Dipalmitoylphosphatidylcholine was used as a marker for the lipid phase and [^{3}H]inulin was used to monitor the aqueous phase. Two size ranges were taken into consideration: small multilamellar vesicles (SLMV) of 0.15–0.7 μm and large multilamellar vesicles (LMLV) of 0.3–2.0 μm. At 24 hr after administration with the large liposomes the highest fraction of drug remaining at the injection site was found. A constant ^{3}H/^{14}C ratio indicated stability of the remaining vesicles. With SMLV a larger fraction of the administered dose of insulin was present in the inguinal lymph node than with LMLV. As no inulin was found in the lymph node after administration of a free inulin solution (145), it was concluded that small liposomes were taken up via the lymphatics to a greater extent than the LMLV. Smaller size range liposomes of two different lipid compositions were used by Tümer et al. (146). They investigated the fate of liposomes of two lipid compositions and of different size: small (30–60 nm) and large (about 400 nm) liposomes composed of phosphatidylcholine (PC) or sphingomyelin (SM) and equimolar cholesterol. Liposomes containing carboxyfluorescein (CF) and ^{111}In-labeled bleomycin were injected in the footpad of rats. Monitoring latent CF concentrations in the plasma, it was concluded that small liposomes consisting of PC and SM entered the blood circulation in intact form. In the case of small liposomes hepatic uptake of ^{111}In for SM/Chol vesicles was lower than for PC/Chol vesicles. From radioactivity measurements in the (whole) footpad at 48 hr after injection it was derived that the large liposomes were retained in the footpad to a greater extent than the small liposomes.

Fluidity

Data showing that bilayer fluidity is an important variable in liposome behavior after i.m. injection were provided by Schreier et al. (147). ^{125}I- and ^{3}H-labeled gentamicin sulfate (GS) containing liposomes were administered i.m. to mice. Phospholipids from two different sources, either egg phosphatidylcholine (EPC) or soy phosphatidylcholine (SPC), were selected as major bilayer components because of known differences in their fatty acid composition, i.e., degree of unsaturation and chain length. The rate of release of encapsulated drug as well as the erosion of liposomes at the injection site were found to be a function of the fluidity of the lipid membranes. The release of GS from EPC liposomes was about 7 times slower than from SPC liposomes. The authors suggested that drug absorption rates may be controlled (within certain limits) by using lipids with different degrees of bilayer "fluidity."

Charge

The role of liposome charge was investigated by Stevenson et al. (148). Insulin was entrapped in either neutral liposomes composed of phosphatidylcholine and cholesterol (10:2) or negatively charged liposomes (phosphatidylcholine/cholesterol/dicetyl phosphate in a ratio of 10:2:1 by weight). These liposomes were injected subcutaneously in dogs that were made to be diabetic. The average liposome diameter was 98 nm (range 30–270 nm). Compared to "free" insulin given admixed without empty liposomes, a threefold prolonged hypoglycemic effect in diabetic dogs resulted from the s.c. administration of negatively charged liposome encapsulated insulin. When neutral liposomes were administered a much weaker biological response was obtained. In experiments in rats with liposomally entrapped [^{125}I] insulin or [^{14}C]inulin, the depot at the s.c. injection site was excised at various time intervals after administration. For the first hours the injectate appeared diffuse, but thereafter the liposomes aggregated to form a distinct pellet. Plasma was found to contain ^{125}I activity associated with intact liposomes. The presence of intact liposomes in plasma could be explained by absorption of small liposomes in the general circulation, whereas large liposomes remained at the depot injection site. Since the liposome size distribution was not accurately defined in this study, it was not possible to obtain information concerning the extent to which this process contributed to total absorption.

Route of Administration and Injected Dose

Titulaer et al. (149) were the first to mention the influence of the administration route (i.m. or s.c.) on the release pattern of a liposomal depot. Free (0.6 mg) and liposome-encapsulated (0.6 and 3.0 mg) chloroquine were administered i.m. and s.c. to mice. Chloroquine concentrations in blood as well as organs were determined. After i.m. and s.c. injection the chloroquine liposomes provided a slower release from the s.c. layer than from muscular tissue. In addition, increasing the liposomal dose resulted in an additional retardation of the release. This phenomenon was investigated in more detail by Kadir et al. (150). The kinetics and antimalarial activity of increasing doses of liposomal chloroquine (0.6, 1.3, 2.0, 2.6, and 3.3 mg) were investigated in the same animal model (*Plasmodium berghei* mouse model) as the one used by Titulaer et al. Detectable chloroquine concentrations were present for 8 days in case of the lowest dose (0.6 mg) and 15 days in case of the highest dose (3.3 mg). Changes in the areas under the curve ($AUC_{0-8\ days}$ and $AUC_{0-15\ days}$), mean residence times (MRT), and absorption rate constants (k_a) were dose-related. In the same study the prophylactic efficacy was tested. Parasitemia was determined after infection of the mice with *P. berghei* parasites. Chloroquine prophylactic efficacy increased with its concentration in blood at the time of infection (Table 2). In the case of the injection of 3.3 mg liposomal chloroquine enough chloroquine had been released to protect the mice at day 21 after injection.

Injection Volume

The influence of the injection volume on the sustained release properties of liposome-encapsulated chloroquine was recently investigated (151). In this study mice were given 0.6 mg liposomal chloroquine dispersed in 20, 40, 60, 80, and 100 μl either i.m. or s.c. The drug profiles again showed that these liposomes were able to provide prolonged release (Figure 8). In most cases chloroquine concentrations in blood were above the detection limit up to 8 days after i.m. injection. The sustained-release properties were more pronounced when the chloroquine liposomes were injected into smaller injection

Table 2 Number of Mice Not Developing a Parasitemia 7 Days After Injection and the Chloroquine Concentrations in Blood at the Time of Infection for Increasing Intramuscular Doses of Liposomal Chloroquine (Lip-CQ).

Dose lip-CQ (mg)	Mean (± sd) concentration (μg/L)	Number of noninfected mice/initial number of mice
0.64	85 ± 12	0/6
1.32	110 ± 11	2/6
1.98	120 ± 42	6/6
2.63	183 ± 42	6/6
3.30	237 ± 48	6/6

Note: Gel state liposomes (0.29 ± 0.01 μm) composed of distearoylphosphatidylcholine (DSPC) : dipalmitoylphosphatidylcholine (DPPC) : cholesterol (Chol), 10:1:10 molar ratio were used. Mice were infected with *Plasmodium berghei* (strain K173) by i.p. administration of 10^5 infected red blood cells (for further details of the experimental model see ref. (150), (151)).

volumes. In the case of s.c. injection of 80 and 100 μl, chloroquine could be detected for a period of only 4 days.

Lymphatic Uptake of Intact Liposomes

Muscles are well supplied by capillary blood vessels and few lymph vessels exist in muscle tissue. Lymphatic capillaries are usually located in the fascial planes around the muscle bundles. Subcutaneous tissue, on the other hand, is known to be well supplied with both capillary blood vessels and lymphatic vessels (83,152,153).

The lymphatic network forms a closed system separated by an endothelial layer from the tissue spaces. The wall of lymphatic capillaries consists of a thin endothelium lacking a basement membrane around its wall. The intercellular junctions between endothelial cells play an important role in uptake and retention of all kinds of materials (e.g., bacteria, lymphocytes, etc.) into lymphatic vessels.

After being taken up in the lymph vessels intact, liposomes either enter the blood circulation through the ductus thoracicus or they may become entrapped in the lymph nodes and release the drug from these sites. These findings indicate that liposomes can serve as systems for site-directed and controlled delivery of drugs to, say, diseased lymph nodes. Pathological conditions such as inflammation of the interstitial tissue or stages of development of tumours also affect the localization of liposomes in the lymph nodes after i.m. or s.c. injection (144, 154-157). Uptake and localization of liposomes into the lymphatics strongly depends on factors such as size, lipid composition, anatomical site of injection, and physical condition. Small-diameter

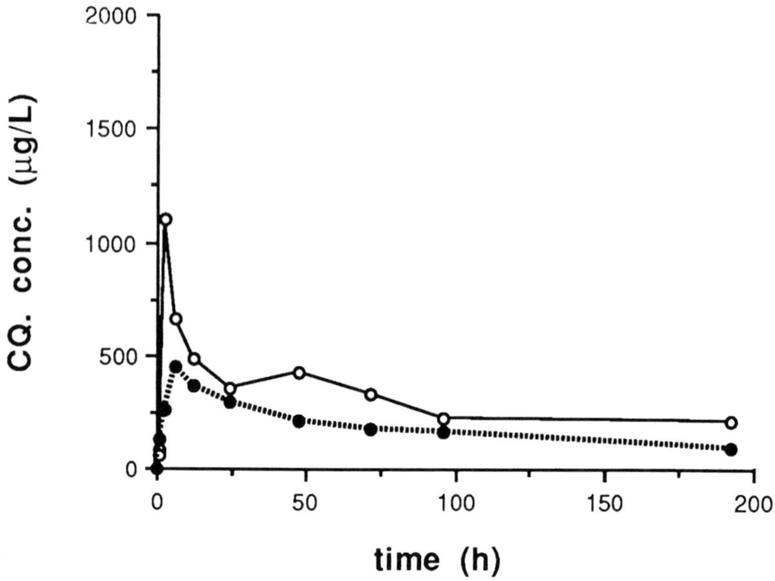

Figure 8 Mean chloroquine (CQ) concentration in blood as a function of time after i.m. or s.c. administration of 0.6 mg liposome-encapsulated chloroquine in (a) 20 μl and (b) 100 μl. Gel state liposomes (0.30 ± 0.07 μm) were composed of distearoylphosphatidylcholine (DSPC): dipalmitoylphosphatidylcholine (DPPC): cholesterol (Chol), 10:1:10 molar ratio. Each time point represents pooled data from $n = 3$ to $n = 6$ mice. (From Ref. 151.)

particles are mobilized from the site of injection more quickly and to a higher extent than large particles. Liposomes smaller than 100 nm are easily taken up by the lymphatics while those larger than 100 nm probably tend to stay at the injection site (148,156,158-160). The properties of the drug involved also play a role. Patel et al. (171) performed experiments with ^{99m}Tc- labeled liposomes containing [^{14}C]cholesterol with entrapped [^{125}I]polyvinylpyrrolidone (MW 30,000–40,000). The liposomes were injected interstitially in rats and homogenates of the regional lymph nodes were analyzed. The results suggested that 60–70% of the liposomes localized in the lymph were intact, but they lost most of the membrane marker (^{99m}Tc) to some components of the lymph. Injection of low molecular weight liposomal substances ([^{3}H]methotrexate, MW 454) in a similar study showed that during their passage through the lymphatics the liposomal structure was not totally destroyed but became leaky, in particular to low molecular weight substances.

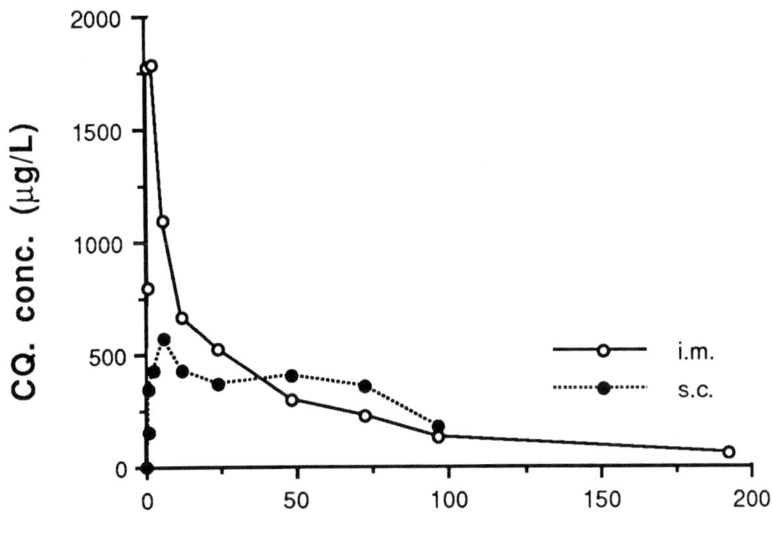

(b)

Studies on the uptake and localization of liposomes into the lymphatics must be interpreted with great prudence because the outcome may vary with the choice of the liposomal marker used. For example, Osborne et al. (156,169) found that ^{99m}Tc-labeled positively charged liposomes were localized in primary lymph nodes to a higher extent than ^{99m}Tc-labeled negative liposomes. In another study, Stevenson and others (148,157) found that using an aqueous liposomal marker (^{125}I-labeled polyvinylpyrrolidone) negative liposomes were localized in the lymph nodes to a greater extent than positive liposomes. Lipid markers may dissociate and bind or exchange with components of the lymph or plasma membrane, while aqueous markers may leak out of the liposomes either at the site of injection or during their passage through the lymphatics. Thus the results may lead to false interpretation of the raw experimental data if these possibilities are not recognized in data analysis and evaluation. Results that can be interpreted in a more straightforward way can be obtained by the use of double-labeled liposomes. With both an aqueous phase label and a lipid marker the ratio of these labels can be monitored and hence the in vivo degradation of liposomes.

In conclusion, lymphatic uptake of liposomes injected subcutaneously or intramuscularly offers opportunities for delivery of materials to the blood capillaries and the draining lymph nodes. Potential therapeutic applications for lymph node accumulation include delivery of anticancer drugs for treatment of lymph node metastases and delivery of radioimaging agents as a diagnostic tool in monitoring lymph node architecture (161-164).

Discussion

The route of administration has a great impact on the tissue distribution of liposome-encapsulated agents. Intramuscularly or subcutaneously administered liposomes show a different tissue distribution pattern from that of i.v. administered liposomes. As shown above, the ability of liposomes to deliver encapsulated drugs in a sustained way upon i.m. and s.c. administration has drawn considerable attention.

Liposome size is an important variable in determining the absorption characteristics, disposition, and ultimate fate of liposomes in the body. Therefore, liposome size should be clearly defined. Unfortunately, many studies pay little or no attention to good characterization of liposome size. The influence of pH and osmolarity of the internal aqueous phase on the stability of the encapsulated drug and the lipid constituents are rarely investigated. Most investigators assume that these kind of interactions are minimal, but little or no evidence is available to confirm this.

The choice of test drugs and markers in studies concerning liposomal fate in vivo is of great importance. Conclusions concerning the fraction of the original liposome dose retained at a particular tissue will vary depending on whether a membrane-associated marker entrapped in the aqueous phase is being used. The membrane-associated markers will provide information on retention of liposomal lipids, but they do not answer the question of whether the lipids are in the form of intact liposomes retaining material entrapped in the internal aqueous compartments. The presence of non-bilayer-interacting entrapped label (e.g., carboxyfluorescein) retained in tissue gives a better indication about liposomal integrity. Data from studies with two markers (i.e., both a lipid marker and an aqueous phase marker) provide the most detailed view on the fate of liposomes. If the ratio between the concentration of the labels is constant, liposomes are probably intact. But even then one should be aware of the possibility of liposomal lipids to exchange with lipids of the surrounding cells and simultaneous leakage of the aqueous phase marker.

When studying the therapeutic effects of liposome-encapsulated agents, it may be important to measure biological effects of "empty" liposomes of identical composition and prepared according to the same protocol as the drug-containing liposomes. The use of these "control" liposomes is important as liposomes are known to possess immunoadjuvant activity. Impurities from the lipids or excipients or impurities introduced via the equipment used can in principle be responsible for triggering immunological responses. For extrapolation of the findings to application in the clinic one should take into account the disease state of the test animal or human volunteer. Data may not be directly applicable because of changed physiological conditions in the pathological state. For example, pores in capillary and lymphatic endothelium are dynamic structures that enlarge dramatically at sites of inflammation or in cases of increased motion. Thus, the permeability of blood vessels increases significantly during inflammation (38,165).

One aspect of liposomes that has gained little attention so far in the literature is their tolerability. Most of the standard lipids (PC, PG) are generally considered to be nontoxic because they are composed of natural body constituents that are biodegradable. However, in particular for the intramuscular and subcutaneous route of injection, few in-depth studies are available concerning the toxic side effects. Most investigators focus on kinetic and therapeutic improvements rather than on fundamental studies concerning liposome safety. Available data on adverse effects associated with in vivo use of liposomes was recently given by Storm et al. (76). The authors reviewed results obtained with empty liposomes (no entrapped drug). Toxic side effects can result from liposomes themselves, from the drug they contain or from synergistic action of the carrier and its content. Especially when drugs that are known to have irritating properties are entrapped in liposomes, continuous release of the active agent might result in increased irritation at the injection site. In order to adequately evaluate the usefulness of liposomes as a drug carrier system, adverse side effects need to be carefully studied as well.

The lipid composition of the liposome bilayer can vary widely resulting in a wide spectrum of physical properties. More rigid structures show a low permeability through the bilayer for the entrapped agent. The regularly used bilayer-forming lipids used in in vivo studies are egg and synthetic PC or semisynthetic derivatives such as DSPC, DPPC, or DMPC plus cholesterol. These lipids have been studied extensively. Their behavior is relatively well understood. Nevertheless, some more exotic constituents such as gangliosides, diacetylenic phospholipids, methacrylate group containing derivatives of PC, carbamyl PC derivative, etc., have been reported as well. These "foreign" lipids might not be readily biodegradable and therefore show toxic effects. When used, biodegradability and safety aspects (including immunogenicity) should be considered carefully when interpreting the data and deciding on future studies.

It has been well established that there is a considerable variability in the extent of entrapment in liposomes. While lipid- soluble compounds can show high encapsulation rates, entrapment of water-soluble drugs, limited to the fraction in the internal aqueous phase, is rather low. This does not apply if "remote loading" techniques can be used (166). Low encapsulation efficiencies may limit the total amount of drug to be administered and possibly the duration of action. Besides, they may pose technological and economical problems. Therefore, other carrier systems are presumably superior to liposomes in cases where the drug is poorly entrapped and high doses of drug are required. Alternative carrier systems should be preferred in these cases.

An important feature of liposomes that should be stressed is the presence of water inside the liposomes and the fact that entrapment within liposomes does not require the formation of specific (covalent) bonds. The presence of water makes them excellent delivery systems for (glyco)proteins, peptides, or other endogenous compounds that are sensitive to damage induced when water is absent in the system, as is the case with most other particulate carrier systems. Under those circumstances liposomes may be preferred to other long-acting sustained-release formulations such as the poly(lactide-*co*-glycolide) polymer matrices.

Another potential field of application of liposomes which has not been mentioned thus far is their use as immunological adjuvants. After i.m. or s.c. injection liposomes can stimulate the induction of humoral and cell-mediated immunity to antigens. Recently, Gregoriadis (167) reviewed the potentials of liposomes as immunological adjuvants. A large collection of data from immunization experiments was summarized and it was concluded that liposomes have a modulating effect on the immune response. Liposome adjuvanticity depends on several of the system's characteristics such as lipid composition, vesicle size, surface charge, fluidity of the bilayer, type of coadjuvants, and lipid to antigen mass ratio. These properties determine the fate in the biological environment and thus the mode of adjuvanticity. To conclude, taking into account the relatively low toxicity of liposomes they can be considered as promising candidates in the area of vaccinology especially with regard to the development of future subunit vaccines that are known to be weakly immunogenic (168).

CONCLUDING REMARKS

Among other particulate systems, liposomes have been investigated widely as potential drug carriers. Sufficient experimental observations on liposome behavior in vivo have now become available to draw some general conclusions regarding the potential utility of liposomes as drug delivery systems. Liposomes are able to function as a depot for sustained drug release. When injected s.c. or i.m., liposomes are cleared slowly from the injection site. With large liposomes, most of the encapsulated drug is absorbed in the form after being released from the liposomal depot. The continuous and slow release from the i.m. or s.c. injection sites can generate therapeutic levels of drugs over significantly longer periods of time than free drugs. Smaller sized liposomes can be absorbed intact into the lymphatics and eventually enter the bloodstream. Lymphatic uptake of small liposomes injected via the i.m. or s.c. route offers opportunities for delivery of drug to the draining lymph node as well. A promising application of this last feature is the delivery of radioimaging agents as a diagnostic tool in monitoring lymph node architecture and treatment of metastatic tumor cells infiltrated in lymph nodes (169,170-172).

Release of the drug after i.m. or s.c. injection is a complex process. A scheme describing the different release pathways and liposomal fate after i.m. injection has been postulated. The exact mechanism is not completely clarified yet. From the reviewed studies it might be concluded that the scheme also applies to the s.c. route of administration.

For a number of liposome-encapsulated drugs the therapeutic advantages of liposome encapsulation have been proven. However, one should be very careful when comparing data from different studies. Variables such as the nature of the lipids and their fluidity, the method of preparation, the number of lipid bilayers per particle, particle size, dose, potential aggregation of particles prior to injection or in situ, and physiological activity in different experimental animals or experimental conditions vary widely. These parameters are known to influence liposome stability in vitro and in vivo and therefore provide the means to manipulate within certain limits the pharmacokinetic profile of encapsulated drugs. Unfortunately, in most studies the liposomal systems under investigation are not characterized adequately. Nevertheless, on the basis of current experience with animals and patients, therapies with liposome-entrapped drugs appear to be promising enough to warrant further efforts.

Studies concerning the i.m. and s.c. administration of drugs with liposomes have been concentrated on specific therapeutic areas. The attention is mainly focused on anticancer and antimicrobial therapies and immunological applications. Most antitumor drugs have a narrow therapeutic window and the use of liposomes may help to improve therapeutic utilization of existing anticancer drugs. Research on liposome-entrapped antitumor agents is still proceeding and clinical investigations are currently underway. Promising results have already been attained (173,174). A number of investigators have demonstrated the increased effectiveness of several different antibiotics (gentamicin, streptomycin, penicillin, cefradin, ampicillin, rifampin, etc.) when the drugs were given in liposomal form. In almost all studies incorporation of the drug in liposomes increased the therapeutic index. These studies have been reviewed extensively (117,175,176). Liposomes are also effective in potentiating the immune response of a variety of antigens. The major advantage of liposomes as an antigen carrier and an immunopotentiating agent is their biodegradability and presumably nontoxic features. Consequently, presently liposome based vaccine preparations are being developed.

Although liposomes consisting of regularly used (phospho)lipids such as PC, PG, and cholesterol appear to be relatively nontoxic when administered as single injections, more information is needed concerning their safety when administered repeatedly at frequent intervals over several weeks or months. Several other issues remain to be addressed before liposomes can be widely used clinically. Among these is the need to further develop cost-effective methods for large-scale preparation of homogeneous formulations that are stable for relatively long periods of time (177).

In defining the prospects for the use of liposomes as i.m. or s.c. administered drug carriers, both pharmaceutical (production of large and reproducible batches, appropriate methods for characterization, sterilization of the liposome formulations, stability, etc.) and safety (toxicity of constituents) issues must be considered. Fortunately, good progress has been made lately in the large-scale production of pharmaceutical liposomal formulations and the number studies with i.m. and s.c. administered liposomes is growing.

REFERENCES

1. J. Zuidema, F. A. J. M. Pieters, and G. S. M. J. E. Duchateau, *Int. J. Pharm.*, *47*, 1 (1988).
2. J. R. Robinson, *Sustained and Controlled Release Drug Delivery Systems*, Marcel Dekker, New York, 1978.
3. C. H. Kleinbloesem, *Clin. Pharmacol. Therapeut.*, *35*, 742 (1984).
4. R. Alex and R. Bodmeier, *J. Microencapsulation*, 7, 347 (1990).
5. J. Kreuter, *J. Microencapsulation*, *5*, 115 (1988).
6. N. A. Armstrong and K. C. James, *Int. J. Pharm.*, *6*, 185 (1980).
7. P. Catellani, G. Vaona, P. Plazzi, and P. Colombo, *Acta Pharm. Technol.*, *34*, 38 (1988).
8. R. Baker, *Controlled Release of Biologically Active Agents*, John Wiley and Sons, New York, 1987.
9. R. L. Juliano, *Drug Delivery Systems*, Oxford University Press, New York, 1980.
10. T. Kissel, *Acta Pharm. Technol.*, *29*, 221 (1983).
11. M. J. Poznansky and R. L. Juliano, Pharmacol. Rev., 36, 277 (1984).
12. E. Tomlinson, *Int. J. Pharm. Technol. Prod. Mfr.*, *4*, 49 (1983).
13. M. Bornschein, P. Melegari, C. Bismarck, and S. Keipert, *Pharmazie*, *44*, 585 (1989).
14. R. H. Müller, *Colloidal Carriers for Controlled Drug Delivery and Targeting*, CRC Press, Stuttgart, 1991, p. 145.
15. G. S. Banker & C. T. Rhodes, *Modern Pharmaceutics*, Marcel Dekker, New York, 1979.
16. E. Tomlinson, J. J. Burger, J. G. McVie, and K. Hoefnagel, in *Recent Advances in Drug Delivery Systems* (J. M. Anderson and S. W. Kim, eds.), Plenum Press, New York, 1983.
17. D. A. Wood, *Int. J. Pharm.*, 7, 1 (1980).

18. M. M. P. Paulssen and B. C. Van Pelt, *Lancet*, 1310 (1981).
19. M. K. Bijsterbosch and T. J. C. van Berkel, *Adv. Drug Deliv. Rev.*, *5*, 231 (1990).
20. A. L. Weiner, S. S. Carpenter-Green, E. C. Soehngen, R. P. Lenk, and M. C. Popescu, *J. Pharm. Sci.*, *74*, 922 (1985).
21. R. E Counsell and R. C Pohland, *J. Med. Chem.*, *25*, 1115 (1982).
22. R. S. Munford, J. M. Andersen, and J. M. Dietschy, *J. Clin. Invest.*, *68*, 1503 (1981).
23. C. De Duve, T. De Barsy, B. Poole, A Trouet, P. Tulkens, and F. Van Hoof, *Biochem. Pharmacol.*, *23*, 2495 (1974).
24. A. T. Forence, *Pharm. J.*, *229*, 291 (1982).
25. D. Gal, P. C. MacDonald, J. C. Porter, and E. R. Simpson, *Int. J. Cancer*, *28*, 325 (1981).
26. M. Tsakala, J. Gillard, M. Roland, F. Chabot, and M. Vert, *J. Controlled Release*, *5*, 233 (1988).
27. D. H. Lewis, in *Biodegradable Polymers as Drug Delivery Systems* (M. Chasin and R. Langer, eds.), Marcel Dekker New York, 1990, p. 1.
28. M. J. D. Eenink, J. Feijen, J. Olijslager, J. H. M. Albers, J. C. Rieke, and P. J. Greidanus, *J. Controlled Release*, *6*, 225 (1987).
29. C. Jones, M. A. Burton, and B. N. Gray, *J. Pharm. Pharmacol.*, *41*, 813 (1989).
30. T. K. Lee, T. D. Sokoloski, and G. P. Royer, *Science*, *213*, 233 (1981).
31. W. E. Longo, H. Iwata, T. A. Lindheimer, and E. P. Goldberg, *J. Pharm. Sci.*, *71*, 1323 (1982).
32. S. S. Davis, S. N. Mills, and E. Tomlinson, *J. Controlled Release*, *4*, 293 (1987).
33. D. L. Wise, J. D. Gresser, and G. J. McCormick, J. Pharm. Pharmacol., 31, 201 (1979).
34. L. M. Sanders, G. McRae, K. Vitale, R. Burns, P. Hoffman, and E. Shek, *J. Pharm. Sci.*, *78*, 888 (1989).
35. Y. Ogawa, M. Yamamoto, H. Okada, T. Yashiki, and T. Shihamoto, *Chem. Pharm. Bull.*, *36*, 1103 (1988).
36. M. Szekerke, L. Wade, and M. E. Whisson, *Neoplasma*, *20*, 163 (1973).
37. M. Tsakala, J. Gillet, and J. Gillard, *J. Pharm. Belg.*, *45*, 5 (1990).
38. C. G. Pitt, in *Biodegradable Polymers as Drug Delivery Systems* (M. Chasin and R. Langer, eds.), Marcel Dekker, New York, 1990, p. 71.
39. Y. Shoji, Y. Mizushima, A. Yanagawa, and T. Yonaha, *Drugs Exp. Clin. Res.*, *9*, 601 (1985).
40. M. Y. Wu, J. Johansson, N. H. Altman, D. H. Haynes, and V. Fiserova Bergerova, *J. Controlled Release*, *9*, 1 (1989).
41. H. Onishi and T. Nagai, *Chem. Pharm. Bull.*, *34*, 2561 (1986).
42. M. Hashida, T. Yoshioka, S. Muranishi, and H. Sezaki, *Chem. Pharm. Bull.*, *28*, 1009 (1980).
43. H. Sezaki, M. Hashida, and S. Muranishi, in *Optimization of Drug Delivery* (H. Bundgaard, A. B. Hansen, and H. Kofod, eds.), Munksgard, Copenhagen, 1982, p. 316.
44. T. K. Lee, T. D. Sokoloski, and G. P. Royer, *Science*, *213*, 233 (1981).

45. Y. Tsukoda, Y. Kato, N. Umemoto, Y. Takeda, T. Hara, and H. Hirai, *J. Natl. Cancer Inst.*, *73*, 721 (1984).
46. H. J. P. Ryser and W. C. Shen, *Proc. Natl. Acad. Sci. USA*, *75*, 3867 (1978).
47. L. J. Arnold, A. Dagan, and N. O. Kaplan, *Polym. Biol. Med.*, *2*, 89 (1983).
48. A. Manzzoni, R. A. Gambetta, F. Trave, and F. Zunino, *Cancer Drug Deliv.*, *3*, 163 (1986).
49. R. F. Tuma, J. O. Forsberg, and B. Agerup, *Cancer*, *50*, 1 (1982).
50. P. Couvreur, B. Kante, L. Grislain, M. Roland, and P. Speiser, *J. Pharm., Sci.*, *71*, 790 (1982).
51. M. El-Samaligy and P. Rohdewald, *Pharm. Acta Helv.*, *57*, 201 (1982).
52. J. Kreuter, *Pharm. Acta Helv.*, *58*, 217 (1983).
53. P. G. Waser, U. Müller, J. Kreuter, S. Berger, K. Munz, E. Kaiser, and B. Pfuger, *Int. J. Pharm.*, *39*, 213 (1987).
54. J. W. McGinity, A. Martin, G. W. Cuff, and A. B. Combs, *J. Pharm. Sci.*, *70*, 372 (1979).
55. Y. F. Leung, G. M. O'Shea, M. F. A. Goosen, and A. M. Sun, *Artif. Organs*, *7*, 208 (1983).
56. W. T. K. Stevenson and m. V. Sefton, *Biomaterials*, *8*, 449 (1987).
57. R. Bodmeier and O. Paeratakul, *J. Pharm. Sci.*, *78*, 964 (1989).
58. R. Duncan, I. C. Hume, P. Kopeckova, K. Ulbrich, J. Strohalm, and J. Kopecek, *J. Controlled Release*, *10*, 51 (1989).
59. P. Edman and I. Sjöholm, *J. Pharm. Sci.*, *71*, 576 (1982).
60. N. Ogata, K. Sanui, H. Nakamura, and H. Kishi, *J. Polym. Sci. Polym. Chem.*, *18*, 933 (1980).
61. V. W. Yeung and J. R. Nixon, *J. Microencapsulation*, *5*, 331 (1988).
62. Y. Machida, T. Nagai, M. Abe, and T. Sannan, *Drug Design Deliv.*, *1*, 119 (1986).
63. D. Knorr and M. Daly, *Process Biochem.*, 48 (April 1988).
64. M. Mizushima, *J. Pharm. Pharmacol.*, *35*, 398 (1983).
65. M. R. Mauk, R. C. Gamble, and J. D. Baldeschwieler, *Science*, *207*, 309 (1980).
66. L. C. Collins-Gold, R. T. Lyons, and L. C. Bartholow, *Adv. Drug Deliv. Rev.*, *5*, 189 (1990).
67. O. von Dardel, C. Mebius, T. Mossberg, and B. Svensson, *Br. J. Anaesth.*, *55*, 41 (1983).
68. H. Steffen, *Roche Magazin*, *20*, 1 (1984).
69. C. Tanford, *The Hydrophobic Effect: Formations of Micelles and Biological Membranes*, John Wiley and Sons, New York, 1980.
70. M. Ostro, *Liposomes: From Biophysics to Therapeutics*, Marcel Dekker, New York, 1987.
71. F. Szoka and D. Papahadjopoulos, *Ann. Rev. Biophys. Bioeng.*, *9*, 467 (1980).
72. G. Gregoriadis, *Liposome Technology*, Vols. 1–3, CRC Press, Boca Raton, 1984.
73. G. Gregoriadis, *Liposomes as Drug Carriers: Recent Trends and Progress*, John Wiley and Sons, New York, 1988.
74. D. Lichtenberg and Y. Barenholz, in *Methods of Biological Analysis* (D. Glick, ed.) John Wiley and Sons, New York, 1988, p. 337.

75. U. K. Nässander, G. Storm, P. A. M. Peeters, and D. J. A. Crommelin, in *Biodegradable Polymers as Drug Delivery Systems* (M. Chasin and R. Langer, eds.), Marcel Dekker, New York, 1990, p. 261.
76. G. Storm, C. Oussoren, and P. A. M. Peeters, in *Membrane Lipid Oxidation* (C. Vigo-Pelfrey, ed.), CRC Press, 1991, p. 239.
77. I. D. Steger and R. J. Desnick, *Biochim. Biophys. Acta*, *464*, 530 (1977).
78. F. C. Szoka and D. Papahadjopoulos, *Ann. Rev. Biophys. Bioeng.*, *9*, 467 (1980).
79. U. K. Nässander, G. Storm, P. A. M. Peeters, and D. J. A. Crommelin, in *Biodegradable Polymers as Drug Delivery Systems* (M. Chasin and R. Langer, eds.), Marcel Dekker, New York, 1990, pp. 261–338.
80. D. I. Abramson, H. V. Sparks, L. P. Thompson, and D. R. Bell, in *Blood Vessels and Lymphatics in Organ Systems* (D. I. Abramson and P. B. Dobrin, eds.), Academic Press, Orlando, 1984, p. 640.
81. F. L. S. Tse and P. G. Welling, *J. Parent. Drug Assoc.*, *34*, 409 (1980).
82. L. Illum, I. M. Hunneyball, and S. S. Davis, *Int. J. Pharm.*, *29*, 53 (1986).
83. M. Simionescu and N. Simionescu, in *Handbook of Physiology*, Vol 4 (E. M. Renkin and C. C. Michel, eds.), American Physiological Society, Bethesda, Maryland, 1984, p. 467.
84. Newton and Newton, *J. Am. Pharm. Assoc.*, *17*, 685 (1977).
85. Y. W. Chien, *J. Parent. Sci., Techn.*, *35*, 106 (1981).
86. S. W. Jacob and C. A. Francone, *Structure and Function in Man*, 2nd ed. W. B. Saunders, Philadelphia, 1970.
87. S. Feldman, *Bull. Parent. Drug Assoc.*, *28*, 53 (1974).
88. F. Kadir, J. Zuidema, A. Pijpers, A. Vulto, and J. H. M. Verheyden, *Int. J. Pharm.*, *64*, 171 (1990).
89. F. Kadir, C. B. J., Seijsener, and J. Zuidema, *Int. J. Pharm.* (submitted for publication).
90. J. W. Dundee, J. A. S. Gamble, and R. A. E. Assaf, *Lancet*, *2*, 1461 (1974).
91. E. S. M. Modderman, F. W. H. M. Merkus, J. Zuidema, H. W. Hilbers, and T. Warndorff, *Int. J. Leprosy*, *51*, 359 (1985).
92. J. H. Senior, *Crit. Rev. Ther. Drug Carrier Syst.*, *3*, 123 (1986).
93. K. G. Hwang, in: *Liposomes: From Biophysics to Therapeutics* (M. J. Ostro, ed.), Marcel Dekker, New York, 1987, p. 109.
94. J. Hernandez, J. Esterlich, M. T. Montero, and O. Valls, *Int. J. Pharm.*, *57*, 211 (1989).
95. A. L. Schroit, J. Madsen, and R. Nayar, *Chem. Phys. Lipids*, *40*, 373 (1986).
96. R. E. Pagano and J. N. Weinstein, *Ann. Rev. Biophys. Bioeng.*, 7, 435 (1978).
97. L. B. Magrolis, A. V. Victorov, and L. D. Bergelson, *Biochim. Biophys. Acta*, *720*, 259 (1982).
98. G. Scherphof, R. Roerdink, M. Waite, and Parks, *Biochim. Biophys. Acta*, *542*, 296 (1978).
99. L. S. S. Guo, R. L. Hamilton, J. Goerke, J. N. Weinstein, and R. J. J. Hovel, *Lipid Res.*, *21*, 993 (1980).
100. V. V. J. Ranade, *Clin. Pharmacol.*, *29*, 685 (1989).

101. R. L. Juliano, in *Liposomes* (M. J. Ostro, ed.), Marcel Dekker, New York, 1983, p. 53.
102. L. B. Margolis, B. A. Baibakov, A. A. Neyfakh, A. V. Victorov, S. I. Gelkima, and L. D. Bergelson, *Biochim. Biophys. Acta*, *804*, 23 (1984).
103. S. L. Cook, S. R. Bouma, and W. R. Huestis, *Biochemistry*, *19*, 4601 (1980).
104. R. E. Pagano, D. K. Struck and A. J. Schroit, in *Liposomes and Immunobiology* (B. Tom and H. Six, eds.), Elsevier, Amsterdam, 1980, p. 193.
105. R. M. Straubinger, K. Hong, D. S. Fried, and D. Papahadjopoulos, *Cell*, *32*, 1067 (1983).
106. G. Poste, in *Liposomes in Biological Systems* (G. Gregoriadis and A. C. Allison, eds.), John Wiley and Sons, New York, 1980, p. 101.
107. F. Szoka, K. Jacobson, Z. Derzko, and D. Papahadjopoulos, *Biochim. Biophys. Acta*, *600*, 1 (1980).
108. A. J. Schroit, J. Madsen, and R. Nayer, *Chem. Phys. Lipids*, *40*, 373 (1986).
109. A. Huang, S. J. Kennel, and L. Huang, *J. Biol. Chem.*, *258*, 14034 (1983).
110. J. Dijkstra, J. W. M. van Galen, C. E. Hulstaert, D. Kalicharan, F. H. Roerdink, and G. L. Scherphof, *Exp. Cell Res.*, *150*, 161 (1984).
111. H. K. Kimelberg and E. G. Mayhew, *Crit. Rev. Toxicol.*, *6*, 25 (1978).
112. T. D. Heath, N. G. Lopez, and D. Papahadjopoulos, *Biochim, Biophys. Acta*, *820*, 74 (1985).
113. T. Ohsawa, Y. Matsukawa, Y. Takakura, M. Hashida, and H. Sezaki, *Chem. Pharm. Bull.*, *33*, 5013 (1985).
114. M. Arrowsmith, J, Hadgraft, and I. W. Kellaway, *Int. J. Pharm.*, *17*, 91 (1983).
115. E. Arakawa, Y. Imai, H. Kobayashi, and H. Sezaki, *Chem. Pharm. Bull.*, *23*, 2218 (1975).
116. C. A. Hunt, *Biochim, Biophys. Acta*, *719*, 450 (1982).
117. C. E. Swenson, M. C. Popescu, and R. S. Ginsberg, *Crit. Rev. Microbiol.*, *15*, S1 (1988).
118. R. R. C. New, M. L. Chance, and S. Heath, *J. Antimicrob. Agents Chemother.*, *8*, 371 (1981).
119. C. Tremblay, M. Barza, C. Fiore, and F. Szoka, *Antimicrob Agents Chemother.*, *26*, 170 (1984).
120. P. Pierson, R. Steiger, and A. Trouet, *Biochem. Pharmacol.*, *31*, 3501 (1982).
121. C. R. Alving, E. A. Steck, W. L. Chapman, V. B. Waits, L. D. Hendricks, G. M. Swartz, and W. L. Hanson, *Life Sci.*, *26*, 2231 (1980).
122. K. Egbaria and N. Weiner, *Adv. Drug Deliv. Rev.*, *5*, 287 (1990).
123. A. Rahman, A Kessler, N. More, et al., *Cancer Res.*, *40*, 1532 (1980).
124. F. Olson, E. Mayhew, D. Maslow, Y. Rustum, and F. Szoka, *Eur. J. Cancer Clin. Oncol.*, *18*, 167 (1982).
125. A. Gazion, D. Goren, Z. Fuks, A. Meshorer, and Y. Barenholz, *Br. J. Cancer*, *51*, 681 (1985).
126. R. L. Juliano and G. Lopez-Berestein, *Pharm. Technol.*, *6*, 164 (1982).
127. G. Gregoriadis, *Liposomes as Drug Carriers*, John Wiley and Sons, New York, 1988.
128. Y. L. Kao and R. L. Juliano, *Biochem. Biophys. Acta*, *453*, 6787, (1981).

129. F. H. Roerdink and A. M. Kroon, *Drug Carrier Systems*, John Wiley and Sons, New York, 1989.
130. M. Woodle, G. Storm, M. S. Newman, L. R. Collins, F. Szoka, and F. J. Martin, Prolonged delivery of vasopressin by long circulating liposomes, *Proceed. Int. Symp. Controlled Release Bioact. Mater.*, Amsterdam, 18, 1991, pp. 103–104.
131. M. S. Wu, J. C. Robbins, R. L. Bugianesi, M. M. Ponpipom, and T. Y. Shen, *Biochim. Biophys. Acta*, *674*, 19 (1981).
132. G. Poste, R. Kirsh and T. Koestler, in *Liposome Technology*, Vol 3 (G. Gregoriadis, ed.), CRC Press, Boca Raton, 1984, p. 1.
133. G. Poste, *Biol. Cell*, *47*, 19 (1983).
134. M. Arrowsmith, J. Hadgraft, and I. W. Kellaway, *Int. J. Pharm.*, *20*, 347 (1984).
135. S. Kim and S. B. Howell, *Cancer Treat. Rep.*, *71*, 705 (1987).
136. D. Papahadjopoulos, *Ann. Rep. Med. Chem.*, *14*, 250 (1979).
137. G. Gregoriadis, *Nature*, *283*, 814 (1980).
138. F. Baillet, M. Housset, A. M. Michelson, and K. Puget, *Free Rad. Res. Commun.*, *1*, 387 (1986).
139. D. A. Eppstein, M. A. van der Pas, C. A. Gloff, and K. F. Soike, *J. Infect. Dis.*, *159*, 616 (1989).
140. I. Rutenfranz, A. Bauer, and H. Kirchner, J. Interferon Res., 10, 337 (1990).
141. R. Roy and S. Kim, Multivesicular liposomes (MLV) containing bleomycine (B) for subcutaneous (SC) administration, *Proc., Am. Assoc. Cancer Res.*, San Diego, 1989, p. 604.
142. C. E. Swenson, K. A. Stewart, and S. Minchey, J. Liposome Res., 1, 379 (1989–90).
143. O. Zierenberg and H. Betzing, *Arzneim. Forsch./Drug Res.*, *29*, 494 (1979).
144. A. J. Jackson, *Drug Met. Dispos.*, *9*, 535 (1981).
145. A. J. Jackson, *Res. Commun. Chem. Pathol. Pharmacol.*, *27*, 293 (1980).
146. A. Tümer, C. Kirby, J. Senior, and G. Gregoriadis, *Biochim, Biophys. Acta*, *760*, 119 (1983).
147. H. Schreier, M. Levy, and P. Milhako, *J. Controlled Release*, *5*, 187 (1987).
148. R. W. Stevenson, H. M. Patel, J. A. Parsons, and B. Ryman, *Diabetes*, *31*, 506 (1982).
149. H. A. C. Titulaer, W. M. C. Eling, D. J. A. Crommelin, P. A. M. Peeters, and J. Zuidema, *J. Pharm. Pharmacol.*, *42*, 529 (1990).
150. F. Kadir, J. Zuidema, and D. J. A. Crommelin, *J. Controlled Release* (Submitted for publication).
151. F. Kadir, J. Zuidema, and D. J. A. Crommelin, *J. Controlled Release* (in press).
152. D. Shepro and P. A. D'Amore, in *Handbook of Physiology* (E. M. Renkin and C. C. Michel, eds.), American Physiology Society, Bethesda, 1984. p. 103.
153. B. E. Ballard, *J. Pharm. Sci.*, *57*, 357 (1968).
154. R. McDougall, J. K. Dunnick, and M. L. Goris, *J. Nucl. Med.*, *16*, 488 (1975).
155. A. W. Segal, G. Gregoriadis, and C. D. V. Black, *Clin. Sci., Mol. Med.*, *49*, 99 (1975).

156. M. P. Osborne, V. J. Richardson, K. Jeyasingh, and B. E. Ryman, *Int. J. Nucl. Med. Biol.*, *6*, 75 (1979).
157. R. W. Stevenson, H. M. Patel, J. A. Parsons, and B. E. Ryman, *Diabetes*, *31*, 506 (1983).
158. M. N. Croll, L. W. Brady, and S. Dadpavvar, *Sem. Nucl. Med.*, *13*, *4* (1983).
159. J. N. Weinstein, in *Rational Basis for Chemotherapy* (B. Chabner, ed.), Alan R. Liss, New York, 1983, p. 441.
160. L. Bergqvist, S. E. Strand, and B. R. R. Persson, *Sem. Nucl. Med.*, *XIII*, 9 (1983).
161. J. Khato, E. R. Priester, and S. M. Sieber, *Cancer Treat. Rep.*, *66*, 517 (1982).
162. P. Hirnle and R. Schubert, *Int. J. Pharm.*, *72*, 259 (1991).
163. V. J. Richardson, B. E. Ryman, R. F. Jewkes, K. Singh, M. N. H. Tattersall, E. S. Newlands, and S. B. Kaye, *Br. J. Cancer*, *40*, 35 (1979).
164. V. I. Kaledin, N. A. Matienko, V. P. Nikolin, Y. V. Gruntenko, V. G. Budker, and T. E. Vakhrusheva, *J. Natl. Cancer Inst.*, *69*, 67 (1982).
165. G. Gabbiani and G. Majno, in *Microcirculation* (G. Kaley and B. M. Altura, eds.), University Park Press, Baltimore, 1980, p. 143.
166. L. D. Mayer, M. B. Bally, M. J. Hope, and P. R. Cullis, *Chem. Phys. Lipids*, *40*, 333 (1986).
167. G. Gregoriadis, *Immunol. Today*, *11*, 89 (1990).
168. I. J. Fidler, R. Nayar, and A. J. Schroit, in *Liposomes as Drug Carriers: Recent Trends and Progress*, John Wiley and Sons, New York, 1988, p. 115.
169. M. P. Osborne, V. J. Richardson, K. Jeyasingh, and B. E. Reyman, *Int. J. Nucl. Med. Biol.*, *9*, 47 (1982).
170. J. Khato, E. R. Priester, and S. M. Sieber, *Cancer Treat. Rep. 66., 517 (1982).*
171. H. M. Patel, K. M. Boodle, and R. Vaughan-Jones, *Biochim. Biophys. Acta*, *801*, 76 (1984).
172. V. J. Caride, *CRC Crit. Rev. Ther. Drug Carrier Syst.*, *1*, 121 (1985).
173. S. S. Daoud, L. R. Hume, and R. L. Juliano, *Adv. Drug Deliv. Rev.*, *3*, 405 (1989).
174. J. Treat and A. Rahmann, *J. Cell. Biochem.*, *12 B*, 252 (1988).
175. R. L. Juliano, in *Drug Carrier Systems* (F. H. D. Roerdink and A. M. Kroon, eds.), John Wiley and Sons, New York, 1989, p. 249.
176. C. R. Alving, in *Targeting of Drugs* (G. Gregoriadis, J. Senior, and A. Truoet, eds.), Plenum Press, New York, 1981, p. 337.
177. F. J. Martin, in *Specialized Drug Delivery Systems: Manufacturing and Production Technology* (P. Tyle, ed.), Marcel Dekker, New York, 1990, p. 267.

8

Subcutaneous Liposomal Tumor Antigen Vaccines: Preclinical and Clinical Studies

Nigel C. Phillips*

Montreal General Hospital Research Institute and McGill University Montreal, Quebec, Canada

LIPOSOMAL TUMOR ASSOCIATED ANTIGENS AS VACCINES FOR CANCER

Active Immunotherapy in Cancer

The treatment of metastatic cancer has relied extensively on chemotherapeutic, radiotherapeutic, and surgical approaches. Immunotherapy, using agents such as BCG or *C. parvuum*, has been used in those patient groups at risk from recurrent disease (1–4). Such intervention has proven to be of marginal effectiveness in changing the overall prognosis. The identification of surface membrane antigens preferentially expressed by tumor cells has provided a target area and a more rational approach to immunotherapy. Immunization with inactivated tumor cells or crude tumor cell–derived products has been largely unsuccessful in inducing the regression of transplantable tumors in animal models (5). However, immunization with such material can be successful in decreasing metastatic tumor growth when given immediately after removal of the primary tumor (6) or with additional immune adjuvants such as BCG (7). Adoptive immunotherapy in such animal models is successful when large numbers of syngeneic lymphoid cells are transferred from hyperimmunized animals and suppressor cells in the host animal have been

**Current affiliation*: Centre D'Immunologie et de Biologie Parasitaire, Institut Pasteur de Lille, Lille, France

inactivated (8). Preliminary clinical studies using tumor-associated antigens (TAA) for the specific immunotherapy of malignant melanoma (9) or of colorectal cancer (10) have indicated that patients with minimal residual tumor burden are able to recognize and respond to the antigens. Other approaches that have attempted to augment the immunogenicity of tumor cells via the use of vaccines containing vesicular stomatitis virus (11), viral oncolyzates (12), HLA mismatched cultured tumor cells (13), or by the heterogenization of tumor cells with unrelated antigens such as PPD (8) have resulted in varying (and largely disappointing) degrees of success with respect to the subsequent immunological and clinical response of the immunized patients. Active specific therapy of melanoma using allogenic antigen preparations and detoxified lipid A as adjuvant (14,15) appears to be capable of eliciting immunological responses in patients with a significant tumor load.

Suppressor Activity in Cancer Patients

The prospect of immunizing patients with advanced cancer who possess a significant degree of immune suppressor activity with tumor antigens has been recently addressed. Bystryn and coworkers reported that patients with stage II melanoma respond more frequently to vaccination with a polyvalent melanoma vaccine than do patients with stage III melanoma (16). Prior or concomitant administration of low-dose cyclophosphamide to patients known to be anergic results in enhanced immunological responsiveness to antigens such as keyhole limpet hemocyanin (17). The prior administration of low-dose cyclophosphamide has been shown to potentiate the immune response against autologous melanoma cell vaccines; in contrast, patients receiving the vaccine alone showed impaired immune reactivity (18). Two of the patients in this study were reported to have significant antitumor responses to treatment with cyclophosphamide and melanoma cell vaccine. This observation was proposed as being due to the elimination or inhibition of excess suppressor T cells responsible for the state of anergy (19,20). The target cell for cyclophosphamide has been subsequently shown to be a $CD4^+$, $2H4^+$, suppressor-inducer T cell (21). The induction of this cell population would appear to be dependent on the presence of the disease rather than on the progression of metastasis. The conflicting efficacy of low-dose cyclophosphamide in conjunction with tumor cell vaccine therapy (22–24) may well be due to differing levels of this T-cell subpopulation in different patient populations.

Tumor Cell Membrane Antigens

The problem of obtaining sufficient autologous TAA for the evaluation of immunization procedures requires the use of vaccine formulations designed to maximize their impact on the immune system. Experimental data indicate that the use of cell lines as the source of antigen requires the presence of cross-reactive antigens (8): while tumor cells lines may provide a long-term solution, the identification and preparation of antigens expressed by a majority of tumors is most readily addressed by cloning procedures. Until such material is available in sufficient quantities and purity to enable clinical trials to take place, alternative approaches must be directed toward enhancing the impact of the antigenic material via its mode of presentation. In a series of studies using a panel of immunologically distinct methylcholanthrene-induced murine fibrosarcomas, it was observed that butane-1-ol–extracted tumor cell antigens were capable of inducing prophylactic immunity against tumor cell challenge (22). The ability of butane-1-ol to extract tumor cell antigens would appear to be a general phenomenon applicable to human tumor cells: antigens have been prepared from human sarcomas (25), melanomas and colonic adenocarcinomas (23) using butane-1-ol. The treatment of viable tumor cells with butane-1-ol releases a series of plasma membrane components that are not contaminated with cytoplasmic or transmembrane components (24). These components appear to be membrane-associated glycoproteins whose hydrophobic anchor has been destabilized by the chaotropic action of butane-1-ol.

Immunization procedures that result in the presentation of such antigens in a manner similar to that found in the intact tumor cell membrane can be expected to have potential for generating immunological responsiveness against relevant antigen epitopes. In experimental models of cancer, butane-1-ol–extracted tumor cell membrane antigens have been successfully used in active, specific immunotherapy (22).

Immune Adjuvants for Potentiating Immune Responses Against Tumor-Associated Antigens

Immunoadjuvants

Immunization strategies directed against actively dividing tumor cells and/or significant tumor-induced immunosuppression will almost certainly require the use of immunoadjuvants other than alum, BCG, or *C. parvuum* if protective responses are to be achieved. Two groups of immunoadjuvants that are capable of stimulating both cell-mediated and antibody responses have received considerable attention in a number of disease areas. These are liposomes (26–38) and muramyl dipeptides (39–52). Liposomes, small concentric vesicles composed of one or more phospholipid bilayers enclosing aqueous spaces, possess immunoadjuvant activity toward encapsulated or associated protein or peptide antigens (26,33,53). Attachment of protein antigens to the surface (54) or within the aqueous phase of the liposome (55) appears to be effective for immunoadjuvant activity, with internalized antigen being the most effective (27). Encapsulation appears to be more effective in stimulating IgG isotype production, whereas surface coupling favors both IgG and IgM (56). A number of putative liposome vaccine formulations for infectious diseases where the organism possesses intrinsic resistance to chemotherapeutic treatment or has antigens of low antigenicity have been described (28,57–66). The immunoadjuvant activity of liposomes has been shown to be augmented by the simultaneous incorporation of other immunostimulants such as lipid A (31,32) or interleukin-2 (IL-2) (37).

N-Acetylmuramyl-L-alanyl-D-isoglutamine (MDP) is the minimal structural unit of bacterial cell wall peptidoglycan possessing immunoadjuvant activity (39–41). MDP and other derivatives have been found to be effective adjuvants for a variety of antigens (43,44,46–48,60). The immunoadjuvant activity of lipophilic MDP derivatives appears to be significantly enhanced after liposomal incorporation (30,52,67–70), whereas the immunoadjuvant activity of hydrophilic MDP derivatives is reduced (38,67). The poor retention characteristics of hydrophilic MDP derivatives in liposomal carriers (71,72) may be responsible for these observations.

The antigen-presenting function of the macrophage has been implicated in the enhanced immunogenicity of liposomal antigens (73). Liposome-bound antigen is significantly more effective than free antigen when macrophages are used as presenting cells in vitro (74,75). The ability of MDP to stimulate a wide range of macrophage functions (76,77) may significantly affect macrophage processing and presentation of liposome-associated antigens.

Liposomal TAA

Liposomal encapsulation of TAA has been shown to be effective in eliciting protective immune responses in animal models (29,78,79). Our studies on the effect of liposomal encapsulation of the purified oncofetal marker carcinoembryonic antigen (CEA) (80) on autologous TAA derived from a metastatic murine melanoma (81) and in phase I clinical trials (82-84) indicates that liposomal adjuvants, especially with incorporated muramyl dipeptides (85), may be effective in eliciting protective humoral (antibody) and cellular responses.

Other Immunotherapeutic Approaches

Considerable attention has recently focused on the use of liposomes to deliver immunostimulants to macrophages as a means of inducing macrophage-mediated tumoricidal activity in situ, especially in organs rich in resident macrophages (70–72,86–87). While such immunotherapy may be applicable for those sites possessing considerable cellular traffic (such as the lung), it may be inappropriate for organs such as the liver where tumor foci in the parenchyme are largely inaccessible to the resident macrophage population

(Kupffer cells) in the sinusoids (88,89). Passive immunotherapy of solid tumors with immunoconjugates is not always successful (90,91) and can lead to the induction of antibodies against the targeting antibody (92,93). Under such circumstances it may be more appropriate to consider active immunization with TAA as a means of controlling tumor growth.

STRATEGIC APPROACH

The ability of liposomes to act as immune adjuvants for incorporated or associated protein antigens provided the starting point for a research program aimed at identifying the impact of liposomal preparations of autologous TAA as vaccines for human melanoma. In order to characterize their potential use in the treatment of metastatic cancer, three main areas of concern have been evaluated:

1. The efficacy and potency of free versus liposomal TAA in eliciting humoral and cellular immunity in animal models
2. The protective activity of liposomal TAA in an animal model of melanoma
3. The clinical acceptability and immunogenicity of liposomal TAA preparations obtained from autologous tumors of patients with stage III melanoma.

LIPOSOMAL FORMULATION AND TOXICITY

Liposome Formulation and Toxicity

The systemic administration of liposomes results initially in a specific concentration within cells of the monocyte/macrophage lineage. Our initial studies were therefore concerned with identifying liposomal formulations that showed minimal toxicity toward macrophages. Such a consideration is not trivial given the potential for high concentrations of phospholipid (>10 mol/ml) at subcutaneous (sc) immunization sites and the subsequent interaction with Langerhans cells, the major antigen-processing cell initially encountered with this route of administration. Using the release of lactate dehydrogenase from murine peritoneal macrophages and the proliferation of a partially differentiated macrophage cell line as indicators, liposomes formulated from DPPC/DMPG in the molar ratio range 7:3–10:1 were identified as having minimal toxicity (Figure 1). All subsequent studies on the effect of liposomal encapsulation on TAA immunogenicity were carried out using this formulation.

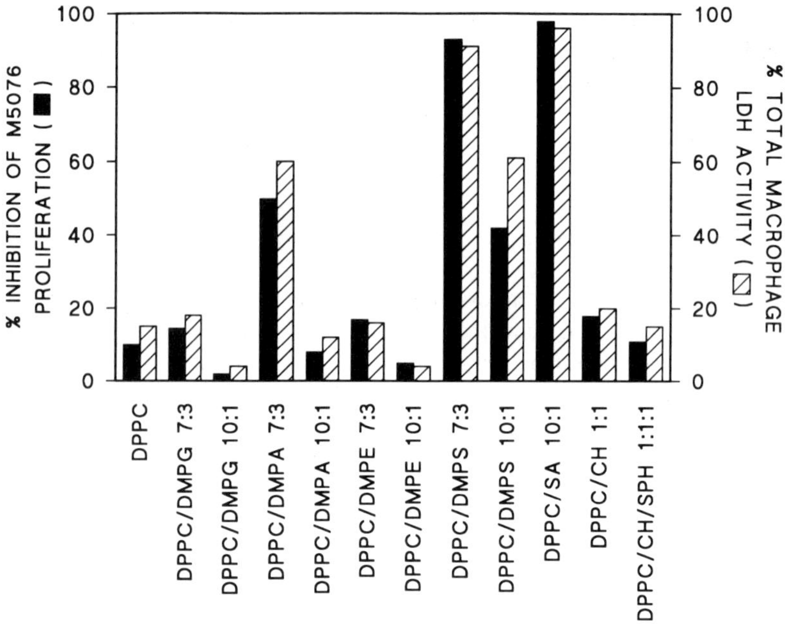

Figure 1 Effect of liposome formulation on toxicity toward macrophages or macrophage-like cells. Macrophages or M5076 cells (10^5 cells in 200 μl minimal essential medium supplemented with 10% fetal calf serum) were incubated with 1 μmol liposomes for 24 hr at 37°C. An aliquot of the supernatant was removed, and lactate dehydogenase (LDH) activity determined using a commercial clinical biochemistry kit. Total LDH was determined by lysing the cells with 0.5% Tween 20. Proliferation of M5076 cells was determined by measuring the incorporation of [^{3}H]methylthymidine (0.1 μCi/10^5 cells) present during the incubation period. The correlation between LDH release and proliferation was highly significant ($r = .995$, $p < .001$).

Enhanced Liposomal Association of a Defined Tumor Antigen, Carcinoembryonic Antigen (CEA), via Palmitation

Incorporation of large hydrophilic molecules such as proteins into multilamellar liposomes is normally inefficient (water entrapment ratios of 2.5 μl/μmol phospholipid). Derivitization of CEA with palmitoyl residues (2–5 molecules/molecule of CEA) using the method of Huang et al. (94) resulted in a significant enhancement (four- to fivefold) of CEA incorporation within MLV liposomes (Figure 2) as well as improved retention (95). Such a procedure allows some modification of the phospholipid/antigen ratio independent of the water entrapment efficiency of the liposome formulation.

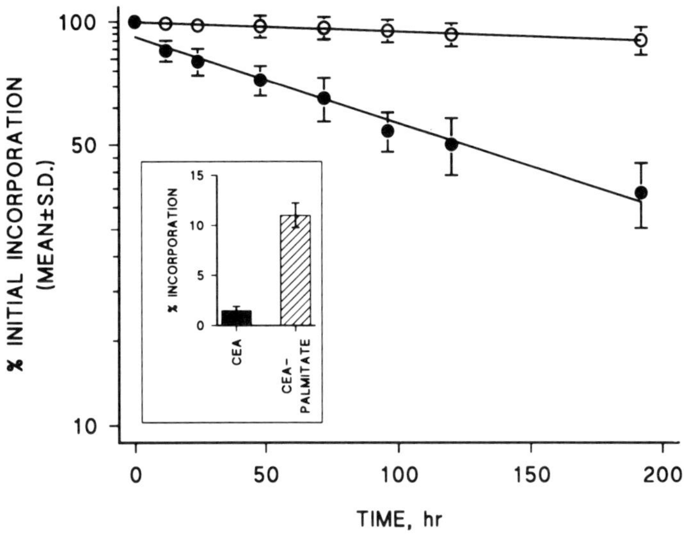

Figure 2 Effect of palmitoylation on CEA incorporation and retention in liposomes. ^{125}I-labeled CEA (●) or CEA palmitate (○) was incorporated within multilamellar liposomes using a hydration volume of 1.0 ml:10 μmol phospholipid. After removal of unincorporated CEA by centrifugal washing, the percent incorporation and retention after incubation at 20°C was determined. The results shown are mean ± SD of three experiments.

Phospholipid Affinity of Butane-1-ol-Derived Proteins

Incubation of [^{125}I]labeled proteins derived from butane-1-ol treatment of human melanoma tumor cell line HTB-70 (ATCC tumor cell repository, Rockville, MD; HTB70 is representative of a primary human melanoma) resulted in a time-dependent association with liposomal membranes (Figure 3). The requirement for palmitation to enhance liposomal association, as was observed for CEA, was not required for plasma membrane TAA because of their intrinsic lipophilic nature. Liposomal TAA preparations were subsequently incubated for 2 hr at 20°C prior to injection to ensure that adequate interaction of the antigens with the phospholipid bilayers would occur.

LIPOSOMAL ENHANCEMENT OF TAA IMMUNOGENICITY

Effect of Liposomal Encapsulation on the Immunogenicity of CEA

CEA is a heavy glycosylated oncofetal marker for carcinomas of the gastrointestinal tract (96-98). Our preliminary studies were designed to evaluate

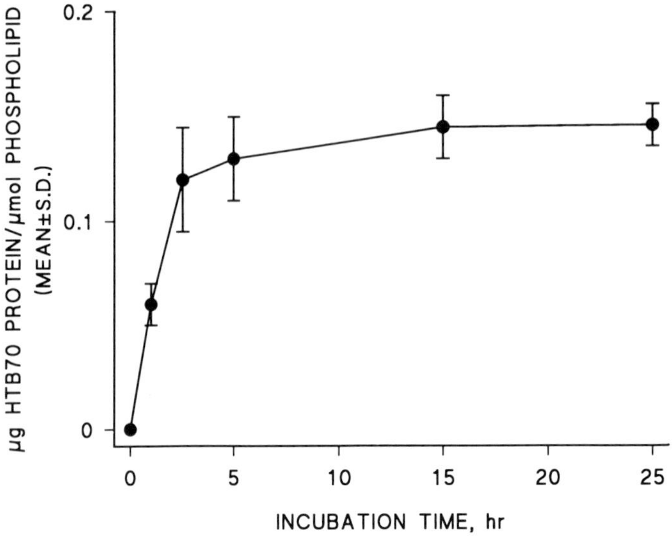

Figure 3 Interaction of butane-1-ol–prepared TAA with preformed liposome membranes. HTB-70 TAA was iodinated with ^{125}I by the Iodogen method (Pierce Chemical Co., Rockford, IL) and incubated with preformed multilamellar DPPC/DMPG liposomes (10:1 molar ratio; average diameter 0.38 μm) in phosphate-buffered saline at 20°C. Aliquots were removed at the indicated times and liposome-associated TAA was determined after centrifugal washing (15,000 *g*/15 min/4°C).

the impact of liposomes on antibody responses toward a pure TAA, as well as the cellular immunity response, as defined by the lymphocyte stimulation index (LSI) in vitro. Immunization and boosting of A/J mice with CEA in the dose range 5–500 μg/mouse via the sc route resulted in a dose-related induction of anti-CEA antibodies (Figure 4). Immunization with CEA emulsified in Freund's complete adjuvant (FCA) or coadministered with MDP resulted in a significant enhancement of the antibody titer. Liposomal encapsulated CEA was administered in the dose range 0.05–5μg/mouse and resulted in a significant enhancement of the antibody titer compared with CEA/FCA or CEA/MDP (Figure 4). A further increase in antibody titer was observed when the lipophilic muramyl dipeptide derivative MDP-glyceroyldipalmitate (MDP-GDP) was incorporated within liposomes containing CEA (Figure 4).

Palmitoylated CEA was recognized by monoclonal antibodies against CEA, and liposomes containing palmitoylated CEA had the same potency and ability to induce IgG antibodies as liposomal CEA. Such results demonstrate that enhanced lipophilicity via palmitoylation does not change the

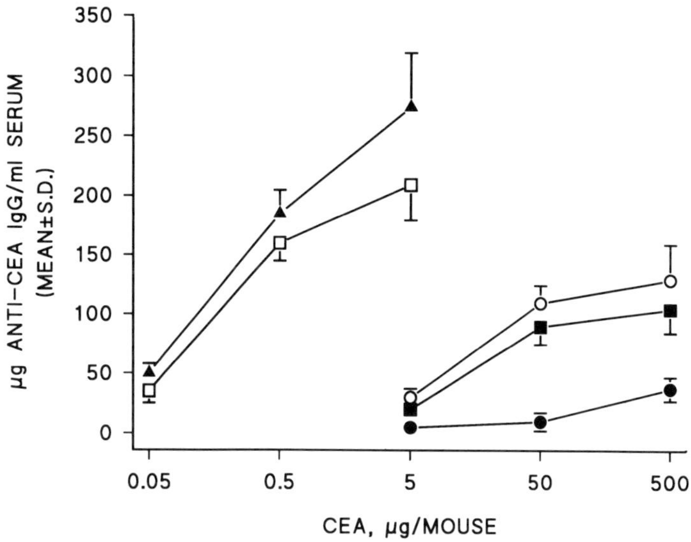

Figure 4 Immunogenicity of CEA in mice. Groups of five C57BL/6 mice were immunized with CEA by sc injection on days 0 and 14. Serum was obtained on day 28, and anti-CEA IgG antibodies determined by ELISA with peroxidase-labeled goat anti-mouse IgG antibody. ●, CEA alone; ■, CEA/50 μg MDP; ○, CEA/FCA; □, liposomal CEA (1 μmol phospholipid); ▲, liposomal CEA containing 10 μg MDP-GDP.

antigenic epitopes or subsequent immunogenicity of the TAA. Isolation of splenic lymphocytes from mice immunized with CEA, CEA/FCA, or liposomal CEA demonstrated that liposomes were significantly more effective than FCA in eliciting cellular immunity as determined by the LSI (Figure 5).

Effect of Liposomal Encapsulation on the Immunogenicity of Butane-1-ol-Derived Melanoma Tumor Antigens

The initial characterization of the effect of liposomal encapsulation on the immunogenicity of TAA was carried out using a TAA preparation derived from human melanoma tumor cell lines maintained in tissue culture. Treatment of HTB-70 melanoma tumor cells maintained in serum-free media for 48 hr with butane-1-ol (3% v/v for 3 min) released 18–20 plasma membrane-associated proteins without affecting the viability of the tumor cells. Concentration of these proteins 100-fold by membrane filtration (MW cutoff

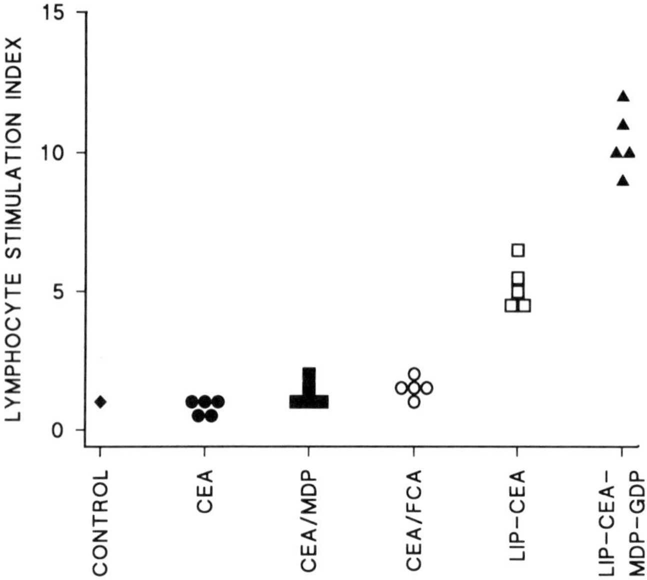

Figure 5 Lymphocyte stimulation index following sc immunization with CEA in mice. Groups of five C57BL/6 mice were immunized with CEA by sc injection on days 0 and 14. Spleen lymphocytes were isolated on day 28, and 10^5 cell in 200 μl RPMI 1650 supplemented with 2.5% fetal calf serum were incubated with 50 μg CEA for 72 hr at 37°C. Mitogenic activity was determined by adding [^{3}H]methylthymidine (0.1 μCi/10^5 cells) for the last 24 hr of the incubation period.

10,000 D) enabled dose–response studies of immunogenicity to be carried out in mice. The administration of HTB-70 TAA in the dose range 0.1–10 μg/mouse resulted in a dose-related production of IgG antibodies (Figure 6). The coadministration of MDP resulted in a slight but significant enhancement of antibody titers. In contrast, liposomal encapsulation resulted in a significant enhancement of antibody titers. The incorporation of the lipophilic muramyl dipeptide derivative MDP-GDP within the liposomes resulted in a further increase in antibody titers (Figure 6).

Antibody Isotype Response

The ability to induce IgG antibodies of defined isotype, specifically cytotoxic antibodies of the IgG2a/2b isotypes (45,99-101), may be of crucial importance in the development of efficient antitumor immunity. Immunization of mice with TAA (CEA or melanoma TAA) or TAA and MDP resulted in the induction of IgM and IgG1 antibodies both pre- and postboost (Table 1).

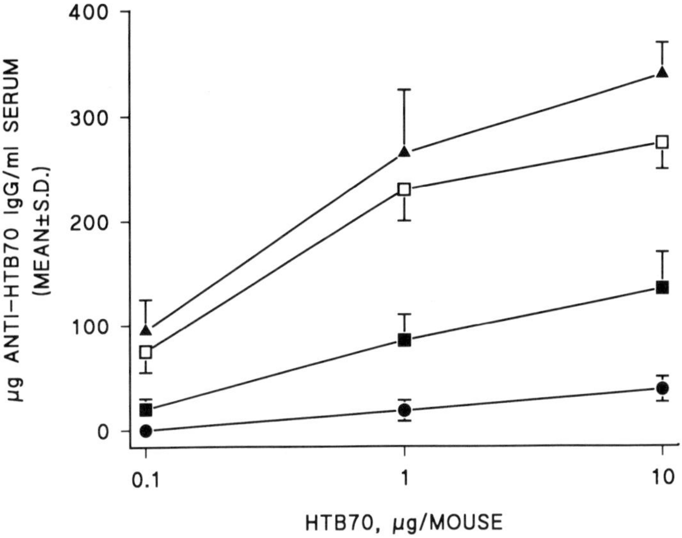

Figure 6 Immunogenicity of human melanoma TAA in mice. Groups of five A/J mice were immunized with HTB-70 TAA by sc injection on days 0 and 14. Serum was obtained on day 28 and anti-HTB IgG antibodies determined by ELISA. ●, HBT-70; ■, HTB-70/50 µg MDP; □, liposomal HTB-70 (1 µmol phospholipid); ▲, liposomal HTB-70 containing 10 µg MDP-GDP.

Immunization with liposomal TAA or liposomal TAA/MDP-GDP resulted in a significant decrease in IgM antibodies as well as the induction of IgG antibodies of all isotypes (IgG1,2a/2b; and 3; Table 1).

SUBCUTANEOUS IMMUNIZATION OF MICE AGAINST SYNGENEIC B16 MELANOMA USING LIPOSOMAL TAA

Immunoadjuvant Activity of Liposomes

The ability of liposomes to augment the immune response of C57BL/6 mice against syngeneic B16 melanoma tumor cells was determined prior to immunization studies. The administration of butane-1-ol–prepared TAA from B16 melanoma cells maintained in tissue culture (maximum of five passages from excision of primary solid tumor) resulted in marginal antibody production (Figure 7) or cellular responsiveness (as determined by lymphocyte stimulation indices; Figure 8) at a dose of 40 µg/mouse. Doses of 10 µg or less were inactive. Incorporation of 0.1 µg TAA within liposomes resulted in a significant enhancement of the LSI and significant antibody production

Table 1 Effect of Immunoadjuvants on Anti-TAA Antibody Isotypes in Mice

	1° Response (%)						2° Response (%)					
Immunization	IgM	IgG	(1	2a	2b	3)	IgM	IgG	(1	2a	2b	3)
CEA[a]												
CEA	90	10	(95	5	5	−)	80	20	(93	3	2	1)
CEA/FCA	80	20	(90	4	4	2)	60	40	(92	2	3	3)
CEA/MDP	78	22	(96	2	2	−)	38	62	(93	2	2	3)
Liposomal CEA	37	63	(50	30	18	2)	31	69	(40	25	25	10)
Liposomal CEA/MDP-GDP	42	58	(44	28	20	8)	30	70	(42	24	26	8)
HTB-70[b]												
HTB-70	62	38	(90	5	4	1)	41	59	(90	3	4	3)
HTB-70/MDP	58	42	(92	3	3	2)	33	67	(90	2	6	2)
Liposomal HTB-70	35	65	(48	20	23	7)	25	75	(45	28	20	7)
Liposomal HTB-70/MDP-GDP	32	68	(46	18	29	7)	23	77	(40	25	30	5)

[a]Groups of five female C57BL/6 mice were immunized via the sc route with CEA (5 μg), CEA emulsified with 50 μg MDP, CEA in liposomes (5 μg/1 mol phospholipid); or CEA in liposomes containing MDP-GDP (10 μg). Mice were immunized on days 0 and 14, and the primary response was determined 13 days after immunization. The secondary response was determined on day 28. IgM and IgG are expressed as a percentage of the total serum Ig, and the IgG isotypes are expressed as a percentage of the total serum IgG. The difference in the IgM/IgG ratio and in the IgG isotype distribution between immunization with CEA, CEA/FCA, or CEA/MDP and the liposomal formulations was highly significant ($p < .001$, ANOVA).

[b]Groups of five female A/J mice were immunized with HTB-70 (5 μg), HTB-70 mixed with 50 μg MDP, HTB-70 in liposomes (5 μg/1 μmol phospholipid), or HTB-70 in liposomes containing MDP-GDP (10 μg). Mice were immunized on days 0 and 14 and the primary response was determined 13 days after immunization. The secondary response was determined on day 28. IgM and IgG are expressed as a percentage of the total serum Ig, and the IgG isotypes are expressed as a percentage of the total serum IgG. The difference in the IgM/IgG ratio and in the IgG isotype distribution between immunization with HTB70 or HTB-70/MDP and the liposomal formulations is highly significant ($p < .001$, ANOVA).

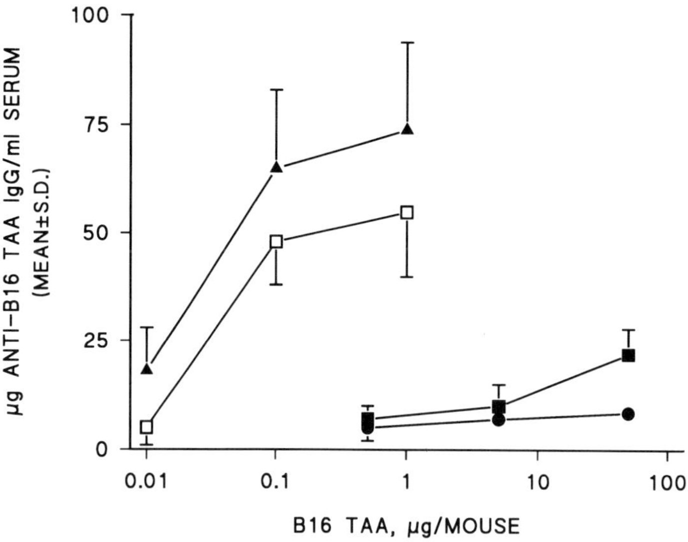

Figure 7 Immunogenicity of B16 melanoma TAA in syngeneic C57BL/6 mice. Groups of five C57BL/6 mice were immunized sc with B16 melanoma TAA on days 0 and 14. Serum was obtained on day 28, and anti-B16 melanoma TAA determined by ELISA. ●, B16 melanoma TAA: ■, B16 melanoma TAA/50 µg MDP; □, liposomal B16 melanoma TAA (1 µmol phospholipid); ▲, liposomal B16 melanoma TAA containing 10 µg MDP-GDP.

(Figures 7 and 8). The concomitant incorporation of MDP-GDP within the liposomal TAA preparation resulted in a further increase in both the LSI and the antibody titer (Figures 7 and 8).

Immunization of C57BL/6 Mice Against B16 Melanoma

The ability of liposomal TAA preparations to induce protective immunity against B16 melanoma was determined by immunizing and boosting C57BL/6 mice with TAA, liposomal TAA, liposomal TAA containing MDP-GDP, or irradiated B16 tumor cells (18,000 rads) corresponding to the amount of B16 TAA administered as free TAA. Immunization with irradiated tumor cells or free TAA failed to control tumor growth on subsequent challenge (Figure 9) and did not prolong survival of the mice (Figure 10). Immunization with liposomal TAA resulted in a significant retardation of tumor growth as well as a significant number of long-term survivors. Immunization with liposomal TAA/MDP-GDP severely retarded tumor growth and resulted in over 60% long-term survivors (>90 days posttumor challenge). Antigen preparations

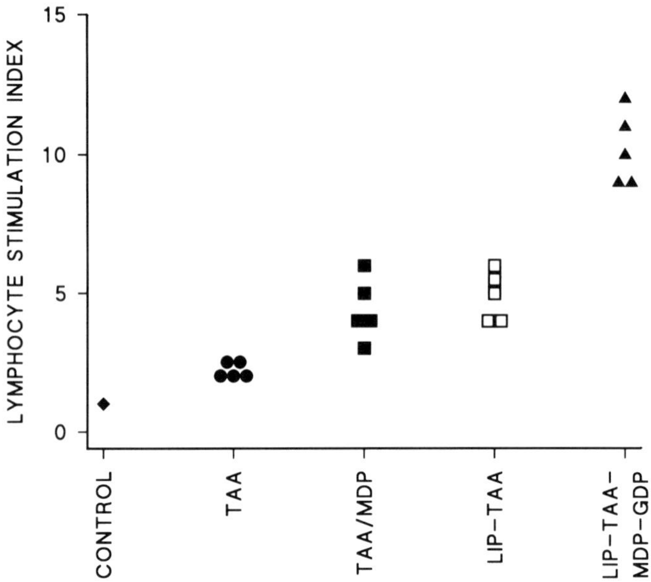

Figure 8 Lymphocyte stimulation index following s.c. immunization with B16 melanoma TAA. Groups of five C57BL/6 mice were immunized with B16 melanoma TAA by sc injection on days 0 and 14. Spleen lymphocytes were isolated on day 28, and 10^5 cells in 200 μl RPMI 1650 supplemented with 2.5% fetal calf serum were incubated with 50 μg B16 melanoma TAA for 72 hr at 37°C. Mitogenic activity was determined by adding [^{3}H]methylthymidine (0.1 μCi/10^5 cells) for the last 18 hr of the incubation period.

derived from B16 melanoma cells maintained in culture for >15 passages resulted in TAA preparations that were not capable of conferring protective immunity when incorporated within liposomal carriers.

CLINICAL SAFETY OF LIPOSOMAL TAA IN PATIENTS WITH MELANOMA

Patient Population

The clinical safety of liposomal melanoma TAA has been determined in patients with stage III melanoma whose disease had proven to be refractory to chemotherapy and/or radiotherapy. All patients entered in the study had multiple metastatic lesions, either subcutaneous or subcutaneous/visceral organ.

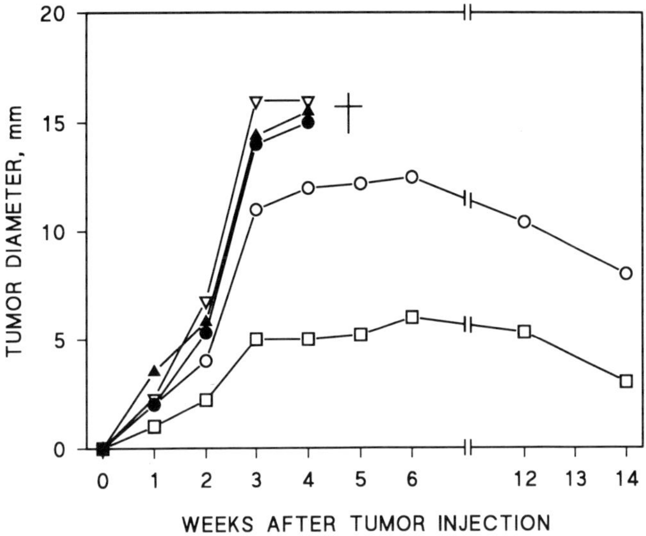

Figure 9 Effect of sc immunization with B16 melanoma TAA on tumor growth. Groups of 10 C57BL/6 mice were immunized with 10^6 irradiated tumor cells (▽), 40 μg TAA (●), control liposomes (▲), liposomal TAA (1 μmol phospholipid containing 0.1 μg) (○), or liposomal TAA/MDP-GDP (10 μg MDP-GDP) (□) on days 0 and 14. The mice were challenged with a sc injection of 10^5 viable B16 melanoma tumor cells in the flank region on day 35. †, animals dead. Treatment with liposomal TAA or liposomal TAA/MDP-GDP was significantly different from control treatment (irradiated tumor cells or liposomes alone: $p < .05$ and 0.001, respectively, Student's *t*-test for unpaired data; $p < .01$, ANOVA, treatment versus tumor diameter). Mean results are shown, SD being <10%.

Preparation of Autologous Melanoma TAA

Our preclinical studies with TAA derived from B16 melanoma cells had demonstrated that prolonged culture of the tumor cells resulted in TAA preparations that were not protective. The clinical safety of liposomal melanoma TAA was therefore determined with antigen preparations derived from freshly excised solid tumors rather than tumor cell lines derived from the patients. Treatment of single-cell preparations obtained from subcutaneous nodules resulted in sufficient TAA to conduct phase I clinical trials.

Immunization Protocol

Patients were immunized subcutaneously with liposomes prepared from DPPC/DMPG, 10:1 molar ratio, and autologous TAA. Immunization was carried

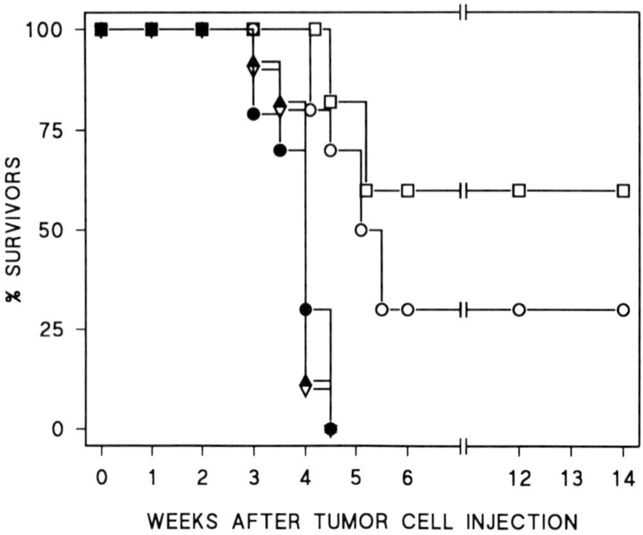

Figure 10 Effect of sc immunization with B16 melanoma TAA on survival following tumor challenge. Legend as for Figure 9. Treatment with liposomal TAA or liposomal TAA/MDP-GDP was significantly different from control treatment (irradiated tumor cells or liposomes alone: $p < .05$, χ^2 analysis with Yates correction).

out at 2- to 4-week intervals, with a starting dose of 35 mg phospholipid and escalation (50%) every 4 weeks. No patient was receiving steroids or NSAID agents on entry or during the trial. All patients were monitored regularly by physical examination, hematological profile, liver and renal function tests, and radiological examination where necessary.

Toxicity

A mild transient tenderness at the site of injection was reported. No other local or systemic reactions were observed in any patient (follow-up period of 9–10 weeks). There was no evidence of systemic toxicity as determined by hematological and biochemical analysis.

Clinical Responses

Of the 13 patients entered into the study, 3 had complete responses and 2 had mixed responses (Table 2). The remaining 8 patients showed no response and continued to have disease progression.

Table 2 Patient Characteristics and Summary of Therapy

No.	Age/sex	Metastatic distribution[a]	Injections	Period (weeks)	Response
1	42/M	sc,Li,Lu,Sb	3	11	Mixed[b]
2	66/M	sc,Li,Lu,B	3	36	None
3	52/F	sc	20	110	Complete
4	22/M	sc,Lu	4	14	None
5	36/F	sc,Sk,Lu,B	7	27	Mixed
6	61/M	sc,Li,Lu	6	32	None
7	66/F	sc	15	102	Complete
8	46/M	sc,Li,Lu	3	12	None
9	34/F	sc,Sk,Rp	4	9	None
10	42/F	sc,Lu	4	15	None
11	46/M	sc	8	26	Complete
12	67/M	sc,Li	4	8	None
13	41/M	sc	4	8	None

[a]B, bone; Li, liver; Lu, lung; Rp, retroperitoneum; Sb, small bowel; sc, subcutaneous; Sk, skin.
[b]Complete, all metastases disappeared; mixed, significant change in subcutaneous metastases but not in visceral metastases; none, no change or increase in metastatic size or number of lesions.

Natural Killer Cell Activity

All patients with normal natural killer (NK) cell activity before entry showed a progressive decrease in activity during the course of treatment. Those patients with low NK cell activity on entry showed no significant change during the course of treatment. There was no correlation between NK activity or clinical responsiveness/immune reactivity.

Lymphocyte Responsiveness to Autologous TAA and Concanavalin A

There was a significant enhancement of the ability of the autologous TAA preparations to stimulate peripheral blood lymphocyte proliferation in vitro in those patients showing a reduction in tumor mass (Figure 11). The proliferative response to the T-cell mitogen concanavalin A (Con A) was maintained or increased in these patients. No lymphocyte proliferative changes were seen in those patients who did not respond clinically to liposomal TAA treatment, and there was a significant decrease in the mitogenic response to Con A.

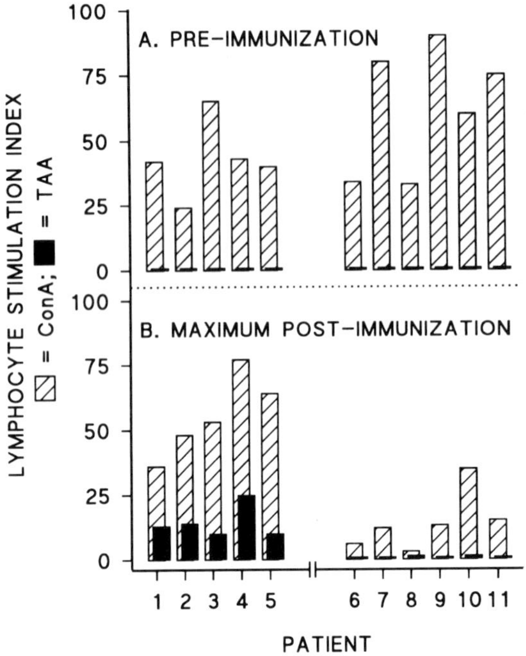

Figure 11 Lymphocyte mitogenic responses to Con A or autologous TAA following liposomal TAA immunization. Peripheral blood lymphocytes (10^5 cells in 200 μl RPMI 1650 medium containing 2.5% fetal calf serum) were incubated with Con A (4 μg/ml) or autologous TAA (5–10 μg/ml) for 72 hr at 37°C. Mitogenic activity was determined by adding [^{3}H]methylthymidine (0.1 μCi/10^5 cells) for the last 18 hr of the incubation period.

Lymphocyte Tumor Cell Stasis Activity

Peripheral blood lymphocytes from patients showing a reduction in tumor mass were able to significantly inhibit the growth of heterologous melanoma tumor cells (HTB-70) in vitro (Figure 12). In contrast, the ability of lymphocytes from clinical nonresponders to inhibit tumor cell growth in vitro did not change or decreased during the course of treatment. The difference in cytostatic activity between responder and nonresponder patients was not significant at the start of treatment ($p > .3$). The difference at the end of treatment was highly significant ($p < .002$). There was a positive correlation between TAA-induced LSI, lymphocyte-mediated tumor stasis activity, and clinical responder status (Figure 13). These results indicate that successful

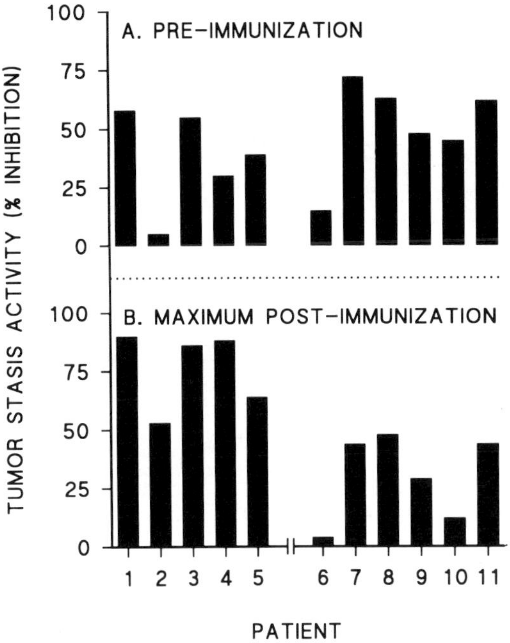

Figure 12 Lymphocyte tumor stasis activity. Peripheral blood lymphocytes (10^5 cells) were added to 5×10^3 HTB-70 cells in the wells of microtiter plates in a volume of 200 μl minimum essential medium containing 10% fetal calf serum. Incubation was carried out for 72 hr at 37°C. Inhibition of tumor cell proliferation was determined by adding [^{3}H]methylthymidine (0.1 μCi) the last 18 hr of the incubation period.

immunization using liposomal TAA will only occur in those patients capable of responding immunologically to the antigenic stimulation. Clinical anergy would appear to be a major limiting factor for positive immunization.

Anti-TAA Antibodies

The presence of IgG anti-TAA antibodies was determined retrospectively in two clinical and two nonclinical responder patients. Anti-TAA IgG antibodies were detected in both clinical responders after the third immunization: anti-TAA antibodies were not detected in the two nonresponders throughout the treatment period (Figure 14).

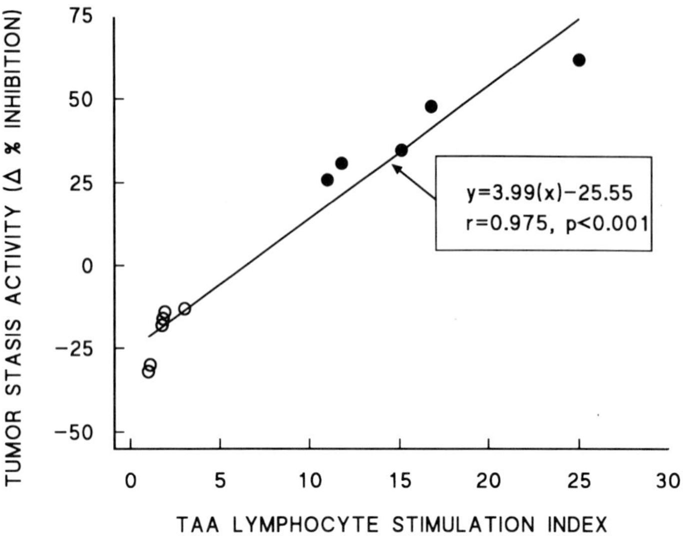

Figure 13 Correlation between TAA mitogenic activity, tumor stasis activity, and clinical responder status. The correlation between TAA mitogenic activity and tumor stasis activity was highly significant ($p < .001$). ●, clinical responder; ○, clinical nonresponder.

Concomitant Treatment with Low-Dose Cyclophosphamide

Patient 5 was initially refractory to liposomal TAA treatment, and was anergic as determined by standard antigen skin testing and peripheral blood lymphocyte responses to Con A or TAA. In an attempt to overcome immunosuppression, low-dose cyclophosphamide (100 mg iv/m^2 body area) was administered 3 days prior to sc vaccination with liposomal TAA. Significant changes in lymphocyte mitogenic responses to Con A and TAA were observed (Figure 15), followed by clinical responsiveness (Table 2).

DISCUSSION

The potential for utilizing immune responses directed against tumor-associated antigens or differentiation antigens as a means of controlling tumor growth and metastasis would appear to depend on two main criteria: relevance of the antigen preparation and host responsiveness. The demonstration of prophylactic or therapeutic antitumor activity using TAA is significant because few (if any) true tumor-specific antigens have been described. In contrast,

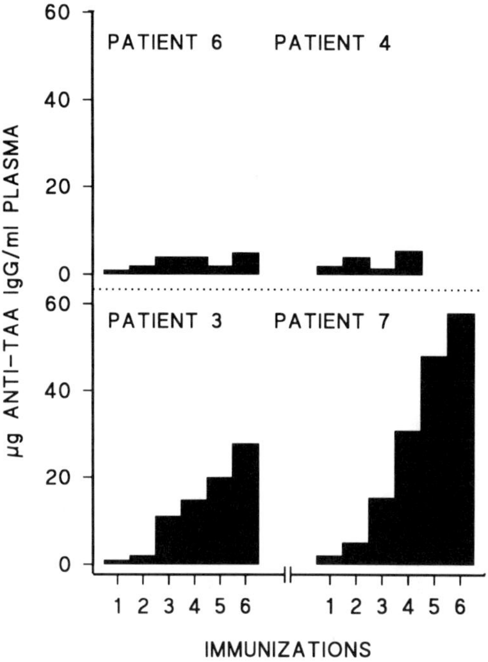

Figure 14 Anti-TAA IgG responses after sc immunization. The presence of anti-TAA IgG antibodies was determined by ELISA using peroxidase-labeled goat anti-human IgG antibody. Two patients who had positive clinical response had detectable IgG antibodies after the third immunization. Two clinical nonresponders failed to produce anti-TAA IgG antibodies following four and six immunizations.

numerous antigens having a high level of expression on tumor cells but only a restricted and low expression on normal cells have been described. Host responsiveness would appear to be governed primarily by the degree of immunosuppression present at the time of immunization. The use of adjuvants capable of eliciting tumor-induced immune responses in the presence of immunosuppression or concomitant therapy to reduce immunosuppressive cell populations would appear to be the major factors in overcoming TAA-specific anergy.

The incorporation of autologous TAA or purified CEA within liposomes results in a significant enhancement of both cellular and humoral immunity directed against the antigens. The use of liposomes as immunoadjuvants would appear to differ from other adjuvants in that IgG isotypes that are recognized as being capable of participating in antibody-dependent cellular cytotoxicity reactions (99–102) are elicited. The demonstration of protective

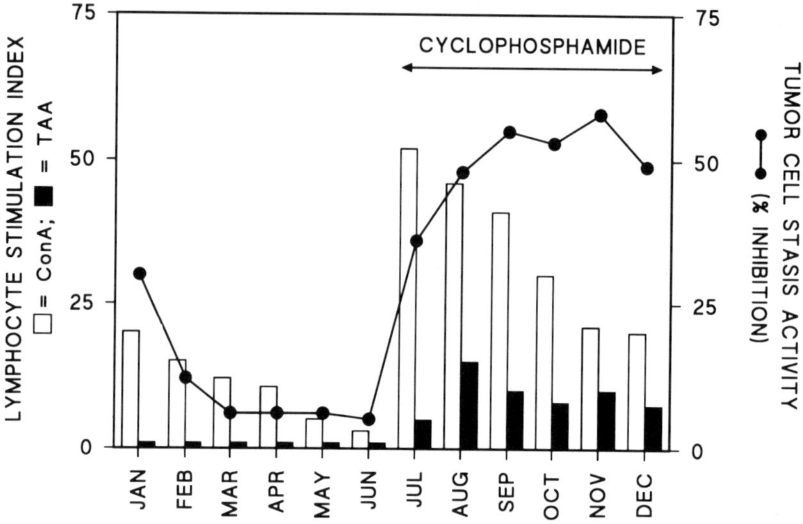

Figure 15 Effect of concomitant treatment with cyclophosphamide on response to liposomal TAA. Immunizations from January through June 1988 failed to elicit any immunological response. Commencing July, the patient was treated 3 days before each vaccination with cyclophosphamide (100 mg/m^2 body surface). A significant increase in Con A and TAA lymphocyte stimulation indices and in lymphocyte tumor stasis activity was observed, and corresponded to a change from clinical nonresponder to clinical responder status (patient 5, Table 2).

immunity induced by liposomal TAA in a murine model of malignant melanoma indicates a correlation between the induction of such antibodies and inhibition of tumor growth. Such a phenomenon is not the result of antigen stimulation alone, as demonstrated by the ineffectiveness of irradiated tumor cells or free antigen. The presence of an additional immunoadjuvant within the liposome, MDP-GDP, resulted in a further increase in protective activity that was accompanied by an increase in cellular reactivity. The difference in activity between liposomal TAA and liposomal TAA/MDP-GDP requires further investigation.

Our clinical evaluation of liposomal TAA in patients with stage III melanoma demonstrated that such a vaccination procedure can be used without side effects. The presence of positive clinical responses in a number of these patients, which correlated with changes in cellular immunity, indicates that such therapy may be of use in controlling metastatic disease that is refractory to conventional treatments with chemotherapy and radiotherapy. The major limitation to immunological intervention at the level of TAA would appear to be one of disease-induced (or prior chemotherapy-induced) immunological anergy. Our (limited) data suggest that this should be evaluated prior to and during immunization, and that the use of cytotoxic agents acting at the level of T-suppressor cells may be one way of eliciting immune responses.

The phase I clinical trial was carried out using tumor antigens derived from patients with considerable tumor mass: immunization of patients with stage II melanoma, who have minimal tumor load and for whom immunosuppression is not a major problem, will require the use of cloned or synthetic antigens. The observation that two patients had mixed responses is of interest with respect to the site of tumor used for TAA preparation. In both of these patients immunization was carried out using TAA prepared from subcutaneous tumors: clinical responses were observed in skin and subcutaneous tumors whereas there was no effect on visceral metastases. These results indicate that heterogeneity of melanoma tumor cell antigens at different metastatic sites may be an additional and potentially limiting factor for therapeutic immunization. Any immunization procedure using synthetic or cloned melanoma antigens will have to consider this heterogeneity in terms of antigenic specificity.

SUMMARY

Liposomal incorporation of tumor antigens results in a significant enhancement of immunogenicity, resulting in specific cellular and humoral immunity. In contrast to other immunoadjuvants such as Freund's complete adjuvant or muramyl dipeptide, liposomal adjuvants elicit IgG isotypes that are capable of participating in antibody-dependent cellular cytotoxicity reactions.

Liposomal tumor antigen preparations are capable of conferring protective immunity in an animal model of melanoma. Irradiated tumor cells or tumor antigen preparations alone were without activity. The concomitant administration of a liposomal-associated muramyl dipeptide resulted in enhanced protection against tumor challenge.

Liposomal preparations of autologous tumor antigens are well tolerated in patients with stage III melanoma and are capable of eliciting positive clinical changes. The clinical responsiveness correlates with the induction of cellular immunity as well as tumor antigen-specific IgG antibody. Clinical anergy is a limiting factor that may be overcome by the elimination of suppressor T cells by low-dose cyclophosphamide treatment. Heterogeneity of tumor antigen expression at different metastatic sites may be a limiting factor for liposomal tumor antigen vaccines.

ACKNOWLEDGMENTS

The financial support of the Medical Research Council of Canada, the National Cancer Institute of Canada, and the Cancer Research Society is gratefully acknowledged.

REFERENCES

1. R. C. Bast, Jr., B. Zbar, T. Borsos, and H. J. Rapp, *N. Engl. J. Med.*, *290*, 1413 (1974).
2. C. W. Aungst, J. E. Sokal, and B. V. Jager, *Ann. Int. Med.*, *82*, 666 (1975).
3. W. D. Terry, *N. Engl. J. Med.*, *303*, 1174 (1980).
4. C. J. Wikstrand and D. D. Bigner, *Am. J. Pathol.*, *98*, 517 (1980).
5. S. Nomi, N. R. Pellis, and B. D. Kahan, *J. Natl. Cancer Inst.*, *73*, 943 (1984).
6. G. Steele, Jr., P. J. Harte, A. A. Rayner, J. M. Corson, J. Madara, A. E. Munroe, V. P. King, and R. E. Wilson, *J. Immunol.*, *128*, 7 (1982).
7. R. J. Y. Ho, R. L. Burke, and T. C. Merigan, *Antiviral Res.*, *13*, 187 (1990).
8. P. J. Lachmann, R. M. Grant, L. S. Freedman, K. Sikora, and N. M. Bleehen, *Br. J. Cancer*, *51*, 415 (1985).
9. A. Hollinshead, M. Arlen, R. Yonemoto, M. Cohen, K. Tanner, W. D. Kundin, and J. Scherrer, *Cancer*, *49*, 1387 (1982).
10. A. Hollinshead, E. G. Elias, M. Arlen, B. EΣuda, M. Mosley, and J. Scherrer, *Cancer*, *56*, 480 (1985).
11. P. O. Livingston, A. P. Albino, T. J. Chung, F. X. Real, A. N. Houghton, H. F. Oettgen, and L. J. Old, *Cancer*, *55*, 713 (1985).
12. C. G. Ioannides, C. D. Platsoucas, C. A. O'Brian, R. Patenia, J. M. Bowen, J. T. Wharton, and R. S. Freedman, *Anticancer Res.*, 9, 535 (1989).
13. D. E. Bullard, D. G. Thomas, J. L. Darling, C. J. Wikstrand, J. V. Diengdoh, R. O. Barnard, J. G. Bodmer, and D. D. Bigner, *Br. J. Cancer*, *51*, 283 (1985).
14. M. S. Nitchell, J. Kan-Mitchell, R. A. Kempf, W. Harel, H. Shau, and S. Lind, *Cancer Res.*, *48*, 5883 (1988).
15. M. S. Mitchell, *Biotechnol. Ther.*, 1, 163 (1900).
16. J. C. Bystryn, R. Oratz, M. N. Harris, D. F. Roses, F. M. Golomb, and J. L. Speyer, *Cancer*, *61*, 1065 (1988).
17. D. Berd, H. C. Maguire, Jr., and M. J. Mastrangelo, *Cancer Res.*, *44*, 5439 (1984).
18. D. Berd, H. C. Maguire, Jr., and M. J. Mastrangelo, *Cancer Res.*, *46*, 2572 (1986).
19. P. O. Livingston, X. Kaelin, C. M. Pinsky, H. F. Oettgen, and L. J. Old, *Cancer*, *56*, 2194 (1985).
20. P. O. Livingston, S. Cunningham-Rundles, G. Marfleet, C. Gnecco, G. Y. Wong, G. Schiffman, W. E. Enker, and M. K. Hoffman, *J. Biol. Response Mod.*, 6, 392 (1987).
21. D. Berd and M. J. Mastrangelo, *Cancer Res.*, *48*, 1671 (1988).
22. S. J. LeGrue, *Cancer Metast. Rev.*, 4, 209 (1985).

23. S. K. Liao, J. W. Smith, and P. C. Kwong, *Cancer Immunol. Immunother.*, *17*, 95 (1984).
24. S. J. LeGrue, J. P. Allison, C. M. Macek, N. R. Pellis, and B. D. Kahan, *Cancer Res.*, *41*, 3956 (1981).
25. J. A. Roth and R. A. Wesley, *Cancer Res.*, *42*, 3978 (1982).
26. A. C. Allison and G. Gregoriadis, *Nature*, *252*, 252 (1974).
27. P. N. Shek and B. H. Sabiston, *Immunology*, *47*, 627 (1982).
28. H. Snippe, J. E. G. van Dam, A. J. van Houte, J. M. N. Willers, J. P. Kamerling, and J. F. G. Vliegenthart, *Infect. Immun.*, *42*, 842 (1983).
29. A. J. Schroit and M. E. Key, *Immunology*, *49*, 431 (1983).
30. S. Kotani and H. Takada, *Yakugaku Zasshi*, *103* (1983).
31. G. J. W. J. Zigterman, M. Jansze, H. Snippe, and J. M. N. Willers, *Int. Arch. Allergy Appl. Immunol.*, *81*, 245 (1986).
32. C. R. AlYing, R. L. Richards, J. Moss, L. I. Alving, J. D. Clements, T. Shiba, S. Kotani, R. A. Wirtz, and W. T. Hockmeyer, *Vaccine*, *4*, 166 (1986).
33. G. Gregoriadis, D. Davis, and A. Davies, *Vaccine*, *5*, 145 (1987).
34. G. J. W. J. Zigterman, M. Jansze, H. Snippe, and J. M. N. Willers, *Immunology*, *65*, 37 (1988).
35. W. E. Varmier and S. L. Snyder, *Immunol. Lett.*, *19*, 59 (1988).
36. H-M. Thérien and E. Shahum, *Immunol. Lett.*, *22*, 253 (1989).
37. L. Tan and G. Gregoriadis, *Biochem. Soc. Trans.*, *17*, 693 (1989).
38. G. Gregoriadis and C. Panagiotidi, *Immunol. Lett.*, *20*, 237 (1989).
39. F. Ellouz, A. Adam, R. Ciorbaru, and E. Lederer, *Biochem. Biophys. Res. Commun.*, *59*, 1317 (1974).
40. C. Merser, P. Sinaÿ and A. Adam, *Biochem. Biophys. Res. Commun.* *66*, 1316 (1975).
41. P. Lefrancier, J. Choay, M. Derrien, and I. Lederman, *Int. J. Peptide Protein Res.*, *9*, 249 (1977).
42. C. Carelli, F. Audibert, and L. Chedid, *Infect. Immun.*, *33*, 312 (1981).
43. F. Audibert, G. Przewlocki, C. Leclerc, M. E. Jolivet, H. Gras-Masse, A. L. Tartar, and L. A. Chedid, *Infect. Immun.*, *45*, 261 (1984).
44. E. Telzak, S. M. Wolff, C. Dinarello, T. Conlon, A. El Kholy, G. Bahr, J. P. Chaoy, A. Morin, and L. Chedid, *J. Infect. Dis.*, *153*, 628 (1986).
45. A. C. Allison and N. E. Byars, *J. Immunol. Meth.*, *95*, 157 (1986).
46. M. E. Jolivet, F. N. Audibert, H. Gras-Masse, A. L. Tartar, D. H. Schlesinger, R. Wirtz, and L. A. Chedid, *Infect. Immun.*, *55*, 1498 (1987).
47. N. E. Byars and A. C. Allison, *Vaccine*, *5*, 223 (1987).
48. N. Khuller, C. M. Gupta, and S. Sehgal, *Immunol. Invest.*, *17*, 1 (1988).
49. K. Onozuka, H. Shinomiya, N. Cho, T. Saito-Taki, and M. Nakano, *Int. J. Immunopharmacol.*, *11*, 781 (1989).
50. H. Manabe, H. Utsumi, T. Kusama, and A. Hamada, *J. Biochem. (Tokyo)*, *105*, 861 (1989).
51. A. J. Morgan, A. C. Allison, S. Finerty, F. T. Scullion, N. E. Byars, and M. A. Epstein, *J. Med. Virol.*, *29*, 74 (1989).
52. M. Tsujimoto, S. Kotani, T. Okunaga, T. Xuko, H. Takada, T. Shiba, S. Kusumoto, T. Takahashi, Y. Goto, and F. Kinoshita, *Vaccine*, 7, 39 (1989).

53. Q. Xiao, G. Gregoriadis, and M. Ferguson, *Biochem. Soc. Trans.*, *17*, 695 (1989).
54. M. W. C. Hatton, S. L. Moar, and M. Richardson, *J. Lab. Clin. Med.*, *115*, 356 (1990).
55. N. Van Rooijen and R. van Nieuwmegen, *Immunol. Commun.*, *9*, 243 (1980).
56. E. Shahum and H.-M. Thérien, *Immunology*, *65*, 315 (1988).
57. A. Steinberger, G. M. Slutzky, J. El-On, and C. L. Greenblatt, *Exp. Parasitol.*, *58*, 223 (1984).
58. J. V. Desiderio and S. G. Campbell, *Infect. Immun.*, *48*, 658 (1985).
59. R. L. Gregory, S. M. Michalek, G. Richardson, C. Harmon, T. Hilton, and J. R. McGhee, *Infect. Immun.*, *54*, 780 (1986).
60. U. Sengupta, S. Sinha, G. Ramu, R. K. Lavan, and C. M. Gupta, *Int. J. Leprosy*, *56*, 45 (1988).
61. G. F. A. Kersten, T. Teerlink, H. J. G. M. Derks, A. J. Verkleij, T. L. Van Wezel, D. J. A. Crommelin, and E. C. Beuvery, *Infect. Immun.*, *56*, 432 (1988).
62. R. L. Richards, M. D. Hayre, W. T. Hockmeyer, and C. R. Alving, *Infect. Immun.*, *56*, 682 (1988).
63. G. F. A. Kersten, A-M. van de Put, T. Teerlink, E. C. Beuvery, and D. J. A. Crommelin, *Infect. Immun.*, *56*, 1661 (1988).
64. R. L. Richards, M. D. Hayre, W. T. Hockmeyer, and C. R. Alving, *Biochem. Soc. Trans.*, *16*, 921 (1988).
65. R. L. Richards, G. M. Swartz, Jr., C. Schultz, N. D. Hayre, G. S. Ward, W. R. Ballou, J. D. Chulay, W. T. Hockmeyer, S. L. Berman, and C. R. Alving, *Vaccine*, 7, 506 (1989).
66. K. Brynestad, B. Babbitt, L. Huang, and B. T. Rouse, *J. Virol.*, *64*, 680 (1990).
67. M. Jolivet, E. Sache, and F. Audibert, *Immunol. Commun.*, *10*, 511 (1981).
68. M. Tsujimoto, S. Kotani, T. Shiba, and S. Kusumoto, *Infect. Immun.*, *53*, 517 (1986).
69. M. Tsujimoto, S. Kotani, F. Kinoshita, S. Kanoh, T. Shiba, and S. Kusumoto, *Infect. Immun.*, *53*, 511 (1986).
70. S. J. LeGrue, I. Saiki, C. A. Romerdahl, and J. Fidler, *Transplantation*, *43*, 584 (1987).
71. N. C. Phillips, M. L. Moras, L. Chedid, P. Lefrancier, and J. M. Bernard, *Cancer Res.*, *45*, 128 (1985).
72. N. C. Phillips, L. Chedid, J. M. Bernard, M. Level, and P. Lefrancier, *J. Biol. Response Mod.*, *6*, 678 (1987).
73. P. N. Shek, B. Y. K. Yung, and N. Z. Stanacev, *Immunology*, *49*, 37 (1983).
74. P. Dal Monte and F. C. Szoka, *J. Immunol.*, *142*, 1437 (1989).
75. P. R. Dal Monte and F. C. Jr. Szoka, *Vaccine*, 7, 401 (1989).
76. M. Parant, *Springer Sem. Immunopathol.*, *2*, 101 (1979).
77. A. Adam and E. Lederer, *Med. Res. Rev.*, *4*, 111 (1984).
78. S. J. LeGrue, *Cancer Immunol. Immunother.*, *17*, 135 (1984).
79. G. Steele, Jr., T. Ravikumar, D. Ross, V. King, R. E. Wilson, and T. Dodson, *Surgery*, *96*, 352 (1984).
80. N. C. Phillips, P. P. Najor, M. Ishida, and H. Sikorska, *J. Biol. Response Mod.*, *6*, 581 (1987).

81. N. C. Phillips, P. P. Major, and H. Sikorska, *Cancer Detect. Prev.*, *12*, 451 (1988).
82. N. C. Phillips, A. Loutfi, M. Kareem, H. Shibata, and M. Baines, in *Liposomes in the Therapy of Infectious Diseases and Cancer* (G. Lopez-Berestein and I. J. Fidler, eds.), Alan R. Liss, New York, 1989, pp. 15–24.
83. N. C. Phillips, A. Loutfi, A. M. A-Kareem, H. R. Shibata, and M. G. Baines, *Cancer Detect. Prev.*, *14*, 491 (1990).
84. N. C. Phillips, *Prog. Clin. Biol. Res.*, *354A*, 257 (1990).
85. N. C. Phillips, *Proc. West. Pharmacol. Soc.*, *32*, 107 (1989).
86. G. Poste, C. Bucana, A. Raz, P. Bugelski, R. Kirsh, and I. J. Fidler, *Cancer Res.*, *42*, 1412 (1982).
87. E. S. Kleinerman, J. L. Murray, J. S. Snyder, J. E. Cunningham, and I. J. Fidler, *Cancer Res.*, *49*, 4665 (1989).
88. N. C. Phillips, *Cancer Metast. Rev.*, *8*, 231 (1989).
89. N. C. Phillips and M.-S. Tsao, *Cancer Immunol. Immunother.*, *33*, 85 (1991).
90. V. S. Byers, R. Rodvien, K. Grant, L. G. Durrant, K. H. Hudson, R. W. Baldwin, and P. J. Scannon, *Cancer Res.*, *49*, 6153 (1989).
91. J. Carlsson, E. Daniel-Szolgay, G. Frykholm, B. Glimelius, A. Hedin, and B. Larsson, *Cancer Immunol. Immunother.*, *30*, 269 (1989).
92. M. J. Losman, H. J. Hansen, R. M. Sharkey, D. M. Goldenberg, and M. Monestier, *Proc. Natl. Acad. Sci. USA*, *38*, 3421 (1991).
93. D. E. Milenic, B. Detrick, J. C. Reynolds, and D. Colcher, *Int. J. Biol. Markers*, *5*, 177 (1990).
94. A. Huang, L. Huang, and S. J. Kennel, *J. Biol. Chem.*, *255*, 8015 (1980).
95. J. D. Rioux, N. C. Phillips, and H. Sikorska, *4th. Int. Congress Cell Biol.* P1.2.27 (1988) (Abstract).
96. P. Gold and S. O. Freedman, *J. Exp. Med.*, *122*, 467 (1965).
97. P. Gold and J. Shuster, *Cancer Res.*, *40*, 2973 (1980).
98. S. Krishnamurthy, J. F. Morris, R. Antonovic, A. Ahmed, W. T. Galey, C. Duncan, and G. T. Krishnamurthy, *Cancer*, *65*, 458 (1990).
99. D. S. Wechsler and P. A. Kongshavn, *J. Immunol.*, *137*, 2968 (1986).
100. F. Skvaril, *Monogr. Allergy*, *19*, 134 (1986).
101. M. S. Kaminski, K. Kitamura, D. G. Maloney, M. J. Campbell, and R. Levy, *J. Immunol.*, *136*, 1123 (1986).
102. B. Holzmann, E. B. Brocker, and J. M. Lehman, *Int. J. Cancer*, *39*, 366 (1987).

9

MEDICAL APPLICATIONS OF INTRAMUSCULARLY ADMINISTERED BROMOCRIPTINE MICROSPHERES

Marcella Montini, Alberto Pedroncelli, Francesca Tengattini, Marina Desiree Pagani, Daniela Gianola, Liana Cortesi, and Giorgio Pagani

Ospedali Riuniti
Bergamo, Italy

Ioana Lancranjan

Sandoz Pharma, Ltd.
Basel, Switzerland

INTRODUCTION

Bromocriptine (2-bromo-α-ergocryptine), an ergot derivative, is a potent dopamine agonist acting on D2 receptors and its activity both in vivo and in vitro mimics the effect of dopamine, causing a rapid suppression of prolactin (PRL) secretion. Unlike dopamine, which has a short-lasting action in vivo and in vitro, bromocriptine has a marked and sustained effect on PRL release (1), which in vivo lasts for several hours after bromocriptine withdrawal.

When it was introduced, bromocriptine (Parlodel, Sandoz Pharma Ltd., Basel, Switzerland) was primarily considered a symptomatic treatment of hyperprolactinemia and its associated symptoms (infertility, amenorrhea, oligomenorrhea, and galactorrhea in women and sexual impotence and infertility in men) (2). Subsequently, important evidence has accumulated showing that Parlodel can also induce marked tumor shrinkage in patients bearing macroprolactinomas, as shown by neuroradiological examination and improvement or normalization of visual field defects. Since Corenblum reported in 1975 the first documented reduction of tumor size in two patients with macroprolactinomas (3), an increasing number of cases of tumor shrinkage associated with suppression of PRL secretion to normal values in patients bearing large PRL-secreting adenomas on treatment with different formulations of bromocriptine, namely, oral Parlodel, Parlodel LA, and Parlodel LAR, have been reported (1,4–31). The tumor shrinkage was documented by tomographs, computed tomography (CT) scans, metrimazide cisternography, and magnetic resonance imaging (MRI). The tumor size reduction results from the involution of adenomatous prolactin cells with reduction in cytoplasmic, nuclear, and nucleolar areas (32,33).

Treatment with bromocriptine or other orally active dopamine agonists may be associated with nausea, vomiting, and postural hypotension at the start of treatment (2). These adverse effects may be particularly severe if full doses are used at the very beginning of treatment. In order to reduce or avoid them, the dose should be built up progressively in 2–3 weeks and this may lead to a critical delay for patients with visual field and/or oculomotor nerve defects.

Therefore, in order to improve the effectiveness (by producing a rapid tumor shrinkage of PRL-secreting tumors) and tolerability of Parlodel, to eliminate difficulty with compliance, and also to offer more comfort of administration, a long-acting depot formulation of bromocriptine, Parlodel LA, was developed for single intramuscular injection. Recently, a new long-acting injectable and repeatable form of bromocriptine, Parlodel LAR, was developed for the long-term treatment of patients with prolactinomas and nontumoral hyperprolactinemia.

PARLODEL LA AND PARLODEL LAR COMPOSITION

Parlodel LA contains 50-mg bromocriptine microspheres prepared according to a spray-drying technique employing polylactic acid as carrier material. The bromocriptine-containing microspheres are suspended in a dextran-containing liquid vehicle by means of a double-chamber syringe immediately before administration. The compound is injected deep intragluteally where these microspheres slowly release the drug. Due to the slow release of bromocriptine from the microspheres, a single intramuscular injection of Parlodel LA produces an inhibition of PRL secretion for 4–6 weeks. Therapeutic plasma concentrations of bromocriptine are reached within the first hours following the injection of Parlodel LA, maximal values being achieved 2–12 hr after injection. The drug release pattern of Parlodel LA is characterized as a two-part parallel process, i.e., an initial fast and an underlying slow-release process. The half-life of the fast-release process is between 10 and 55 min, while the one of the slow-release is between 7 and 22 hr. The polylactic acid, the carrier polymer of Parlodel LA, has a long half-life (approximately 6 months) which makes Parlodel LA suitable for the initial treatment of patients with prolactinomas/hyperprolactinemia and for the prevention/suppression of lactation.

The new injectable form of bromocriptine, Parlodel LAR, employs the glucose-initiated poly(DL-lactide-*co*-glycolide) as carrier polymer. The biodegradation of the carrier material selected for Parlodel LAR was investigated in vitro and in vivo. The mass degradation of the microspheres in vitro determined gravimetrically started after about 15 days and it was almost complete after 61 days, only 5% of the polymer being still present (34). In vivo studies carried out to assess the degradation of the polymer used cages filled with Parlodel LAR or placebo that were implanted intraperitoneally in rats. Biodegradation of Parlodel LAR and placebo was nearly complete 52 days after implantation (35). Similar results were obtained by repeating the experiments in rabbits (36).

The toxicology of Parlodel LAR was investigated in several studies (36,37) aimed at defining both the biocompatibility and the biodegradation of the polymer used as carrier material for the bromocriptine microspheres. The local tolerability and biodegradation of the glucose-initiated poly-(DL-lactide-*co*-glycolide) microspheres with either placebo or Parlodel were assessed in a toxicology study carried out in rats (37). The injection did not produce any local macroscopic findings. Microscopic evaluation showed that there were small groups of microspheres lying in areas of chronic inflammatory infiltrate in either muscle fascicles or interfascicular spaces. The size of these microspheres was variable and most of them were in different stages of biodegradation. Intact microspheres could be seen only occasionally. The local tissue reaction and healing response were of a granulation tissue type and consisted of macrophages, lymphocytes, foreign body multinucleated giant cells, and capillary formations. The tissue reaction was limited to the injection site and the adjacent muscle fibres were not involved. There was no evidence of a capsule formation around the local reaction and inflammation was in no case severe (37). Thus, it has been concluded that single intramuscular injections of both Parlodel-impregnated microspheres and empty microspheres caused minimal to moderate tissue reactions that were limited to the muscle fascicles and interfascicular spaces, while the surrounding tissue remained intact. There was no difference in severity between the Parlodel and placebo microspheres. The type and severity of the muscle reaction resulted similar to those observed in intramuscular tolerance studies using other comparable sustained drug delivery systems previously reported by Gray (38) and Vischer et al. (39).

Polyglycolic acid polymers and polylactic/polyglycolic acid polymers are used as surgical suture materials. Studies on the biocompatibility and biodegradation of these polymers have already been performed in the last decade (40,41). As these polymers are based on substances found in the human body, i.e., glycolic acid and lactic acid, they are not completely foreign materials and all the clinical experience accumulated hitherto supports their good tolerability (40,41).

Therefore, as a result of these biocompatibility and biodegradation studies performed in both rats and rabbits, it was concluded that Parlodel LAR can be administered repeatedly.

PHARMACOKINETIC/PHARMACODYNAMIC AND CLINICAL EFFICACY STUDY COMPARING PARLODEL LAR TO ORAL PARLODEL

The aim of this study was to assess (a) the efficacy in suppressing PRL secretion and improving clinical symptoms/signs of hyperprolactinemia of either Parlodel LAR 50 mg or oral Parlodel administered daily for 28 days, (b) any differences between these two forms of bromocriptine as to circulating bromocriptine levels, and (c) tolerability both local and systemic.

The study was carried out in a double-blind double-dummy fashion in parallel groups of hyperprolactinemic patients. Twenty-five female patients, aged 17–43 years (mean age 31.2 ± 7 years), received a single im injection of either Parlodel LAR 50 mg dose and placebo for oral Parlodel (13 patients) or placebo for Parlodel LAR and the active oral compound at the full dose of 2.5 mg three times a day (tid) gradually built up in a week (12 patients). All patients received placebo or active oral Parlodel tid for 28 days.

Blood was sampled for PRL and bromocriptine assessment at 08:00, 09:00, 10:00, 11:00, 12:00, 14:00, 16:00, and 20:00 hr on days 0 (baseline), 1 (first day of either treatment), 3, 7, 14, 28, 35, and 42. Serum PRL was measured by immunoradiometric assay, using as standard MRC 75/504. Normal serum PRL concentrations are taken as below 400 mU/L with a midrange interassay coefficient of variation of 3.8%. The limit of detection for the assay is 9 mU/L.

Plasma bromocriptine was measured by a radioimmunological assay with a limit of sensitivity of 10.1 pg/ml (42,43).

Prior to the treatment, patients presented with galactorrhea (18/25 patients), amenorrhea (16/25 women), oligomenorrhea (3/25 women), libido disturbances (8/25 patients), and headache (11/25 patients).

All patients had high-resolution CT scans (General Electric 9800 scanner) demonstrating an appearance compatible with the presence of a microadenoma in 10 of them.

The mean PRL concentrations at 8 o'clock in the morning on the first day of treatment administration were similar between the two groups. The decrease in the mean serum PRL levels was more marked in the Parlodel LAR group than in the oral Parlodel one on the first treatment day (Figure 1). Serum PRL concentrations decreased to within the normal range in

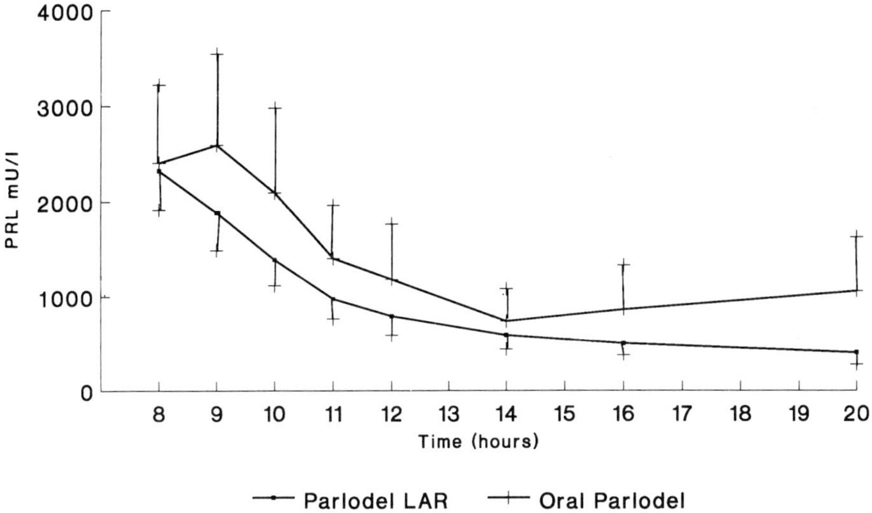

Figure 1 Serum PRL profile on the first day of drug administration in both groups of patients.

12/13 patients in the Parlodel LAR group, in 10/12 patients 12 hr following the injection of Parlodel LAR, and in 2/12 patients after 3 days, whereas in the oral Parlodel group, 7/12 patients reached normal PRL values, 5/7 patients after 12 hr, 1/7 after 7 days, and 1/7 after 14 days.

In the Parlodel LAR group, a constant suppression of PRL secretion to within the normal range was observed in 7/13 patients on day 14 and in 5/13 patients on day 28. On postinjection day 35, PRL secretion was suppressed to normal values in 2/13 patients whereas on day 42 all patients had serum PRL levels above the normal range.

The patients who received oral Parlodel showed a more gradual decrease in serum PRL when compared to the Parlodel LAR group. Among this group of patients, serum PRL concentrations were consistently within the normal range in five patients on day 7 and six patients on days 14 and 28. None of the patients had their serum PRL suppressed to normal values on either days 35 or 42, i.e., 7 and 14 days, respectively, after the discontinuation of the oral treatment.

The mean plasma concentrations of bromocriptine after administration of Parlodel LAR 50 mg reached a value of 1921 ± 467 pg/ml (mean ± SEM) at 3 hr and further increased to 2151 ± 336 pg/ml at 12 hr (Figure 2) and they remained at approximately this level until day 3. The plasma bromocriptine levels declined faster during days 3 to 14 and more slowly there-

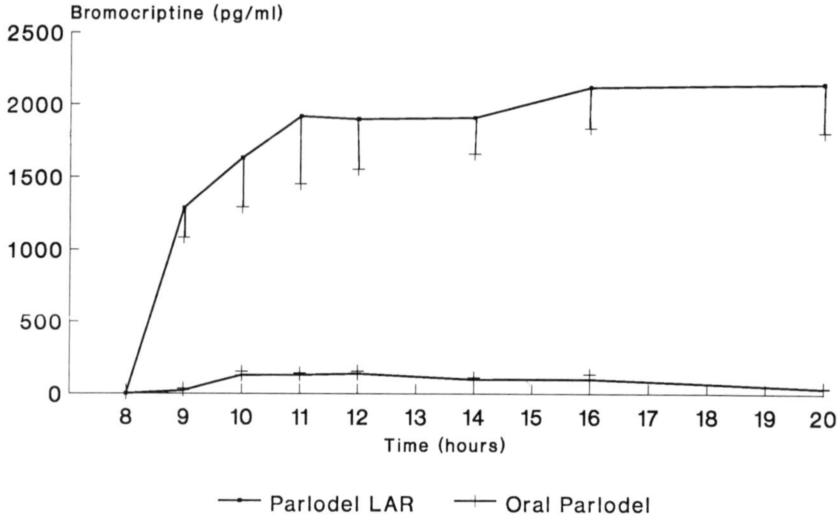

Figure 2 Plasma bromocriptine profile on the first day of drug administration in both groups of patients.

after. The mean plasma concentrations were 346 ± 71 pg/ml on day 28, decreasing to 108 ± 16 pg/ml on day 42 (Figure 3). Taking into considerations the median value, peak plasma concentrations of bromocriptine following the administration of Parlodel LAR occurred between days 1 and 2 (Table 1). Circulating bromocriptine was thus detectable and therapeutically effective up to day 42, pointing out a slow release of the Parlodel LAR preparation. The release half-life calculated using the mean plasma profile was 10 days.

The mean plasma concentrations of bromocriptine following the oral administration of the first dose of Parlodel 2.5 mg reached a value of 128 ± 14 pg/ml at 3 hr, decreasing to 37 ± 6 pg/ml at 12 hr (Figure 2). On day 3 mean bromocriptine levels increased to 106.0 ± 21.3 pg/ml (dose of oral Parlodel: 5 mg daily) and they further raised to 179.87 ± 16.8 pg/ml on day 7 (dose of oral Parlodel: 7.5 mg daily). On day 28 the mean bromocriptine concentrations had a further increase to 202.2 ± 17.3 pg/ml, falling to values below the limit of detection on postinjection days 35 and 42 (Figure 3). Peak plasma concentrations of bromocriptine after oral administration of the first dose of Parlodel 2.5 mg occurred at around 4 hr (Table 2).

Of the 12 patients in the Parlodel LAR group reporting abnormal menstrual cycles at baseline, 9 resumed normal menses after a single administration of Parlodel LAR 50 mg. Also 4/7 patients with abnormal menses at

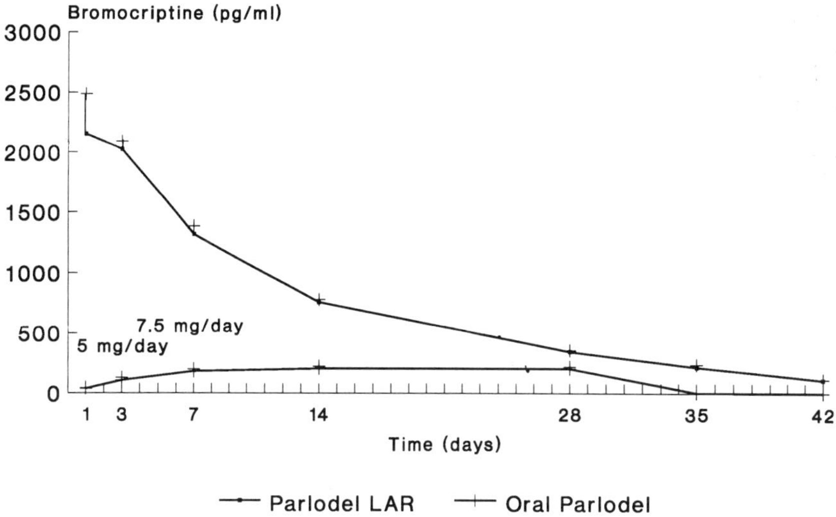

Figure 3 Plasma bromocriptine levels over the 42-day study period in both groups of patients.

Table 1 Mean ± SD and (Median) Values of the Pharmacokinetic Parameters of Bromocriptine After Administration of Parlodel LAR

Parameter	Parlodel LAR group
t_{max} (days)	4.9 ± 10.7 (1.8)
C_{max} (pg/ml)	2800 ± 1300 (2500)
AUC (0–42 days) (pg/ml/days)	31,800 ± 5800 (33,600)

baseline in the oral Parlodel group resumed normal menses within 28-day treatment. Libido normalized in all patients who had complained of it in both groups of treatment. Galactorrhea disappeared in 6/9 patients in the Parlodel group and in 5/9 patients treated with oral Parlodel. Headache was relieved in 6/7 patients in the Parlodel LAR group and in 2/4 patients in the oral Parlodel group (Table 3).

Table 2 Mean ± SD and (Median) Values of the Pharmacokinetic Parameters of Bromocriptine Calculated on the First Day of Administration of Oral Parlodel 2.5 mg

Parameter	Oral Parlodel
t_{max} (hr)	4.0 ± 2.0 (4.0)
C_{max} (pg/ml)	200 ± 100 (200)
AUC (0–12 hr) (pg/ml/hr)	1000 ± 500 (1000)

Adverse events in both groups were mostly mild and transient. The most commonly reported side effects were nausea, few episodes of vomiting, and orthostatic symptoms. All symptoms occurred mostly on the first day of treatment and the majority of them disappeared spontaneously. There was no difference between the two groups as to incidence, severity, and duration. No pain at the injection site was reported by any patient.

Although both formulations of bromocriptine, i.e., Parlodel LAR and oral Parlodel, were very efficacious in lowering serum PRL levels, a higher number of patients in the Parlodel LAR group (12/13) than in the oral Parlodel group (7/12) reached normal PRL values during the 42-day observation period. This was more marked on the first day of treatment: 10 of the 13 patients who received Parlodel LAR achieved normal serum PRL levels at 12 hr compared to 5 of the 12 patients who received oral Parlodel.

The plasma concentrations of bromocriptine were much higher for the Parlodel LAR formulation as compared to the oral preparation, which could be due to incomplete absorption and metabolism. Mean plasma bromocriptine levels were within the therapeutic range during the whole 42-day study period in the Parlodel LAR group.

DOSE PROPORTIONALITY STUDY WITH DIFFERENT DOSES OF PARLODEL LAR, NAMELY, 25 mg, 50 mg, 100 mg: PHARMACOKINETIC, PHARMACODYNAMIC, AND CLINICAL DATA

This study was undertaken to assess the relationship of different doses of Parlodel LAR, i.e., 25 mg, 50 mg, and 100 mg, in terms of (a) the degree and duration of PRL suppression, (b) circulating bromocriptine levels, and (c) tolerability. It was carried out in a double-blind randomized fashion in

Table 3 Symptoms/Signs of Hyperprolactinemia Prior to and After Parlodel LAR 50 mg or Oral Parlodel 2.5 mg tid

Pt.	Treatment	Menstrual disturbance/ resumption of normal menses	Galactorrhea/ disappearance	Headache/ disappearance
1	Parlodel LAR	Yes/–	No	Yes/28 days
2	Parlodel LAR	Yes/–	No	Yes/14 days
3	Oral Parlodel	Yes/–	No	No
4	Oral Parlodel	Yes/–	Yes/–	Yes/–
5	Oral Parlodel	No	No	Yes/–
6	Oral Parlodel	Yes/35 days	Yes/14 days	No
7	Parlodel LAR	Yes/28 days	Yes/28 days	No
8	Parlodel LAR	Yes/–	Yes/–	Yes/42 days
9	Oral Parlodel	Yes/28 days	Yes/28 days	No
10	Parlodel LAR	Yes/35 days	Yes/–	Yes/–
11	Oral Parlodel	No	Yes/–	No
12	Parlodel LAR	Yes/28 days	Yes/28 days	No
13	Parlodel LAR	No	No	No
14	Parlodel LAR	Yes/7 days	Yes/28 days	Yes/28 days
15	Parlodel LAR	Yes/28 days	Yes/28 days	No
16	Oral Parlodel	Yes/35 days	Yes/28 days	No
17	Parlodel LAR	No	Yes/7 days	No
18	Parlodel LAR	Yes/35 days	Yes/28 days	Yes/28 days
19	Oral Parlodel	Yes/35 days	Yes/28 days	No
20	Oral Parlodel	No	Yes/–	Yes/28 days
21	Oral Parlodel	No	Yes/42 days	Yes/42 days
22	Parlodel LAR	Yes/28 days	Yes/–	Yes/28 days
23	Parlodel LAR	Yes/28 days	No	No
24	Oral Parlodel	Yes/–	Yes/–	No
25	Parlodel LAR	Yes/28 days	No	No

21 female patients (7 patients/dose) with both prolactinomas (15 patients) and nontumoral hyperprolactinemia (6 patients).

Prior to the study, galactorrhea was recorded in 16/21 patients and oligomenorrhea/amenorrhea in 18/21 patients. Headaches were reported by 7/21 patients.

Blood was sampled for PRL and bromocriptine assessment at 08:00, 09:00, 10:00, 11:00, 12:00, 14:00, 16:00, and 20:00 hr prior to, on days 1, 14, and 28, while on days 2–7, 21, 42, and 60 one blood sample for PRL and bromocriptine estimation was collected at 8 o'clock in the morning.

Serum PRL was measured by immunoradiometric assay, using as standard MRC 75/504. Normal serum PRL is taken below 400 mU/L with a midrange interassay coefficient of variation of 3.8%. The limit of detection for the assay is 9 mU/L.

Plasma bromocriptine was measured by a radioimmunological technique with a limit of sensitivity of 10.1 pg/ml (42,43).

Serum PRL levels normalized in 19/21 patients, in 18/19 after 12 hr and 1/19 on day 3 following the administration of Parlodel LAR. The suppression of PRL secretion lasted 28 days in 4/7 patients treated with either the 25 mg or 50 mg dose and in 5/7 patients treated with the 100 mg dose. In 5/7 patients treated with Parlodel LAR 100 mg, PRL secretion was suppressed to normal values for 60 days (Table 4).

Table 4 Effect of Treatment with Parlodel LAR 25, 50, or 100 mg on Serum PRL Concentrations

Pt.	Mean PRL basal levels (mU/L)	Nadir PRL levels (mU/L)	Time to get nadir PRL levels (days)	Mean PRL levels on day 28 (mU/L)	Duration of normalized PRL levels (days)	Dose given (mg)
1	1860	114	0.5	380	28	25
2	840	44	0.5	158	28	25
3	7080	840	3.0	4460	0	25
4	1600	88	0.5	560	7	25
5	900	124	0.5	260	42	25
6	1140	206	0.5	740	14	25
7	700	24	0.5	320	28	25
8	3660	328	0.5	1200	7	50
9	1780	220	0.5	88	42	50
10	940	48	0.5	68	42	50
11	1080	220	0.5	520	14	50
12	640	90	0.5	102	28	50
13	680	246	0.5	600	14	50
14	2920	266	3.0	270	28	50
15	1520	88	0.5	48	60	100
16	2220	300	0.5	480	1	100
17	2340	580	0.5	1060	0	100
18	2280	260	0.5	18	60	100
19	1180	68	0.5	78	60	100
20	940	112	0.5	300	60	100
21	760	120	0.5	200	60	100

A comparison between the effects of the 25-mg, 50-mg, and 100-mg doses of Parlodel LAR showed that the maximal PRL-suppressing effect was attained at around 3–4–5 days for all three dosages. The time during which the PRL secretion was more than 50% suppressed was 32 days for the 25-mg dose, 51 days for the 50-mg dose, and 60 days for the 100-mg dose.

The statistical analysis by paired Student's *t* test did not show any significant correlation between the dose of Parlodel LAR and the nadir decrease in serum PRL levels. A positive correlation was conversely seen between the mean length of PRL suppression by various doses of Parlodel LAR. A significant difference by paired Student's *t* test was found only comparing the patients treated with Parlodel LAR 25 mg and Parlodel LAR 100 mg ($p < .05$). Neither the nadir serum PRL levels nor the percentage of patients who experienced a suppression of PRL secretion to normal values was dose-dependent. Conversely, a significant longer lasting suppression of serum PRL concentrations was found in patients treated with the 100-mg dose of Parlodel LAR.

Mean plasma concentrations of bromocriptine reached a peak value of 1609 ± 243.3 pg/ml (mean ± SEM) at 4 hr (12:00 hr) following the injection of Parlodel LAR 25 mg (Figure 4) and they stayed at around the same level for 3 days. Thereafter, circulating bromocriptine steadily declined and it was

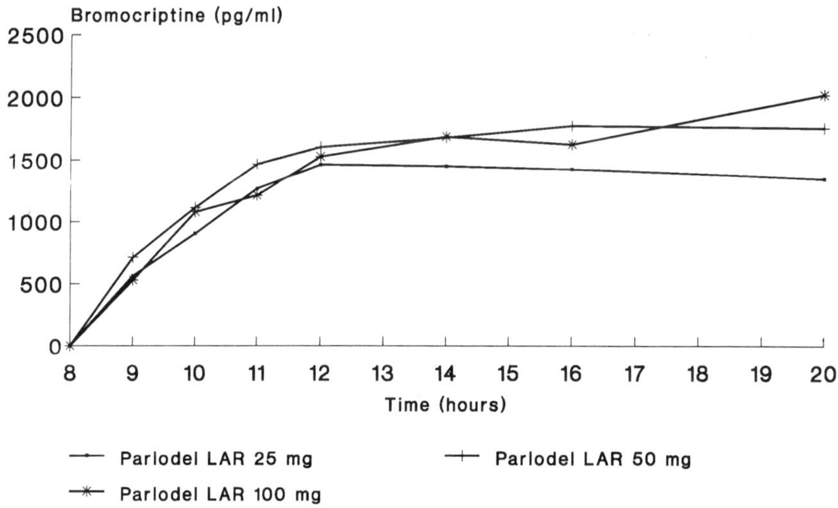

Figure 4 Plasma bromocriptine profile on the first day of administration of Parlodel LAR 25, 50, or 100 mg.

665.6 ± 85.9 pg/ml on day 7, decreasing to 229.9 ± 28.28 pg/ml on day 28. The mean plasma bromocriptine concentrations were below the limit of detection on day 60 (Figure 5).

Mean plasma bromocriptine levels, after the administration of Parlodel LAR 50-mg dose, reached a peak value of 1779 ± 342.4 pg/ml at 6 hr (14:00 hr) (Figure 4) and they stayed at around the same level until day 6. Subsequently, they steadily decreased and they were 1171 ± 188.1 pg/ml on day 7, further declining to 524 ± 104 pg/ml on day 28. Circulating bromocriptine was slightly below the therapeutic threshold on day 60, mean bromocriptine concentrations being 86.17 ± 37.7 pg/ml (Figure 5).

After the administration of Parlodel LAR 100-mg dose, mean plasma concentrations of bromocriptine reached a peak value of 2031 ± 253 pg/ml at 12 hr (20:00 hr) (Figure 4). They further increased to 2381 ± 423.9 pg/ml on day 2 and stayed at around the same level for 7 days. Thereafter, they steadily declined and were 945.7 ± 155.5 pg/ml on day 21, decreasing to 440.8 ± 83.26 pg/ml on day 42. The mean plasma concentrations were 295 ± 81.37 pg/ml on day 60, so that circulating bromocriptine was still therapeutically effective even after 60 days following administration (Figure 5).

Peak of plasma concentrations of bromocriptine after administration of Parlodel LAR 25-mg dose was reached around day 1 and around day 2 following the injection of 50 mg Parlodel LAR. With regard to the 100-mg

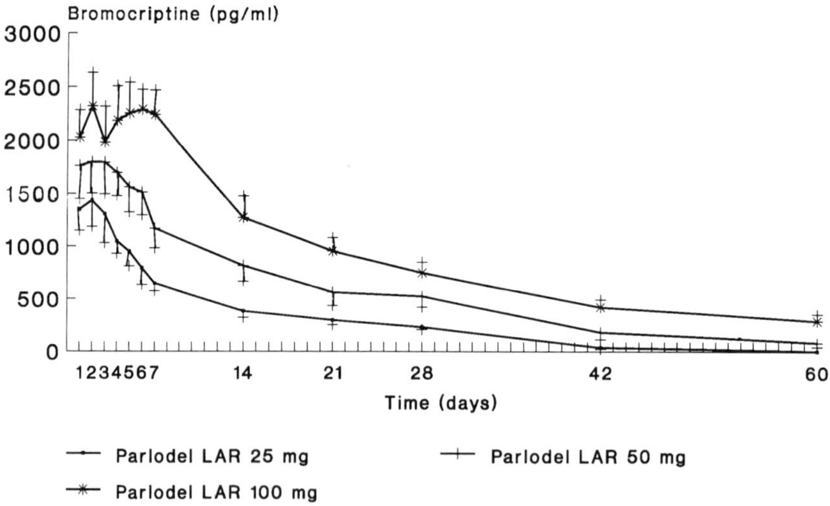

Figure 5 Plasma bromocriptine levels over the 60-day study period in all groups of patients.

dose, peak of plasma concentrations of bromocriptine occurred between days 2 and 3 (Table 5).

Galactorrhea disappeared in 4/16 patients and 16/18 patients had the resumption of normal menses after the first dose of Parlodel LAR. Headaches were relieved in 3/7 patients (Table 6).

Adverse events were mostly mild and transient: nausea was reported by 14/21 patients for 1 hr to 2 days, one to four vomiting episodes occurred in 10/21 patients on the injection day, and postural hypotension occurred in 11/21 patients on days 1 to 3. No significant difference in the incidence, severity, or duration of adverse reactions was observed between the three dosages.

One should note that 8/21 patients included in this study were known as intolerant of oral dopamine agonists. Although all these patients experienced adverse events after the injection of Parlodel LAR, they were transient and mostly mild in all but one case. The treatment with Parlodel LAR by monthly injections was thereafter continued in all patients previously intolerant of oral dopamine agonists.

The results of this study give evidence of the high effectiveness of Parlodel LAR in lowering serum PRL levels in patients with prolactinomas or nontumoral hyperprolactinemia. Normalization of PRL secretion attained in 19/21 patients lasted for at least 28 days in 13/21 patients. One patient treated with the 25-mg dose, two with the 50-mg dose, and five with the 100-mg dose showed a longer lasting, i.e., 42 days, normalization of serum PRL values; indeed, in 5/7 patients who received 100 mg of Parlodel LAR, PRL secretion was suppressed to normal values for the whole 60-day observation period.

Clinical data show a rapid regression of clinical symptoms/signs of hyperprolactinemia in almost all patients.

Although the incidence of side effects is higher in this study than in those previously reported (19,22,25,27), this fact was probably due to the great number of patients known as intolerant of dopamine agonists (8/21

Table 5 Mean ± SD Values of the Pharmacokinetic Parameters of Bromocriptine After a Single Administration of Parlodel LAR, 25, 50, or 100 mg

Parameter	Parlodel LAR 25 mg	Parlodel LAR 50 mg	Parlodel LAR 100 mg
t_{max} (days)	0.87 ± 0.77	1.95 ± 1.66	2.55 ± 2.20
C_{max} (pg/ml)	1,782 ± 664	2,260 ± 858	2,726 ± 767
AUC (0–60 days) (pg/ml/days)	16,328 ± 4,435	32,875 ± 9,687	50,647 ± 11,544

Table 6 Symptoms/Signs of Hyperprolactinaemia Prior to and After One Injection of 25, 50 or 100 mg Dose of Parlodel LAR

Pt	Pituitary tumor	Menstrual disturbance/ resumption of normal menses (days)	Galactorrhea/ disappearance (days)	Headache/ disappearance (days)
1	Microadenoma	Yes/32 days	Yes/–	No
2	–	Yes/28 days	No	No
3	Macroadenoma	Yes/–	Yes/–	Yes/28 days
4	–	Yes/18 days	Yes/–	Yes/–
5	Microadenoma	Yes/35 days	Yes/28 days	Yes/–
6	Microadenoma	No	Yes/–	No
7	Microadenoma	Yes/16 days	Yes/–	No
8	Microadenoma	Yes/30 days	Yes/–	No
9	–	Yes/30 days	Yes/42 days	No
10	–	Yes/16 days	No	Yes/–
11	Microadenoma	Yes/58 days	Yes/–	No
12	Microadenoma	No	Yes/–	No
13	Microadenoma	Yes/32 days	No	No
14	Microadenoma	Yes/28 days	Yes/32 days	No
15	Microadenoma	Yes/21 days	No	Yes/60 days
16	–	Yes/17 days	Yes/–	No
17	Residual tumour	Yes/–	Yes/–	No
18	Macroadenoma	Yes/35 days	Yes/28 days	No
19	–	Yes/37 days	Yes/–	No
20	Microadenoma	Yes/58 days	No	Yes/28 days
21	Microadenoma	No	Yes/–	Yes/–

patients) who were included in this study. They were accepted based on data showing that Parlodel LAR is well tolerated in patients intolerant of oral dopamine agonist therapy (44). The lack of relationship between the incidence, severity, and duration of adverse events and the dose used points out the patients' individual sensitivity to dopamine agonists, which seems to be more important than the dose of Parlodel LAR used at the start of therapy.

In conclusion, these data support the excellent efficacy and good tolerability of Parlodel LAR in the treatment of patients with hyperprolactinemia. Moreover, it has been pointed out that the highest dose, i.e., 100 mg, provides better results in terms of efficacy, i.e., longer lasting suppression of PRL secretion.

LONG-TERM TREATMENT WITH PARLODEL LAR IN PATIENTS WITH PROLACTIN-SECRETING MACROADENOMAS: EFFICACY, TOLERABILITY, AND SAFETY RESULTS

Aim of this study was to assess the efficacy of Parlodel LAR (a) in suppressing PRL secretion, (b) improving clinical symptoms/signs of hyperprolactinemia, (c) improving symptoms/signs related to the pituitary tumor, and (d) tolerability, both local and systemic, during long-term treatment in patients with macroprolactinomas.

Patients and Methods

Thirty-nine patients (24 men and 15 women), aged 16–68 years, with prolactin-secreting pituitary macroadenomas were treated with Parlodel LAR. Four of the 39 patients received only one injection of Parlodel LAR, whereas the other 35 patients were included in a long-term study (4–64 months).

The initial dose of Parlodel LAR was 25 mg in one patient, 50 mg in 32 patients, and 100 mg in 6 patients. During the long-term treatment, Parlodel LAR was administered as intramuscular monthly injections at the dose of 50 mg in 14 patients, 75 mg in 1 patient, and 100 mg in 20 patients.

Following the first administration of Parlodel LAR, blood was sampled for PRL assessment at 08:00, 09:00, 10:00, 11:00, 12:00, 14:00, 16:00, and 20:00 hr on days 0 (baseline), 1 (day of the injection), 3, 7, 14, and 28. During the long-term treatment, blood samples for PRL estimation were collected at 08:00, 10:00, 12:00, 16:00, and 20:00 hr on the day of the injection and the day before.

Serum PRL was measured by immunoradiometric assay, using as standard MRC 75/504. Normal serum PRL is taken below 400 mU/L with a midrange interassay coefficient of variation of 3.8%. The limit of detection for the assay is 9 mU/L.

CT scan (General Electric 9800 scanner) or MRI (1.5 Signal Tesa scanner) was performed before starting treatment and after 1, 6, and 12 months of therapy. Thereafter, the neuroradiological examination was conducted yearly.

Visual field perimetry assessed by the Goldman machine was carried out at baseline and it was repeated on postinjection days 1, 3, 5, 7, 14, and 28 after the first injection and then after 3 and 6 months, should any abnormality be present at study entry.

Prior to the treatment, galactorrhea was present in 12/39 patients. At study entry, 3 of the 15 women were at postmenopausal stage. Eleven of the 12 women who did not reach menopause reported amenorrhea and 1/12 oligomenorrhea. Libido disturbances were recorded by 22/39 patients. Potency alterations were also reported by 15/24 men, either slight (2/15 men) or severe (13/15 men). Tumor-related disturbances consisted of headaches in 20/39 patients and visual field defects in 9/39 patients.

Six of the 39 patients had previously undergone neurosurgery and 2 of them had subsequently received radiotherapy. Prior to Parlodel LAR, 4/39 patients had been submitted to neurosurgery followed by oral Parlodel treatment. Pharmacotherapy consisting of orally active dopamine agonists had been previously administered to 7/39 patients.

Hematology and biochemistry safety tests together with ECG recordings were performed every 3–6 months.

Results

Serum PRL Levels

Prior to the injection of Parlodel LAR, mean serum PRL concentrations were 62,706 ± 21,338 mU/L (mean ± SEM). On day 1 after the first injection, serum PRL levels decreased to within the normal range in 2/39 patients (5%) and to more than 50% of the pretreatment levels in 34/39 patients (Figure 6). Mean serum PRL values decreased from 62,706 ± 21,338 mU/L to 15,870 ± 4898 mU/L 12 hr after the administration of Parlodel

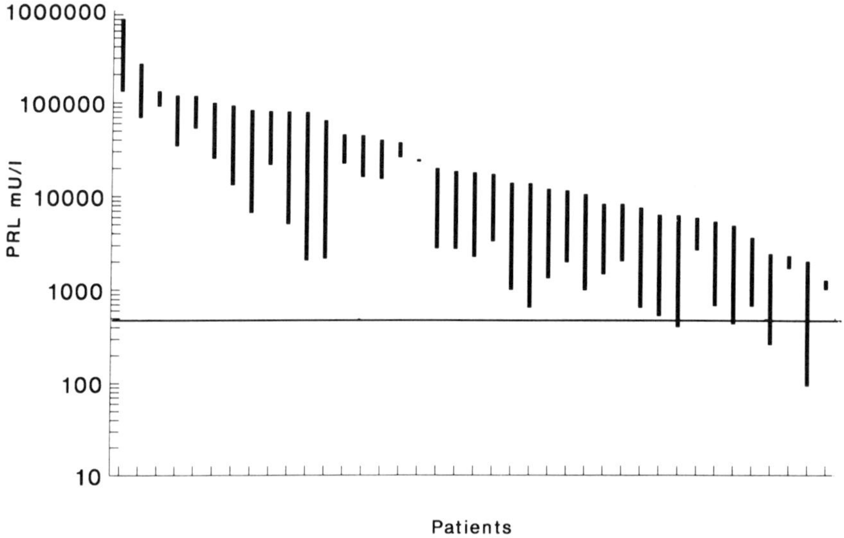

Figure 6 PRL values in patients with macroprolactinomas before and 12 hr after the first dose of Parlodel LAR.

LAR. Twelve of the 39 patients (31%) reached normal PRL values during the 28-day follow-up. A long-lasting suppression of PRL secretion, i.e., up to day 28, to normal values was noted in 6/39 patients (15.4%) after the first injection of Parlodel LAR (Figure 7). Although 27/39 patients had their serum PRL concentrations constantly above the normal range during the 28-day follow-up period, 20/27 patients had their PRL levels suppressed to more than 50% on postinjection day 28.

During the long-term treatment with Parlodel LAR, a long-lasting and constant suppression of PRL secretion to within the normal range, i.e., up to day 28, 35, 42, 49, 60, and 70, was observed in 21/35 patients (60%) after 1–21 injections of Parlodel LAR (Figure 8). Of the 14 patients who did not reach normal PRL values during the long-term treatment, all of them showed a decrease in their PRL levels by more than 50% on day 28 after each subsequent injection when compared to baseline values.

The administration interval could be extended to 35 days in 4/35 patients, 42 days in 2/35 patients, 49 days in 1/35 patients, 60 days in 1/35 patients, and 70 days in another patient (Table 7).

Neuroradiological Examination

A significant tumor shrinkage (20–60%) was documented by CT scan/MRI examination in 20/39 patients after the first injection of Parlodel LAR (Figure

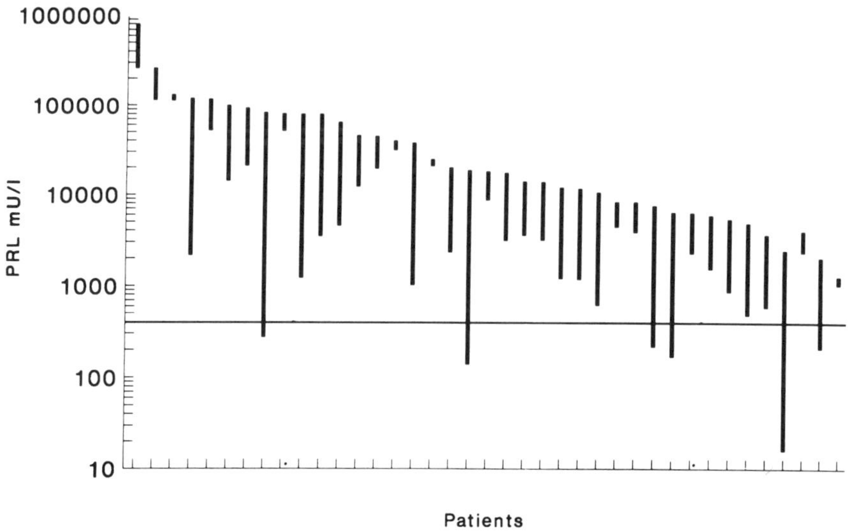

Figure 7 PRL values in patients with macroprolactinomas before and on day 28 after the first dose of Parlodel LAR.

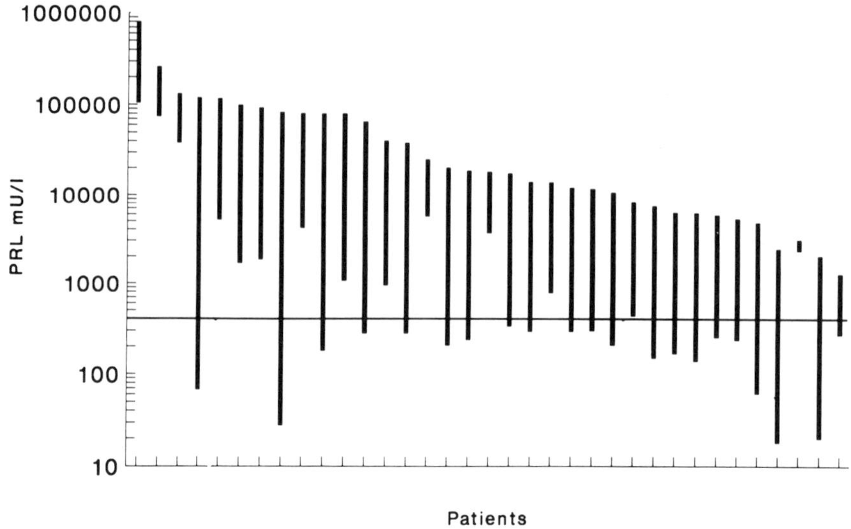

Figure 8 PRL values in patients with macroprolactinomas before and during long-term treatment with Parlodel LAR.

9). Four patients showed a further significant tumor shrinkage during the long-term treatment with Parlodel LAR: 2 patients after 6 months (percentage reduction in tumor height: 56 and 76%, respectively, of the baseline size), 1 patient after 12 months (tumor shrunk by 62% when compared to the pretreatment size), and the other patient after 24 months (tumor shrinkage of 34% with respect to the pretreatment size). Three patients had an empty sella at the CT scan/MRI control performed after 6 months (2 patients) and

Table 7 Administration Interval of Parlodel LAR 50–200 mg in Patients Bearing Macroprolactinomas During Long-Term Treatment

Days	No. patients
28	26/35
35	4/35
42	2/35
49	1/35
60	1/35
70	1/35

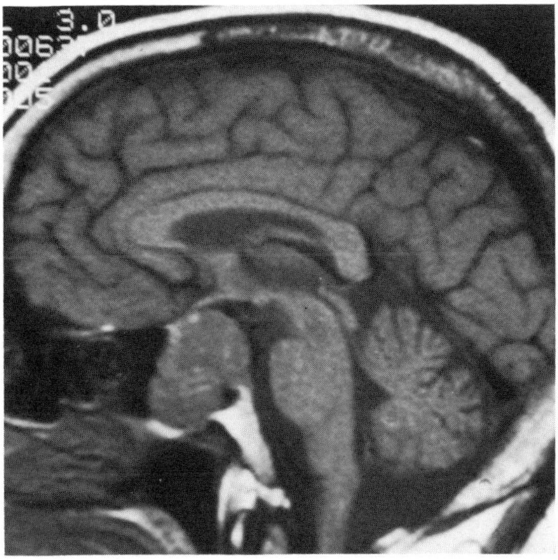

(A)

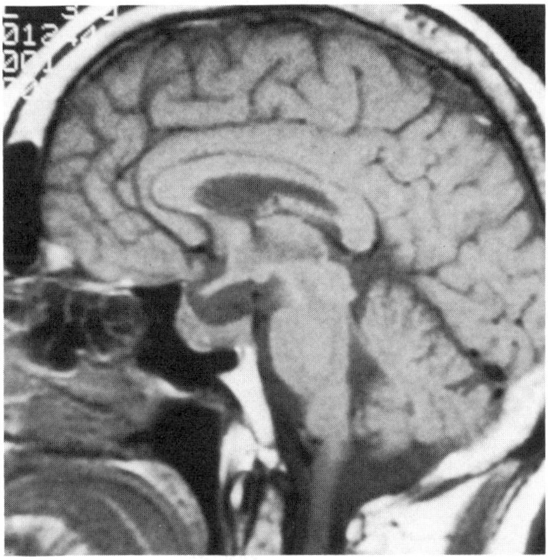

(B)

Figure 9 (A) MRI examination of the pituitary fossa performed prior to the treatment and (B) on day 21 after the first injection of Parlodel LAR in a patient bearing macroprolactinoma.

24 months (1 patient) of treatment, suggesting a radiological disappearance of the pituitary tumor (Figure 10).

Clinical Symptoms/Signs (Table 8)

After the first injection galactorrhea disappeared in 8/12 patients who reported this sign at study entry. In other two patients galactorrhea disappeared after 9 and 15 injections, respectively. Two of the 11 women with amenorrhea prior to the treatment resumed normal menses after the first injection of Parlodel LAR. During the long-term treatment, 7/11 women reported regular menses. Regular menses were also recorded by the only oligomenorrheic woman after 8 months of treatment. Five of the 22 patients who reported diminished libido prior to the start of this study noted a normalization after the first administration of Parlodel LAR. In 14/22 patients, libido normalized after 2–18 injections. Sexual performance was improved in all men who had complained of decreased libido or impotence before treatment. A normali-

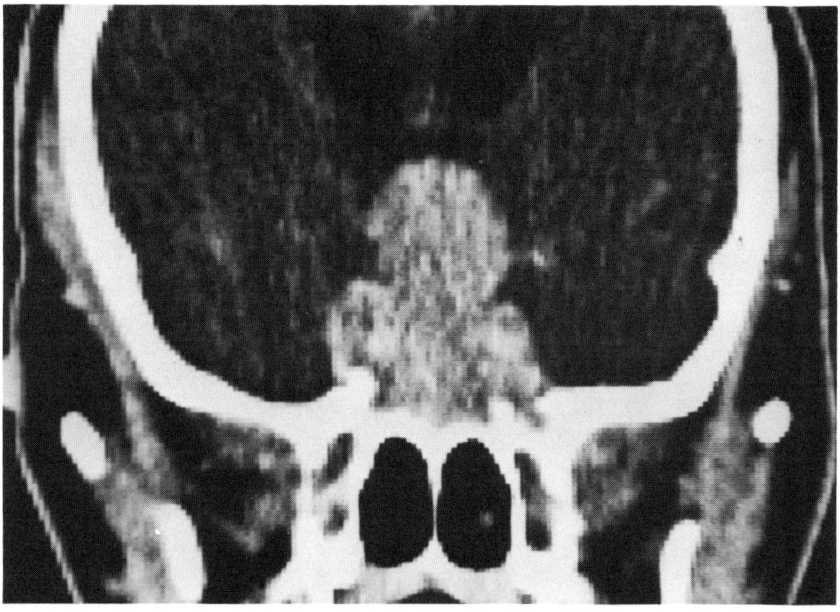

(A)

Figure 10 (A) CT scan examination of the pituitary fossa performed prior to the treatment and (B) after 6-month treatment with Parlodel LAR in a patient bearing macroprolactinoma. Complete radiological disappearance of the pituitary tumor can be seen.

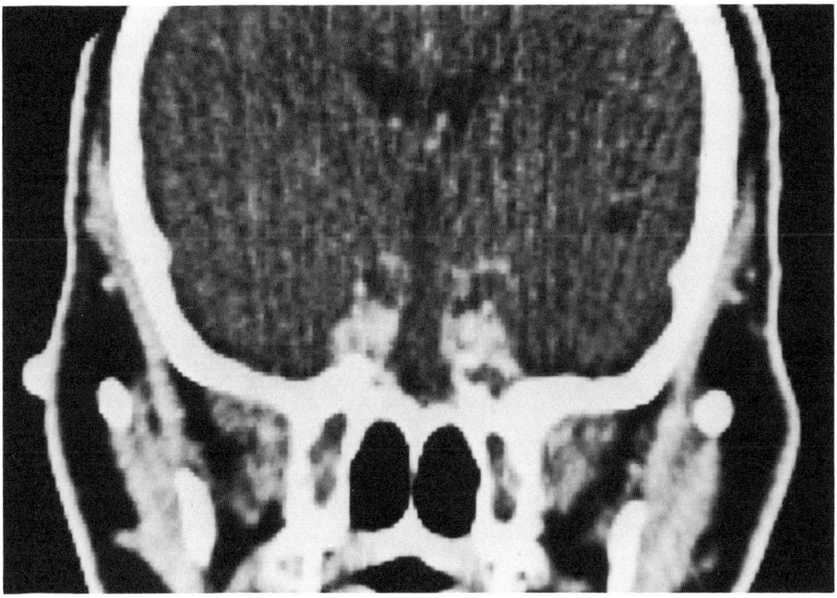

(B)

zation of potency was noted in 10/15 men: in 4/10 after the first injection and in other 6 men after 2–18 doses of Parlodel LAR. Headaches were relieved in 18 of the 20 patients who had suffered them: in 10/18 after the first month and in other 8 patients within 2–12 months. Visual field perimetry normalized in 7/9 patients within 2–14 days after the first injection of Parlodel LAR.

Adverse Events

Adverse events were mostly mild and transient. After the first injection of Parlodel LAR, the most frequently reported side effects were nausea (9/39), vomiting (4/39), and postural hypotension (5/39). They were reported on the day of the injection, they were short-lasting, and they disappeared spontaneously within the first 24 hr following the injection. Anyway, tolerability to Parlodel LAR further improved during the long-term treatment, side effects being reported by only two patients on the day of the injection. In general, tolerability can be rated as good or very good in every patient.

One should note that of the 11 patients who had previously received orally-active dopamine agonists, 6 were intolerant of them. All these patients tolerated Parlodel LAR very well.

Table 8 Symptoms/Signs of Hyperprolactinemia/Prolactinoma Prior to and After First Injection and During Long-Term Treatment with Parlodel LAR in Patients with Macroprolactinomas

			Disappearance	
Clinical symptom/sign	Baseline	Disappearance	After 1st inj.	During long-term treatment (months)
Galactorrhea	12/39 patients	10/12 patients	8/10 patients	2/10 patients (15)
Oligo/amenorrhea	12/12 women	10/12 women	2/10 women	8/10 women (2–15)
Impotence	15/24 men	10/15 men	4/10 men	6/10 men (2–18)
Decreased libido	22/39 patients	19/22 patients	5/19 patients	14/19 patients (2–18)
Headache	20/39 patients	18/20 patients	10/18 patients	8/18 patients (2–12)
Visual field deficit	9/39 patients	7/9 patients	7/7 patients	–

No pain at the injection site was reported by any patient.

During the long-term treatment, no impairment in hematology and biochemistry safety tests was observed. No newly occurring abnormalities in ECG recordings were noted in any patient.

Conclusions

The long-term treatment with Parlodel LAR in patients with macroprolactinomas gives evidence that this new formulation of bromocriptine is very efficacious and well tolerated during the long-term application. A rapid and long-lasting suppression of PRL secretion was observed in the majority of patients who resumed normal menses or potency and showed a progressive disappearance of the clinical symptoms/signs of hyperprolactinemia. Besides, the interval between each subsequent injection could be extended to 35–70 days in 9/39 patients. A rapid shrinkage of the pituitary tumor occurred in most patients within the first month of treatment. Moreover, in 3/39 patients was observed a complete radiological disappearance of their macroprolactinomas. Tolerability was rated as good or very good in all patients; even the patients who had been previously intolerant of oral dopamine agonist agents tolerated Parlodel LAR very well. There were no impairment effects noted on the safety parameters. The compliance of the patients to the long-term treatment, i.e., up to 64 months, was very good.

LONG-TERM TREATMENT WITH PARLODEL LAR IN PATIENTS WITH HYPERPROLACTINEMIC DISORDERS: EFFICACY, TOLERABILITY, AND SAFETY RESULTS

The aim of this study was to assess (a) the efficacy in suppressing PRL secretion and improving clinical symptoms/signs of hyperprolactinemia, (b) tolerability, both local and systemic, and (c) safety of the long-term treatment with Parlodel LAR given as intramuscular monthly injections in hyperprolactinemic patients.

Patients and Methods

Sixty-six patients (65 women and 1 man), aged 17–48 years, entered this long-term study with Parlodel LAR (4–64 months).

Before starting Parlodel LAR, 42/66 patients had received oral bromocriptine at the daily dose of 2.5–20 mg. Two patients had previously undergone neurosurgery followed by oral treatment with bromocriptine 20 mg/day in one case.

CT scan (General Electric 9800 scanner) or MRI (1.5 Signal Tesa) evaluation was carried out prior to the treatment with Parlodel LAR in every patient. The neuroradiological examination showed an appearance compatible with the presence of microadenoma in 38/66 patients and empty sella in 2/66 patients. Twenty-six of the 66 patients had no evidence of pituitary tumor. CT scan/MRI examination of the pituitary fossa was repeated after 6- to 12-month treatment and thereafter at yearly intervals.

Visual field perimetry assessed by the Goldman machine was performed at study entry in patients bearing a PRL-secreting pituitary microadenoma. It was repeated on postinjection day 28 after the first injection of Parlodel LAR and thereafter every 6 months, should any abnormality be present at baseline.

The first dose of Parlodel LAR was 25 mg in 8 patients, 50 mg in 50 patients, and 100 mg in 8 patients. During the long-term treatment, Parlodel LAR was given as intramuscular monthly injections at the dose of 25 mg in 1 patient, 50 mg in 51 patients, 100 mg in 11 patients, and 150 mg in 3 patients.

After the first injection of Parlodel LAR, serum PRL concentrations were assessed at 08:00, 09:00, 10:00, 11:00, 12:00, 14:00, 16:00, and 20:00 hours on day 0 (baseline), 1 (day of the injection), 3, 7, 14, and 28. Thereafter, blood was taken for PRL assessment at 08:00, 10:00, 12:00, 16:00, and 20:00 hr on the day of the injection and the day before.

Serum PRL was measured by immunoradiometric assay using as standard MRC 74/504. Normal serum PRL is considered as below 400 mU/L with a midrange interassay coefficient of variation of 3.8%. The limit of detection for the assay is 9 mU/L.

Hematology and biochemistry safety tests and 12-lead ECG recordings were performed at study entry and then every 3–6 months.

At baseline galactorrhea was present in 49/66 patients. At study entry, 37 of the 65 women reported amenorrhea and 18/65 oligomenorrhea. Decreased libido was recorded by 11/66 patients. Headaches were present in 30/66 patients. No patients had impairment of their visual field perimetry.

Results

Serum PRL Levels

Prior to the first application of Parlodel LAR, mean serum PRL levels were 2200.6 ± 348.4 mU/L (range: 630–11,680 mU/L) in patients bearing microprolactinomas and 1790.4 ± 209.8 mU/L (range: 774–5540 mU/L) in patients with nontumoral hyperprolactinemia (NTH). On day 1, day of the injection, serum PRL concentrations decreased to within the normal range in 26/38 microadenoma patients (68.4%) (Figure 11) and in 24/28 NTH patients (86%) (Figure 12). Fifty-five of the 66 patients (83%) reached normal values during the 28-day follow-up. A long-lasting suppression of PRL secretion, i.e., up to day 28, to normal levels was observed in 19/38 microad-

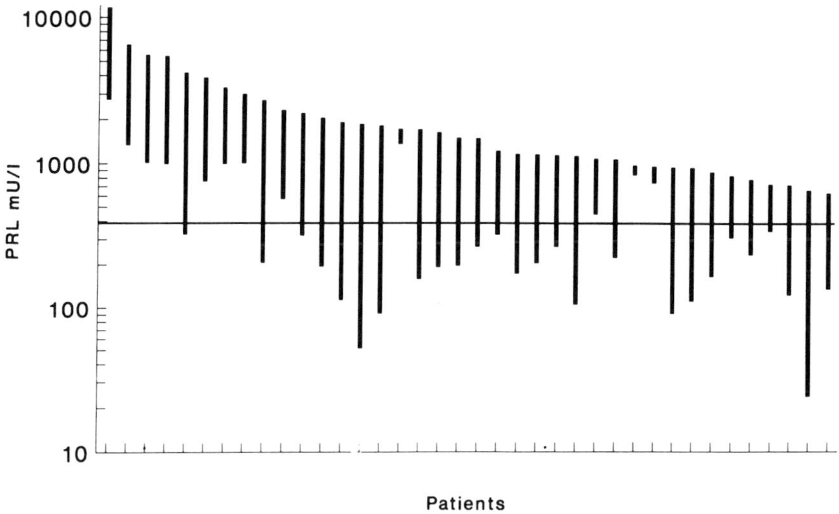

Figure 11 PRL values in patients with microprolactinomas before and 12 hr after the first dose of Parlodel LAR.

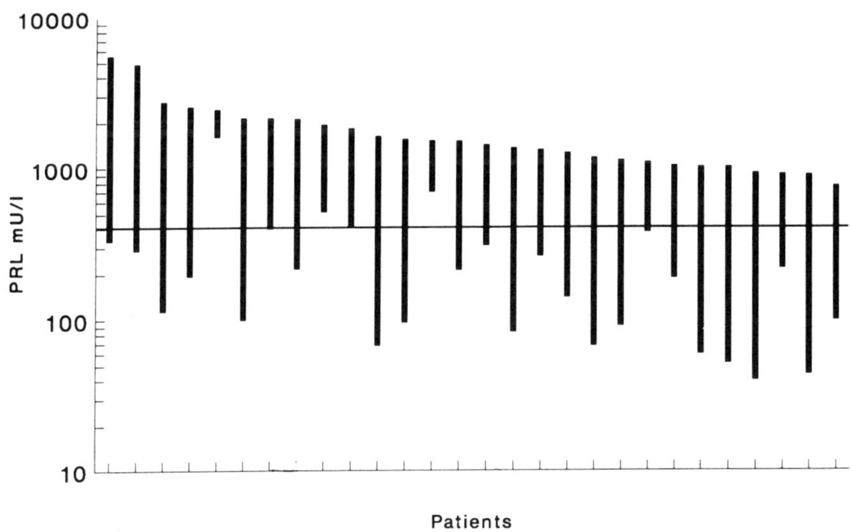

Figure 12 PRL values in patients with NTH before and 12 hr after the first dose of Parlodel LAR.

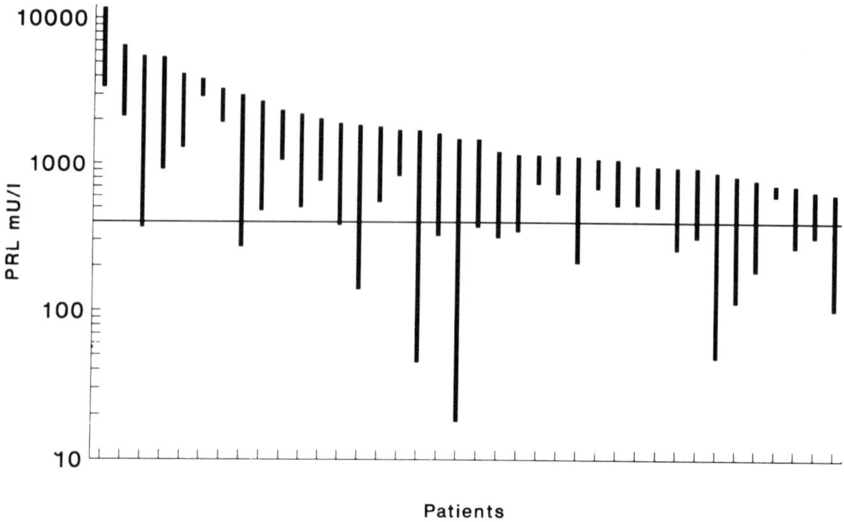

Figure 13 PRL values in patients with microprolactinomas before and on day 28 after the first dose of Parlodel LAR.

enoma patients (50%) (Figure 13) and in 14/28 NTH patients (50%) (Figure 14). Mean serum PRL concentrations decreased from 2200.6 ± 348.4 mU/L to 660 ± 121 mU/L in microadenoma patients and from 1790.4 ± 209.8 mU/L to 474.4 ± 73 mU/L in NTH patients on postinjection day 28.

The long-term treatment with Parlodel LAR was continued in all patients who received a total of 4–64 injections.

A long-lasting suppression of PRL secretion to within the normal range, i.e., up to day 28, 35, 42, 49, 56, 70, and 90, was recorded in 52/66 patients, namely, 29 microadenoma patients (76%) (Figure 15) and 23 NTH patients (82%) (Figure 16), after 1–20 injections of Parlodel LAR. Of the 14 patients who did not reach normal PRL values during the long-term treatment, 12 had their serum PRL suppressed to more than 50% on postinjection day 28, when compared to the pretreatment levels.

The interval between each subsequent injection could be extended to 35 days in 9 patients, to 42 days in 10 patients, to 49 days in 13 patients, to 56 days in 2 patients, to 70 days in 3 patients, to 90 days in 2 patients (Table 9).

Neuroradiological Examination

A complete disappearance of the pituitary microadenoma was documented by MRI examination performed with gadolinium enhancement in 2/38 microadenoma patients after 6–15 months of treatment with Parlodel LAR.

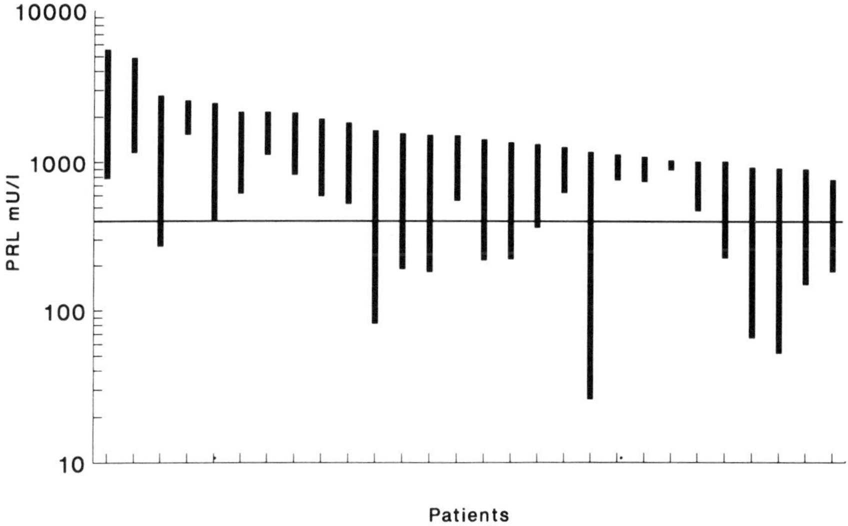

Figure 14 PRL values in patients with NTH before and on day 28 after the first dose of Parlodel LAR.

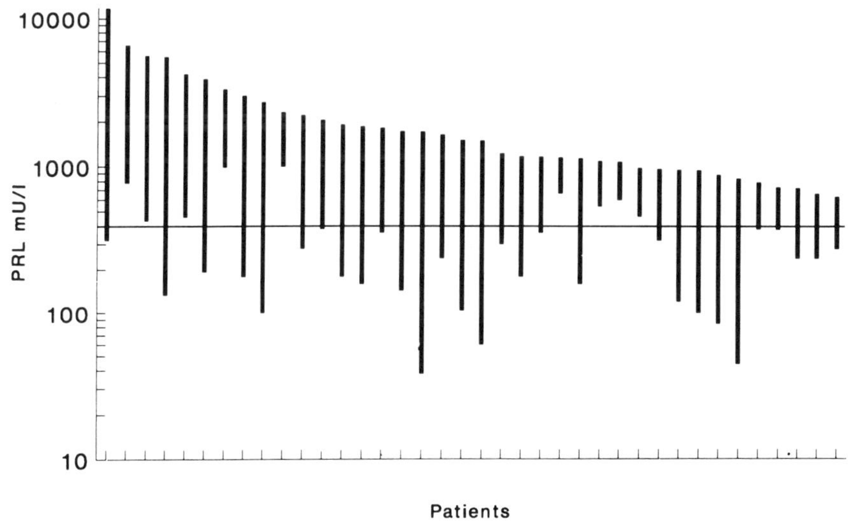

Figure 15 PRL values in patients with microprolactinomas before and during long-term treatment with Parlodel LAR.

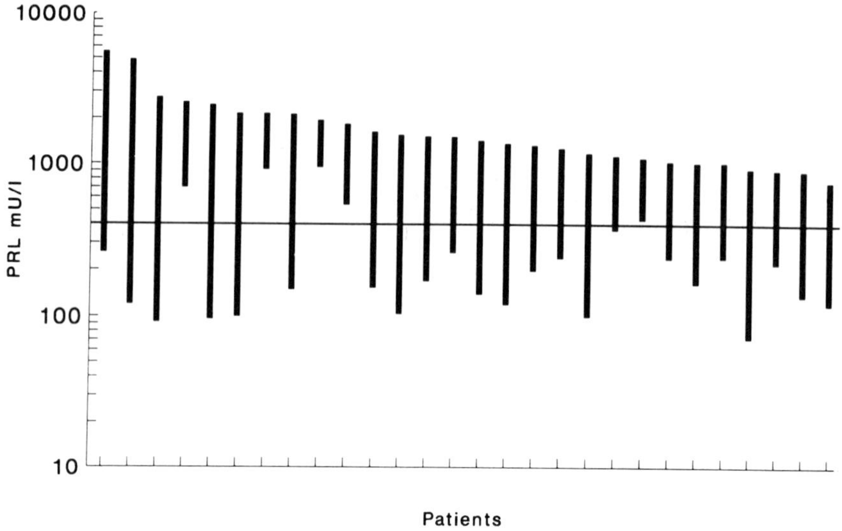

Figure 16 PRL values in patients with NTH before and during long-term treatment with Parlodel LAR.

No changes of the microadenomas were noted on repeated CT scan or MRI examinations in the other patients bearing microadenomas.

Clinical Symptoms/Signs (Table 10)

Galactorrhea, which was present in 49/66 patients at study entry, disappeared in 18/49 after the first injection of Parlodel LAR. In the other 23 patients galactorrhea disappeared after 2–26 months of treatment.

Table 9 Administration Interval of Parlodel LAR 50–200 mg in Patients with Hyperprolactinemic Disorders

Days	No. patients
28	27/66
35	9/66
42	10/66
49	13/66
56	2/66
70	3/66
90	2/66

Table 10 Symptoms/Signs of Hyperprolactinemia Prior to and After First Injection and During Long-Term Treatment with Parlodel LAR in Patients with Hyperprolactinemic Disorders

Clinical symptom/sign	Baseline	Disappearance	Disappearance	
			After 1st inj.	During long-term treatment (months)
Galactorrhea	49/66 patients	41/49 patients	18/41 patients	23/41 patients (2–26)
Oligo/amenorrhea	55/65 women	54/55 women	26/54 women	28/54 women (2–15)
Decreased libido	11/66 patients	11/11 patients	5/11 patients	6/11 patients (2–6)
Headache	30/66 patients	24/30 patients	16/24 patients	8/24 patients (2–10)

Twenty-six of 37 women who reported amenorrhea prior to the treatment resumed menses after the first injection of Parlodel LAR. Menses occurred in another 10 previously amenorrheic women after 2 to 15 months of treatment. Thereafter, regular and normal menses were recorded in 36/37 women during the long-term treatment. Besides, regular menses were also recorded in all the previously oligomenorrheic women (18/65) after 3–4 months of treatment.

Libido normalized in 5/11 patients, who reported diminished libido at baseline, after the first injection and in the remaining 6 patients after 2–6 months of treatment.

Headaches were relieved in 24 of the 30 patients who had suffered them: in 16/24 within 28-day treatment and in other 8 patients after 2–10 injections.

Adverse Events

In all patients the injections of Parlodel LAR caused minimal and short-lasting side effects. After the first injection of Parlodel LAR, adverse effects were reported by 30/66 patients and consisted of nausea (22/66), vomiting (18/66), and postural hypotension (7/66). These adverse events were transient, lasting less than 24 hr in all but two patients. Besides, most side effects disappeared spontaneously. During the long-term treatment tolerability to Parlodel LAR further improved, adverse events being reported by 4/66 patients on the day of the injection. One should note that of the 42 patients who had previously received orally active dopamine agonists, 24 were intolerant of oral bromocriptine. All but two of these patients tolerated Parlodel LAR well or very well.

Local tolerability was very good, no pain or redness at the injection site being reported by any patient.

During the long-term treatment with Parlodel LAR, no changes in hematology and biochemistry safety tests were noted. No newly occurring abnormalities in ECG recordings were observed either.

Conclusions

The long-term treatment with Parlodel LAR was very effective and well tolerated in patients with hyperprolactinemic disorders. A rapid and very long-lasting suppression of PRL secretion together with a rapid disappearance of clinical symptoms/signs of hyperprolactinemia were noted in the great majority of patients. The interval between each subsequent injection could be extended to 35–90 days in 39/66 patients. Moreover, radiological disappearance of pituitary microadenoma was observed in two patients. Tolerability was rated as good or very good in most patients. Even patients previously intolerant of oral bromocriptine tolerated Parlodel LAR well or very well. No impairment of safety parameters was observed in any patient. The compliance of the patients to the long-term treatment with Parlodel LAR, namely, up to 64 months, was very good.

PARLODEL LAR, TREATMENT OF OPTION IN SOME ACROMEGALIC PATIENTS, IN THE LONG-TERM TREATMENT OF ACROMEGALY: EFFICACY, TOLERABILITY, AND SAFETY RESULTS

Conventional therapy of acromegaly is represented by surgery and/or radiation, which are often unsatisfactory (45–51). Generally, growth hormone (GH)–secreting pituitary adenomas are not completely removed by neurosurgery. Radiotherapy at best restores elevated GH levels to normal only after a period of several years. At worst, GH hypersecretion may persist despite surgical intervention and radiation therapy. Pharmacotherapy of acromegaly has therefore received much attention by both researchers and investigators as an alternative or adjunctive therapy to surgery and/or radiotherapy.

At present, the medical treatment of choice in patients with acromegaly is represented by octreotide, a long-acting synthetic analog of somatostatin (52-58). However, as some of the acromegalic patients proved to be resistant or partially resistant to the treatment with octreotide, it has been decided to precede the pharmacotherapy by an acute inhibition test with both octreotide and bromocriptine, a dopamine agonist that paradoxically lowers serum GH levels in acromegalic patients (59-68), as the usual effect of this agent and other dopamine agonists is stimulation of GH secretion in normal individuals.

Patients and Methods

Based on a better suppression of GH secretion by bromocriptine (2.5 mg orally) over ocreotide (50 μg sc) during an acute inhibition test performed before starting pharmacotherapy, we selected four male patients with clinically active acromegaly (Figures 17–20), aged 30–70 years, for the treatment with Parlodel LAR 100 mg im administered every 28 days. One patient received one injection of Parlodel LAR, followed by oral treatment with Parlodel SRO (slow-release oral form of bromocriptine, developed in order to decrease the frequency of administration to once a day and further reduce unwanted side effects, which are thought to be related to the rapid absorption of the drug. The aim of this new formulation is to replace the 2.5-mg two- or three-times-a-day regimen with a well-tolerated 5- or 10-mg once-a-day form). The other three patients were long-term treated with Parlodel LAR, receiving so far 9–48 injections.

Active acromegaly was biochemically confirmed at the time of starting treatment by failure to suppress serum GH concentrations to less than 4 mU/L during a 75-g oral glucose tolerance test.

At study entry, all patients had high-resolution CT scan (General Electric 9800 scanner) or MRI (1.5 Signal Tesa) of the pituitary fossa demonstrating

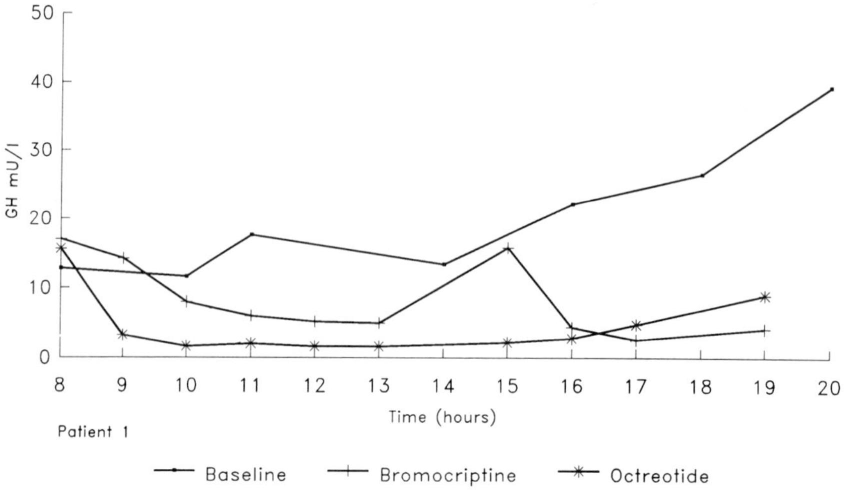

Figure 17 Basal GH curve and acute GH response to octreotide (50 μg sc) or bromocriptine (2.5 mg orally).

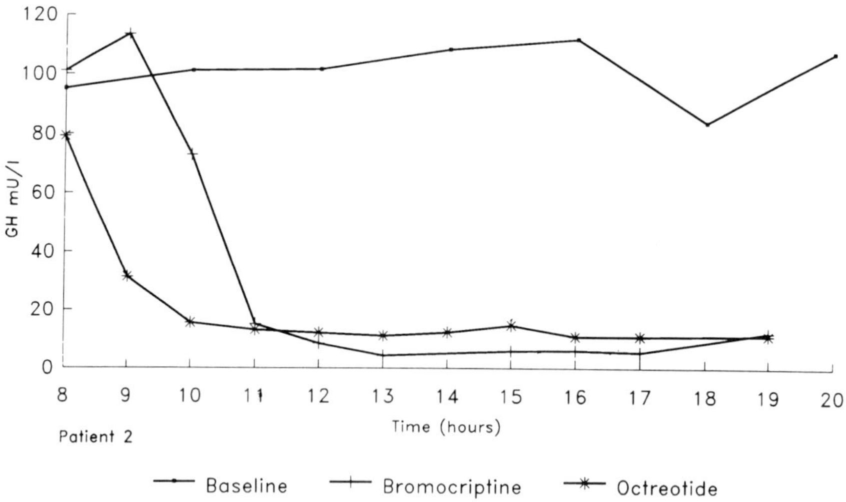

Figure 18 Basal GH curve and acute GH response to octreotide (50 μg sc) or bromocriptine (2.5 mg orally).

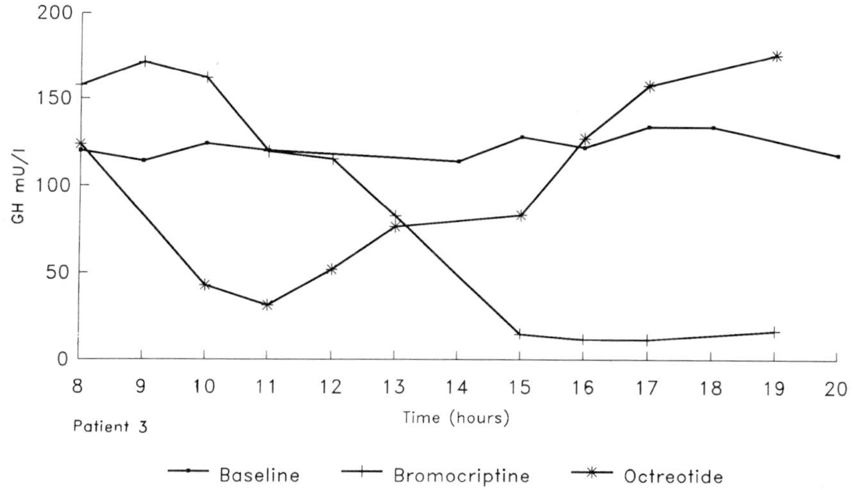

Figure 19 Basal GH curve and acute GH response to octreotide (50 μg sc) or bromocriptine (2.5 mg orally).

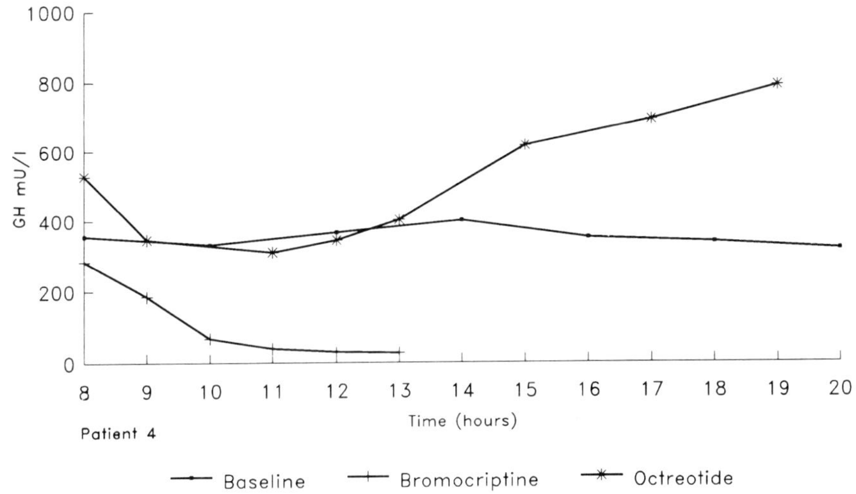

Figure 20 Basal GH curve and acute GH response to octreotide (50 μg sc) or bromocriptine (2.5 mg orally).

an appearance compatible with the presence of a macroadenoma in all patients. In 1/4 patients the macroadenoma was a mixed GH- and PRL-secreting one, while in another 2 patients mild hyperprolactinemia was recorded. The neuroradiological examination was repeated after 6 and 12 months of Parlodel LAR treatment and yearly thereafter. Visual field perimetry performed prior to the start treatment resulted normal in every patient.

Before Parlodel LAR, all patients had had no other treatment for acromegaly.

One patient required appropriate replacement therapy with thyroxine.

The patients were clinically and biochemically assessed every month. In view of the known pulsatility of GH secretion, a mean value during the day was calculated as an index of GH secretion. Blood was sampled for GH assessment at 08:00 (fasting), 10:00, 12:00, 14:00, 16:00, 18:00, and 20:00 hr and the results averaged. Serum GH was assessed at the above time points on days 0 (baseline), 1 (day of the injection), 3, 7, 14, and 28 after the first dose of Parlodel LAR. Thereafter, circulating GH was measured on days 1 and 28 after each subsequent injection of Parlodel LAR.

Serum GH was measured by an immunoradiometric technique with a midrange interassay coefficient of variation of 2.7%. The limit of detection of the assay is 0.04 mU/L. Normal serum GH is taken as below 4 mU/L.

Blood samples for serum IGF1 assessment were taken at 8 o'clock in the morning at baseline and then on day 28 after each dose of Parlodel LAR.

Serum IGF1 was measured by a radioimmunological assay. It uses a double-antibody disequilibrium technique that includes an octadecylsulfate (ODS)-silica extraction procedure for serum samples. Normal serum IGF1 level in adults is considered as below 1.1 U/mL.

Clinical evaluation was performed through systematic enquiry into the symptoms of acromegaly and the subjective changes on Parlodel LAR treatment. In all patients finger size was assessed by measuring the circumference of the fourth finger of each hand using a set of jeweller's rings. Hematology and biochemistry safety tests together with ECG recordings were monitored every 2–3 months.

Results

Serum GH Levels

At study entry the mean serum GH concentrations were 20.4 mU/L (patient 1), 101.8 mU/L (patient 2), 120.8 mU/L (patient 3), and 348 mU/L (patient 4).

Twelve hours after the first injection of Parlodel LAR, mean serum GH levels fell below 4 mU/L in patients 1, 2, and 3. In patient 4 serum GH decreased from 348 mU/L to 22 mU/L (−93.6%) 12 hr after the first injection (Figure 21). On postinjection day 28, circulating GH was markedly suppressed in all patients, mean GH concentrations being 7.0 mU/L (−65.7%), 17.2 mU/L (−83.1%), 30.4 mU/L (−74.8%), and 15.6 mU/L (−95.5%) in the four patients, respectively (Figure 22).

During the long-term treatment GH secretion remained well suppressed in all three patients. In 1/3 patients (patient 4) mean serum GH values fell below 4 mU/L after nine injections of Parlodel LAR. Thereafter, circulating GH remained below 4 mU/L throughout the period of treatment, namely, 48 months. Patient 2 had mean GH levels on treatment of 10 mU/L (−90.1%) when last assessed. With regard to patient 3, mean serum GH concentrations decreased to 8 mU/L (−93.3%) during the long-term treatment, i.e., 34 injections (Figure 23).

Serum IGF1 Levels

Serum IGF1 concentrations decreased in all patients in accordance with the fall in GH.

Serum PRL Levels

At study entry, in the three patients who had hyperprolactinemia, serum PRL levels were 54,620 mU/L (patient 4), 1348 mU/L (patient 3), and 990 mU/L (patient 2). The other patient had normal PRL values. A long-lasting suppression of PRL secretion, namely, up to day 28, was observed after the

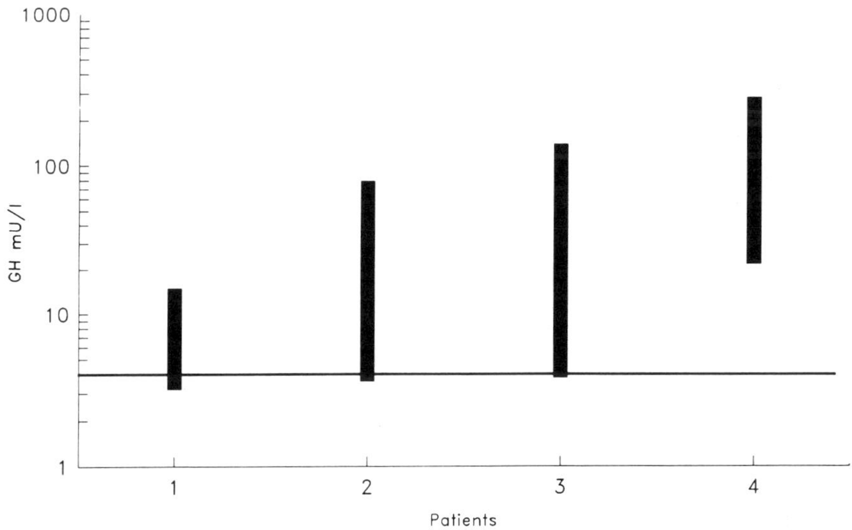

Figure 21 Serum GH values before and 12 hr after the first injection of Parlodel LAR.

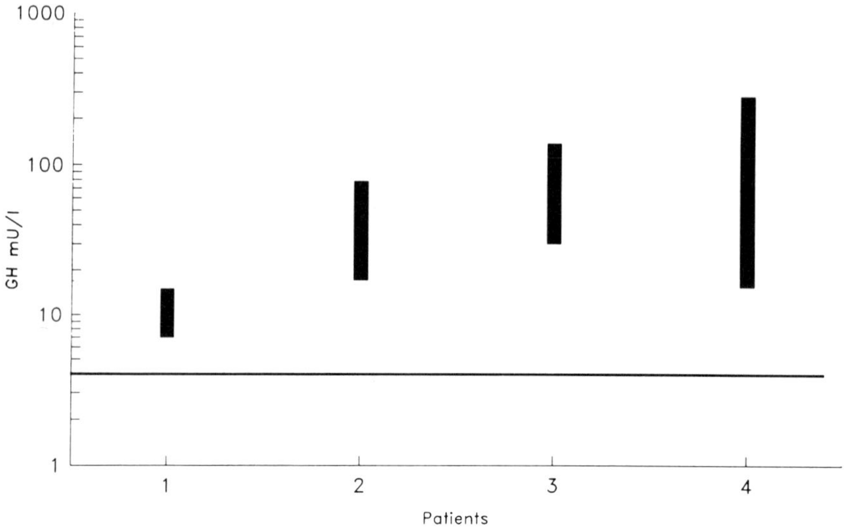

Figure 22 Serum GH values before and on day 28 after the first injection of Parlodel LAR.

first dose in patients 2 and 3 and after seven injections in patient 4. These levels were maintained thereafter in all patients (Figure 24).

Neuroradiological Examination

Significant tumor shrinkage (28–37.5%) was documented with CT scan or MRI examination in 3/4 patients within 6-month treatment. In particular, patient 4 had his tumor shrunk by 30% after the first injection of Parlodel LAR and further tumor size reduction was noted after 6-month therapy. Furthermore, it was observed a significant tumor shrinkage (28%) after the first dose of Parlodel LAR even in a patient with normal serum PRL values at baseline.

Clinical Response

All patients reported excessive perspiration before treatment and this symptom was abolished in all of them. The skin became thinner in all patients and facial features improved in all of them. Ring size decreased in all patients during the long-term treatment. The foot volume decreased in each patient and 2/4 patients could wear shoes at least one size smaller. Improved energy and well-being were reported by all patients.

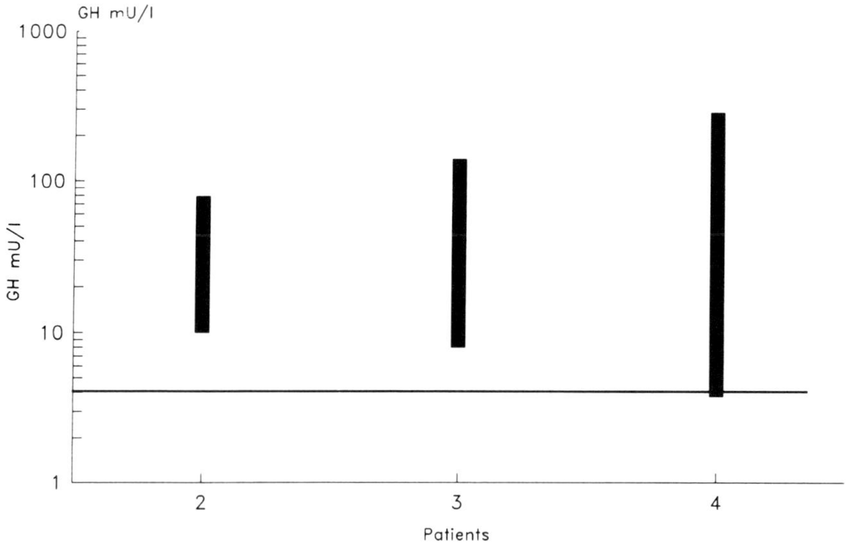

Figure 23 Serum GH values before and during long-term treatment with Parlodel LAR.

Tolerability

Tolerability was rated as very good in all patients. Not a single adverse event was reported by any patient either after the first injection or during the long-term treatment with Parlodel LAR.

Conclusions

Long-term treatment with Parlodel LAR was effective in acromegalic patients responsive to bromocriptine during the acute inhibition test performed prior to the pharmacotherapy. Serum GH levels were markedly reduced in all patients. In patient 4 mean serum GH concentrations decreased to less than 4 mU/L. Although in the other two treated long-term patients circulating GH did not normalize, GH values fell by more than 90% of the pretreatment concentrations.

All patients showed clinical improvement. Excessive perspiration was abolished soon after starting the Parlodel LAR treatment. Thereafter, facial features and skin thickness improved as well. Hand and foot size, too, gradually decreased during the long-term application.

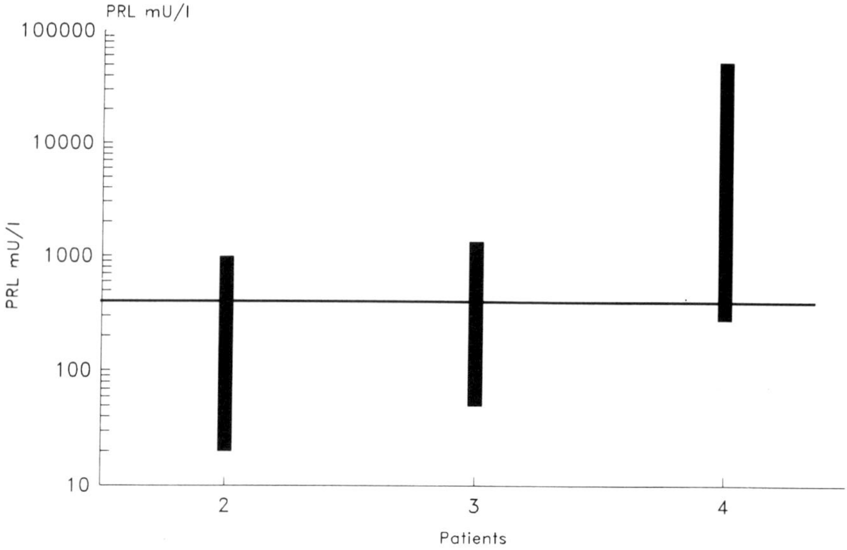

Figure 24 Serum PRL values before and during long-term treatment with Parlodel LAR in patients 2, 3, and 4.

Tumor shrinkage was observed in 3/4 patients, in 2/3 after the first injection. Indeed, one patient with normal serum PRL levels had his tumor shrunk by 28%.

Tolerability was very good in all patients, no side effects at all being reported either after the first injection or during the long-term treatment.

The long-term treatment with monthly injections of Parlodel LAR was successfully accepted by all patients.

EFFECT OF PARLODEL LA AND PARLODEL LAR THERAPY ON NONFUNCTIONING PITUITARY MACROADENOMAS

The natural history of nonsecreting pituitary tumors is little known, but one study (69) found it to be characterized by continued expansion. Conventional initial treatment for these tumors is surgery or radiotherapy or both, but they present a therapeutic dilemma if they recur or if these approaches fail.

Bromocriptine is a well-established treatment for prolactinomas. It can suppress PRL secretion, normalize the clinical picture, and, above all, rapidly reduce the size of many such tumors, providing an early decompression of the visual pathways and resolution of field defects due to its antimitotic effect complemented by its dopaminergic effect (70,71).

The therapeutic effect of bromocriptine in functionless pituitary tumors in which serum PRL is either normal or raised due to tumor compression of the hypophyseal portal vasculature is still uncertain. Although the treatment of choice in patients with mass-related symptoms remains neurosurgery, apparent shrinkage of functionless tumor with dopamine agonists, especially bromocriptine, was reported by some authors (4,72-74). Conversely, some groups showed that nonfunctioning pituitary adenomas do not shrink with dopamine agonists (7,75-77), despite adenomatous cells possessing D2 receptors (77,78).

In view of this uncertainty we have prospectively studied a group of patients with nonsecreting pituitary macroadenomas with normal or slightly elevated PRL levels and have monitored the effects of two new long-acting forms of bromocriptine, namely, Parlodel LA and Parlodel LAR, on tumor size and mass-related symptoms.

Patients and Methods

Twenty-one patients (15 men and 6 women), aged 18–80 years, were included in this study. All the patients beared large pituitary tumors (defined as pituitary tumors greater than 10 mm in diameter with lateral and/or suprasellar extensions) and they had serum PRL concentrations below 1400 mU/L (normal serum PRL values <400 mU/L). Acromegaly and Cushing's disease were excluded by conventional tests.

Three of the 21 patients had previously undergone neurosurgery, followed in two cases by external irradiation. One of the 21 patients had received only radiotherapy (Table 11).

Basal endocrine function was evaluated by assaying serum PRL levels (mean of five blood samples), serum luteinizing hormone (LH) and follicle-stimulating hormone (FSH), serum thyroxine, serum cortisol at 08:00, 08:30, 09:00 hrs and, in male patients, serum testosterone. Dynamic function testing was carried out by means of luteinizing hormone–releasing hormone (LHRH), 100 μg iv), thyrotropin-releasing hormone (TRH), 200 μg iv), and corticotropin-releasing hormone (CRH), 100 μg iv). Appropriate hormone replacement therapy was started according to the results of these tests.

Visual field perimetry, assessed by the Goldman machine, was performed prior to and on days 3, 7, 14, 21, 28 after the first injection of Parlodel LAR and at monthly intervals thereafter.

Radiological investigation by means of CT scan (General Electric 9800 scanner) or MRI (1.5 Signal Tesa) was carried out at study entry and on day 28 following the first injection and thereafter at 6-month intervals.

Table 11 Clinical Details of 21 Patients Bearing Large Pituitary Tumors and Serum Prolactin Below 1400 mU/L

Pt.	Age	Sex	Previous treatment	Symptoms	Visual involvement	Endocrine defects	Serum PRL (mU/L)
1	19	M	None	Visual defect, headache, impaired libido	BH	LH, FSH, ACTH, TSH	1140
2	52	M	None	Visual defect, headache, impotence	BH	LH, FSH, ACTH, TSH	200
3	62	M	None	Impotence	None	LH, FSH, ACTH, TSH	170
4	60	M	RT	None	None	LH, FSH, ACTH, TSH	600
5	80	M	None	Visual defect, headache, impotence	BQ	LH, FSH, ACTH, TSH	500
6	36	M	NS	Visual defect, impotence	TH	None	500
7	63	M	NS + RT	Visual defect	MA	ACTH	124
8	44	F	None	Amenorrhea, headache	None	None	130
9	33	M	None	Headache	None	None	380
10	21	F	None	Headache, amenorrhea	None	None	200
11	71	M	None	Impaired libido, impotence	None	None	1360
12	40	M	None	Visual defect	BH	None	700
13	50	M	NS + RT	Visual defect	TH	LH, FSH, ACTH, TSH	150
14	18	F	None	Visual defect, headache	CN	None	440
15	43	M	None	Headache, impaired libido, impotence	None	LH, FSH, ACTH, FSH	140
16	38	M	None	Impaired libido, impotence	None	LH, FSH, ACTH, TSH	1200
17	64	M	None	Headache	None	LH, FSH, ACTH, TSH	66
18	53	M	None	Impaired libido	CS	None	740
19	44	F	None	Amenorrhea, impaired libido	None	None	300
20	24	F	None	Headache, amenorrhea	None	LH, FSH, ACTH, TSH	350
21	68	F	None	Visual defect	TH	LH, FSH, ACTH, TSH	740

RT, radiotherapy; NS, neurosurgery; BH, bitemporal hemianopia; BQ, bitemporal quadrantinopia; TH, temporal hemianopia; MA, monolateral amaurosis; CN, concentric narrowing; CS, central scotoma.

At study entry, headache was present in 10/21 patients, decreased libido in 6/21 patients. Impotence was reported by 7/15 men. Amenorrhea was recorded in 4/6 women. Visual field defects were recorded in 10/21 patients and they consisted of bitemporal hemianopia in 3/10 patients, bitemporal quadrantinopia in 1/10 patients, temporal hemianopia in 3/10 patients, monolateral amaurosis in 1/10 patients, concentric narrowing in 1/10 patients, and central scotoma in 1/10 patients (Table 11).

At presentation, 11 of the 21 patients were normoprolactinemic; 11 patients had complete loss of their remaining anterior pituitary function, and 1 patient had secondary adrenal insufficiency (Table 11).

All patients received a single im injection of either Parlodel LA 50 mg (6/21 patients) or Parlodel LAR 50 mg (7/21 patients) or 100 mg (8/21 patients) followed by chronic treatment with either Parlodel LAR in 4/21 patients or oral bromocriptine in 11/21 patients.

Results

Twelve hours after Parlodel LAR administration, serum PRL levels decreased to within the normal range in all patients with mild hyperprolactinemia. Prolactin secretion was consistently suppressed to normal values since then.

Eight of the 21 patients had their tumor shrunk by 14–25% with respect to the pretreatment tumor volume within 4 weeks following either Parlodel LA (3/8 patients) or Parlodel LAR (5/8 patients) (Table 12). Of the 8 patients, 1 had a further tumor shrinkage (45% of the pretreatment tumor size) after 24 months of treatment with oral bromocriptine 20 mg daily.

Visual field perimetry became normal in 4/10 patients within the first few days after the first dose of either Parlodel LA (1/4) or Parlodel LAR (3/4) (Table 12).

After the first injection, headaches were relieved in 3/10 patients. Libido and sexual performance was improved in 4/7 men, two of whom had complained of impotence before treatment.

Adverse events were mostly mild and short-lasting. On the day of injection, 8/21 patients reported mild to moderate nausea, 7/21 patients few episodes of vomiting, and 2/21 patients postural hypotension. These side effects disappeared spontaneously within 24 hr. Thereafter, no adverse events were reported by any patient.

During the long-term treatment, no significant changes in tumor size and clinical symptoms/signs were observed in any patient.

Conclusions

The results of this study are in agreement with previous data suggesting that bromocriptine could be of benefit in some patients with functionless pituitary

Table 12 Results of Treatment in 21 Patients with Large Pituitary Tumors Treated with Parlodel LA or Parlodel LAR

Pt.	Drug	CT/MRI pre	CT/MRI post	VF pre	VF post
1	Parlodel LAR	25 mm	NC	BH	NC
2	Parlodel LA	50 mm	NC	BH	Normal
3	Parlodel LAR	16 mm	NC	Normal	–
4	Parlodel LAR	23 mm	18 mm	Normal	–
5	Parlodel LAR	15 mm	NC	BQ	NC
6	Parlodel LA	27 mm	22 mm	TH	NC
7	Parlodel LAR	24 mm	NC	MA	NC
8	Parlodel LAR	12 mm	9 mm	Normal	–
9	Parlodel LA	22 mm	NC	Normal	–
10	Parlodel LAR	10 mm	NC	Normal	–
11	Parlodel LA	14 mm	12 mm	Normal	–
12	Parlodel LA	26 mm	NC	BH	NC
13	Parlodel LAR	25 mm	NC	TH	Normal
14	Parlodel LAR	10 mm	8 mm	CN	NC
15	Parlodel LAR	17 mm	NC	Normal	–
16	Parlodel LAR	30 mm	NC	Normal	–
17	Parlodel LAR	14 mm	NC	Normal	–
18	Parlodel LAR	40 mm	32 mm	CS	Normal
19	Parlodel LAR	10 mm	NC	Normal	–
20	Parlodel LA	18 mm	15 mm	Normal	–
21	Parlodel LAR	20 mm	16 mm	TH	Normal

VF, visual field; NC, no change; BH, bitemporal hemianopia; BQ, bitemporal quadrantinopia; TH, temporal hemianopia; MA, monolateral amaurosis; CN, concentric narrowing; CS, central scotoma.

tumors by reducing tumor size and improving visual defects. The two long-acting forms of bromocriptine, namely, Parlodel LA and Parlodel LAR, known to produce high plasma concentrations of bromocriptine within a few hours following the injection, without evidence of important side effects, could be used to identify the patients with nonfunctioning pituitary tumors who can benefit of the chronic treatment with bromocriptine. Furthermore, these two long-acting depot formulations of bromocriptine constitute a useful treatment of inoperable patients with recurrences of large tumors who already received maximal doses of irradiation.

CONCLUSIONS

Parlodel LA, the first long-acting injectable form of bromocriptine, represents a new galenical form of Parlodel developed for intramuscular administration of bromocriptine contained in polylactic acid microspheres. Due to the slow release of the drug from the polylactic acid microspheres, a single intramuscular injection of Parlodel LA was shown to induce an inhibition of PRL secretion for 4–6 weeks in normal volunteers. In most patients (9-11,16) a single dose of Parlodel LA 50 mg was reported to produce a marked reduction of serum PRL concentrations and rapid tumor shrinkage. Since the polylactic acid, the carrier polymer of Parlodel LA, is slowly degraded (approximately 6 months), this new galenical form of Parlodel is suitable for being administered once only to either initiate treatment of patients with hyperprolactinemia/prolactinomas or to prevent/suppress lactation. Parlodel LAR, the newest injectable form of Parlodel, consists of bromocriptine contained in glucose-initiated poly-DL-lactide-*co*-glycolide microspheres, a polymer that is fully metabolized within 2 months (34-36). Thus, this new long-acting injectable form of bromocriptine can be repeatedly administered at 4-week intervals.

Repeated injections of Parlodel LAR 25–200 mg at 4- to 13-week intervals in patients with hyperprolactinemia/prolactinomas produced a rapid and sustained suppression of PRL secretion and a disappearance of symptoms/signs related to hyperprolactinemia. Moreover, a marked and rapid tumor size reduction was observed in the majority of patients bearing macroprolactinomas within the first month of treatment with Parlodel LAR. Indeed, the complete disappearance of macroprolactinoma was recorded in 3 of the 39 macroadenoma patients.

Based on these results, Parlodel LAR could be considered both as the initial treatment of choice of prolactinomas even in the presence of visual field defects and as the first choice for the long-term treatment of patients with both hyperprolactinemia and prolactinomas. The very good tolerability of Parlodel LAR in comparison with orally active dopamine agonists and the once-a-month administration determined a very good compliance of the patients during the long-term treatment. Even the patients who had been previously intolerant of oral dopamine agonist agents tolerated this new formulation of bromocriptine well or very well.

Besides, Parlodel LAR resulted very effective in acromegalic patients responsive to dopamine agonists. A marked suppression of GH concentrations even to normal values was observed in all the patients investigated; the clinical pattern improved very rapidly and all but one patient showed a significant reduction of the pituitary tumor. The tolerability was very good, with no patients reporting side effects either after the first injection or during the long-term treatment.

Finally, Parlodel LA and Parlodel LAR could be of benefit in some patients with functionless pituitary tumors by reducing the tumor size and improving the visual defects. Parlodel LAR represents a useful treatment for inoperable patients with recurrences of large tumors who have already received maximal doses of radiation.

REFERENCES

1. M. O. Thorner, J. A. Burdman, M. I. Calabrese, C. A. Valdenegro, M. L. Vance, and R. M. McLeod, in *Lisuride and Other Dopamine Agonists* (D. B. Caine, R. Horowski, R. McDonald, and W. Wuttke, eds.), Raven Press, New York, 1983, pp. 213–229.
2. G. M. Besser, J. A. H. Wass, A. Grossman, P. J. A. Moult, and P. Bouloux, in *Lisuride and Other Dopamine Agonists* (D. B. Caine, R. J. McDonald, R. Horowski, and W. Wuttke, eds.), Raven Press, New York, 1983, pp. 239–254.
3. B. Corenblum, B. R. Wester, C. B. Martinez, and C. Ezrin, *Clin. Res. 23*, 614 A (1975).
4. J. A. H. Wass, J. Williams, M. Charlesworth, D. P. E. Kingsley, A. M. Halliday, I. Doniach, L. H. Rees, W. I. McDonald, and G. M. Besser, Bromocriptine in the management of large pituitary tumours. *Br. Med. J. 284*, 1908–1911 (1982).
5. K. von Werder, R. Fahlbusch, and H. K. Rjork, in *Prolactin and Prolactinomas* (G. Tolis, C. Stefarris, T. Mountokalakis, and I. Labrie, eds.), Raven Press, New York, 1983, pp. 415–429.
6. A. Liuzzi, P. G. Chiodini, G. Oppizzi, D. Dellabonzana, B. Spelta, and F. Silvestrini, in *Lisuride and Other Dopamine Agonists* (D. B. Caine, R. Horowski, R. McDonald and W. Wuttke, eds.), Raven Press, New York, 1983, pp. 231–238.
7. D. L. Barrow, G. T. Tindall, K. Kovacs, M. O. Thorner, E. Horvath, and J. C. Hoffman, *J. Neurosurg.*, *60*, 1–7 (1984).
8. M. E. Molitch, R. L. Elton, R. E. Blackwell, B. Caldwell, R. J. Chang, R. Jaffe, G. Joplin, R. J. Robbins, M. O. Thorner, and the Bromocriptine Study Group, *J. Clin. Endocrinol. Metab.*, *60*, 698–705 (1985).
9. A. Grossman, R. Ross, J. A. H. Wass, and G. M. Besser, *Clin. Endocrinol.*, *24*, 231–238 (1986).
10. G. Benker, B. Gieshoff, O. Freundlieb, R. Windeck, H. M. Schulte, I. Lancranjan, and D. Reinwein, *Clin. Endocrinol.*, *24*, 505–513 (1986).
11. M. Montini, G. Pagani, D. Gianola, M. D. Pagani, M. Salmoiraghi, L. Ferrari, and I. Lancranjan, *J. Clin. Endocrinol. Metab.*, *63*, 266–268 (1986).
12. C. Ferrari and P. G. Crossignani, *Hum. Reprod.*, *1*, 507–514 (1986).
13. M. D. Bronstein, C. S. Cardin, and R. Marino, Jr., *Surg. Neurol.*, *28*, 31–37 (1987).
14. A. Zarate, C. Moran, R. Miranda, M. Loyo, M. Medina, and M. E. Fonseca, *J. Endocr. Invest.*, *10*, 233–236 (1987).
15. I. Halperin, M. D. Rodriguez, C. Cardenal, R. Casamitjana, M. J. Martinez Osaba, V. Lienas, and E. Vilardel, *J. Endocr. Invest.*, *10*, 277–282 (1987).
16. I. Lancranjan, in *Gynaecological Endocrinology* (A. R. Genazzani, A. Volpe, and F. Facchinetti, eds.), Parthenon, Carnforth, 1987, pp. 239–252.

17. G. Schettini, G. Lombardi, B. Merola, P. Miletto, C. Fariello, S. Cirillo, R. Fusco, and I. Lancranjan, *J. Endocr. Invest.*, *11*, 47–51 (1988).
18. E. J. van Cutsen and S. W. J. Lamberts, *Neth. J. Med.*, *32*, 112–117 (1988).
19. J. W. van't Verlaat, I. Lancranjan, M. J. Hendriks, and R. J. Croughs, *Acta Endocrinol.*, *119*, 51–55 (1988).
20. H. Schatz, H. Stracke, and G. Hildebrandt, *Path. Res. Practice* (West Germany), *183*, 546–551 (1988).
21. G. Pagani, M. D. Pagani, M. Montini, D. Gianola, P. Dominoni, M. Salmoiraghi, G. Ghilardi, and I. Lancranjan, in *Recent Research on Gynaecological Endocrinology*, Vol. 1 (A. R. Genazzani, F. Petraglia, A. Volpe, and F. Facchinetti, eds.), Parthenon, Carnforth, 1988, pp. 299–304.
22. M. Montini, M. D. Pagani, D. Gianola, P. Dominoni, M. Salmoiraghi, G. Pagani, and I. Lancranjan, in *Recent Research in Gynaecological Endocrinology*, Vol. 1 (A. R. Genazzani, F. Petraglia, A. Volpe, and F. Facchinetti, eds.), Parthenon, Carnforth, 1988, pp. 305–309.
23. M. D. Pagani, M. Montini, I. Lancranjan, D. Gianola, M. Salmoiraghi, G. P. Tonnarelli, F. Sileo, G. Bonaldi, G. Belloni, and G. Pagani, in *Recent Research on Gynaecological Endocrinology*, Vol. 1 (A. R. Genazzani, F. Petraglia, A. Volpe, and F. Facchinetti, eds.), Parthenon, Carnforth, 1988, pp. 293–298.
24. I. Lancranjan, G. Pagani, F. Cavagnini, G. Schettini, and M. Bronstein, in *Pain and Reproduction* (A. R. Genazzani, G. Nappi, F. Facchinetti, and E. Martignoni, eds.), Parthenon, Carnforth, 1988, pp. 299–306.
25. I. Lancranjan, F. Cavagnini, C. Felley, M. D. Pagani, M. Montini, and J. P. Felber, in *Pain and Reproduction* (A. R. Genazzani, G. Nappi, F. Facchinetti, and E. Martignani, eds.), Parthenon, Carnforth, 1988, pp. 555–564.
26. F. Cavagnini, C. Maraschini, M. Moro, M. De Martin, C. Invitti, A. Brumani, and I. Lancranjan, in *New Activities in Dopamine Agonist Field* (G. M. Besser, ed.), Parthenon, Carnforth, 1988, pp. 47–58.
27. E. Ciccarelli, C. Miola, T. Avataneo, F. Camanni, G. M. Besser, and A. Grossman, *Fertil. Steril.*, *52*, 930–935 (1989).
28. A. Kocijancic, J. Prezelj, I. Vrhovec, and I. Lancranjan, *Acta Endocrinol.*, *122*, 272–276 (1990).
29. P. P. Garcia-Luna, A. Leal-Cerro, J. L. Pereira, E. Navarro, F. Trujillo, A. Cortes, F. Villamil, D. Acosta, C. Santos, and M. Revuelta et al., *Med. Clinic*, *94*, 126–129 (1990).
30. M. Montini, D. Gianola, M. D. Pagani, F. Tengattini, P. Dominoni, F. Sileo, L. Ferrari, S. Stroppa, L. Gualteroni, and G. Pagani, *Endocrinol. Exp.*, *24*, 159–166 (1990).
31. G. Schettini, G. Lombardi, B. Merola, A. Colao, P. Miletto, E. Caruso, and I. Lancranjan, *Clin. Endocrinol.*, *33*, 161–169 (1990).
32. S. S. Reganchary, T. Tomita, B. F. Jeffries, et al., *Neurosurgery*, *10*, 242–251 (1982).
33. G. T. Tindall, K. Kovacs, and E. Horvath, et al., *J. Clin. Endocrinol. Metab.*, *55*, 1178–1183 (1982).
34. U. Siemann, P. Koci, and W. Ableidinger, personal communication, 1986.

35. F. Nimmerfall, personal communication, 1987.
36. P. Vit and S. Qureshi, personal communication, 1986.
37. P. Vit and S. Qureshi, personal communication, 1988.
38. J. E. Gray, in *Sustained and Controlled Release Drug Delivery Systems* (R. Robinson, ed.), Marcel Dekker, New York, 1978, pp. 351–405.
39. G. E. Vischer, R. L. Robinson, H. B. Maulding, J. W. Forg, U. E. Pearson, and G. J. Argentieri, *J. Biomed. Mater. Res.*, *19*, 349–365 (1985).
40. H. Autian, in *Fundamental Aspects of Biocompatibility*, Vol. 2 (D. F. Williams, ed.), Marcel Dekker, New York, 1978, pp. 351–405.
41. D. F. Williams, in *Fundamental Aspects of Biocompatibility*, Vol. 2 (D. F. Williams, ed.), CRC Press, Baton Raton, 1981, pp. 129–138.
42. H. F. Schran, S. I. Bhuta, H. J. Schwarz, and M. O. Thorner, in *Neuroendocrine and Neuropsychiatric Aspects* (M. Goldstein, D. Caine, A. Lieberman, and M. O. Thorner, eds.), Raven Press, New York, 1980, p. 125.
43. J. Rosenthaler, H. Munzer, and R. Voges, in *Drug Metabolite Isolation and Determination* (E. Reid and G. P. Leppard, eds.), Plenum Press, New York, 1983, p. 215.
44. K. von Werder, J. Schopohl, and G. Mehltretter, in *Advances in Gynaecological Endocrinology*, Vol. 2 (A. R. Genazzani, F. Petraglia, A. Volpe, and F. Facchinetti, eds.), Parthenon, Carnforth, 1988, pp. 27–38.
45. H. J. Quabbe, *Clin. Endocrinol.*, *16*, 107–119 (1982).
46. H. J. Quabbe, in *Trends in the Diagnosis and Treatment of Pituitary Adenomas* (S. W. J. Lamberts and F. J. H. Tilders, eds.), Proceedings of the 3rd European Workshop on Pituitary Adenomas, Amsterdam, September, 1983.
47. R. C. Eastman, P. Gorden, and J. Roth, *J. Clin. Endocrinol. Metab.*, *48*, 931–940 (1979).
48. C. M. Feek, J. McLelland, J. Seth, A. D. Toft, W. J. Irvine, P. L. Padfield, and C. R. W. Edwards, *Clin. Endocrinol.*, *20*, 401–408 (1984).
49. N. T. Zervas, in *Secretory Tumours of the Pituitary Gland* (Progress in Endocrine Research and Therapy, Vol. 1) (P. M. C. L. Black, N. T. Zervas, E. C. Ridgeway, and J. B. Martin, eds.), Raven Press, New York, 1984, pp. 377–385.
50. J. F. Aloia, M. S. Roginski, and J. O. Archambeau, *Am. J. Med. Sci.*, *267*, 81–87 (1979).
51. B. Bloom and S. Kramer, in *Secretory Tumours of the Pituitary Gland* (Progress in Endocrine Research and Therapy, Vol. 1) (P. M. C. L. Black, N. T. Zervas, E. C. Ridgeway, and J. B. Martin, eds.), Raven Press, New York, 1984, pp. 179–190.
52. W. Bauer, U. Briner, W. Doepfner, R. Haller, R. Huguenin, P. Marbach, T. J. Petcher, and J. Pless, *Life Sci.*, *31*, 1133–1141 (1982).
53. L. B. Barnard, W. G. Granthan, P. Lamberton, T. M. O'Dorisio, and I. M. D. Jackson, *Ann. Intern. Med.*, *105*, 856–862 (1986).
54. I. Jackson, L. D. Barnard, and P. Lamberton, *Am. J. Med.*, *81*(6b), 94–99 (1986).
55. S. W. J. Lamberts and E. Del Pozo, *Scand. J. Gastroenterol.*, *21*(119), 141–149 (1986).

56. S. W. J. Lamberts, R. Oosterom, M. Neufeld, and E. Del Pozo, *J. Clin. Endocrinol. Metab.*, *60*, 1161–1165 (1985).
57. M. Salmoiraghi, M. Montini, M. D. Pagani, D. Gianola, G. Ghilardi, F. Tengattini, G. Previtali, L. Cortesi, A. Pedroncelli, P. Dominoni, and G. Pagani, in *Recent Research in Gynaecological Endocrinology*, Vol. 1 (A. R. Genazzani, F. Petraglia, A. Volpe, and F. Facchinetti, eds.), Parthenon, Carnforth, 1988, pp. 36–40.
58. G. Pagani, M. Montini, D. Gianola, M. D. Pagani, F. Tengattini, P. Dominoni, G. Ghilardi, L. Cortesi, A. Pedroncelli, and G. P. Tonnarelli, *Endocrinol. Exp.*, *24*, 175–185 (1990).
59. A. Liuzzi, P. G. Chiodini, L. Botalla, G. Cremascoli, E. E. Muller, and F. Silvestrini, *J. Clin. Endocrinol. Metab.*, *38*, 910–912 (1974).
60. M. O. Thorner, A. Chait, and M. Aitken, et al., *Br. Med. J.*, *1*, 299–303 (1975).
61. J. A. H. Wass, M. O. Thorner, D. V. Morris, L. H. Rees, A. Stuart-Mason, A. E. Jones, and G. M. Besser, *Br. Med. J.*, *1*, 875–878 (1977).
62. R. N. Clayton, Y. Vrionides, S. S. Lynch, et al., *Acta Endocrinol.*, *89*, 469–482 (1978).
63. R. Pelkonen, R. Yilkahri, and S. L. Karonen, *Clin. Endrocrinol.*, *12*, 219–224 (1980).
64. P. M. Bell, A. B. Atkinson, D. R. Hadden, L. Kennedy, H. Leslie, D. Merret, and B. Sheridan, *Arch. Intern. Med.*, *146*, 1145–1149 (1986).
65. J. W. Nortier, R. J. Croughs, G. H. Donker, J. H. Thijssen, and F. Schwarz, *Acta Endocrinol.*, *106*, 175–183 (1984).
66. J. W. Nortier, R. J. Croughs, J. H. Thijssen, and F. Schwarz, *Clin. Endocrinol.*, *22*, 209–217 (1985).
67. V. Popovic, D. Micic, M. Nesovic, P. Djordjevic, J. Micic, and D. S. Djuric, *Exp. Clin. Endocrinol.*, *85*, 351–357 (1985).
68. D. J. Gross, Y. Halperin, J. M. Gomori, and B. Glaser, *Isr. J. Med. Sci.*, *25*, 256–260 (1989).
69. G. E. Sheeline, *Am. J. Roentgenol., Radium Ther. Nucl. Med.*, *112*, 768–773 (1971).
70. K. von Werder, R. Fahlbush, and H. K. Rjosk, *Neuroendocrinol. Lett.*, *3*, 328 (1981).
71. D. G. Johnston, K. Hall, P. Kendall-Taylor, R. Prescott, and R. Hall, in *Program and Abstracts of the Serono Symposium on Pituitary Hormones and Related Peptides*, Rome, Italy, Abstract 71, 1981.
72. F. Wollesen, T. Anderson, and A. Karle, *Ann. Intern. Med.*, *96*, 281–286 (1982).
73. D. G. Johnston, K. Hall, A. McGregor, W. M. Ross, P. Kendall-Taylor, and R. Hall, *Am. J. Med.*, *71*, 1059–1061 (1981).
74. A. M. McGregor, M. F. Scanlon, R. Hall, and K. Hall, *Br. Med. J.*, *2*, 700–703 (1979).
75. A. Grossman, R. Ross, M. Charlesworth, C. B. T. Adams, J. A. H. Wass, I. Doniac, and G. M. Besser, *Clin. Endocrinol.*, *22*, 679–686 (1985).

76. C. W. Burke, J. S. Bevan, A. J. Molyneux, and C. B. T. Adams, *J. Endocrinol.*, *104*(suppl.), 18, Abstract 4 (1985).
77. J. S. Bevan and C. W. Burke, *Clin. Endocrinol.*, *25*, 561–672 (1986).
78. O. Serri, A. M. Marchisio, R. Collu, J. Hardy, and M. Somma, *Hormone Res.*, *19*, 97–102 (1984).

10

DESIGN AND PROPERTIES OF PARTICULATE CARRIERS FOR INTRAVASCULAR ADMINISTRATION

Karel Petrak

CIBA-GEIGY Pharmaceuticals
Suffern, New York

DESIGN AND PROPERTIES OF PARTICULATE CARRIERS

Rationale for the Use of Particles in Drug Delivery

Drugs produce their effects through interacting reversibly with receptor sites within the body. For the drug to elicit the required therapeutic response, the correct amount of drug must be available at the site at the right time. Also, the presence of the required amount of drug needs to be maintained to achieve the sufficient duration of the therapeutic effect. The distribution of drugs to sites other than those of action and elimination is not only uncalled for but is also a potential source of toxicity.

The advancing knowledge of biological sciences makes it possible for us better to understand the mechanisms of diseases. Identification of specific chemical structures aids in designing and making drugs that can in principle act directly on the effectors of the disease. However, even using the body's own "drugs" (e.g., tissue plasminogen activator, steroid hormones) requires that, for their optimum action, they be presented to the target exclusively and when required. Various particular carriers have been tested as means for achieving the required mode of delivery.

The basic rationale for using particulate carriers in the intravenous application presupposes that the drugs included in the system will be distributed according to the properties of the carrier. Therefore, the carrier is expected to "seek out" the preferred site and consequently the drug could be directed to the intended site of action. For this to happen, the particulate carrier must have an access to the intended site of drug action, and further it must be able to avoid interactions with other sites within the body. At present, no such ideal particulate drug carriers are available.

Depending on the specific indication, an ideal particulate delivery system should facilitate some of the following properties of drugs:

Prolong drug effect due to longer circulation time as compared to free drug
Increase the drug concentration at the required site of action due to a preferential sequestering of the particles by the tissue of the site
Reduce drug toxicity in the tissue that do not accumulate the particles
Protect the drug from, say, metabolism and immune system recognition until it reaches the desired target site
Through an appropriate particle size choice, confine the drug delivery system to the chosen anatomical compartment
Interact selectively with the cells of the target site, if equipped with specific, biologically recognized structural units
Retain the drug within the particle while "in transit," and release the drug at the target site at the appropriate rate
Through participating in adsorptive endocytosis, deliver drugs to the appropriate phagocytic cells

For the drug's performance to be optimal, its quantity, the rate of delivery to the site, the selectivity of distribution, and the knowledge of the pharmacodynamics of the drug action need to be considered in the design of an effective drug delivery system (1). Unfortunately, much of this essential information is frequently not available in practice. Further, an altered pattern of drug distribution can alter the way the drug is metabolized. For all these reasons, "the designer" of a particulate drug delivery system is often unaware of what exactly is required.

To appreciate fully the rationale for employing particles as drug carriers in the vascular compartment it may be useful to remember that blood circulation plays a vital role in supplying the body with nutrients and oxygen, as well as in removing waste products. Blood flow also serves often as the key medium for the distribution of pharmacological agents through the body and as a mediator for delivering it to the sites of action. The rate and extent to which a drug enters various tissues are influenced by the rate with which it is brought to the tissue through circulation, and the ease with which the drug can leave the circulation and enter the tissue and be retained by that tissue. Particulate carriers, being fundamentally physically different from the drug, are expected to alter these rates so that a more favorable drug pharmacokinetics results.

For the drug to enter other than the vascular compartment of the body it has to pass through the capillary membranes. The rate with which drugs can cross the endothelium depends largely on their size and solubility (2). Even the smallest available particulate systems (i.e., nanoparticles having molecular weight in 10^6 D) as a rule do not cross the normal endothelium. Opportunities do exist, however, for particles to cross the endothelium due to local physiological (e.g., hepatic capillaries) or disease-related (e.g., inflammation) variations.

Intravascular Applications for Particulate Carriers

Particulate carrier systems could be utilized for a number of intravascular applications as discussed below.

Capillary Filtration

In this application, particles larger than 10 μm are filtered out of the circulation largely in the first capillary bed encountered after administration. Therefore, by selecting the appropriate port of entry for local administration, a drug-loaded delivery system can be retained in the required organ and tissue (e.g., the lungs) (3).

As Circulating Depot

In this application, particulate carrier must avoid a rapid uptake by the body's mononuclear phagocyte system (MPS); further, it needs to release the drug at a required rate, and needs to be removed from the body without initiating any unwanted response (e.g., immune). The required particle range is 0.1–1.0 μm. Although various indications could be considered (e.g., infection, leukemia, thrombosis), it is doubtful whether a circulating depot is really required where other forms of controlled delivery (e.g., a release from a intramuscular depot) might suffice.

Cellular Targets Within the Vasculature

Once a particular system that can circulate in the vascular compartment becomes available, selective delivery to other targets can be contemplated. Since it is most likely that for an effective delivery to cells the particles will need to enter the interior of the cells, delivery to phagocytic cells is the most obvious application. Therefore, the use of particulate systems for the delivery to diseased macrophages (e.g., parasitic, fungal, or viral invasion; enzyme function disorders; autoimmune disease such as rheumatoid arthritis; or gene therapy) would appear to be worth considering. It is less obvious how cells that do not ingest particles (e.g., erythrocytes), and therefore would need to rely on the delivery of drugs to within the cell volume from the particles adhering to the cell surface, could benefit from the use of such delivery systems.

Cellular Extravascular Targets

In this application, a long-circulating delivery system would additionally need to have means of accessing diseased or denuded endothelia, or would need to cross the endothelial barrier to reach cells (e.g., macrophages) active at an extravascular site (as for example at the site of active rheumatoid arthritis) (4).

Diagnostic Imaging

Water-soluble radiographic contrast materials have been found to improve detection of, say, tumors, but these agents are not ideal (i.e., they eventually reach both the normal and diseased tissue, are rapidly excreted, and can actually obscure some lesions) (5). Insoluble, particulate contrast agents tend to produce higher contrast between normal and pathological tissue, and can maintain this contrast for longer. Therefore, attempts have been made to encapsulate contrast agents into particulate delivery systems.

Structure Requirements for Long-Circulating Particles

From the above comments it is clear that the first, most important requirement on a therapeutically useful particulate delivery system for the intravascular route of administration is for the system to be able to avoid an uptake by the MPS. It has been known for some time that the distribution of colloidal particles, after intravenous administration, depends on interfacial physicochemical events (6-7).

Hydrophobic colloids are avidly cleared from circulation by the MPS, in particular via the liver and spleen (8). Evidence suggests that this event follows adsorption of proteins onto the particle surface (9). The presence at the particle surface of adsorbed or chemically attached hydrated polymers has been shown to reduce protein adsorption, and in consequence also the particle removal by the MPS (10,11). It is known in colloid science in general that the effectiveness of the hydrated polymer shell in stabilizing the particles is determined by the solution physicochemical properties of the polymer, the thickness of the adsorbed layer, and the density of the surface coverage (12).

The conditions that lead to repulsion of proteins from surfaces were recently examined by Andrade et al. (13). Their qualitative conclusions do suggest that high surface density and long chain length, of poly(oxyethylene) in this case, are necessary for protein resistance. More significantly, the surface density is predicted to have a greater effect than chain length on the steric repulsion and van der Waals attractions. The effect of other properties of materials such as surface charge, surface energy, interfacial free energy, and surface motion, together with the effects associated with specific compounds (e.g., albumin, glycoproteins, heparin) on blood compatibility, were also considered separately by Andrade et al. (14).

Based on the rationale of hydrated dynamic surfaces (15), lipid microspheres and other particles have been stabilized by coating with polysaccharides. A marginally increased hemocompatibility, as judged by the length of survival of the particles in circulation, was reported for such materials (16).

It certainly does appear to be reasonable to assert that for materials where interactions between circulating proteins and particles are prevented, there would be an increased likelihood for such materials to be hemocompatible. Silberberg (17) explained very succinctly one approach to this: "Particles which have a saturated layer of macromolecules adsorbed to their surfaces 'see' each other as do these macromolecules in solution. If the macromolecules give stable solution so do the coated particles." Yet further complications arise from the fact that the adsorbed coat is more stable when no free macromolecules are present in the surrounding solution (18). If this is not the case, as it clearly is not in a vascular compartment of the body, exchange between adsorbed and circulating macromolecules leads to changes in the composition of the particle surface.

In the first approximation it can be said that a particle that has an extensively hydrated surface composed of polymers that do not interact with plasma proteins has a very good chance of being hemocompatible in the sense of this application, and is likely to exhibit an extended residence in circulation (19).

Although particles modified by adsorbing hydrophilic polymers to their surface have been useful in demonstrating the above rationale (20), the results obtained have also pointed out the major drawbacks of this approach:

There is a lack of complete and homogeneous surface coverage.
Reversible adsorption occurs.
"Model" synthetic polymers such as polystyrene are not a realistic therapeutic system; adsorption to other materials is even less likely to be successful.

An alternative approach to modification by adsorption is to chemically attach polymers to the particle surface (21). However, even with this approach it is difficult to obtain a fully and homogeneously covered surface, due to the distribution of the reactive sites on the surface and the steric hindrance presented by the excluded volume of already attached polymer chains. Other practical complications are also likely to arise when grafting to polymer particles in the presence of drugs is attempted.

The difficulty of making hemocompatible particles would thus appear to be not just because of the lack of understanding as to what needs to be done but because of the practical problems associated with preparing stable surfaces saturated with suitable macromolecules.

A very direct lesson can be learned about what surface features are needed by examining natural materials. Most appropriately for our discussion, red blood cells remain in circulation with an apparent half-life of about 25 days (22). A detailed analysis of the surface of these blood cells shows it to be composed largely of carbohydrates (which are in turn a part of surface glycoproteins). The glycoproteins produce at the cell surface a densely packed hydration sheet, around 16 nm thick. This hydration layer effectively shields the hydrophobic cell membrane from interacting with circulating vascular components (such as opsonizing proteins). An additional stabilizing effect is also likely imparted to these cells through a net negative charge present at the surface (under the physiological conditions, the sialic acid part of the cell carbohydrate is ionized) (cf. Figure 1).

Therefore, various attempts have been made in recent years to incorporate carbohydrates, and especially sialic acid, into the surface of colloidal drug delivery materials. Geho and Lau (23) patented a method of modifying liposomes that claimed to lead to a longer residence time of liposomes in circulation. These liposomes included in their composition lipids that had a trisialic acid as their headgroup. The rationale for this approach is sound as the role of sialic acid in controlling the extent and timing of "recognition" in biological systems is well known (24). Recently, several attempts were made to utilize compounds containing sialic acid for modifying surfaces of particles and liposomes (25-27).

Most of the results support the rationale that surface modification could lead to increased hemocompatibility of the material. It is also clear that the so far very limited success has been due to mainly two factors. First, there were subsequent changes in the surface of the particles caused by desorption of the surface-modifying species used (e.g., poloxamers). Second, the initial coverage of the particle surface with the modifying agent was both incomplete and insufficient, giving an "imperfect" material (e.g., gangliosides in liposomes). In other words, the correct chemical entities (such as "noninteracting" sugars, e.g., sialic acid) need to be present in a correct physical form (i.e., here as a hydrated, sufficiently thick, preferably negatively charged layer, capable of screening hydrophobic interactions between circulating proteins and the drug-containing particle core).

An elegant way of producing particles having a much more accurately defined surface composition has been described by Rolland et al. (28). Their approach could briefly be summarized as follows: Amphiphilic block copolymers can be considered to be higher molecular weight analogs of nonionic surfactants. Indeed, such block copolymers are known to form micelles, with their degree of association and the apparent molecular weight of the block copolymer micelles increasing with the increasing molecular weight of the copolymer and with the decreasing content and the sequence length of the hydrophilic component of the block copolymer (29). Block copolymers of the structure and composition that gives micelles at sufficiently low critical

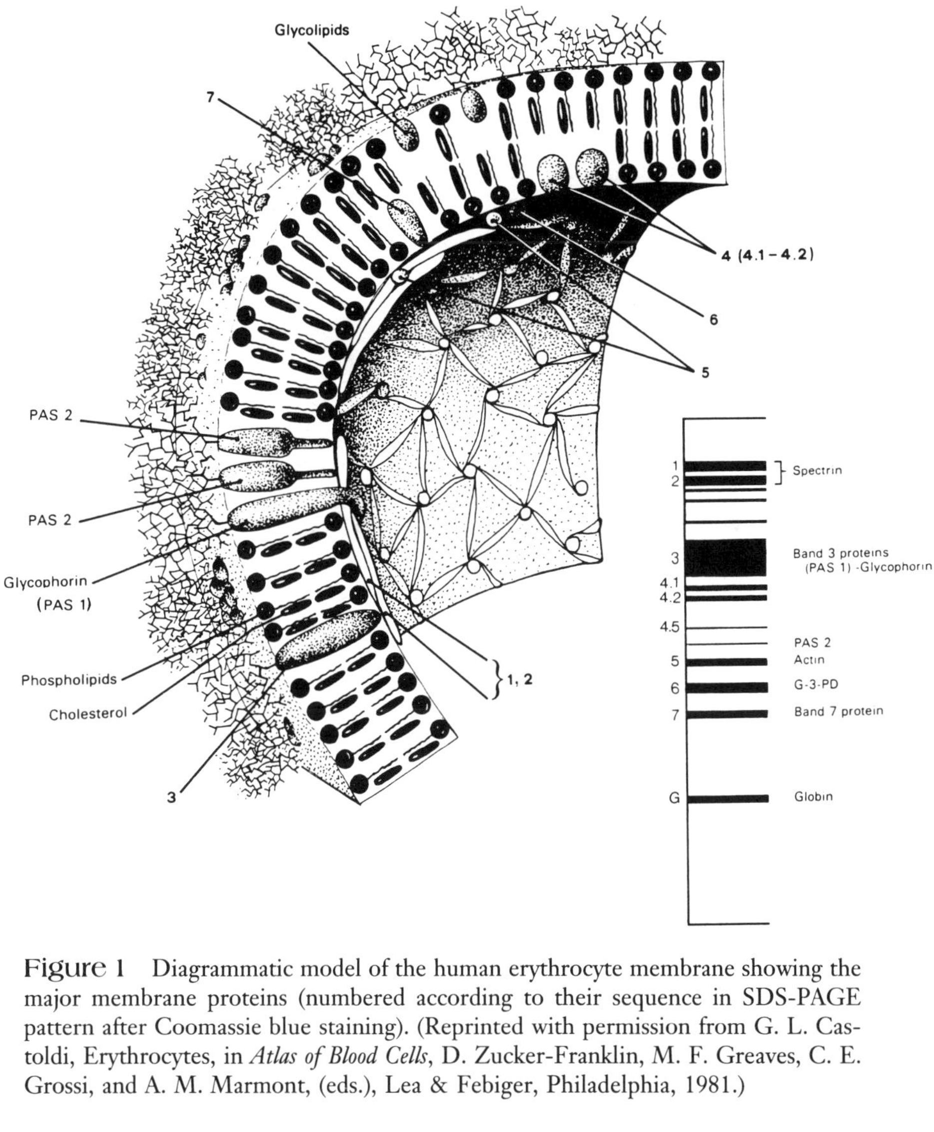

Figure 1 Diagrammatic model of the human erythrocyte membrane showing the major membrane proteins (numbered according to their sequence in SDS-PAGE pattern after Coomassie blue staining). (Reprinted with permission from G. L. Castoldi, Erythrocytes, in *Atlas of Blood Cells*, D. Zucker-Franklin, M. F. Greaves, C. E. Grossi, and A. M. Marmont, (eds.), Lea & Febiger, Philadelphia, 1981.)

micelle concentrations ($\sim 10^{-8}$–10^{-10} M) can thus produce stable particles that do not interact with proteins under physiological conditions. In experimental animals, the micelles remained in circulation with a half-life in excess of 50 hr. This "self-assembly" principle, when utilized together with biodegradable block copolymers, should eventually lead to superior drug delivery systems. The similarity in the physical nature of the surface of such particles and some biological surfaces (such as that of erythrocytes shown in Figure 1) is indicated in the idealized particle depicted in Figure 2.

Further, for particulate carriers to be acceptable they must avoid undesirable interactions at various levels in the body (e.g., blood and cell surface interactions, stimulus–response coupling, host defense interactions) (30). It would appear that blood compatibility could be achieved at least with some

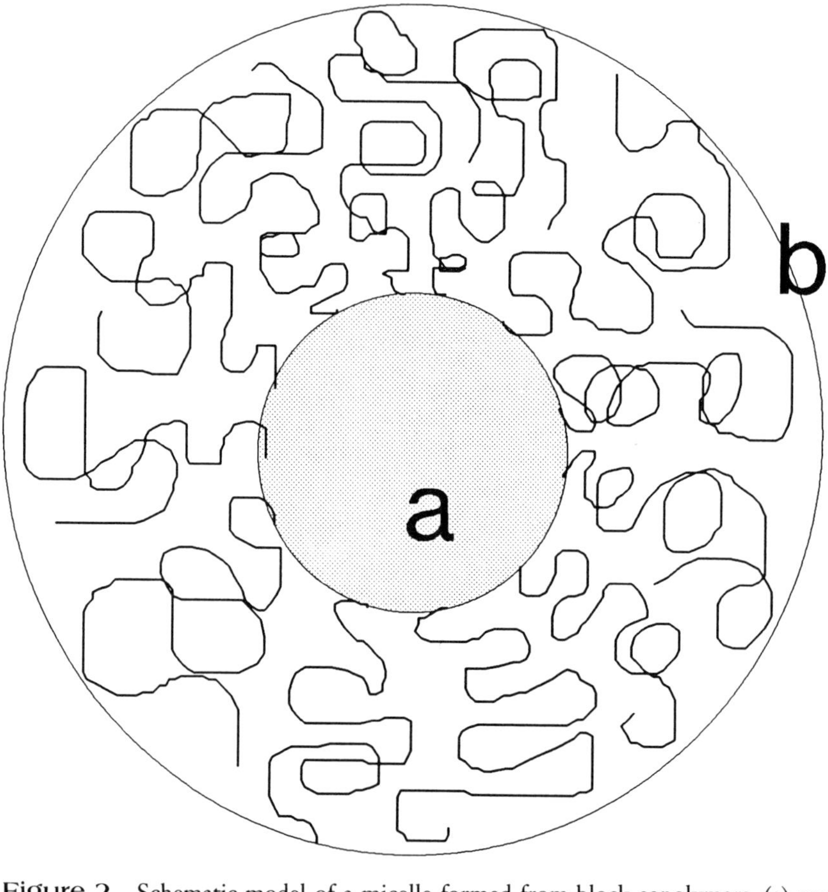

Figure 2 Schematic model of a micelle formed from block copolymers. (a) core formed from hydrophobic blocks also containing the drug; (b) hydrophilic chains extending into the aqueous environment and preventing interactions between drug and circulating proteins.

materials, either endogenous such as serum albumin, natural such as polysaccharides, or entirely synthetic (28). However, the inevitable effects resulting from attaching or encapsulating drugs on the properties of particles has received little attention, but it can be safely assumed that such modifications will often reduce the blood compatibility of the delivery system (107). Further, the actual release of the drug from such particles in any controlled manner remains largely a matter of conjecture.

Clearly an important property of particulate drug-carrying systems is their lack of toxicity. For polystyrene particles, it has been demonstrated that particle-induced, dose-dependent systemic hypotension and myocardial depression depends on particle size (with small, 3.4-μm particles causing a bigger effect than 11.6-μm particles) (31). These effects, believed to be due to delayed deposition of the smaller particles in the body, can be expected to be largely eliminated when particles can degrade in the body at the appropriate rate (e.g., albumin). Although adequate acceptability of albumin particles in general has been reported by several workers (32,33), the toxic effects will need to be examined carefully for any new drug delivery system or drug combination.

Various materials that could be considered as basic elements of drug delivery systems were recently reviewed (34,35). In my view, materials used by the body itself for specific tasks should be examined first of all (mainly carbohydrates, polysaccharides, glycoproteins, proteins, poly(amino acids)). Degradation of the delivery system into harmless, excretable components must occur with any material that is to have a fighting chance of being used in a regular clinical practice.

The particles alone, and even more so particles carrying drug molecules, can be expected to elicit cell responses leading to events such as local inflammation (36) or immune response (37). Even when a carrier itself is nonimmunogenic, it can well serve as a hapten response–inducing carrier once it has been coupled to a drug (38).

Once a generally noninteracting particulate system is available, its desirable interaction with the target site or molecules at the site presents a new, unsolved problem.

APPLICATION OF PARTICULATE SYSTEMS FOR SPECIFIC INDICATIONS

The above ideal particles for delivering drugs to sites of action are still in the realm of science fiction. Much less perfect materials are available for the current experimental medical practice.

Particulates in the Treatment of Infectious Diseases

Given the physical and pharmacological attributes of particulate carriers, and especially of liposomes, it is not surprising that such delivery systems have been tested in vivo mainly in connection with infectious diseases. This application, although something of a "saving grace," follows very naturally from the known pathogenesis of intracellular infections. Most particulate delivery systems, being recognized as foreign in the body, are removed by the cells of the MPS. It is not surprising, therefore, that such delivery systems have been found to be effective against diseases caused by intracellular pathogens, especially those that colonize the cells of the MPS (39).

Drugs that cannot reach intracellular pathogens located in such "pharmacological sanctuary" because of their limited ability to cross the membranes of the host cells can do so with the aid of phagocytosed particles (40). This logical rationale for using particles for intracellular delivery has now been confirmed many times experimentally, e.g., by Bonventre and Gregoriadis (41).

A number of investigators (e.g., see Ref. 42) have observed that the effect of a liposomal drug preparation depended markedly on the liposome composition. In another study, Mehta et al. (43) observed, using erythrocytes as a representative mammalian cell, that whereas free amphotericin B was toxic to both fungal cells and mammalian cells in vitro, the liposomal preparation of the drug was toxic only to fungal cells. Similarly differentiated action was also reported for liposomes acting on macrophages and lymphoid cells (44). The exact mechanism responsible for such selectivity of action is not clear at present. A recently observed behavior that liposomes of certain composition can facilitate transfer of amphotericin B from liposomes to fungal cell membrane but not to mammalian cell membranes might be at least in part connected with the apparently selective drug delivery (45). A theoretical basis, derived from the knowledge of the physical state of lipid membranes, for the construction of vesicles that would transfer its content to some cells and not others does seem to exist (46). However, the phenomenon is not yet fully understood, and the actual in vivo conditions are far too complex, to make this approach alone of much direct practical value. It must not be forgotten that particulate delivery systems, being by definition colloidal in nature, will have their properties determined largely by the nature of the particle surface.

It is not the purpose of this chapter to catalogue all the experiments that have been reported on particulate delivery systems. Several thorough reviews already exist on this and on the potential use of particulate delivery systems (39,47-50), the potential being based on the in vitro and in vivo animal model results (51). As an illustration, sterilization of liver and spleen of mice previously infected with *Listeria monocytogenes* resulted from administration of a liposome-entrapped ampicillin (two doses of 0.27 mg). When the same free drug was administered, complete elimination of bacteria was achieved only after the administration of 48 mg (eight doses of 6 mg) of the drug (52). Similarly, about a 100-fold increase in the effectiveness of methotrexate was observed when the drug was attached to maleyated bovine serum albumin (53). This study again confirmed that an effective delivery of drug to macrophages can be achieved based on the scavenger receptor-mediated endocytic pathway, leading to selective killing of parasites residing in these cells.

Liposome-incorporated antimicrobial agents and 8-aminoquinolines have been reported to be far superior to the free drug against experimental leishmaniasis (54,55). Similarly, Desiderio and Campbell reported that liposome-entrapped cephalothin was more effective than the free drug in the treatment of experimental murine salmonellosis (56,57). The therapeutic index of primaquine was enhanced significantly when presented in a liposomal form. Primaquine phosphate's toxicity toward the lungs, heart, kidneys, and brain was modified through the radically altered distribution of the drug in the body when administered as a liposomal formulation (58). Liposome-encapsulated macrophage activators (immunomodulators) have been shown to be more effective than the free agents in other disease states: antitumor activity (59), herpes (60), hepatitis (61), and fungal infections (62).

In some cases, although liposomes did not increase the efficacy of the drug as compared to its free form, lower drug toxicity due to liposome encapsulation permitted administration of a higher dose, which then resulted in complete cure (63).

Very few reports exist, however, regarding the use of such systems in the treatment of human diseases. The pioneering and now very frequently referred to studies were conducted by Lopez-Berestein et al. (64,65). In the latter study, 46 patients with systemic fungal infections were treated with a liposomal preparation of amphotericin B (40 of these patients were reported to have failed previously to respond to conventional amphotericin B therapy, and the remaining 6 patients suffered severe side effects in response to such conventional therapy). Twenty-four patients showed a complete and positive response to liposomal amphotericin, with 22 patients failing to respond. No short- or long-term toxic effects were observed. The success of liposomal over free-drug preparation has been further demonstrated in other clinical trials (66-68). So, through lucky coincidence, even particles not specifically designed for targeting to sites of disease have been found useful; by extrapolation this approach is very promising. In my view, much will be achieved in this area when various scientific disciplines make a concerted effort to solve this multifaceted problem.

Delivery of Anticancer Drugs

A thorough review of the delivery of anticancer drugs has been published by Zee-Cheng and Cheng (69). The ideal delivery system that would, after being administered in intravascular fashion, deliver the drug in its active form exclusively to the required site (i.e., solid or metastatic tumor) of action while containing the drug inactive in the rest of the body still does not exist. However, as in the case of delivery to sites of infection, some benefits can be derived from using particulate delivery in cancer therapy. Recent innovations in the synthesis of microparticulate drug delivery systems, and their *potential* clinical applications, were recently reviewed by Kerr (70).

Systemic Administration

So far, most of the available experimental systems – liposomes, microspheres of natural and synthetic polymers, soluble polymers, nanoparticles, magnetic particles, and even activated carbon particles – have been tested in clinical investigations. It would appear that any benefits that may derive from the use of such systems are related to the depot action of the anticancer agents carried by the system (71). In experiments with liposome-entrapped adriamycin it was shown that drug (particularly cardio-) toxicity was reduced while the therapeutic effects against a number of different tumors were maintained or enhanced (72-74).

A steady flow of promising preclinical data keeps the expectations about the usefulness of liposomes and other particulate systems high (75). At present, however, particulate antitumor agents have no established role in the treatment of cancer. A number of clinical trials using liposome-encapsulated anticancer drugs have been reported to be in progress (e.g., doxorubicin, phases I and II; liposomal platinum, phase I; muramyl tripeptide phosphatidylethanolamine, phase II; amphotericin (for systemic fungal infections), compassionate IND and phase I; gentamicin (gram-negative infections), phase I; daunorubicin, phase I; and 6-aminochrysene, phase I) (76).

It is clear that the pharmacokinetics of chemotherapeutic agents can be altered by their formulation with particulate carriers. For example, Rolland (108) showed this in hepatoma patients using doxorubicin bound to polymethacrylate nanoparticles (cf. Figure 3). However, the all-important question regards the possible benefits that could result from altered drug distribution.

The current practice in chemotherapy is such that the doses administered are limited only by the currently inevitable side effects. The use of more conservative, lower doses may compromise the cure rate; therefore the preferred approach is to maintain full-dose therapy and consider reducing the total duration of treatment. When high doses or concentrations of anticancer

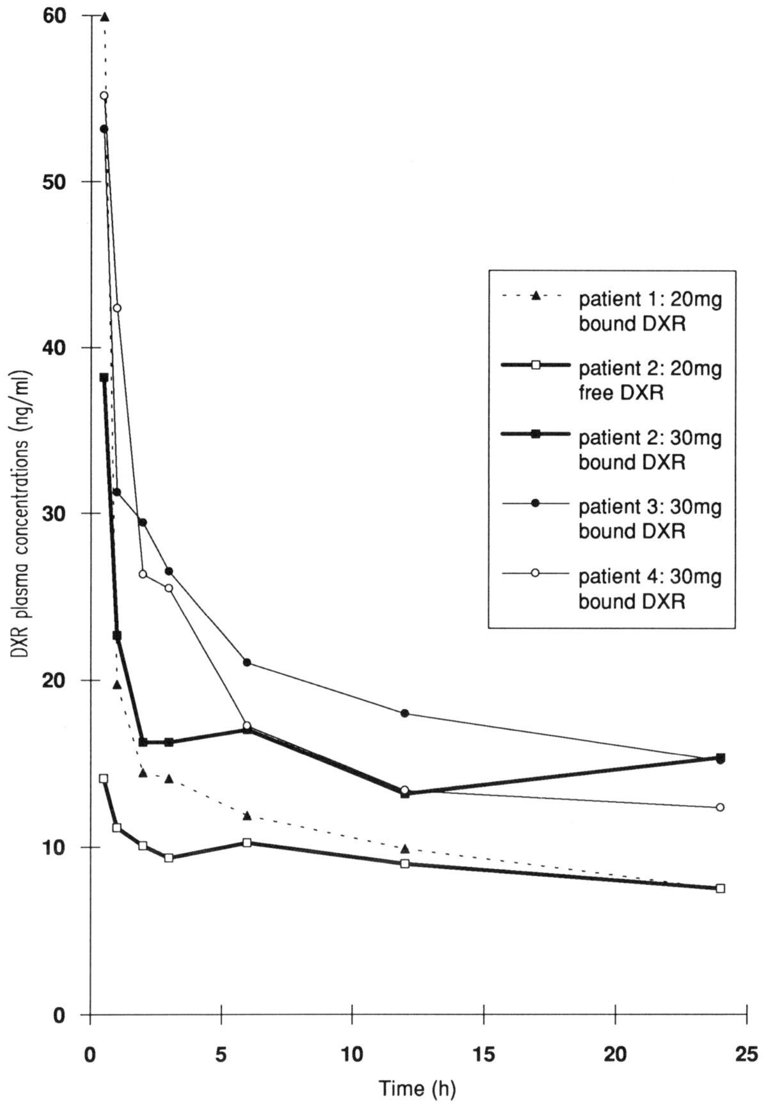

Figure 3 Doxorubicin (DXR) plasma concentration–time profiles obtained after a single intravenous injection of free or nanoparticle-bound doxorubicin in hepatoma patients. (Reproduced with permission from Ref. 108.)

drugs can be achieved, as in regional infusion applications, there is in general clear evidence that a substantial increase in response rate occurs (109).

Further, it is known that the dose–response curve for most chemotherapeutic agents is steep. This is related to the sensitivity of the tumor to a given drug. Consequently, when high concentrations of antitumor drugs can be achieved in the tumor, as in some cases of regional infusion studies, a substantially increased beneficial therapeutic response can be seen (110).

The toxicity of the currently available antitumor agents is such that any approach resulting in (at least) equal effectiveness of lower doses is appealing. There is some indication that relatively low-dose maintenance therapy for some type of cancer (e.g., testicular, lymphoma) may prolong the duration of remission without necessarily increasing the cure rate (109). In other cases, toxicity has been reduced without compromising the effectiveness of therapy (e.g., adriamycin in breast cancer and sarcomas) (111).

Therefore, a clear benefit should be derived from using particulate delivery systems as long as (a) therapeutically meaningful concentrations of the drug can be sustained in the tumor while (b) the plasma concentration of the drug is reduced significantly (i.e., below the levels of limiting toxicity).

It should be kept in mind also that exposing tumor cells to sublethal concentrations of chemotherapeutic agents can lead to the development of resistance of tumor to drug (112). An empirical rule derived from radiotherapy also applies to chemotherapy: if the same total dose is administered by reducing the dose rate while increasing the duration of treatment correspondingly, the cure is substantially reduced in experimental systems.

Nevertheless, such systems remain to have much potential as long as more determined effort is made to include the physiological, pharmacokinetic, and medical considerations in their design instead of repeatedly trying to apply what is readily available. For example, instead of using liposomes made from conventional phospholipids, novel glycolipids designed and made specifically for drug delivery systems should be used (77). With the clinical perspective in mind, it is very likely that a successful treatment of systemic metastatic disease will require the development of liposome formulations able to access extra-MPS tumor sites (78).

Local Drug Delivery and Chemoembolization

Local drug delivery has today become common practice in cancer chemotherapy, e.g., as hepatic arterial infusion and intrathecal administration. To obtain any meaningful gain in local drug concentration by local drug delivery as compared to systemic administration, the clearance of the drug by the body should be much higher than the blood flow through the target. Similarly, for a peritarget delivery, i.e., administration into the space (cavity) that surrounds the target, as in administration of drugs into the cerebrospinal fluid, the body clearance should be much higher than the efflux from the peritarget space (79). Hepatic arterial chemotherapy with hepatic arterially administered yttrium-90 microspheres appears to be a promising way to more effective therapy (80). This treatment is directed to primary and metastatic liver cancers, and might eradicate efficiently (i.e., during a 1- to 2-month course of treatment) and effectively (a complete response rate of >80% being reported) surgically unresectable tumors. Combining potent radiosensitizers with yttrium-90 microspheres appeared to generate improved selectivity of treatment.

Additional advantage in local drug administration can be derived from chemoembolization. Particles of appropriate size (>10 μm diameter) will be filtered out from the bloodstream by the first available capillary bed. Depending on the in vitro half-life of the particles and the dose, a decrease in blood flow through the capillary bed will result. The combination of decreased blood flow and release of the drug from entrapped particles may lead to an increased local drug concentration (81-83).

Following the administration of 15- to 35-μm albumin microspheres through the artery supplying hypervascular liver tumors, the carrier particles were reported to be retained predominantly by the tumor vasculature (84). Fujimoto et al. (85) used mitomycin C–loaded albumin microspheres for the treatment of inoperable hepatic cancer. The patient survival was claimed by the authors to show a statistical improvement.

An additional method for retaining drug-loaded particles at the site is offered by including ultrafine particles of magnetite (Fe_3O_4) into, e.g., a human serum albumin particle matrix (86). Albumin particles suffer from the same fate as most other colloids – they are taken up rapidly by the MPS (the plasma circulating half-life ~8–10 min). However, when magnetically responsive particles are infused locally via an artery, they can be retained within the capillaries of tumor by an externally applied magnetic field. Assuming that the drug release kinetics and the general pharmacokinetics of the released drug are favorable, this can lead to the drug being present at the site in high concentration possibly causing tumor cell death (87,88). However, magnetically responsive (albumin) particles suffer the fate of all the other tested systems – methods to produce, on a large scale, particles with uniform physicochemical properties and the ability to avoid uptake by the MPS need to be developed (89). Further, this approach is suitable only for well-defined tumors, with little or no applicability to metastatic neoplasms. Information is still needed on relative drug localization between tumor and normal sites within a given target volume, and on extracellular and intracellular distribution of both the target and drug molecules within the lesional foci, following administration of drug-loaded magnetic particles.

Any successful treatment of disseminated metastases must overcome the difficulty arising from neoplastic cellular heterogeneity and the development of resistance of tumor cells to therapy. Largely in animal models so far, macrophages have been activated in situ through interactions with liposomes containing various immunomodulators. Cancer metastases in several tumor system have been eradicated through repeated administration of such liposomes. The immunomodulators (e.g., muramyl dipeptide, muramyl tripeptide phosphatidylethanolamine, lymphokines such as macrophage-activating factor and γ-interferon) all suffer from a very short circulation life (90). The availability of liposomes has facilitated a way of activating macrophages in situ (91). There appear to be at least two limiting factors using this approach. The first is to some extent due to the removal of the liposomes by the MPS, resulting in the activation of macrophages away from the tumor site. The activated macrophages (blood monocytes) may not readily identify and act on the metastases. Second, the tumor burden may exceed the destructive capacity of the available macrophages. For this reason, a combined therapy consisting of macrophage activation, chemotherapy, and local irradiation has been used. Results obtained using human blood monocytes are now also available (92). The use of liposomes for a more general macrophage "suicide" technique using dichloromethylene diphosphonate as the encapsulated drug was described by Van Rooijen (93).

Overall, the currently available particulate delivery systems do not deliver drugs exclusively to the intended site of action. At best the delivery system accumulates in the organ or tissue where the tumor is located, and the local release of drug from particles might temporarily increase drug concentration at the site. Alternatively, the particulate system functions as a slow-release formulation, altering the drug plasma concentration profile from which some benefits (e.g., decreased systemic toxicity) might ensue (94,108). This situation is similar to that observed for polymer-bound anticancer prodrugs (95).

At present, out of all tested delivery systems, liposomes appear to offer the best option for systemic drug delivery. Even with these "particles," a huge effort and a major conceptual breakthrough is still needed to overcome the existing technological difficulties before a pharmaceutically acceptable liposome preparation becomes available. It is also essential that in seeking a new effective therapy, experimental approaches and clinical strategies with a sound rationale are adopted.

Diagnostic Imaging

For the currently available diagnostic imaging techniques (e.g., gamma camera, nuclear magnetic resonance imaging, computed tomography) it is important that the technique be able to resolve and contrast tissues which, though histologically different, appear to be similar to the technique (e.g., magnetically similar). Contrast-enhancing agents have therefore been developed to help differentiate among different tissues. Particulate carriers have also been considered for this application, and the question of delivering a contrast agent simply becomes a question of delivering site-selectively a pharmacologically inactive contrast agent (96). A number of metal ions and compounds can be used as such agents, with the problems of particle body distribution remaining the same as for the delivery of drugs. For example, supermagnetic iron oxide particles have been found to be very effective contrast agents, but their destination in the body is predominantly the liver and spleen (97). This type of distribution has been utilized in the clinic for detecting by magnetic resonance hepatic lesions in the liver (98).

Although antibodies are not the type of "particle" discussed in this chapter, the use of antibodies for delivering imaging agents is instructive. In principle, the use of antibodies that have been well established to target to tumors in connection with contrast agents should facilitate imaging. Thus, monoclonal antibody MAb 19.9 (directed against colon adenocarcinoma) was attached to 16 molecules of dimeglumine chelated to gadolinium (99). The antibody retained its biological and immunological properties after labeling; however, the degree of labeling used was considered unlikely to lead to a sufficient contrast enhancement of the tumor when magnetic resonance imaging is used. It has been estimated that as many as 100 or more chelated paramagnetic ions per MAb may be required to achieve sufficient contrast enhancement in vivo (100). Using direct conjugation to antibodies it is unlikely that with this number of chelated ions present the antibody's biological properties would not alter. One approach to solve this problem is to use polymeric carriers (e.g., polylysine) for attaching the chelating ligand to MAbs (101). Although it has been claimed that over 90% of the MAb activity was retained in vivo, it is not clear whether the body distribution of the whole conjugate remains equally favorable. The development of agents that would be tissue-specific and would be taken up selectively by only the normal or the pathological tissue and be retained long enough to facilitate imaging is still required. One strategy is to develop contrast agents with particulate properties (102). An alternative is to encapsulate contrast agents into a particulate delivery vesicle. The particle itself would need to have no image-enhancing properties but would serve to alter the biodistribution of the entrapped compound. Liposomes have been developed for carrying gadolinium diethylenetriaminepentaacetic acid. Although the magnetic relaxivity of the entrapped contrast agent was less than that of the free compound, these encapsulated contrast agents improved detection of liver metastases in experimental animals as compared to the free agent. Enhanced imaging of other tissue has not been reported (103,104). Again, the pathology of disease may facilitate accumulation of particles in the non-MPS tissues. Encouraging results have been obtained in delivering imaging agents to ischemic, inflamed, and tumor tissues (105). [Although this chapter deals only with the intravenous applications of particles, it should not be forgotten that the use of different routes of administration is likely to produce different results. For example, the subcutaneous administration of encapsulated imaging agents tends to lead to an enhanced visualization of lymph nodes, and can demonstrate very accurately any involvement of lymph nodes in cancer (106).]

Again, several critical issues, similar to those relevant to the use of liposomes for drug delivery, must be resolved before particulate carriers can find a wide and successful use in this application.

CONCLUSIONS

The rationale of using a drug delivery system, whether administered through the vascular compartment or otherwise, is sound as long as the delivery system alters the pharmacokinetics of the drug in such a way that a therapeutic benefit is accrued. It could be argued that there are likely to be very few unique chemical entities that exhibit only desirable effects in the body. To limit the undesirable effects, when these are due to the drug acting at the wrong place at the wrong time, requires the use of an "intelligent" delivery system. Such systems do not yet exist, but could be designed by "copying" the body's own means of controlling the spacial distribution of endogenous effectors (such as hormones). To this end, particulate systems need to be made that can

Avoid indiscriminate interactions with the MPS;
Selectively enter the desired cell population;
Release the active component at the appropriate time and rate; however, it should be also kept in mind that the overall "performance" of the system will depend on the pharmacokinetics of the drug after it has been released from the particle; and
Be eliminated from the body without causing any other response.

For administering particulate delivery systems into the vascular compartment of the body, there is very little reason other than site-selective drug delivery. Creating a circulating depot is of little appeal, since local depots can be utilized more readily (although exceptions might exist, e.g., an intramuscular depot of an anticoagulant might cause local bleeding that would not occur with a circulating intravascular system).

At present the most promising approach appears to be based on the self-assembly of amphiphilic compounds, into either liposomes or lipid core particles. Synthesis of novel glycolipids, or block copolymers consisting of carbohydrates and other biodegradable macromolecules, is the required first step before particulate delivery systems that satisfy the above requirements can be made.

REFERENCES

1. L. Aarons, A. Boddy, and K. Petrak, Pharmacokinetic evaluation of site-specific drug delivery systems, in *Novel Drug Delivery* (L. F. Prescott and W. S. Nimmo, eds.), John Wiley and Sons, New York, pp. 121–127.
2. K. Petrak and P. Goddard, *Adv. Drug Deliv. Rev.*, *3*, 191–214 (1989).
3. M. J. A. P. Daemen, H. T. M. Vervoort-Peters, and H. H. W. Thijssen, *J. Pharm. Pharmacol.*, *40*, 283–285 (1988).
4. M. M. Weiss, *Sem. Arth. Rheum.*, *19*(1), 9–21 (1989).
5. S. Seltzer, *Invest. Radiol.*, *23(Suppl.1)*, S122–S125 (1988).
6. D. J. Wilkins and P. A. Myers, *Br. J. Exp. Pathol.*, 7, 568–576 (1966).
7. D. J. Wilkins, *J. Colloid Interf. Sci.*, *25*, 84–89 (1967).
8. P. Couvreur, *CRC Crit. Rev. Ther. Drug Systems*, *5*(1), 1–20 (1988).
9. J. W. B. Bradfield, in *Microspheres in Drug Therapy* (S. S. Davis, L. Illum, J. G. McVie, and E. Tomlinson, eds.), Elsevier, Amsterdam, 1984, pp. 25–37.
10. J. E. O'Mullane, K. Petrak, L. E. F. Hutchinson, and E. Tomlinson, *Int. J. Pharma.*, *63*, 177–180 (1990).
11. K. Petrak, *Br. Polym. J.*, *22*, 213–219 (1990).
12. D. Napper, *J. Colloid Interf. Sci.*, *58*, 390 (1977).
13. (a) S. I. Jeon, J. H. Lee, J. D. Andrade, and P. G. de Gennes, *J. Colloid Interf. Sci.*, *142*(1), 149–158 (1991). (b) S. I. Jeon and J. D. Andrade, *J. Colloid Interf. Sci.*, *142*(1), 159–166 (1991).
14. J. D. Andrade, S. Nagaoka, S. Cooper, T. Okano, and S. W. Kim, *ASAIO*, *10*(2), 75 (1987).
15. S. Nagaoka, Y. Mori, H. Tanzawa, Y. Kikuchi, F. Inagaki, Y. Yokota, and Y. Noishiki, *ASAIO*, *10*(2), 76 (1987).
16. A. Carlsson, T. Sato, and J. Sunamoto, *Bull. Chem. Soc. Jpn.*, *62*(3), 791–796 (1989).
17. A. Silberberg, *J. Colloid & Interface Sci.*, *111*(2), 486–497 (1986).
18. P. G. de Gennes, *Advances Colloid Interf. Sci.*, 27, 189–209 (1987).
19. L. Illum, S. S. Davis, R. H. Mueller, E. Mak, and P. West, *Life Sci.*, *40*, 367–374 (1987).
20. L. Illum and S. S. Davis, *J. Pharm. Sci.*, 72, 1086–1089 (1983).
21. P. Artursson, L. Brown, J. Dix, P. Goddard, and K. Petrak, *J. Polym. Sci., Part A; Polym. Chem.*, *28*, 2651–2663 (1990).
22. Geigy Scientific Tables (C. Lentner, ed.), Vol. 3, CIBA-GEIGY, 1984, p. 209.
23. B. Geho and J. Lau, U.S. Patent 4,501,728 (1983).
24. J. D. Aplin and R. C. Hughes, *Biochim. Biophys. Acta*, 694, 374–418 (1982).
25. T. M. Allen, C. Hansen, and J. Rutledge, *Biochim. Biophys. Acta*, 981, 27–35 (1989).
26. T. Allen and A. Chonn, *FEBS Lett.* *223*(1), 42–46 (1987).
27. T. D. Heath, N. G. Lopez, and D. Papadahopoulos, *Biochim. Biophys. Acta*, 74, 820 (1985).
28. A. Rolland, J. O'Mullane, P. Goddard, L. Brookman, and K. Petrak (accepted for publication).
29. C. Price, *Pure Appl. Chem.*, *55*, 1563 (1983).

30. P. Artursson and J. O'Mullane, *Adv. Drug Delivery Rev.*, *3*, 165–189 (1989).
31. J. D. Slack, M. Kanke, G. H. Simmons, and P. P. DeLuca, *J. Pharm. Sci.*, *70*, 660–664 (1981).
32. G. P. Royer, T. K. Lee, and T. D. Sokolski, *J. Parent. Sci. Technol.*, *37*, 34–37 (1983).
33. J. H. Ratcliffe, I. M. Hunneyball, A. Smith, C. G. Wilson, and S. S. Davis, *J. Pharm. Pharmacol.*, *36*, 431–436 (1984).
34. T. St. Pierre and E. Chiellini, *J. Bioact. Compat. Polym.*, *1*(4), 467–497 (1986) (part I); *2*(1), 4–30 (1987) (part II); *2*(3), 238–257 (1987) (part 3).
35. P. Goddard and K. Petrak, *J. Bioact. Compat. Polym.*, *4*, 372–402 (1989).
36. J. C. Foreman and L. M. Lichtenstein, *Biochim. Biophys. Acta*, *629*, 587–603 (1980).
37. M. Kanke, I. Sniecinski, and P. P. DeLuca, *Proc. 3rd Int. Conf. Pharm. Technol.*, Paris, 1983, pp. 162–169.
38. B. K. Park, M. C. Tingle, P. S. Grabowski, J. W. Coleman, and N. R. Kitteringham, *Biochem. Pharmacol.*, *36*(5), 591–599 (1987).
39. R. L. Juliano, in *Drug Carrier Systems; Horizons in Biochemistry and Biophysics*, Vol. 9 (F. H. D. Roerdink and A. M. Kroon, eds.), John Wiley and Sons, New York, 1989, pp. 249–279.
40. J. D. Johnson, W. L. Hand, J. B. Francis, N. King-Thompson, and R. W. Corwin, *J. Lab. Clin. Med.*, *95*, 429–439 (1980).
41. P. F. Boventre and G. Gregoriadis, *Antimicrob. Agents Chemother.*, *13*, 1049–1051 (1978).
42. M. Stevenson, A. J. Baillie, and R. M. E. Richards, *Antimicrob. Agents Chemother.*, *24*, 742–749 (1983).
43. R. Mehta, G. Lopez-Berestein, R. Hopfer, K. Mills, and R. J. Juliano, *Biochim. Biophys. Acta*, *14*, 230–234 (1984).
44. R. Mehta, G. Lopez-Berestein, R. L. Hopfer, K. Mehta, R. A. White, and R. L. Juliano, *Antimicrob. Agents Chemother.*, *28*, 511–513 (1985).
45. R. L. Juliano, C. W. Grant, K. R. Barber, and M. A. Kalp, *Mol. Pharm.*, *31*, 1–11 (1987).
46. J. Bolard, *Biochim. Biophys. Acta*, *864*, 257–304 (1986).
47. F. Emmen and G. Storm, *Pharmaceutisch Weekblad* (Scientific Edition), *9*, 162–171 (1987).
48. A. Coune, *Infection*, *16*(3), 141–147 (1988).
49. G. Lopez-Berestein, *Antimicrob. Agents Chemother.*, *31*(5), 675–678 (1987).
50. C. E. Swenson, M. C. Popescu, and R. S. Ginsberg, *Crit. Rev. Microbiol.*, *15(Suppl.1)*, S1–S32 (1988).
51. I. A. J. M. Bakker-Woudenberg, F. H. Roerdink, and G. Scherphof, *Liposomes as Drug Carriers: Trends and Progress*, John Wiley and Sons, Chichester, 1987.
52. I. A. J. M. Bakker-Woudenberg, A. F. Lokerse, F. H. Roerdink, D. Regts, and M. F. Michel, *J. Infect. Dis.*, *151*, 917–924 (1985).
53. A. Mukhopadhyay, G. Chaudhuri, S. K. Arora, S. Sehgal, and S. K. Basu, *Science*, *244*, 705–707 (1989).

54. C. R. Alvig, E. A. Steck, W. L. Chapman, Jr., V. B. Waits, L. D. Hendricks, G. M. Swartz, Jr., and W. L. Hanson, *Life Sci.*, *26*, 2231–2238 (1980).
55. C. R. Alvig, E. A. Steck, W. L. Hanson, P. S. Loizeaux, W. L. Chapman, Jr., and V. B. Waits, *Life Sci.*, *22*, 1021–1026 (1978).
56. J. V. Desiderio and S. G. Campbell, *J. Infect. Dis.*, *148*, 563–570 (1983).
57. J. V. Desiderio and S. G. Campbell, *J. Reticuloendothel. Soc.*, *34*, 279–287 (1983).
58. P. Pirson, R. Steiger, and A. Trouet, *Biochem. Pharmacol.*, *31*, 3501–3507 (1982).
59. G. Lopez-Berestein, L. Milas, N. Hunter, K. Mehta, E. M. Hersh, C. G. Kurahara, M. Vanderpas, and D. A. Eppstein, *Clin. Exp. Metastas.*, *2*, 127–137 (1984).
60. W. C. Koff and I. J. Fidler, *Antiviral Res.*, *5*, 179–190 (1985).
61. C. LaBonnardiere, *Ann. Microbiol.* (Paris), *129A*, 397–402 (1978).
62. E. B. Fraser-Smith, D. A. Epstein, M. A. Larsen, and T. R. Matthews, *Infect. Immun.*, *39*, 172–178 (1983).
63. P. Pirson, R. F. Steiger, A. Trouet, J. Gillet, and F. Herman, *Ann. Trop. Med. Parasitol.*, *74*, 383–391 (1980).
64. G. Lopez-Berestein, *Antimicrob. Agents Chemother.*, 31(5), 675–678 (1987).
65. G. Lopez-Berestein, G. P. Bodey, V. Fainstein, M. Keating, L. S. Frankel, B. Zeluff, L. Gentry, and K. Mehta, *Arch. Intern. Med.*, *149*, 2533–2536 (1989).
66. G. Lopez-Berestein, V. Fainstein, R. Hopfer, K. Mehta, M. P. Sullivan, M. Keating, M. G. Rosenblum, R. Mehta, M. Luna, E. M. Hersh, J. Reuben, R. L. Juliano, and G. P. Bodey, *J. Infect. Dis.*, *151*, 704–710 (1985).
67. G. Lopez-Berestein, G. P. Bodey, L. S. Frankel, and K. Metha, *J. Clin. Oncol.*, *5*, 310–317 (1987).
68. J. P. Sculier, A. Coune, F. Meunier, C. Brassinne, C. Laduron, C. Hollaert, N. Collette, C. Heymans, and J. Klastersky, *J. Cancer Clin. Oncol.*, *24*, 527–538 (1988).
69. R. K.-Y. Zee-Cheng and C. C. Cheng, *Meth. Find. Exp. Clin. Pharmacol.*, *11*(7-8), 439–529 (1989).
70. D. J. Kerr, *Cancer Drug Delivery*, *4*(1), 55–61 (1987).
71. E. Mayhew, Y. Rustum, and F. Szoka, in G. Gregoriadis, J. Senior, A. Trouet (eds.), *Targeting of Drugs*, NATO Adv. Study Inst., Series A, Life Sci., Plenum Press, New York, 1982, pp. 249–260.
72. A. Gabizon, A. Dagan, D. Goren, Y. Barenholz, and Z. Fuks, *Cancer Res.*, *42*, 4734–4739 (1982).
73. E. Mayhew, Y. Rustum, and W. J. Vail, *Cancer Drug Delivery*, *1*, 43–58 (1983).
74. G. Storm, F. H. Roerdink, P. A. Steerenberg, W. H. de Jong, and D. J. A. Crommelin, *Cancer Res.*, *47*, 3366–3372 (1987).
75. J. H. Senior, *CRC Crit. Rev. Ther. Drug Carrier Systems*, *3*(2), 123–193 (1987).
76. M. J. Ostro and P. R. Cullis, *Am. J. Hosp. Pharm.*, *46*, 1576–1587 (1989).
77. D. Papahadjopoulos and A. Gabizon, *Horizons in Membrane Biotechnology*, Wiley-Liss, New York, 1990, pp. 85–93.
78. A. Gabizon, in *Drug Carrier Systems*, Vol. 9 (F. H. D. Roerdink and A. M. Kroon, eds.), John Wiley and Sons, New York, 1989, pp. 185–211.

79. J. F. M. Smits and H. H. W. Thijssen, in *Rate-Controlled Drug Administration and Action* (H. A. J. Struyker-Boudier, ed.), CRC Press, Boca Raton, 1986, pp. 83–114.
80. W. Ensminger, *Cancer Chemother. Pharmacol.*, *23* (*Suppl.*), S68–S73 (1989).
81. S. Dakhil, W. D. Ensminger, and K. Cho, *Cancer*, *50*, 631–635 (1982).
82. J. W. Gyves, W. D. Ensminger, and D. Vanharken, *Clin. Pharmacol. Ther.*, *34*, 259–265 (1983).
83. W. D. Ensminger, J. W. Gyves, P. Stetson, and S. Walker-Andrews, *Cancer Res.*, *45*, 4464–4467 (1985).
84. J. J. Burger and J. G. McVie, *Pharmaceutish Weekblad*, 7, 27 (1985).
85. S. Fujimoto, M. Miyazaki, F. Endoh, O. Takahashi, K. Okui, and Y. Morimoto, *Cancer*, *56*, 2404–2410 (1985).
86. K. J. Widder, A. E. Senyei, and B. Sears, *J. Pharm. Sci.*, *71*, 107–111 (1982).
87. K. Sugibayashi, M. Okumura, and Y. Morimotor, *Biomaterials*, *3*, 181–186 (1982).
88. M. Sako and S. Hirota, *Gan-to-Kagaku-Ryoho*, *13*, 1618–1624 (1986).
89. P. K. Gupta and C. T. Hung, *Life Sci.*, *44*, 175–186 (1989).
90. I. J. Fidler, Metastasis (Ciba Foundation Symposium 141), John Wiley and Sons, Chichester, 1988, pp. 211–222.
91. A. J. Schroit and I. J. Fidler, *Cancer Res.*, *42*, 161–167 (1982).
92. E. S. Kleinerman and M. M. Hudson, in *Liposomes in Treatment of Infection Diseases and Cancer* (I. J. Fidler and G. Lopez, eds.), Alan R. Liss, New York, 1988.
93. N. Van Rooijen, *J. Immunol. Meth.*, *124*, 1–6 (1989).
94. A. Hagiwara, T. Takahashi, and N. Oku, *Crit. Rev. Oncol. Hematol.*, *9*(4), 319–350 (1989).
95. L. W. Seymour, K. Ulbrich, J. Strohalm, J. Kopecek, and R. Duncan, *Biochem. Pharmacol.*, *39*, 1125 (1990).
96. L. Lex, *Acta Biochim. Biophys. Hung.*, *24*(3), 265–281 (1989).
97. S. Majumdar, S. Zoghbi, C. P. Pope, and J. C. Gore, *Radiology*, *169*, 653–655 (1988).
98. D. D. Stark, R. Weissleder, G. Elizondo, P. F. Hahn, S. Saini, L. E. Todd, J. Wittenberg, and J. T. Ferrucci, *Radiology*, *168*, 297–301 (1988).
99. C. Curtet, C. Tellier, J. Bohy, M. L. Conti, J. C. Saccavini, P. Thedrez, J. Y. Douillard, J. F. Chatal, and H. Koprowski, *Proc. Natl. Acad. Sci. USA*, *83*, 4277–4281 (1986).
100. Y. Manabe, C. Longley, and P. Furmanski, *Biochem. Biophys. Acta*, *883*, 460–467 (1986).
101. V. P. Torchilin, A. L. Klibanov, N. D. Nossiff, M. A. Slinkin, H. V. Strauss, E. Haber, V. N. Smirnov, and B. A. Khaw, *Hybridoma*, *6*, 229–240 (1987).
102. S. E. Seltzer, *Radiology*, *171*, 19–21 (1989).
103. C. Tilcock, *Radiology*, *171*, 77 (1989).
104. E. C. Unger, *Radiology*, *171*, 19 (1989).
105. V. J. Caride, *Nucl. Med. Biol.*, *17*(1), 35–39 (1990).
106. I. Levy, J. Roberts, and M. Baum, *Postgrad. Med. J.*, *63*, 829–833 (1987).

107. L. E. Ambler, L. Brookman, J. Brown, P. Goddard, and K. Petrak, (submitted).
108. A. Rolland, *Int. J. Pharm.*, *54*, 113–121 (1989).
109. E. Frei and G. P. Canellos, *Am. J. Med.*, *69*, 585–594 (1980).
110. T. Kato, R. Nemoto, H. Mori, M. Takahashi, and M. Harada, *Cancer*, *48*, 674–680 (1981).
111. R. Benjamin, S. Legha, B. Mackay, M. Ewer, S. Wallace, M. Valdivieso, S. Rasmussen, G. Blumenschein, and E. Freireich, *Proc. AACR*, *22*, 179 (1981).
112. B. W. Fox and M. Fox (eds.), *Handbook of Experimental Pharmacology*, *Vol. 72, Antitumor Drug Resistance*, Springer-Verlag, New York, 1984.

11

Intravenous Liposomal Therapy of Retroviral Infections: A Strategic Approach

Nigel C. Phillips* and Christos M. Tsoukas

Montreal General Hospital Research Institute and McGill University Montreal, Quebec, Canada

POTENTIAL FOR LIPOSOMAL THERAPY IN RETROVIRAL DISEASE

Human AIDS

It is clear that the multifaceted pathology of human immunodeficiency virus (HIV)–induced acquired immunodeficiency syndrome (AIDS) infection will preclude the majority of, if not all, single therapeutic approaches currently being used or developed. HIV infection induces profound qualitative and quantitative changes in the innate and acquired immune systems (1). These defects are responsible for the susceptibility to opportunistic bacterial and viral infections, neoplasia, and autoimmune phenomena (1,2). HIV-infected patients have diminished immunological responses to foreign antigens associated with the progressive depletion of $CD4^+$ lymphocytes (3). Moreover, significant defects in immune function have been observed in HIV-infected asymptomatic individuals before depletion of $CD4^+$ lymphocytes (4). A direct effect of inactivated HIV-1 or viral components on both T- and B-lymphocyte function in vitro has been demonstrated (5–8), as well as selective signaling

Dr. Nigel C. Phillips and Dr. Christos M. Tsoukas are both members of the McGill Centre for AIDS Research.

**Current affiliation*: Centre D'Immunologie et de Biologie Parasitaire, Institut Pasteur de Lille, Lille, France

defects in T cells (9). HIV infection of dendritic cells (10) may result in impaired antigen presentation. A number of effects on monocyte/macrophage function have been noted (11–14), and the macrophage has been implicated in the pathogenesis of HIV infection (15,16).

Contemporary strategies for the treatment of HIV infection are strongly orientated toward inhibiting infection and replication of the virus. There is an urgent need to identify procedures capable of reducing the toxicity and/or enhancing the activity of agents able to inhibit the replication of HIV, especially nucleoside analogs. In the short term, these inhibitors provide the most promising therapeutic approach. Effective therapy will ultimately depend on vaccination of asymptomatic individuals as a means of arresting viral replication at both the individual and population levels.

Antiviral Therapy with AZT

3′-Azido-3′-deoxythymidine (AZT) inhibits the infectivity and cytopathic effects of HIV (17). AZT was initially shown to be of benefit to patients with HIV infection (18), but longer treatment periods have identified significant AZT-induced bone marrow toxicity (19–25). Defects in the proliferation and differentiation of bone marrow progenitor cells in HIV-infected individuals have been observed (26) and may contribute to the toxicity of AZT. Concomitant granulocyte-monocyte–stimulating factor therapy reduces AZT toxicity (27,28) but is itself associated with dose-limiting side effects (29,30). AZT-resistant variants of HIV have been isolated from patients on long-term Zidovudine (Burroughs Welcome) therapy (31), and interruption of AZT therapy may lead to enhanced viral production and infectivity (32). More recent studies demonstrated that a dose reduction may result in antiviral activity with reduced bone marrow toxicity (23,33,34). Combination therapy with other nucleoside analogs such as 2′,3′-dideoxycytidine (35,36) or other antiviral agents such as acyclovir (37) may reduce the toxicity of AZT. Decreased susceptibility of HIV to AZT following in vivo (38–41) or in vitro (42) exposure has been reported. Chronic exposure of cell lines infected with HIV to AZT results in a decreased production of reverse transcriptase (RT), p24, and infectious viral particles (43). Removal of the AZT results in the expression of infectious viral particles within a period of days, indicating inhibition of intact virus production rather than eradication of integrated virus.

The active metabolite of AZT, which is AZT triphosphate (AZTTP), inhibits HIV RT with an IC_{50} of 2.3 μM. The IC_{50} for cellular DNA polymerase-α is 260 μM, thus showing that at therapeutic concentrations AZT selectively inhibits HIV RT (44). This specificity is not observed for other enzymes. AZTTP competitively inhibits DNA polymerase-γ (45) and thymidylate synthase (46). AZT inhibits UV-induced DNA strand break rejoining in normal human lymphocytes (47), indicating that inhibition of DNA-related enzymic activity is expressed on a functional level.

Mouse Retroviral Diseases

A number of retroviral infections in mice are known to induce immunosuppression and leukemia (e.g., Molony virus, Rauscher virus, Friend virus). A retroviral infection of mice that results in acute lymphoproliferative disease and profound immunosuppression was recently described (48). LP-BM5 murine leukemic virus, a mixture of ectropic and mink cell focus-inducing retroviruses, contains a 4.8-kilobase replication-defective retrovirus with large deletions in the polymerase (*pol*) and envelope (*env*) genes (49,50). The defective retrovirus alone may be oncogenic rather than immunosuppressive (51). Rescue of the defective retrovirus with nonpathogenic ectropic retrovirus induces a disease that has many of the characteristics of human AIDS, including lymphadenopathy, hypergammaglobulinemia, immunosuppression of cellular and humoral immunity, B-cell lymphoma (48), and susceptibility to opportunistic infection (52). Functional T lymphocytes are required for susceptibility to the disease (53), supporting the concept that an initial polyclonal activation of T cells is necessary for the subsequent abnormal stimulation of B cells (53).

Following the initial polyclonal stimulation of B cells, specific antigen responses to both T-dependent or T-independent antigens are abolished (54). Selective signaling defects in helper T cells/antigen-presenting cells have been observed in LP-BM5-infected mice (55), as well as significant changes in the ratio of $L3T4^+/CD8^+$ lymphocytes (56). Unlike human AIDS, murine AIDS (MAIDS) is associated with elevated $L3T4^+$ and depressed CD8+ numbers (57). Nonfunctional $L3T4^+$ and activated/cytotoxic CD8+ cells in MAIDS (57) may result in a pathological process very similar to human AIDS. Macrophages, as in human AIDS, appear to be a primary site of infection (58). A major association between the H-2 haplotype (major histocompatibility complex, or MHC) of the mouse strain and sensitivity or resistance to MAIDS has been identified (59,60). Susceptibility to MAIDS in mouse strains appears to be acute (H-2b, H-2p, H-2j) or chronic (H-2a, H-2d, H-2k) in onset (59,60).

Treatment of Murine AIDS with AZT

Treatment of LP-BM5 infection with AZT inhibits virus-induced changes in lymphoid cell populations (56), prevents immunosuppression (61), prolongs life span and reduces viral titer (62) only if therapy commences within a week of infection. Intermittent and alternating dose regimens of AZT and dideoxycytidine (63) or treatment with antineoplastic drugs (64) also seems to be effective in preventing the development of MAIDS when treatment is started at the time of infection.

The use of such models will ultimately enable the relationship between immunological responsiveness (cellular responses, immunoglobulin isotype expression) and acquired immunodeficiency to be quantified.

STRATEGIC APPROACH

The profound involvement of the macrophage in the pathogenesis of HIV infection (15,16) presents a unique opportunity for evaluating cell-specific drug targeting via the use of liposomal carriers. As a prerequisite to the initiation of clinical studies using liposomes containing antiviral agents and/or immunomodulators in patients with HIV infections, we have investigated the impact of liposomes on retroviral disease in three immunopathological areas:

The amelioration of the therapeutic index of AZT, especially with respect to the bone marrow

The effect of AZT and liposomal AZT on retrovirus-induced immunosuppression, on changes in $CD4^+$ T cells, and on viral replication (viremia)

Specific interaction of liposomes with cells at risk from HIV infection, namely $CD4^+$ T cells and monocytes, via the use liposome surface-associated antibody (immunoliposomes)

AMELIORATION OF THE THERAPEUTIC INDEX OF AZT BY LIPOSOMAL THERAPY

Liposomal Incorporation and Retention of AZT

AZT is a small molecule with a slight degree of lipophilicity. However, liposomal incorporation of AZT corresponded to inclusion within the aqueous phase of the liposome. Using AZT at 10 mg ml^{-1} to hydrate dried films of DPPC/DMPG and DSPC/DMPG* and prepare multilamellar vesicles (MLVs) (average diameter 0.8 μm, range 0.2–2.0 μm), liposomal incorporation was 250 ± 12 and 240 ± 15 μg/10 μmol phospholipid, respectively (mean ± SD from four experiments). Incubation of these liposomes at 4°C did not result in the loss of AZT from the liposomes (Figure 1). Incubation of DPPC/DMPG liposomes at 37°C resulted in a 96% loss of AZT over a period of 24 hr; in contrast, DSPC/DMPG liposomes incubated at 37°C lost 64% of incorporated

*DPPC, dipalmitoylphosphatidylcholine; DMPG, dipalmitoylphosphatidylglycerol; DSPC, distearoylphosphatidylcholine.

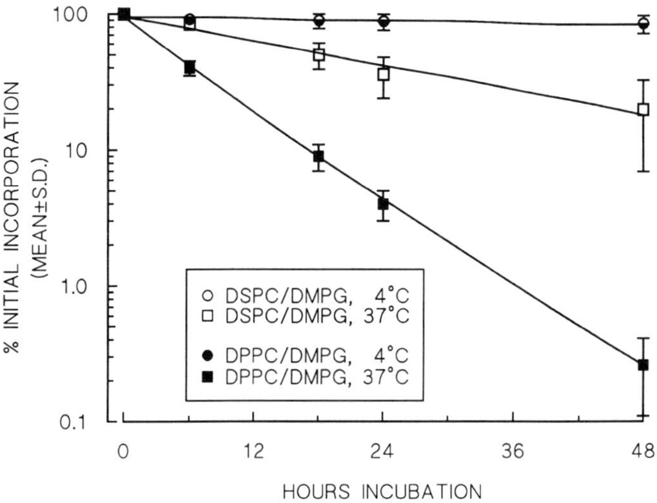

Figure 1 Liposomal retention of AZT. DSPC/DMPG or DPPC/DMPG liposomes containing AZT were incubated at 4 or 37°C. Liposomal AZT was determined at the indicated times. Mean result from four experiments.

AZT (Figure 1). All of our subsequent experiments with liposomal AZT were therefore carried out using liposomes formulated from DSPC/DMPG.

Bone Marrow Localization of Free and Liposomal AZT

AZT (as [^{3}H]methyl AZT) levels in the bone marrow remained constant 1 hr after the injection of 10 mg/kg body weight into male CD1 mice and were associated with the cellular fraction (0.018 ± 0.004% injected dose/10^7 cells). Liposomal AZT at the same dose was not detectable in the bone marrow (<0.00001%). Analysis of plasma AZT concentrations showed that there was a 10-fold difference between the apparent volumes of distribution (V_d) of AZT and liposomal AZT (0.595 and 0.047 L^{-1}, respectively), presumably the result of liposomal localization within organs of the reticuloendothelial system.

Hematological Toxicity of Free and Liposomal AZT

Treatment of male CD1 mice with free AZT iv in the dose range of 0.4–10 mg/kg for 5 days resulted in a dose-dependent reduction of bone marrow cellularity, red blood cells (RBC), and leukocytes (Figure 2). Significant reductions in RBC and leukocytes were seen at doses of 10 and 2 mg/kg ($p <$ .01 and .05 respectively; Student's two-tailed t test), but not at 0.4 mg/kg. The reduction in bone marrow cellularity was significant at all AZT doses

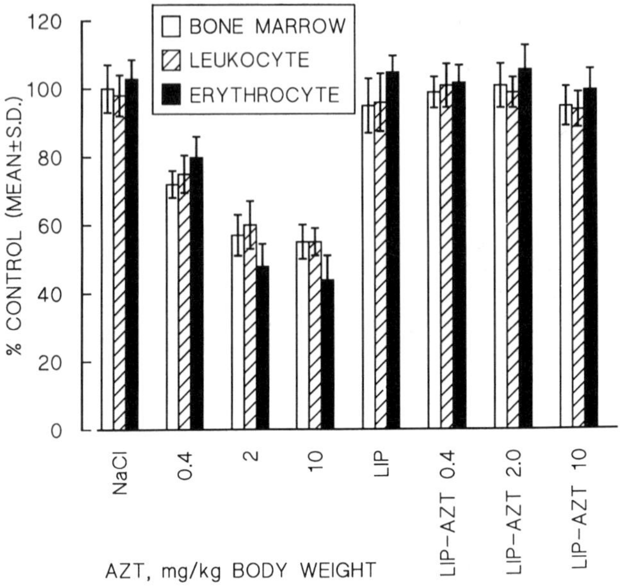

Figure 2 Hematopoietic toxicity of free and liposomal AZT. Groups of five CD1 male mice were treated with free AZT, liposomes (LIP), or liposomal AZT (LIP-AZT) for 5 days. Bone marrow cellularity, erythrocyte and leukocyte numbers were determined on day 6. AZT treatment at 0.4, 2, and 10 mg/kg body weight resulted in a significant decrease in bone marrow cellularity as well as, erythrocyte and leukocyte numbers compared to NaCl treatment ($p < .01$, three-way ANOVA). Treatment with liposomal AZT in the same dose range had no effect on bone marrow cellularity, erythrocyte or leukocyte numbers. Mean ± SD of three experiments.

($p < .01-.05$). Control liposomes or liposomal AZT at 2 mg/kg given iv had no effect on bone marrow cellularity, RBC, or leukocytes (Figure 2).

Inflammatory Neutrophil Response

The bone marrow toxicity of agents such as AZT may result in impaired nonspecific host responses, such as neutrophil production, toward infection. The iv administration of free AZT resulted in a significant and dose-related reduction in the neutrophil inflammatory response against thioglycollate (Student's t test, $p < .01$; Figure 3). Control liposomes or liposomal AZT given iv had no effect on the neutrophil inflammatory response (Figure 3).

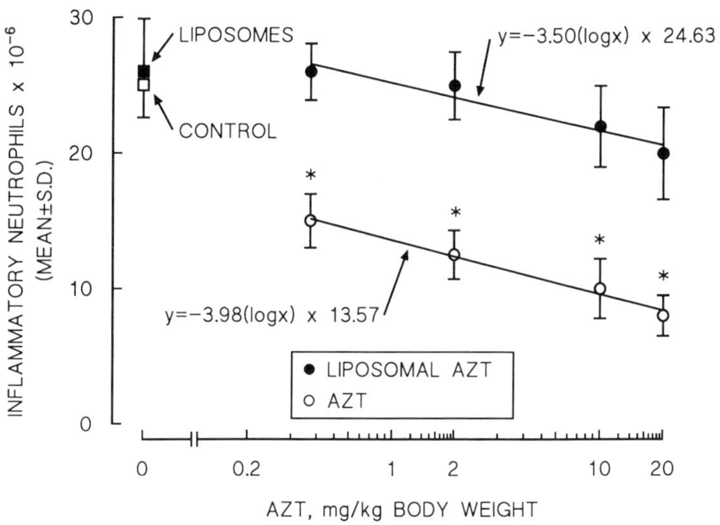

Figure 3 Neutrophil inflammatory response following free or liposomal AZT treatment. Groups of five CD1 male mice were treated with free AZT, liposomes, or liposomal AZT for 5 days. AZT treatment at 0.4–20 mg/kg body weight resulted in a significant decrease in the neutrophil inflammatory response. *, Significantly different from control neutrophil inflammatory response ($p < .01$, Student's t test for unpaired data). Liposomal AZT did not significantly impair the neutrophil inflammatory response.

Abrogation of Bone Marrow Toxicity of AZT by Free and Liposomal Muramyl Dipeptide

Granulocyte-macrophage colony-stimulating factors (GM-CSFs) stimulate the differentiation of granulocyte and monocyte bone marrow precursors (65). Recombinant GM-CSFs have been used in the salvage of patients receiving high-dose chemotherapy (66). Such therapy is limited by the fact that the therapeutic index is approximately 1 (67) and may well be self-limiting due to the induction of control cytokine networks (68). The ability of liposomes containing hydrophilic or lipophilic derivatives of muramyl dipeptide (MDP; *N*-aetylmuramyl-L-alanyl-D-isoglutamine) to induce macrophage-mediated cytotoxic activity against tumor cell targets in vitro and in vivo has been extensively studied (69–72). Liposomal MDPs are also more effective than free muramyl dipeptides in inducing GM-CSF activity in vivo (73) and interleukin-1 (IL-1) secretion in vitro and in vivo (74). Intravenous treatment of mice with either free or liposomal MDP prior to treatment with

AZT results in the maintenance of normal bone marrow cellularity (Figure 4), normal blood leukocyte levels (Figure 4), and normal inflammatory neutrophil responses (75). The relative efficacy of free and liposomal MDP in protecting bone marrow cells against AZT-induced toxicity correlates well with their ability to induce CSFs in vivo (73). Using this procedure, it has been possible to administer AZT via the iv route at doses of up to 50 mg/kg/day without significant myeloid bone marrow toxicity.

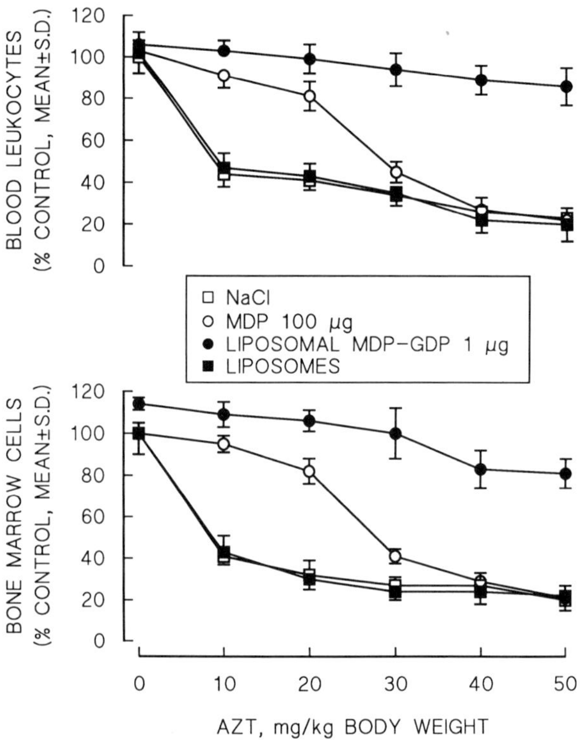

Figure 4 Effect of MDP or liposomal MDP-GDP on AZT toxicity. Groups of five CD1 male mice were treated intravenously with MDP (100 μg/mouse) or liposomal MDP-GDP (1 μmol phospholipid containing 1 μg/mouse) 5 hr before treatment with AZT in the dose range 10–50 mg/kg body weight intravenously. The effect of treatment on peripheral blood leukocytes and bone marrow cells was determined after 5 days.

ACTIVITY OF FREE AND LIPOSOMAL AZT IN MURINE AIDS

Splenocyte Mitogenic Responses in LP-BM5–Infected Mice

One of the consequences of MAIDS is a rapid suppression of the ability of lymphocytes to respond to mitogenic stimulation (60). Our studies have demonstrated that splenocyte mitogenic responses to the lymphocyte mitogens concanavalin A (Con A), lipopolysaccharide (LPS), and phytohemagglutinin (PHA) are significantly reduced compared to control, noninfected mice at 4 and 7 weeks postinfection (Table 1). Intravenous treatment with AZT at 2 mg/kg three times weekly commencing a week after infection resulted in normal mitogenic responses at 4 weeks postinfection (Table 1): the protective effect of AZT treatment was lost at 7 weeks postinfection, where a significant reduction in mitogenic response was observed (Table 1). Intravenous treatment with liposomal AZT at 2 mg/kg three times weekly resulted in mitogenic responses within the normal range at both 4 and 7 weeks postinfection (Table 1). Control liposomes were protect against LP-BM5–induced suppression of mitogenic responsiveness.

Table 1 Inhibition of LP=BM5-Induced Lymphocyte Immunosuppression by AZT and Liposomal AZT[a]

	Stimulation Index					
	4 weeks postinfection			7 weeks postinfection		
Treatment	Con A	PHA	LPS	Con A	PHA	LPS
Normal, uninfected	34 ± 8	12 ± 4	17 ± 5	38 ± 6	15 ± 6	22 ± 5
NaCl	17 ± 6[b]	4 ± 3[b]	6 ± 4[b]	16 ± 8[b]	4 ± 2[b]	3 ± 1[b]
AZT, 2 mg/kg	25 ± 5	7 ± 3	8 ± 5	13 ± 9[b]	3 ± 2[b]	4 ± 3[b]
Liposomal AZT, 2 mg/kg	35 ± 9	11 ± 3	15 ± 6	36 ± 8	14 ± 4	19 ± 5
Liposomes	15 ± 8[b]	5 ± 3[b]	6 ± 4[b]	14 ± 6[b]	3 ± 1[b]	4 ± 2[b]

[a]Groups of five male C57BL/6 mice were injected with LP-BM5 retrovirus ip. Treatment with AZT (2 mg/kg body weight iv) or liposomal AZT (33 mmol phospholipid, 2 mg/kg body weight iv) was started 1 week after infection and was continued two times weekly for 3 or 6 weeks. The mitogenic response of spleen lymphocytes to concanavalin A (Con A, T-cell mitogen), lipopolysaccharide (LPS, B-cell mitogen), or phytohemaglutinin (PHA, T-cell mitogen) was determined *in vitro* by the incorporation of $[^3H]$methylthymidine using standard procedures. Stimulation index = proliferation in the presence of mitogen/control proliferation.

[b]Significantly different from mitogenic response of control mice ($p < .01$, ANOVA).

Effect of AZT and Liposomal AZT on Plasma RT Activity in LP-BM5–Infected Mice

RT is an accepted marker of retrovirus proliferation (76) and has been used to follow the effect of AZT on viremia in LP-BM5–infected mice (61). Plasma RT activity increased 35-fold over background activity during LP-BM5 infection (Figure 5). AZT treatment at 2 mg/kg iv three times weekly retarded but did not inhibit RT activity, indicating a partial suppression of the development of viremia (Figure 5). Liposomal AZT at 2 mg/kg iv three times weekly suppressed the development of RT activity to within background levels (Figure 5).

Effect of AZT and Liposomal AZT on L3T4$^+$-T cells in LP-BM5–Infected Mice

There was a transient increase in the proportion of Thyl.2$^+$-L3T4$^+$ T cells 7 days following LP-BM5 infection (Figure 6). Thyl.2$^+$-L3T4$^+$ T cells were significantly reduced 21 and 35 days postinfection ($p < .001$, ANOVA).

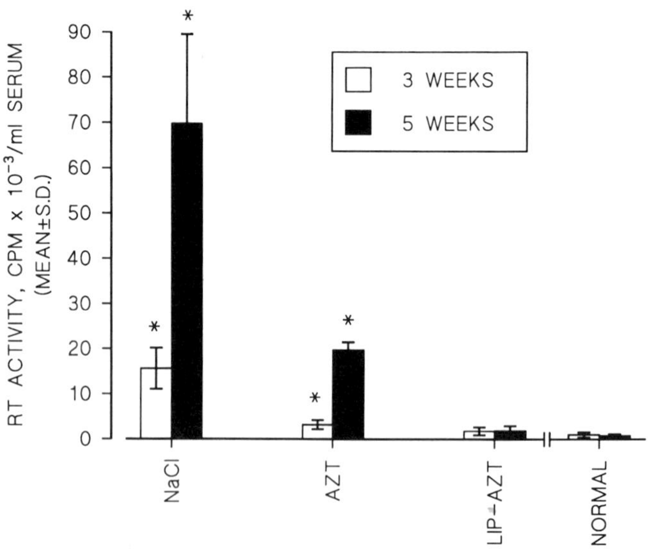

Figure 5 Effect of AZT treatment on plasma reverse transcriptase activity in LP-BM5 infected mice. Groups of five C57BL/6 female mice were infected with LP-BM5 leukemia virus. Treatment with AZT (2 mg/kg body weight iv three times weekly) or liposomal AZT (2.5 μmol phospholipid containing 2 mg/kg body weight iv three times weekly) was started on day 7, and plasma reverse transcriptase activity determined after 3 and 5 weeks. *, Significantly different from control, uninfected mice ($p < .01$, Student's t test for unpaired data).

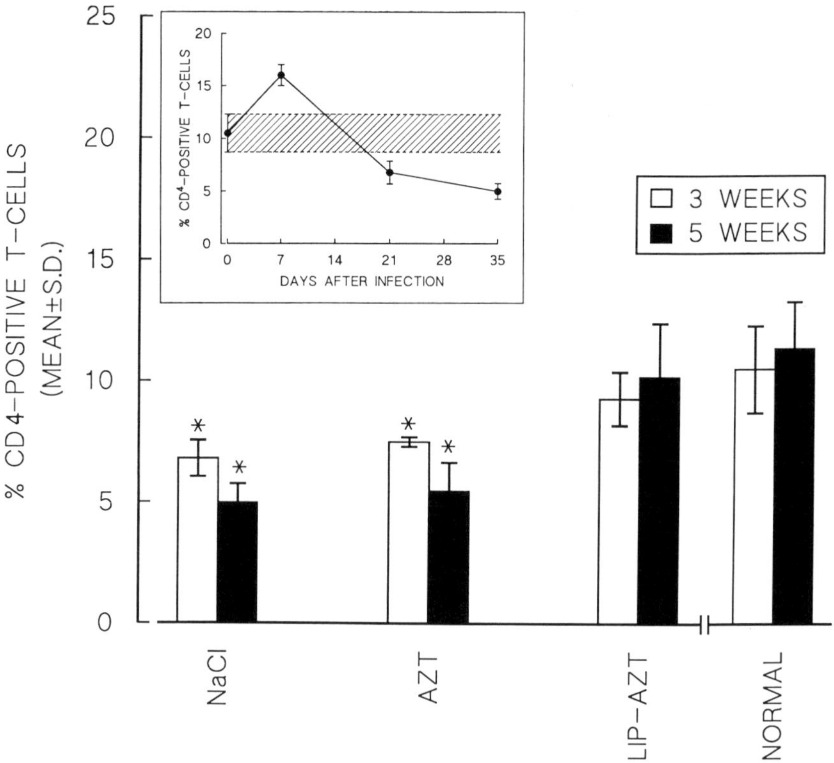

Figure 6 Effect of AZT treatment on Thy 1.2^+-$CD4^+$ T cells in LP-BM5-infected mice. Groups of five C57BL/6 female mice were infected with LP-BM5 leukemia virus. Treatment with AZT (2 mg/kg body weight iv three times weekly) or liposomal AZT (2.5 μmol phospholipid containing 2 mg/kg body weight iv three times weekly) was started on day 7, and splenic Thy 1.2^+-$CD4^+$ T cells determined by FACS analysis after 3 and 5 weeks. *, Significantly different from control, uninfected mice ($p < .01$, Student's t test for unpaired data). Inset: Effect of LP-BM5 infection on Thy 1.2^+-$CD4^+$ T cells during the course of infection.

Treatment with AZT at 2 mg/kg iv three times weekly did not prevent the LP-BM5 induced reduction in Thyl.2^+-L3T4^+ T cells (Figure 6), whereas liposomal AZT treatment at 2 mg/kg iv three times weekly resulted in normal Thyl.2^+-L3T4^+ T-cell levels (Figure 6).

LIPOSOME TARGETING TO CD4 CELLS

The initial tropism of HIV for cells bearing the $CD4^+$ antigen prompted studies on the suitability of directed drug delivery to such cells via immunological targeting. Liposomes bearing anti-CD4$^+$-antibody (immunoliposomes) interact with both monocytes and T-helper lymphocytes (Figure 7) in purified peripheral blood lymphocyte preparations and in whole blood (77). The use of such immunoliposomes to target antiviral agents to both monocytes and T-helper lymphocytes may have a significant effect on infectivity and transmission of HIV. The inability of such liposomes to interact with lymphocytes or monocytes from patients with AIDS (Table 2) almost certainly results from the downregulation of CD4 expression and reduced CD4$^+$ T-cell numbers in these individuals. Immunoliposome therapy may therefore only be applicable during the early stages of the disease when

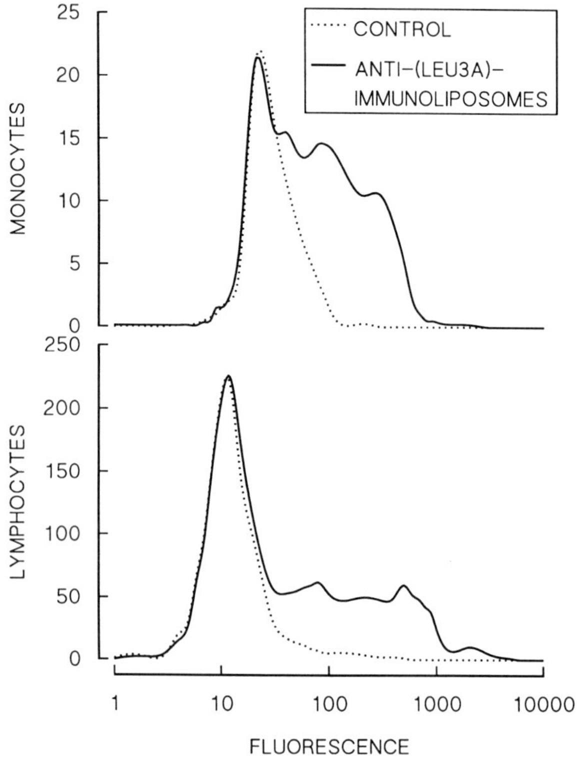

Figure 7 Immunoliposome targeting to CD4$^+$ leukocytes. Peripheral blood leukocytes from normal individuals were incubated with NBD-PE–containing immunoliposomes possessing surface-bound anti-Leu3A (anti-CD4). Binding to monocytes or lymphocytes was determined by FACS analysis of the respective cell populations. Control incubations contained NBD-PE liposomes without surface anti-CD4.

Table 2 Binding of Anti-CD4 Immunoliposomes to Leukocytes[a]

	% Gated cells binding			
	Immunoliposomes		FITC-Anti-Leu3A	
Population	Monocytes	Lymphocytes	Monocytes	Lymphocytes
Normal (n = 10)	91 ± 8	34 ± 11	89 ± 6	33 ± 9
AIDS (n = 6)	33 ± 4[b]	8 ± 2[b]	20 ± 5[b,c]	9 ± 2[b,c]

[a]Peripheral blood leukocyte populations were isolated from normal individuals or from individuals with AIDS by Ficoll Hypaque density centrifugation. After incubation for 60 min with fluorescent NBD-PE liposomes possessing surface-bound anti-Leu3A (anti-CD4) antibody (1 μmol phospholipid/10^6 cells) cell populations were analyzed on a Becton Dickenson FACStar. The percentage of gated monocytes or lymphocytes binding immunoliposomes or FITC-labeled anti-Leu3A was determined using a total cell population of 10,000. Results are the mean ± SD.

[b]Significantly different from normal individuals ($p < .01$, Student's t test for unpaired data).

[c]Significantly different from FITC-anti-Leu3A (Student's t test for unpaired data). The difference represents NBD-PE liposomes taken up by phagocytosis during the incubation period.

$CD4^+$ T cells are at or near normal levels or against cellular antigens that are not affected by HIV infection.

CLINICAL APPLICATION OF LIPOSOMAL AZT

The toxicological profile of intravenous liposomal AZT in mice in conjunction with the enhanced activity against LP-BM5 retroviral infection has lead us to initiate clinical trials with liposomal AZT in HIV-positive individuals. An ongoing phase I clinical trial is currently evaluating the safety of liposomal AZT in patients with AIDS who are unable to continue conventional oral Zidovudine treatment because of bone marrow intolerance. A 31-year-old caucasian male with hemophilia and hemophilic arthropathy was diagnosed as having HIV infection by Western blot analysis in 1983. Hepatosplenomegaly and generalized lymphadenopathy were noted in 1984. The patient remained asymptomatic until spring 1988 when he developed severe fatigue. He developed significant immunodeficiency, manifested by a severe depletion of $CD4^+$ T cells (40/mm^3), and had evidence of HIV disease progression (serum p24 antigen). The patient was given Zidovudine (1200 mg/daily po) in September 1988. Treatment was stopped after 1 month because of moderate headache and nausea. The patient refused to restart treatment until January 1990. The dose of Zidovudine was reduced from 1200 mg daily to 100 mg daily by March 1990 because of anemia. Treatment was discontinued in April 1990 because of progressive anemia fatigue and abnormal liver function. Because of ongoing clinical disease progression and oral Zidovudine intolerance, the patient agreed to participate in a phase I clinical trial of liposomal azidothymidine.

Treatment was started in October 1990 at a dose of 100 mg phospholipid/15 mg AZT by iv transfusion over a period of 3 hr. Treatment was continued weekly for 6 weeks until a dose of 600 mg phospholipid/90 mg AZT was reached. The patient experienced fever (38.2°C) 2–8 hr after each transfusion that was controlled by the administration of 250 mg acetaminophen. The patient had a sense of well-being and maintained a Karnoksi score of 90 throughout the treatment. Although severe leukopenia was present prior to treatment, it did not deteriorate further and there was a trend of improving leukocyte values with time (Figure 8).

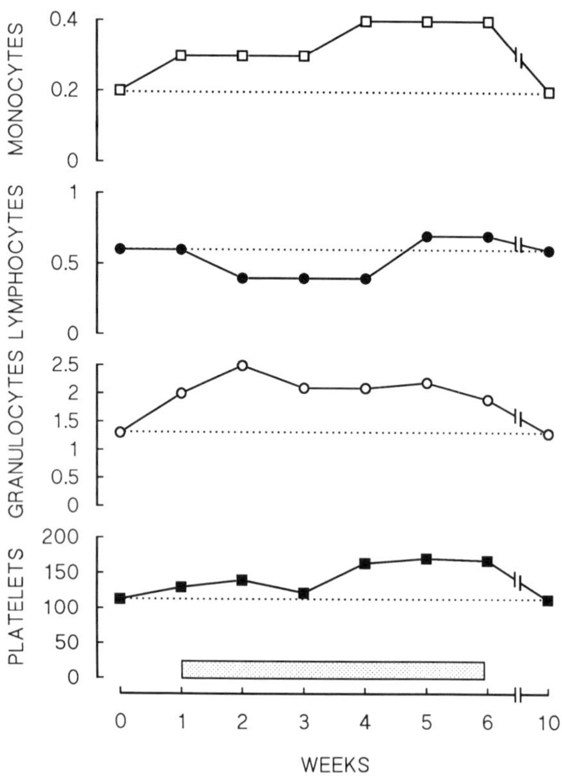

Figure 8 Hematological changes following intravenous liposomal AZT treatment of an HIV-positive individual with AIDS. Platelet, granulocyte, lymphocyte, and monocyte numbers ($\times\ 10^9/L^{-1}$) were determined during liposomal AZT treatment (☐).

There was no change in the leukocyte differential. The patient remained moderately anemic, with no changes in hemoglobin hematocrit or RBC indices with time. Interestingly, the platelet count increased to within the normal range. This treatment effect is frequently found with oral Zidovudine treatment (78). Lymphocyte immunophenotyping revealed a decline in NK cells with treatment, but no substantial changes in total T and B cells. The $CD4^+$ T-cell subset did not change and the patient remained severely immunocompromised with an abnormal CD4/CD8 ratio. Data obtained from two additional patients treated with liposomal AZT has indicated that doses of 60–90 mg of liposomal AZT are well tolerated and do not induce fever.

DISCUSSION

Successful therapy of retroviral infections will require an intervention regimen that deals effectively with the underlying physiopathology. Concomitant liposomal drug therapy and immunoadjuvant activity against viral antigens, with the ability to intrinsically target one of the major cell types involved in retroviral infections, the macrophage, may be an important component of such a therapeutic approach. Such an approach does not attempt to eliminate the virus but rather seeks to control virus-induced immunosuppression in order that effective therapeutic vaccination may be achieved. The ability to target AZT (or other antiretroviral drugs) as well as vaccine antigens at the same time and to the same cell may be more effective than immunization alone for HIV-positive individuals. AZT would appear to be well retained in liposomes formulated from either DPPC/DMPG or DSPC/DMPG and stored at 4°C. Incubation of the liposomes at 37°C showed that DSPC/DMPG liposomes retained AZT better than DPPC/DMPG liposomes ($t_{1/2}$ of 13 versus 4 hr, respectively). These times are in agreement with previously published studies on the liposomal retention of water-soluble molecules (69,71). This formulation, while appropriate for preclinical and phase I clinical studies, will ultimately benefit from the use of lipophilic antiviral agents or derivatives (liposomal prodrug formulation) that are compatible with the phospholipid phase of the liposome rather than the aqueous phase. Our choice of this liposomal formulation is based on its low toxicity toward macrophages (N. C. Phillips and A. Emili, unpublished observations), and by its acceptability in clinical trials (79–81), rather than its ability to retain small water-soluble molecules.

The results of our studies clearly demonstrate that AZT has profound acute toxicity against myelopoiesis and erythropoiesis in mice (82). The short-term dosing regimen used reduced the number of RBC and circulating monocytes and neutrophils, presumably as a result of toxicity toward their precursor cells in the bone marrow. The encapsulation of AZT within liposomal carriers abrogated the hematopoietic toxicity of AZT, resulting in normal RBC and circulating neutrophil and monocyte levels. These observations are consistent with the low targeting efficiency of liposomal AZT to the bone marrow. The encapsulation of AZT within liposomal carriers is therefore an effective way in which to reduce its toxicity against bone marrow cells. The quantity of AZT associated with the bone marrow cells indicates that AZT-induced hematopoietic toxicity may only be associated with a small number of actively dividing precursor cell populations rather than a more nonspecific effect against mature cell populations. In contrast to free AZT, the administration of liposomal AZT was not associated with the incorporation of AZT by bone marrow cells. The inability of liposomes to traverse the blood–brain barrier may mean that this therapeutic approach is not appropriate for treating retroviral infection of the central nervous system. Preliminary pharmacokinetic data would indicate that there is no significant difference between AZT and liposomal AZT in terms of drug localization in the brain (unpublished observations). Whether this is due to AZT leakage or to liposomal localization in brain capillaries remains to be determined.

Therapeutic treatment of MAIDS with liposomal AZT clearly demonstrated its enhanced efficacy over treatment with AZT in terms of retrovirus-induced immunosuppression, changes in $CD4^{+}$pl T cells, and viral replication, as determined by plasma RT activity (83). The initial results of the phase I clinical study to evaluate the safety, tolerance, and toxicity of liposomal AZT indicate that this treatment may be beneficial in patients who are intolerant toward AZT because of bone marrow toxicity. In view of these findings we are currently evaluating the remaining cohort of the study. An alternative approach, namely, that of pretreatment with liposomal MDP, was also shown to be capable of inducing significant protection of myelopoiesis (but not erythropoiesis) against high doses of AZT in mice. The ability to administer such high doses (corresponding to >3.5 g/day in humans) may, however, have some benefit in the effective control of viremia and immunosuppression prior to vaccination. MDPs, either in free form or in liposomes, are capable of inhibiting HIV replication in human monocytes in vitro (84,85). Preliminary studies in mice infected with LP-BM5 retrovirus indicate that disease-related immunosuppression and polyclonal immunoglobulin production can be controlled with a range of MDPs administered systemically (M. Parant and N. C. Phillips, unpublished observations); the possibility that synergistic interactions will occur when AZT and MDPs are both incorporated within liposomes is highly likely and is currently under investigation.

The ability to target cells at risk from HIV infection via the use of immunoliposomes may be limited in its applicability because of the potential downregulation of specific cell antigens such as CD4 during infection and the induction of antibodies directed against the antibody portion of the immunoliposome (86). We are currently evaluating whether manipulation of the liposomal composition in conjunction with the use of restricted portions of the IgG molecule will reduce the immunogenicity. However, it is our belief that the use of immunoliposomes will generate anti-idiotypic antibodies as the rule rather than as the exception. The observation that liposomal adjuvant preparations are active in MAIDS (87), where there is a significant degree of immunosuppression, suggests that immunological reactions will also take place in HIV-positive individuals.

SUMMARY

Liposomal incorporation of AZT would appear to be an effective means of reducing bone marrow toxicity and enhancing antiviral activity. The use of liposomal AZT may be indicated in those patient populations displaying intolerance toward conventional AZT therapy. The potential for combining antiviral therapy with immunomodulator activity via the coencapsulation of muramyl dipeptides as a means of controlling viral replication and immunosuppression may ultimately be necessary for effective therapeutic immunization to take place.

ACKNOWLEDGMENTS

Support for the research programs on liposomal AZT has come in part from the Medical Research Council of Canada and from Health and Welfare Canada. The authors would like to thank Dr. David Yeowell, Burroughs Wellcome Ltd., Research Triangle Park, North Carolina, for the supply of Zidovudine used in the phase I clinical trial, and Dr. Louis Chedid, VacSyn S.A., Paris, for the supply of MDP and MDP-GDP used in the preclinical studies.

REFERENCES

1. R. N. de Martini and J. W. Parker, *J. Clin. Lab. Anal.*, *3*, 56 (1989).
2. R. J. L. Klaassen, J. W. Mulder, A. B. J. Vlekke, J. K. M. Eeftinck Schattenkerk, H. M. Weigel, J. M. A. Lange, and A. E. G. K. Von dem Borne, *Clin. Exp. Immunol.*, *81*, 11 (1990).
3. P. Spornraft, M. Fröschl, J. Ring, M. Meurer, F.-D. Goebel, H. W. L. Ziegler-Heitbrock, G. Riethmüller, and O. Braun-Falco, *Br. J. Dermatol.*, *119*, 1 (1988).
4. V. J. P. Teeuwsen, K. H. J. Siebelink, F. De Wolf, J. Goudsmit, F. G. C. M. Uyt de Haag, and A. D. M. E. Osterhaus, *AIDS*, *4*, 77 (1990).

5. S. Pahwa, R. Pahwa, R. A. Good, R. C. Gallo, and C. Saxinger, *Proc. Natl. Acad. Sci. USA*, *83*, 9124 (1986).
6. D. L. Mann, F. Lasane, M. Popovic, L. O. Arthur, W. G. Robey, W. A. Blattner, and M. J. Newman, *J. Immunol.*, *138*, 2640 (1987).
7. A. Amadori, G. P. Faulkner-Valle, A. De Rossi, and P. Zanovello, *Clin. Immunol. Immunopathol.*, *46*, 37 (1988).
8. D. C. Diamond, B. P. Sleckman, T. Gregory, L. A. Lasky, J. L. Greenstein, and S. J. Burakoff, *J. Immunol.*, *141*, 3715 (1988).
9. G. P. Linette, R. J. Hartzman, J. A. Ledbetter, and C. H. June, *Science*, *241*, 573 (1988).
10. S. Patterson and S. C. Knight, *J. Gen. Virol.*, *68*, 1177 (1987).
11. J. L. Lathey, J. M. Agosti, J. A. Nelson, L. Corey, S. A. Gregory, J. H. Morrissey, T. S. Edgington, and M. B. A. Oldstone, *Clin. Immunol. Immunopathol.*, *54*, 1 (1990).
12. D. L. Mann, S. Gartner, F. LeSane, W. A. Blattner, and M. Popovic, *Clin. Immunol. Immunopathol.*, *54*, 174 (1990).
13. G. T. Spear, H. A. Kessler, L. Rothberg, J. Phair, and A. L. Landay, *Clin. Immunol. Immunopathol.*, *54*, 184 (1990).
14. A. Jewett, J. V. Giorgi, and B. Bonavida, *J. Immunol.*, *145*, 4065 (1990).
15. M. S. Meltzer, D. R. Skillman, P. J. Gomatos, D. C. Kalter, and H. E. Gendelman, *Ann. Rev. Immunol.*, 8. 169 (1990).
16. M. S. Meltzer, M. Nakamura, B. D. Hansen, J. A. Turpin, D. C. Kalter, and H. E. Gendelman, *AIDS Res. Hum. Retroviruses*, *6*, 96, (1990).
17. H. Mitsuya, K. J. Weinhold, P. A. Furman, S. St. Clair, S. Nussinoff-Lehrman, R. C. Gallo, D. Bolognesi, D. W. Barry, and S. Broder, *Proc. Natl. Acad. Sci. USA*, *82*, 7096 (1985).
18. R. Yarchoan, K. J. Weinhold, E. Lyerly, E. Gelmann, R. N. Blum, G. M. Shearer, H. Mitsuya, J. M. Collins, C. E. Myers, R. W. Klecker, P. D. Markham, D. T. Durack, S. Nussinoff-Lehrman, D. W. Barry, M. A. Fischl, R. C. Gallo, D. P. Bolognesi, and S. Broder, *Lancet*, *1*, 575 (1986).
19. M. A. Fischl, D. D. Richman, M. H. Grieco, M. S. Gottlieb, P. A. Volberding, O. L. Laskin, J. M. Leedom, J. E. Groopman, D. Mildvan, R. T. Schooley, et al., *N. Engl. J. Med.*, *317*, 185 (1987).
20. D. D. Richman, M. A. Fischl, M. H. Grieco, M. S. Gottleib, P. A. Volberding, O. R. Laskin, J.M. Leedom, J. E. Groopman, D. Mildvan, M. S. Hirsch, G. G. Jackson, D. T. Durack, and S. Nussinoff-Lehrman, *N. Engl. J. Med.*, *317*, 192 (1987).
21. E. Dournon, W. Rozenbaum, C. Michon, C. Perrone, P. De Truchis, E. Bouvet, M. Levacher, S. Matheron, S. Gharakhanian, P. M. Girard, D. Salmon, C. Leport, M. C. Dazza, and B. Regnier, *Lancet*, *2*, 1297 (1988).
22. P. A. Pizzo, J. Eddy, J. Falloon, F. M. Balis, R.PI. Murphy, H. Moss, P. Wolters, P. Brouwers, P. Jarosinski, M. Rubin, S. Broder, R. Yarchoan, A. Brunetti, M. Maha, S. Nussinoff-Lehrman, and D. G. Poplack, *N. Engl. J. Med.*, *319*, 889 (1988).
23. D. Stambuk, M. Youle, D. Hawkins, C. Farthing, D. Shanson, R. Farmer, A. Lawrence, and B. Gazzard, *Quart. J. Med.*, *70*, 161 (1989).

24. P. S. Gill, M. Rarick, R. K. Brynes, D. Causey, C. Loureiro, and A. M. Levine, *Ann. Int. Med.*, *107*, 502 (1987).
25. C. A. Kennedy, H. S. Griffith, and G. E. Mathison, *Ann. Int. Med.*, *114*, 250 (1991).
26. A. Ganser, O. G. Ottmann, H. Von Briesen, B. Volkers, H. Rubsamen-Waigmann, and D. Hoelzer, *Res. Virol.*, *141*, 185 (1990).
27. H. C. Lane, *AIDS*, *3 Suppl. 1*, S181 (1989).
28. K. Bhalla, M. Birkhofer, S. Grant, and G. Graham, *Exp. Hematol.*, *17*, 17 (1989).
29. A. Delmer, M. Karmochkine, M. Cadiou, H. Gerhartz, and R. Zittoun, *Am. J. Hematol.*, *34*, 73 (1990).
30. G. J. Lieschke, D. Maher, M. O'Connor, M. Green, W. Sheridan, M. Rallings, E. Bonnem, A. W. Burgess, K. McGrath, R. M. Fox, and G. Morstyn, *Cancer Res.*, *50*, 606 (1990).
31. K. Gelmon, J. S. G. Montaner, M. Fanning, J. R. M. Smith, J. Falutz, C. Tsoukas, J. Gill, G. Wells, M. O'Shaughnessy, M. Wainberg, and J. Reudy, *J. AIDS*, *3*, 555 (1989).
32. M. A. Wainberg, J. Falutz, M. Fanning, J. Gill, K. Gelmon, J. S. G. Montaner, M. O'Shaughnessy, C. Tsoukas, and J. Ruedy, *J. Am. Med. Assoc.*, *261*, 856 (1989).
33. P. A. Volberding, S. W. Lagakos, M. A. Koch, C. Pettinelli, M. W. Myers, D. K. Booth, H. H. Balfour, Jr., R. C. Reichman, J. A. Bartlett, M. S. Hirsch, R. L. Murphy, W. D. Hardy, R. Soeiro, M. A. Fischl, J. G. Bartlett, T. C. Merigan, N. E. Hyslop, D. D. Richman, F. T. Valentine, L. Corey, and AIDS Clinical Trial Group of NIAID, *N. Engl. J. Med.*, *322*, 941 (1990).
34. M. A. Fischl, D. D. Richman, N. Hansen, A. C. Collier, J. T. Carey, M. F. Para, W. D. Hardy, R. Dolin, W. G. Powderly, J. D. Allan, B. Wong, T. C. Merigan, V. J. McAuliffe, N. E. Hyslop, F. S. Rhame, H. H. Balfour, Jr., S. A. Spector, P. Volberding, C. Pettinelli, J. Anderson, and AIDS Clinical Trials Group of NIAID, *Ann. Int. Med.*, *112*, 727 (1990).
35. S. A. Bozzette and D. D. Richman, *Am. J. Med.*, *88 Suppl. 5B*, 24S (1990).
36. T-C. Meng, M. A. Fischl, and D. D. Richman, *Am. J. Med.*, *88 Suppl. 5B*, 27S (1990).
37. H. Hollander, A. R. Lifson, M. Maha, M. R. Blum, G. W. Rutherford, and S. Nussinoff-Lehrman, *Am. J. Med.*, *87*, 628 (1989).
38. S. Land, G. Treloar, D. McPhee, C. Birch, R. Doherty, D. Cooper, and I. Gust, *J. Infect. Dis.*, *161*, 326 (1990).
39. B. A. Larder, G. Darby, and D. D. Richman. *Science*, *243*, 1731 (1989).
40. R. Rooke, M. Tremblay, H. Soudeyns, L. DeStephano, X-J. Yao, M. Fanning, S. J. G. Montaner, M. O'Shaughnessy, K. Gelmon, C. Tsoukas, J. Gill, J. Ruedy, and M. A. Wainberg, *AIDS*, *3*, 411 (1989).
41. M. S. Smith, E. L. Brian, and J. S. Pagano, *J. Virol.*, *61*, 3769 (1987).
42. P. Reiss, J. M. A. Lange, C. A. Boucher, S. A. Danner, and J. Goudsmit, *Lancet*, *1*, 421 (1988).
43. R. Rooke, M. Tremblay, and M. A. Wainberg, *Virology*, *176*, 205 (1990).
44. P. A. Furman, J. A. Fyfe, M. H. St. Clair, K. Weinhold, J. L. Rideout, G. A. Freeman, S. Nussinoff-Lehrman, D. P. Bolognesi, S. Broder, H. Mitsuya, and D. W. Barry, *Proc. Natl. Acad. Sci. USA*, *83*, 8333 (1986).

45. H. König, E. Behr, J. Löwer, and R. Kurth, *Antimicrob. Agents Chemother.*, *33*, 2109 (1989).
46. J. M. Dzik, M. Bretner, T. Kulikowski, J. Ciesla, J. M. Ciesla, W. Rode, and D. Shugar, *Biochem. Biophys. Res. Commun.*, *155*, 1418 (1988).
47. B. Munch-Peterson, *Biochem. Biophys. Res. Commun.*, *157*, 1369 (1988).
48. D. E. Mosier, R. A. Yetter, and H. C. Morse III, *J. Exp. Med.*, *161*, 766 (1985).
49. D. C. Aziz, Z. Hanna, and P. Jolicoeur, *Nature*, *338*, 505 (1989).
50. S. K. Chattopadhyay, H. C. III. M :se, M. Nakino, S. K. Ruscetti, and J. W. Hartley, *Proc. Natl. Acad. Sci. USA*, *86*, 3862 (1989).
51. M. Huang, C. Simard, and P. Jolicoeur, *Science*, *246*, 1614 (1989).
52. R. M. L. Buller, R. A. Yetter, T. N. Frederickson, and H. C. Morse III, *J. Virol.*, *61*, 383 (1987).
53. D. E. Mosier, R. A. Yetter, and H. C. Morse III, *J. Exp. Med.*, *165*, 1737 (1987).
54. D. M. Klinman and H. C. Morse III, *J. Immunol.*, *142*, 1144 (1989).
55. T. Mizuochi, J. Mizuguchi, T. Uchida, K. Ohnishi, M. Nakanishi, Y. Asano, T. Kakiuchi, Y. Fukushima, K. Okuyama, H. C. Morse III, and K. Komuro, *J. Immunol.*, *144*, 313 (1990).
56. D. Portnoi, A. M. Stall, D. Schwartz, T. C. Merigan, L. A. Herzenberg, and T. Basham, *J. Immunol.*, *144*, 1705 (1990).
57. H. C. Morse, III, R. A. Yetter, C. S. Via, R. R. Hardy, A. Cerny, K. Hayakawa, A. W. Hugin, M. N. Miller, K. L. Holmes, and G. M. Shearer, *J. Immunol.*, *143*, 844 (1989).
58. D. E. Mosier, R. Gulizia, D. Spector, R. A. Yetter, and H. C. Morse III, *FASEB J.*, *2*, A831 (1988).
59. J. W. Hartley, T. N. Fredrickson, R. A. Yetter, M. Makino, and H. C. Morse III, *J. Virol.*, *63*, 1223 (1989).
60. D. Hamelin-Bourassa, E. Skamene, and F. Gervais, *Immunogenetics*, *30*, 266 (1989).
61. H. Ohnota, Y. Okada, H. Ushijima, T. Kitamura, K. Komuro, and T. Mizuochi, *Antimicrob. Agents Chemother.*, *34*, 605 (1990).
62. T. Basham, C. D. Rios, T. Holdener, and T. C. Merigan, *J. Infect. Dis.*, *161*, 1006 (1990).
63. T. Basham, T. Holdener, and T. Merigan, *J. Infect. Dis.*, *163*, 869 (1991).
64. C. Simard and P. Jolicoeur, *Science*, *251*, 305 (1991).
65. R. E. Dornsife, M. H. St. Clair, A. T. Huang, T. J. Panella, G. W. Koszalka, C. L. Burns, and D. R. Averett, *Antimicrob. Agents Chemother.*, *35*, 322 (1991).
66. T. M. Allen, G. A. Austin, A. Chonn, L. Lin, and K. C. Lee, *Biochim. Biophys. Acta Bio-Membr.*, *1061*, 56 (1991).
67. A. Halbreich, *Biochem. Int.*, *22*, 859 (1990).
68. P. G. Calkhoven, M. Aalbers, V. L. Koshte, R. W. Griffioen, J. C. Van Nierop, D. Van der Heide, and R. C. Aalberse, *Clin. Exp. Allergy*, *21*, 91 (1991).
69. N. C. Phillips, M. L. Moras, L. Chedid, P. Lefrancier, and J. M. Bernard, *Cancer Res.*, *45*, 128 (1985).
70. N. C. Phillips, M. L. Moras, L. Chedid, J. F. Petit, J. P. Tenu, E. Lederer, J. M. Bernard, and P. Lefrancier, *J. Biol. Response Mod.*, *4*, 464 (1985).
71. N. C. Phillips, L. Chedid, J. M. Bernard, M. Level, and P. Lefrancier, *J. Biol. Response Mod.*, *6*, 678 (1987).

72. N. C. Phillips and L. Chedid, in *Liposomes as Drug Carriers* (G. Gregoriadis, ed.), John Wiley and Sons, Chichester, 1988, pp. 243–259.
73. A. Galelli, B. Charlot, N. C. Phillips, and L. Chedid, *Cancer Res.*, *49*, 810 (1989).
74. N. C. Phillips, in *Monokins and Other Non-lymphocytic Cytokines* (M. C. Powanda, J. J. Oppenheim, M. J. Kluger, and C. A. Dinarello, eds.), Alan R. Liss, New York, 1988, pp. 141–144.
75. N. C. Phillips, C. Tsoukas, and L. Chedid, in *Immunotherapeutic Prospects of Infectious Diseases* (N. K. Masihi and W. Lange, eds.), Springer-Verlag, Berlin, 1990, pp. 135–139.
76. D. Baltimore, *Nature*, *226*, 1209 (1970).
77. N. C. Phillips and C. Tsoukas, *Cancer Detect. Prev.*, *14*, 383 (1990).
78. J. Montaner, T. Le, M. Fanning, K. Gelmon, C. Tsoukas, J. Falutz, M. O'Shaughnessy, M. Wainberg, and J. Ruedy, *AIDS*, *4*, 565 (1990).
79. N. C. Phillips, A. Loutfi, M. Kareem, H. Shibata, and M. E. Baines, in *Liposomes in the Therapy of Infectious Diseases and Cancer* (G. Lopez-Berestein and I. J. Fidler, eds.), Alan R. Liss, New York, 1989, pp. 15–24.
80. N. C. Phillips, *Prog. Cli. Biol. Res.*, *354A*, 257 (1990).
81. N. C. Phillips, A. Loutfi, A. M. A-Kareem, H. R. Shibata, and M. G. Baines, *Cancer Detect. Prev.*, *14*, 491 (1990).
82. N. C. Phillips, E. Skamene, and C. Tsoukas, *J. AIDS*, *4*, 959 (1991).
83. N. C. Phillips and C. Tsoukas, *Blood*, *79*, 1137–43 (1992).
84. K. N. Masihi, W. Lange, B. Rohde-Schulz, and L. Chedid, *AIDS Res. Hum. Retroviruses*, *6*, 393 (1990).
85. J. K. Lazdins, K. Woods-Cook, M. Walker, and E. Allteri, *AIDS Res. Hum. Retroviruses*, *6*, 1157 (1990).
86. N. C. Phillips and E. Emili, *Immunol. Lett.*, *30*, 291 (1991).
87. N. C. Phillips, *FEMS Microbiol. Immunol.*, *89*, 209–18 (1992).

12

CLINICAL APPLICATIONS OF MICROSPHERES IN EMBOLIZATION AND CHEMOEMBOLIZATION: A COMPREHENSIVE REVIEW AND PERSPECTIVES

P. M. J. Flandroy, Christian Grandfils, and R. J. Jérôme

University of Liège
Liège, Belgium

INTRODUCTION

Nowadays new drug delivery systems are actively designed with two main objectives in mind (1,2). The first aim is nothing but the old concept of the "missile drug." It consists of targeting the biologically active molecule to its site of action, thus avoiding a random distribution throughout the body. The second purpose is to control the delivery of the drug over an extended period of time. The kinetics of release must be adequate to the pharmacokinetics and biological characteristics of the drug, e.g., a constant rate delivery (zero-order kinetics) and a pulse or feedback release.

Microparticulate drug carriers, such as microcapsules and microspheres, can serve both purposes. Indeed, their systemic or local injection in the vessels ensures the passive targeting of the drug. Furthermore, the drug release can be controlled by diffusion through the polymer matrix and/or by erosion of the polymer (3,4). The role of microspheres and microcapsules primarily depends on their size and injection route (5,6). Microparticles of a diameter smaller than 2 μm can be injected in an intravenous, intra-arterial, and intraperitoneal manner in order to target the reticuloendothelial system. The intravenous injection of microspheres ranging from 3 to 12 μm is intended to block the capillaries of lungs, liver, and spleen (7). Finally, vessels can be hyperselectively embolized with particulate materials of more than 10 μm loaded with a drug or not (8,9). Such a technique will be discussed in the present chapter.

Embolization is a widely acknowledged form of endovascular therapy. It consists of delivering an embolic material locally through a catheter that has been previously inserted in the vessels supplying the pathological area. The endovascular technique causes either a reversible or an irreversible ischemia of the lesion. The outcome of the intervention actually depends on the embolic agent that determines the characteristics of the occlusion (i.e., density, homogeneity, and lifetime). Depending also on the nature of the embolic material, perilesional edema, necrosis, inflammation, and local rupture of the vascular endothelium may occur with the anoxia.

Embolization has proved to be effective in the treatment of numerous diseases:

Hemorrhage blockage (gastrointestinal, pulmonary)
Vascular malformations (aneurysm, arteriovenous fistula or AVF, and arteriovenous malformation or AVM)
Vascularized primary or secondary tumors

Accordingly, it has been recommended as a preoperative, palliative, and curative method.

Although a large variety of materials have been reported in the medical literature during the last decade, only a few of them are commercially available. In a late section a comparison is drawn between the main characteristics of the spherical and nonspherical particulate embolic materials that have been proposed recently in order to fulfill specific requirements of geometry, biodegradability, biocompatibility, and adhesiveness. Later sections review the various experimental testings of the embolic materials on animals as well as their clinical use. The last part of this chapter is devoted to the future improvements of the embolic materials and to their perspectives of application.

Chemoembolization was first introduced by Kato et al. (10–12) in 1978. It involves the selective arterial embolization of a tumor together with a simultaneous or subsequent local delivery of chemotherapeutic agents. The advantage of such a proceeding lies in the fact that it provides a synergistic effect of embolization and local chemotherapy, thereby increasing the therapeutic index of the antimitotic drugs that is comparatively low in systemic injections. The first clinical reports published by Kato and coworkers (10) mentioned the use of nondegradable microcapsules loaded with mitomycin C for the treatment of primary and secondary carcinomas of kidney, liver, bone, and intrapelvic organs. Their results showed an extended average period of reprieve for affected patients, together with a decrease in the usual side effects of systemic chemotherapy.

Up to now, however, this therapeutic technique was applied in oncology to the treatment of unresectable tumors only. In the particular case of the liver, the specific vascular pattern tends to provoke a reversible ischemia, so that applications mainly concern unresectable hepatocellular carcinoma (HCC) (13,14).

This new endovascular treatment has great potentialities since the drug uptake by embolized tissues could be significantly enhanced by:

A higher drug concentration at the site
A longer drug/tissue contact time
An increased vascular permeability to the drug by modification of the environment

The respective roles of the first two characteristics are easily assessed by the pharmacokinetic equation established by Chen and Gross (15) in the particular case of regional chemotherapy. Equation (1) shows that the decrease in the local blood perfusion rate (Q_t) caused by the embolization is responsible for an increase in R_d, i.e., a relative coefficient proportional to the local delivery by arterial infusion:

$$R_d = 1 + K_{app}/Q_t \qquad (1)$$

The last term of Eq. (1) (K_{app}) highlights the importance of the total apparent drug clearance from the body, except for the targeted region. It is the reason why the selection of drug(s) is of fundamental importance.

Despite the expected pharmacokinetic benefit of chemoembolization, the clinical results that have been made available up to now are not definitely convincing because:

The technique appears to be most appropriate for nonresectable, chemoresistant, and highly vascularized tumors.
The hyperselective catheterism that has to be carried out is still a sophisticated technique reserved to a limited number of specialized medical units.
The number of arteries feeding a tumor should be as limited as possible.
An exceedingly high drug concentration is a possible source of local toxicity.
There is not necessarily a direct dependence between drug level and rate of response.
The injection of the "microemboli" in the selected arteries is generally followed by their distal migration (16,17). This is expected to decrease the selectiveness of the therapy or even to cause deleterious side effects.
The local infusion of a drug may not systematically improve the activity of that drug toward a chemoresistant tumor.

These possible limitations might raise the question of the usefulness of chemoembolization compared to the more classical chemotherapeutic techniques of oncology. However, before any final conclusion is drawn, intense research efforts have to be dedicated to

The testing of new drugs selected for regional efficacy in embolized tissues, i.e., a particular hypoxic environment associated with low-energy conditions and acidosis

The optimization of the characteristics of particulate "emboli" in terms of embolization efficiency, rate of biodegradation, and kinetics of drug release

It is worth mentioning that the association of a drug with a carrier is in fact a new pharmacological entity.

ENDOVASCULAR DELIVERY SYSTEMS

Endovascular Devices

According to the targeting method, the selective endovascular approach of a lesion has to be carried out step by step. The damaged organ is first to be reached, then the lesion itself, and finally the cells (18,19). For chemoembolization to be effective, the delivery system should be supplied at the first level (organ), the particulate carriers at the second level (lesion), and drug(s) should be released "in situ" (third level). The reality may of course turn out to be quite different. The vessels that feed a lesion provide the best way for a delivery system to reach and suppress that lesion. Although the method is in many aspects an interesting one, it might be hindered by both the structural features of the vessels and the physical properties of the delivery system, namely, the intravascular catheter.

Indeed, congenital or acquired characteristics of some vessels can prevent an organ to be reached, even though the targeting of a local lesion is expected to be straightforward. Atherosclerotic irregularities, stenosis, and sinuosities of the arteries can generate such difficulties. These potential problems have to be kept in mind since chemoembolization is most often applied to elderly patients. If they occur, they are a severe limitation since the targeted organ must be reached selectively in order to avoid any undesirable interference of the embolic materials with other organs.

Very sophisticated and powerful catheter systems are nowadays available. Indeed, the digital subtraction angiographic units have been significantly improved and minute-sized catheters have been designed that are made of shafts of variable stiffness with a very flexible and supple distal extremity. They provide the best approach to the lesion. The external diameter of the newly designed microcatheters ranges from 1.5 to 2.7 F (1 F = 1/3 mm = 1.313 in.). A gold or platinum ring is inserted in the tip that increases the opacity under fluoroscopy. The available microcatheters are equipped with a steerable microguide wire for torque advancement and thus offer the opportunity to reach up to the sixth division branches of arteries. Intravascular navigation using such devices makes regions accessible that were thought to be out of reach a few years ago (e.g., brain cortical artery).

However, the situation is not yet an ideal one, regarding, for instance, the size and flexibility of the microcatheters. Indeed, their distal aperture is narrow and only liquids or particles of a diameter up to 350 μm can easily be injected. Spherical particles of a 200- to 300-μm diameter size readily pass through the device when correctly dispersed and sufficiently diluted. The flow of nonspherical particles of a similar size distribution is more difficult to control due to a higher friction and the possible clogging of the microcatheter. If clogging is to occur, the catheter has to be withdrawn. Indeed it cannot be unblocked by passing the guide wire through it without compacting the particles further. The 200- to 300-μm microspheres are undeniably very convenient for chemoembolization. The presence of an abnormal arteriovenous shunting might of course generate problems, since these small spherical particles would then have the opportunity to reach the lungs. Fluid "emboli" that harden instantaneously remain most suited to these hemodynamic conditions. Microspheres would be more secure if they were prone to a quick endothelial adhesion or a fast swelling when delivered.

Since a high flexibility is required, hyperselective catheters must be of a coaxial type. They consist of:

A guiding catheter (8 or 9 F size) inserted into a main artery (e.g., the primitive carotid artery)
Microcatheter with or without the steerable microguide wire
Hemostatic valve that provides a tight seal between the guiding catheter and the microcatheter, so that a continuous heparin flush can be used to prevent blood clotting

Such a system is set up backward in a tract placed in a femoral artery. The guiding catheter allows the angiography of the vascular lesion and the observation of anastomoses, arteriovenous shunts, etc., before starting the embolization procedure. A thorough preoperative angiography is mandatory in order to identify the arteries most appropriate for catheterization and thus to decide on the strategy to enforce. It is also possible to check the quality of the occlusion. This time-consuming manipulation of coaxial delivery systems requires a trained experience in catheterizations, all the more so as lesions usually feed on several arteries. The microcatheter and the guide wire must be handled carefully since they are very expensive. Moreover, they can be used only once.

Since the place reached by the endovascular delivery system determines the selectivity of the injection, the question is whether the catheter adjustment can improve the efficiency of the endovascular procedure. Of course, a lesion that is fed by tortuous arteries is expected to be more accessible if the endovascular material is actively moved rather than passively pushed. Thus, the system should take advantage of the hemodynamic conditions of the lesion, e.g., of an increased blood flow at the site as can be promoted by hypervascularity and arteriovenous shunting. This flow could help the microcatheter to be guided throughout the lesion. The flow sensitivity of the microcatheter could, for instance, be increased by a calibrated leak balloon attached to the distal extremity.

If the blood perfusion rate is not abnormally increased in the lesion, an alternative guiding technique might be provided by an external electromagnetic field acting on the microcatheter. In practice such a methodology raises technical problems and has been experimented with only partial success under experimental conditions (20).

To conclude, the first level of targeting (organ) can be achieved successfully in most cases. This is mainly due to the rapid progress in the tailoring of new endovascular devices. If the situation is favorable, the lesions also can be targeted efficiently. Except for liver and lung, which possess a dual arterial and venous vascularization, the endovascular procedures can be safely applied if the second level of targeting is carried out by the embolic material itself.

Characteristics of the Embolic Materials

Embolization

CLASSIFICATION OF THE USUAL EMBOLIC AGENTS. The design of new delivery systems has largely contributed to the efficiency of endovascular techniques in the preoperative and curative treatment of vascularized tumors or vascular malformations. Table 1 shows the most currently used embolic agents and their clinical applications.

Two main types of embolic materials should be brought forward: (a) particulate agents for the embolization of tumors and (b) liquid agents for the embolization of arteriovenous malformations. Although extremely simplified, that classification agrees with the daily working conditions of most physicians around the world (21). Mechanical systems are known to be effective in the treatment of arteriovenous fistulas (e.g., traumatic carotid-cavernous fistulas) and arterial aneurysms; however, these are beyond the scope of this chapter.

The most currently used liquid materials are commonly named "glu." One of the most representative agents of that category is the *N*-butyl-2-cyanoacrylate or Histoacryl (A. B. Braun, A. G. Melsungen). Its use requires

Table 1 Classification of the Usual Embolic Materials: Nature and Clinical Applications

Materials	Indications
(1) Particulate	
a. Degradable:	
Gelatin sponge (Gelfoam)	Hemorrhage blockage
b. Nondegradable:	
Duramatter Polyvinylalcohol (Ivalon)	Tumoral conditions
(2) Liquid	
N-butyl-2-cyanoacrylate (Histoacryl)	Vascular malformation conditions, e.g., arteriovenous malformations
(3) Mechanical systems	
Coils	Vascular malformation conditions, e.g., arterial aneurysms
Detachable balloons	Arteriovenous fistulas

some skillfulness since it hardens instantaneously when in contact with an ionic solution (e.g., contrast material and blood), thus risking that the catheter might get stuck in the vessel. Some agents, such as iodine-containing lipophilic contrast agents, have the ability to slow down the polymerization of Histoacryl, but the addition of these agents does not lead to highly controlled and reproducible conditions for embolization. The use of such a potentially excellent obliterating material awaits its further improvement.

Whenever the particulate materials are concerned, the commercially available ones are of an irregular shape and their use is usually disappointing: clogging of the well-placed microcatheter, proximal arterial occlusion, short lifetime of the vessel obstruction, etc. It is clear that ideal embolic agents are not yet available. This deficiency accounts for the large range of materials that were tested over the last decade. Recently, research teams of the universities of Liège (Belgium), Lille and Paris (France) have established a multidisciplinary Group for the Evaluation of Microspheres in Embolization (GEME). Each team produces and uses microspheres of various natures. The experimental results are regularly discussed in order to evaluate the potentialities and limitations of microspheres in embolization and chemoembolization. Since the authors of this chapter participate in GEME, they will refer to some observations, comments, and conclusions that have been shared by the GEME members.

According to Beaujeux (22), the characteristic features of an ideal embolic agent would be the following:

Biocompatibility
A spherical shape and a smooth surface
Precise calibration
Adhesiveness
Resistance to degradation
Mechanical resistance and suppleness
Easiness of production, sterilization, and conditioning
Easiness of manipulation (dispersion, injection, and visualization)

Two of these characteristics are of critical importance: the shape of the particles and the distribution of the particle size. The nature and degradability of the embolic material are of secondary importance since what is needed is the sealing of the arteries either in the core of the tumoral lesion or in the arteriovenous nidus of a vascular malformation. For instance, the gelfoam powder (Upjohn, Kalamazoo, MI), size 40 μm, produces deep ischemia and necrosis although it is rapidly degraded (within a few days). Conversely, polyvinylalcohol particles (Ivalon, Unipoint, High Point, NC), size 150–250 μm, are not degradable and their immobilization is more proximal than expected, which allows collateral vessels to feed the targeting tissue. Thus the wrong choice of an embolic material can lead to a proximal occlusion that is as uneffective as a surgical ligature.

The right particle size is a necessary condition but not a sufficient one for achieving a distal and homogeneous occlusion. In addition to the size, the shape of the particles has a decisive effect. This is proved by the poor efficiency of the commercially available embolic material of an uneven shape compared to spherical microspheres. A perfectly spherical shape is undoubtedly mandatory.

NONSPHERICAL MATERIALS. They can have either a natural or a synthetic origin. Some of them are sensitive to enzymes and resorbed in the organism. Blood clots, fragments of gelatin sponge, and partially crosslinked collagen fibrils belong to that category. They provide a temporary occlusion and they are suitable either for preoperative treatments or for repetitive chemoembolization if their degradation is perfectly controlled. Except for crosslinked collagen fibrils, the effectiveness of which has only been evaluated recently (23,24), all the other polymorphic materials have the same shortcomings:

A highly irregular shape with numerous bumps.
The heterogeneous calibration of the particles: the size distribution is generally broad. For instance, the proportion of the smallest particles in a polyvinylalcohol material has been reported to be 15 times higher than that of the biggest particles (25).
The ability to aggregate, which is partly due to the shape and size of the particles and to their higher density compared to blood.

SPHERICAL MATERIALS. A spherical shape together with a narrow size distribution is a prerequisite for an embolic material to occlude perfectly blood vessels. However, the capability of the spherical materials described in the literature to perform a homogeneous, precise, and complete occlusion is at variance. Spherical embolic materials can be classified on the basis of their interaction with the biological environment. They belong to two major groups:

Inert materials, such as silicone, glass, dextran, and polylactide (PLA) microspheres

Hydrogels, such as poly(2-hydroxyethyl methacrylate) (PHEMA), trisacryl gelatin, dextran gelatin, and polyvinylalcohol microspheres

Inert spherical microspheres can be responsible for a very distal occlusion of arteries due to their hydrophobic nature and protracted migration. Dion et al. (26) and Laurent et al. (27) reported that the postmigration of the "emboli" can occur for several days until an inflammatory granuloma appears. The mobility of the particles is currently explained by the occurrence of such factors as an increasing blood pressure, the reactional vasoplegia subsequent to embolization, and the elasticity of the vascular walls. The efficiency of the inert materials in mechanically occluded vessels is limited by either a possible restoring of the blood flow or the migration of the spheres through arteriovenous shunts. As polylactide, polystyrene, or silicone polymers are highly hydrophobic, they do not swell and are unreactive toward their biological environment; namely, the endothelial surface and the main blood components.

When the embolic material swells in the blood vessels, it forms a hydrogel. Hydrogel microspheres have been reported by Horak et al. (28–30) and Thanoo and Jayakrishnan (32,33). They are typically soft and permeable to water. Interestingly enough, trisacryl microspheres can be coated by a chemotactic agent, such as gelatin, which increases their ability to interact with the endothelium (27). A major advantage of that class of embolic agents is the formation of a "thrombus" immediately after their injection in the bloodstream. Thus, in addition to a mechanical occlusion, a thrombogenic effect contributes to a long-lasting endovascular occlusion. Finally, the porosity and surface properties of hydrogels may be an incentive for blood cells either to adhere to the microspheres or to engulf in them if the porosity is appropriate. In this regard, the spontaneous adsorption of thrombocytes is strongly favorable to a good embolization, since it promotes an increase in both the particle size and the mechanical strength of the emboli. In fact, fibrinogen is first adsorbed on the hydrogel surface, which is known to stimulate the adsorption and aggregation of thrombocytes. An interconnected network of macropores is also necessary for cell recolonization to occur and for the embolic material to be permanently attached to the vascular lumen. The porous character of the embolic material has a critical effect on cell formation and growth, and particularly on the emergence of a foreign body reaction and streaks of fibrous tissues.

In conclusion, spherical embolic agents, either inert or swollen, are suitable materials for effective occlusions, provided that they are well calibrated. However, their potential drawbacks are poles apart. There is a risk of distal migration whenever inert microspheres are used, particularly if used in excess, which might cause a reflux. A proximal closure can occur if a hydrogel swells too quickly in large arteries, which allows the lesion to be vascularized again by collateral vessels.

Chemoembolization

ESSENTIAL CHARACTERISTICS OF PARTICLES FOR CHEMOEMBOLIZATION. The chemoembolization of tumors relies on two basic principles (34): (a) the embolization of the vessels that feed the tumor and (b) the kinetics of drug release from the emboli introduced by an endovascular route. The first principle has been discussed previously, and spherical particles have thus to be recommended for their achievement of distal, homogeneous, and effective occlusions. Microspheres now have a twofold action since they are not only embolic agents but also drug carriers. The originality of chemoembolization has to be found in an improvement of the local drug uptake (35). That very desirable effect depends on the local drug delivery, a high drug concentration, and a sustained drug action (36–40). Indeed, when the microspheres are supplied in the tumor, all the drug they contain has a potential to reach the targeted tissues. The local drug concentration is accordingly higher and the systemic dose lower. The tumor is in a lengthy contact with the drug since the blood flow can be severely slowed down by the drug carrier itself. At last, the occlusion of the tumoral arteries leads to ischemia and to a decrease of partial oxygen pressure and pH that might increase the permeation of the drug.

Only a few embolic materials have been tested in chemoembolization, namely, ethylcellulose microcapsules, cetylalcohol, PLA polymers, poly(lactide-glycolide) copolymers (PLGA), carnauba wax, and protein (serum albumin, gelatin) microspheres. Table 2 sums up the main characteristics of these materials.

The kinetics of drug release from the emboli and the biodegradation rate of the drug carrier are of major importance, since they determine the amount of drug released per unit of time and, accordingly, the bioavailability of the drug and its antitumor effect. The biodegradation of the embolic agent allows the treatment to be repeated according to a precise schedule. For the application of a sequential treatment to be applied, microspheres must indeed

Table 2 Characteristics of the Microspheres Adopted in Chemoembolization

Polymer	Mean size	Shape	Degradation rate	Drug	Kinetic drug release	Ref.
Ethylcellulose	200–300 μm	Irregular	Nondegradable	Cisplatin, mitomycin C	"Burst effect"	10–12,34,41–47
PLGA[a]	100–200 μm	Spherical	Degradable: few days to months	Cisplatin, adriamycin, aclarubicin	Sustained release	9,49–52
Proteins						
albumin	7–80 μm	Spherical	Degradable, few days	Adriamycin, mitomycin C	"Burst effect"	14,18,38–40,55,56
gelatin	45 ± 8 μm					
Cetylalcohol	300 μm	Spherical	Degradable: ±20 days	CCNU	"Burst effect"	48
Wax	125–800 μm	Spherical	Nondegradable	5-FU	First-order release	54
Chitosan-chitin	150–200 μm	Spherical	Degradable	Cisplatin	Sustained release	55

[a]PLGA, poly (lactide-glycolide) copolymer; CCNU,1-(2-chloroethyl)-3-cyclohexylnitrosourea; 5-FU, 5-fluorouracil.

disappear before the foreign body reaction occurs. According to Madoule et al. (17), this reaction is observed between the seventh and the tenth day after the embolization. The optimum degradation time of the carrier should thus be about 10 days.

The control of drug release in vivo is still a challenge (50). It depends on the nature and the internal structure of the carrier. The former characteristic determines the possible erosion of the particles and the latter feature has a decisive effect on the drug diffusion. As a rule, drug release systems show an initial fast efflux of drug (diffusion) followed by a protracted phase of release (diffusion and/or biodegradation) (39). The initial efflux is known as the "burst effect." It must be avoided since it works as an arterial chemoinfusion with two major drawbacks:

Drug side effects consecutive to a high plasmatic concentration
Reduced drug efficacy due to a shortened drug–tissue contact time

Ideally, the burst effect should be as small as possible and the rate of drug release as constant as possible.

It is the reason why the biodegradability of the carrier and a constant protracted efflux make up the best situation. In this respect, the use of PLA and PLGA copolymers and chitosan has to be recommended. A nondegradable carrier combined with a burst effect is the worst situation, which in fact tends to dominate when ethylcellulose is used. Ethylcellulose microcapsules are furthermore difficult to inject since they often occlude the catheters. This drawback results from an irregular shape and from the aggregation of the particles in the saline injection medium. The drug loading of the ethylcellulose particles is usually very high (up to 80%). This advantage is, however, counterbalanced by the huge burst effect. Biodegradable proteins and cetylalcohol, which favor a burst effect, are acceptable candidates as drug carriers. Carnauba wax is not degradable and is poorly effective, although a relatively constant protracted efflux might be observed.

Since PLA and PLGA copolymers seem to be one of the best materials, the authors of this chapter have focused on microspheres of this type for chemoembolization purposes.

PLA AND PLGA MICROSPHERES FOR CHEMOEMBOLIZATION

Control of the Biodegradation Rate In contrast to natural polymers, synthetic macromolecules are highly versatile materials, the main properties of which—such as mechanical properties and chemical stability—can be easily changed. A recent progress in macromolecular chemistry allows each step of the polymerization process to be carefully controlled to such an extent that each molecular characteristic of the polymer is predictable: molecular weight, tacticity, nature of the end groups, and molecular weight distribution. These properly tailored macromolecules, like pieces of a construction set, can then be used to design new polymeric materials. This macromolecular engineering was first applied to conventional materials; it has now been extended to synthetic biomaterials (58–62).

Increasing attention is being paid to polymers of lactic and glycolic acids for the development of new galenic formulations. These materials are well known for their biocompatibility and biodegradability (63). They have found early use as biodegradable sutures. Their degradation is related to the hydrolytic instability of the ester bonds. The degradation rate is deeply affected by the chain configuration that can be of the LL, DD, or DL form depending on the structure (stereoisomerism) of the lactide precursor (Figure 1). The poly–LL and DD lactides are semicrystalline, in contrast to the poly-DL-lactide, which is totally amorphous. The latter polymer is more sensitive to degradation in a biological environment due to a faster penetration of the surrounding liquid. The biodegradation rate of polylactides is usually enhanced by copolymerizing lactides with glycolide. According to Miller et al. (64), the lifetime lies between 1 week and at least 5 months when the lactide/glycolide ratio of the copolymer is modified.

Due to a low cristallinity and a high hydrophilicity, the random poly(DL-lactide-*co*-glycolide) of a 50/50 (w/w) composition exhibits the shortest degradation time: about a week. The homogeneous hydrolytic rupture of the ester bonds of polylactides and polyglycolide theoretically follows a first-order rate kinetics. In practice, Visscher et al. (63) and Vert et al. (65) showed, in accordance with our observations, that a heterogeneous erosion mechanism prevails when implants or microspheres of polylactide are concerned. The comparison of the external and internal morphology of matrices eroded in vitro and in vivo reveals an unexpected course in the biodegradation process.

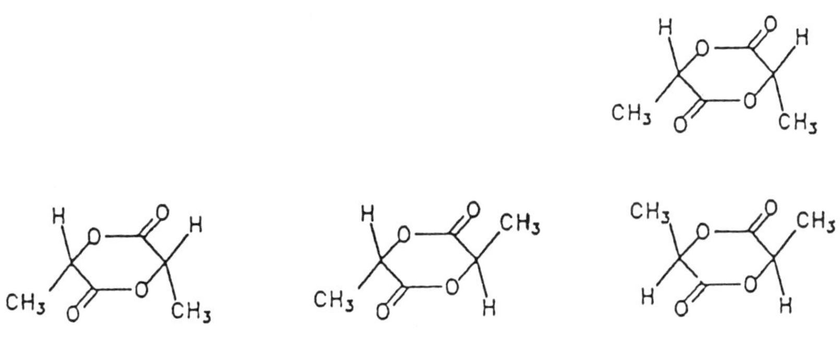

L-lactide MESO-lactide (D,L)-lactide

Figure 1 Chemical structure of the diastereoisomer forms of the lactide monomers.

Scanning electron microscopy (SEM) clearly shows that the initial breakdown of the microparticles occurs at the center whereas no alteration of the outer layers is visible. Although a difference in crystallinity between the outer and inner shells cannot be ruled out, it might be suggested that the protons released by the polymer degradation have to diffuse throughout the matrix, which accordingly generates a gradient of pH. In contrast to all expectations, poly(lactide-*co*-glycolide) (50/50, w/w) microspheres appear to be unmodified after a 1-month incubation in a saline buffer at 37°C. The main conclusion of these experimental observations is that even when the most degradable polyester is used, the microspheres remain mechanically too stable for a repetitive chemoembolization to be performed.

In order to make microspheres biodegradable within a few days, we have combined two poly DL-lactide samples of a different molecular weight. The basic concept relies on the capability of a high molecular weight component (M_n 50,000) to provide the microspheres with a high mechanical strength, whereas a low molecular weight partner (M_n 2500) should decrease the lifetime of the particles very significantly. These two-component microspheres have been prepared by the well-known emulsion-solvent evaporation process. A solution of the two polymers in CH_2Cl_2 is dispersed in an aqueous phase and the microdroplets are then solidified by the slow evaporation of the organic solvent. As discussed in a separate paper, that technique is very convenient to control the size distribution of the microspheres with precision (66). Moreover, it allows for the immobilization of a large variety of drugs and accordingly offers the opportunity to combine the advantages of chemotherapy and embolization. Hydrophobic drugs can indeed be entrapped by using an oil-in-water (o/w) emulsion while hydrophilic drugs require an oil-in-oil (o/o) emulsion or a water-in-oil-in-water (w/o/w) emulsion. Watts and collaborators (67) recently reviewed such a method and emphasized the main parameters that control the encapsulation process and the potential clinical applications of the drug-loaded microspheres.

As reported in Table 3, we prepared the microspheres of the two previously mentioned poly-DL-lactides in various ratios. After a careful purification, the microspheres were incubated at 37°C in a saline medium buffered at pH 7.4, under a slow tangential agitation in the presence of a bacteriostatic agent. The time dependence of their morphology was investigated by SEM. Figures 2–5 refer to microspheres of various compositions and support that the mechanical stability of the particles can be modified on purpose in a narrow span of time. Indeed, under the present experimental conditions, the 50/50 labeled microspheres are broken up after 10 days, whereas only the surface of the 75/25 ones is slightly eroded. The breakdown of the microspheres is observed when a critical fraction of the oligomer has been hydrolyzed. It is worth noting that the two-component poly-DL-lactide micro-

Table 3 Composition of Different Poly-(DL)-lactide Blends Adopted in Order to Modulate the Biodegradation Rate of the Microspheres[a]

	Composition of the organic dispersed phase	
Blend designation	High molecular weight PLA (%)	Low molecular weight PLA (%)
75/25	5.62	5.0
6634	4.95	13.6
58/42	4.35	16.8
50/50	3.70	20.0

[a]The different % relate to the weight composition (% w/w) of the two polymers in the organic phase.

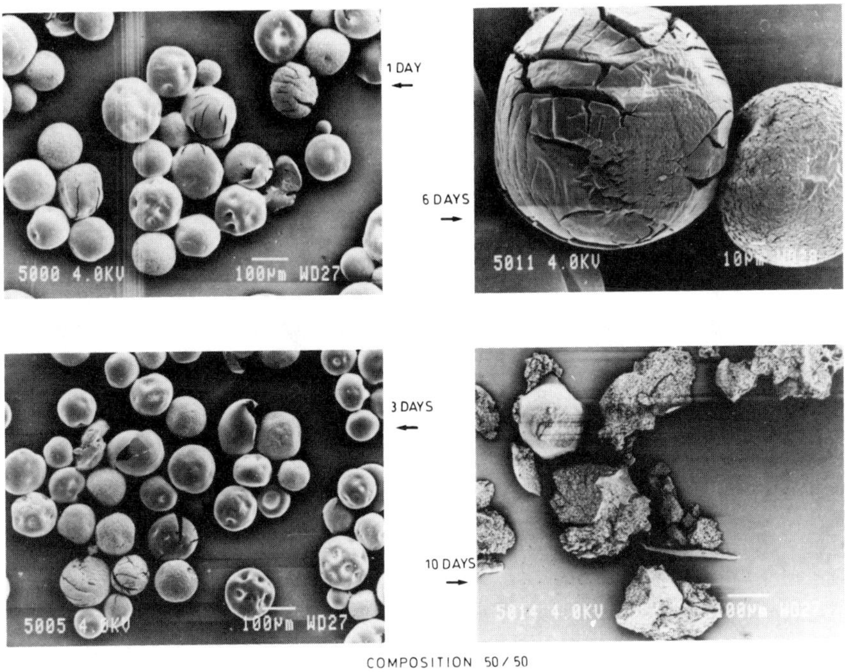

Figure 2 Microscopic characterization of the biodegradation kinetics of the polylactide microspheres (composition: 50/50) observed in vitro at 37°C in a saline medium buffered at pH 7.4.

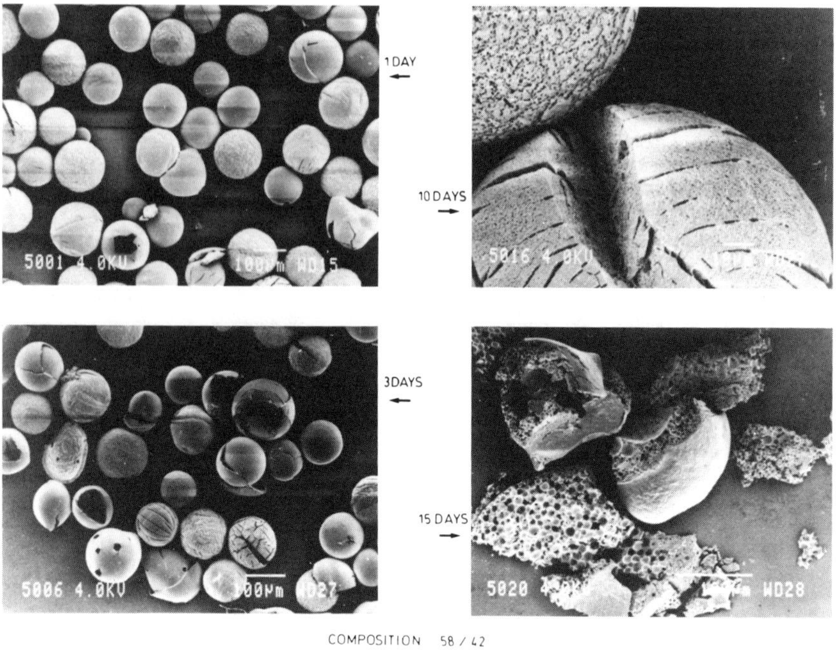

Figure 3 Microscopic characterization of the biodegradation kinetics of the polylactide microspheres (composition: 58/42) observed in vitro at 37°C in a saline medium buffered at pH 7.4.

spheres breaks down although the degradation of the high molecular weight chains is not complete.

Cisplatin Encapsulation and Kinetics of Release Cisplatin (or cisdiaminodichloroplatinum, CDDP) is an antineoplastic agent highly active toward a variety of tumors. Its cytotoxicity seems to be related to its capability of inhibiting selectively the DNA synthesis. Several embolic materials, including polylactic acid (51–53), have been loaded with that powerful antimitotic agent. Poly-DL-lactide microspheres made up of various proportions of a low and high molecular weight component, respectively, have been prepared as previously described and loaded with 5% (w/w) CDDP.

Figure 6 shows that the drug release kinetics is a two-step process whatever the microsphere composition might be. A rapid, although progressive, release is observed for the 10–24 first hours, in a direct relation to the composition of the polymeric matrix. Cisplatin is then very slowly released according to a quasi-linear time dependence. In contrast to the biodegradation rate of the micros-

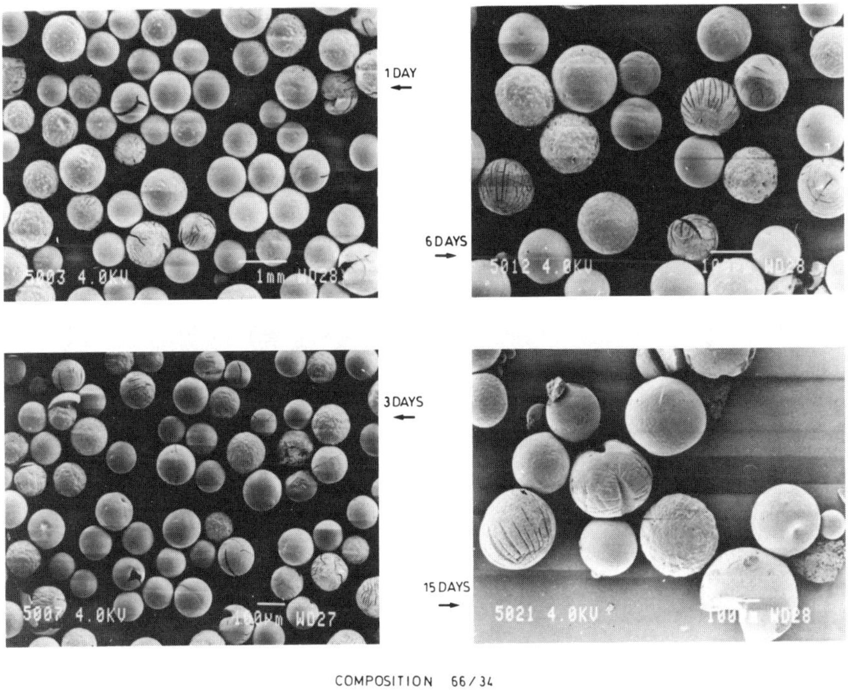

Figure 4 Microscopic characterization of the biodegradation kinetics of the poly-lactide microspheres (composition: 66/34) observed in vitro at 37°C in a saline medium buffered at pH 7.4.

pheres, the kinetics of the CDDP release is hardly affected by the presence of low molecular weight poly-DL-lactide chains, except for the relative importance of the two steps of release. That observation might be explained by the distribution of the CDDP crystalline particles within the polymeric matrix. Indeed, microspheres originate from an organic dispersed phase, the viscosity of which decreases as the content of the low molecular weight poly-DL-lactide increases. Accordingly, the CDDP particles that are present in the organic phase migrate more easily toward the outer shells of the particles when the solvent is removed under stirring. A larger amount of drug can thus contribute to the first step of release. Furthermore, the low molecular weight component decreases the glass transition temperature (T_g) of the poly-DL-lactide matrix (68). If the plasticization effect is strong enough, then the water penetration can be enhanced and contribute to a faster drug release.

In conclusion, the blending of poly-DL-lactide of at least two different molecular weights can allow the microspheres to be biodegraded within a

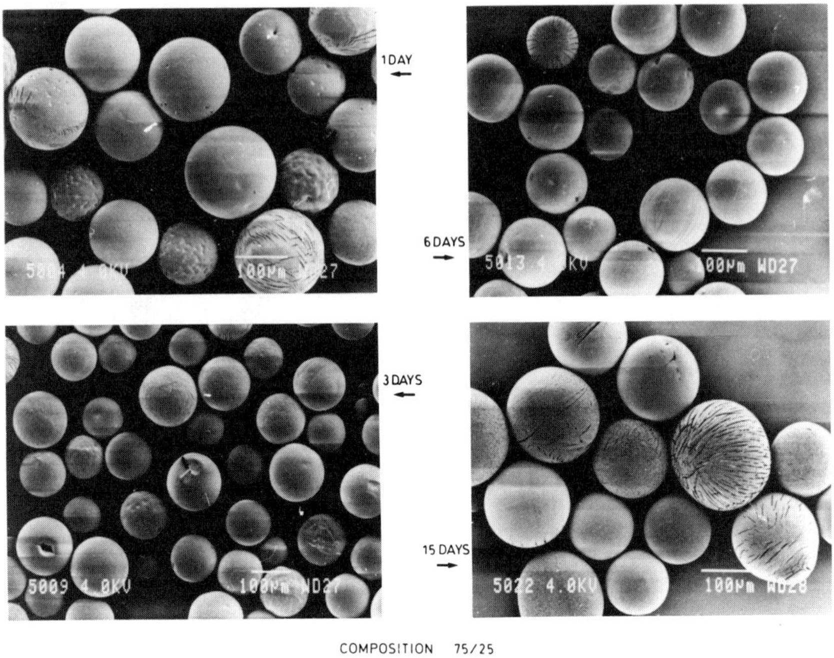

Figure 5 Microscopic characterization of the biodegradation kinetics of the polylactide microspheres (composition: 75/25) observed in vitro at 37°C in a saline medium buffered at pH 7.4.

time scale appropriate to a repetitive chemoembolization. Interestingly enough, the drug is released in a sustained, although initially rapid, manner.

Selection of Drugs

Antimitotic Drugs

When they are used in a systemic way, the anticancer drugs have a low therapeutic index. This drawback must be attributed to the plasmatic distribution of the drug and to the related systemic toxicity and lack of specificity. When a lesion is successfully targeted by a particulate carrier, then the drug must be released and reach the tumoral cells. Since the tumoral response is proportional to the local drug concentration, such a requirement is of critical importance. It depends on the contact time of the drug with the tumoral cells as well as on the drug extraction by the organ and on the arterial flow in the tumor compared to that in the host organ (69). If the regional delivery is high enough, then the local toxicity might be a dose-limiting factor (36).

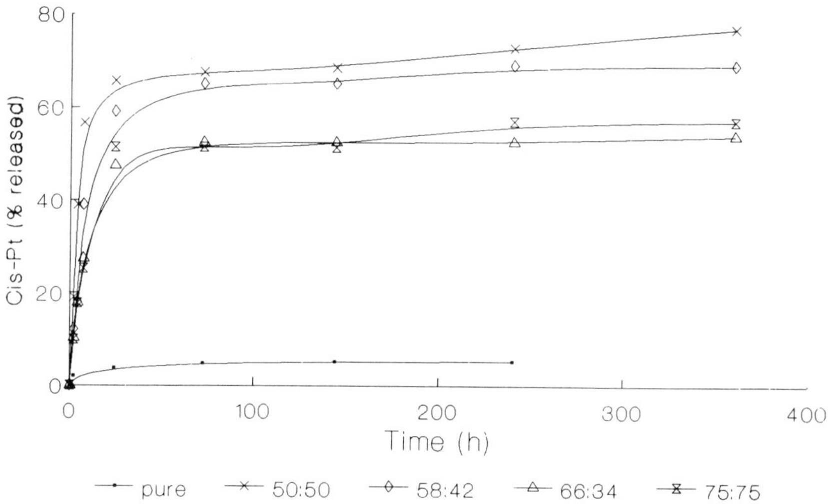

Figure 6 Kinetic profiles of cisplatin released in vitro from the different poly-DL-lactide microspheres. Composition 50:50, 58:42, 66:34, and 75:25 are the same as quoted in Table 3. "Pure" refers to microspheres composed only of the high molecular weight component. The released kinetics experiment was performed in a saline buffer (pH 7.4) at 37°C under slow agitation.

Of course, the local drug concentration, and thus the drug released per unit of time, must be sufficiently high to overcome the resistance of the tumoral cells (70).

The drug must be selected in relation to the target organ and the particulate carriers. A favorable drug–carrier affinity might prevent a burst effect from occurring. The drug must also be compatible with the environment. Indeed, the vascular occlusion can induce changes in the physiological microenvironment and perturb the drug uptake and metabolism (71). Accordingly, drugs of an increased efficiency in the presence of hypoxic cells should be recommended. As an example, mytomycin C and adriamycin are preferentially toxic to cells under hypoxic conditions (71,72). In contrast, 1,3-bis(2-chloroethyl)-1-nitrosourea (BCNU), 1-(2 chloroethyl)-3-cyclohexyl-nitrosourea (CCNU), cisplatin, and 5-fluorouracil (5-FU) do not show any preferential toxicity under these conditions. Even worse, bleomycin, procarbazine, streptonygrin, actinomycin D, and vincristine are preferentially toxic to cells under oxygenated conditions. This is the reason why mitomycin C has been encapsulated within ethylcellulose microcapsules in protein and starch microspheres. Similarly, PLA microspheres and protein microparticles have been loaded with adriamycin. Cisplatin, which shows an intermediate toxicity under hypoxic conditions, has been incorporated within ethylcellulose microcapsules and PLA microspheres (Table 2). The ability of the drug to penetrate tissues is of more critical importance when chemoembolization rather than systemic administration is concerned. As pointed out by Teicher et al. (71), it is essential that therapeutic amounts of the drug penetrate the tissues of the hypoxic cell of the tumor. Since the hypoxic and hypoenergetic conditions of the embolized tissues might adversely affect the drug diffusion, the availability of highly diffusing drugs seems to be mandatory for effective chemoembolization. That requirement, however, is still to be assessed. The variability of the turnover rate of the tumoral cells should also be taken into account. There is a need for drugs active toward noncycling or slowly progressing cells and toward very dynamic ones. Therefore, it is not surprising that combinations of several drugs have been proposed and used in systemic injections. However, the technique is not know to have been extended to clinical chemoembolization. Although it may seem unrealistic to combine several drugs into a unique embolic carrier, a cocktail of microparticles loaded with different drugs is feasible and would deserve interest.

In conclusion, a drug active in a systemic administration is not necessarily appropriate to chemoembolization. Drugs must be active under hypoxic conditions. This requires that the drug efficiency be tested under these specific tissular conditions and that not only classic but also new antimitotics should be considered.

Other Pharmaceutical Agents

α-THROMBIN

Introduction The efficiency of chemoembolization based on antimitotic agents could be increased by the release of additional pharmacological agents, such as a coagulation factor, α-Thrombin is a proteolytic enzyme that plays a decisive role in blood coagulation, since it has a catalytic effect on the conversion of fibrinogen into fibrin. Sugitachi et al. (73) explored the potentialities of the local administration of α-thrombin in cancer chemotherapy. They produced gelatin microparticles as a chemoembolization material capable of releasing anticancer drugs combined with α-thrombin and factor XIII. That material has been used as a thrombogenic agent in the preoperative arterial embolization of localized breast cancers. The authors report very encouraging results concerning the efficiency of the vessel occlusion and the slow release of adriamycin or mitomycin C.

The combination of an anticancer drug with a blood coagulation factor has been patented as an active material in chemoembolization and needle therapy (74).

More recently, D'Amico et al. (75) investigated the antimetastatic effect of α-thrombin directly injected into a rat's tumor. Sealing of some tumor vessels appears to be effective in preventing dissemination from a primary tumor. However, the benefit of that therapeutic strategy is limited by the short lifetime of α-thrombin in the bloodstream. It should thus be of interest to slow down the release of that coagulating agent.

α-Thrombin Microencapsulation by Coacervation We have immobilized bovine α-thrombin in polylactide (PLA) and poly(lactide-glycolide) (PLGA) microparticles by a coacervation technique. This phase separation method is currently used to encapsulate hydrophilic drugs and was recently tested for the encapsulation of peptides. Figure 7 schematizes the four main steps of coacervation (76):

Dispersion of the drug, in a solid or liquid form, in a coating polymer solution.

Partial desolvation of the coating polymer to the point where a phase separation occurs together with a formation of droplets of a polymer-rich phase (or coacervate). This key step can be induced by the addition of a solute that decreases the solubility of the polymer.

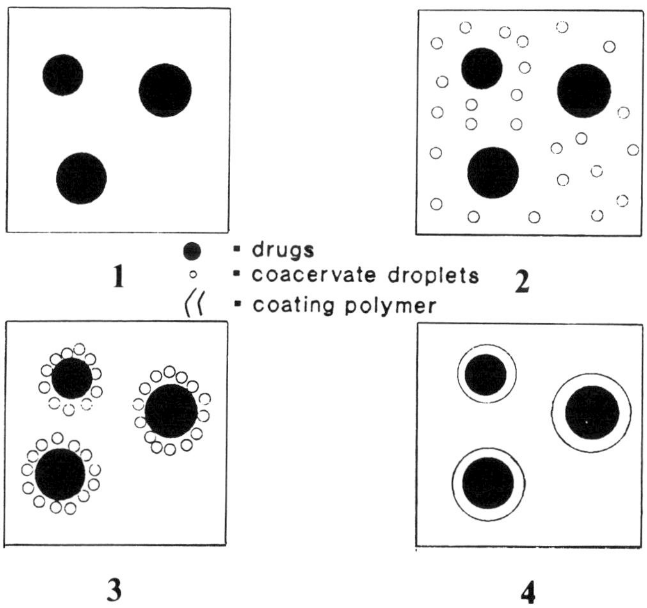

Figure 7 Schematic description of the four main steps involved in the coacervation microencapsulation process.

Deposition of the coacervate all around the drug particles (solid or liquid) with formation of embryo microcapsules.
Further desolvation of the coating polymer in order to harden the microparticles before isolating them by sedimentation.

PLA or PLGA microparticles have thus been prepared by using solutions of the coating polymer in CH_2Cl_2. Coacervation has been initiated by the progressive addition of silicone oil under stirring. If the drug is a protein, there is of course a risk of its denaturation by the organic solvent and by the hydrophobic polymer. Kinetics of the α-thrombin release extend over more than 5 days (Figure 8). However, an initial burst effect is observed, the magnitude of which depends on the viscosity of the silicone oil used in the coacervation process. Later on the α-thrombin release agrees satisfactorily with a zero-order kinetic.

Figure 9 shows the time dependence of the percentage of the released α-thrombin that is active. These experimental data have been calculated from the total amount of the released protein and from the analysis of its activity by a fluorescent binding technique (77). The results clearly show that the release of an essentially inactive protein is associated with the burst effect. Thus a protein that is immobilized in the very outer shells of the microspheres

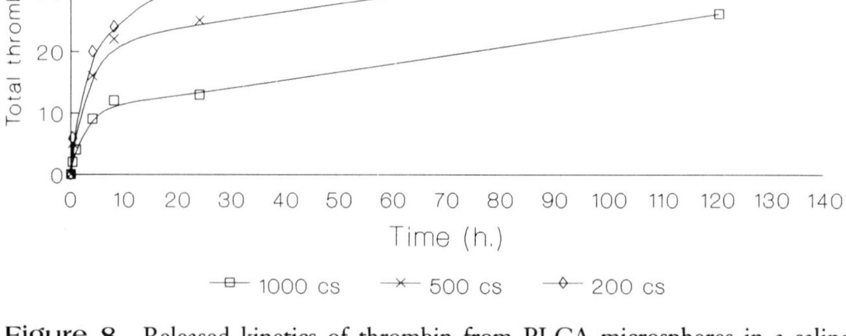

Figure 8 Released kinetics of thrombin from PLGA microspheres in a saline buffer at 25°C. Polymer precipitation has been performed by addition of silicone oil: 1000, 500, or 200 cs. The activity of the protease has been related to its ability to bind a fluorescent ligand (77).

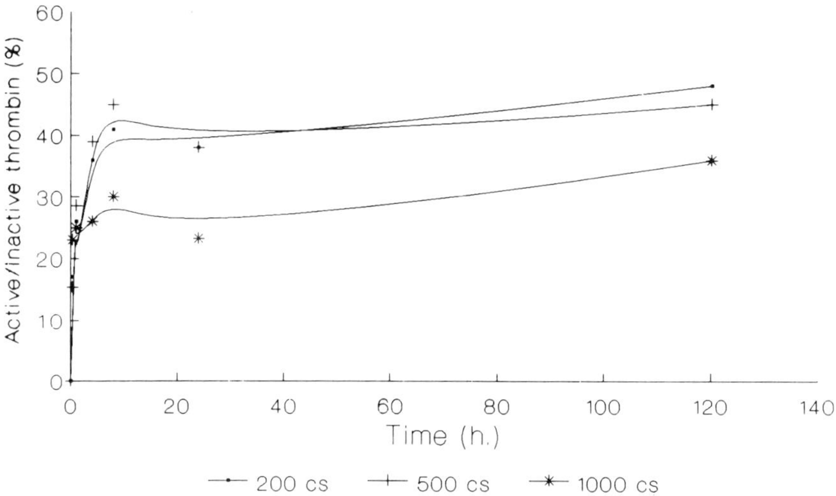

Figure 9 Time dependence of the percentage of the released thrombin (= ratio active/inactive form) from PLGA microspheres prepared in the presence of the incompatible polymer, silicone oil, with a viscosity of 200, 500, or 1000 cs.

is inactive, most likely due to a direct contact with the organic solvent and the coacervation agent. Nearly half the released α-thrombin is active when the zero-order kinetics is observed (78).

Since the release is very fast compared to the lifetime of the coating polymer in vitro, diffusion has to be the dominant mechanism for the protein release. The effect of the molecular weight of silicone oil on the release kinetics might be related to a change in the porosity of the polymer matrix. It is finally worth noting that data of Figure 9 are only valid for the PLGA microparticles. Indeed, the most hydrophobic PLA causes the irreversible denaturation of the protein.

SEX HORMONES. For several years, sex hormones, such as estrogens, have been used in oral or selective arterial treatments of many lesions and malformations, e.g., Rendu–Osler disease, juvenile angiofibroma (79), and dural AVF (80).

Similarly to the antitumoral chemoembolization, embolic particles could be loaded with an exogenous estrogen, so that it might be fruitful to combine embolization with local hormonotherapy. For instance, there is an 85% risk for the Rendu–Osler disease, i.e., the haemorrhagic hereditary telangiectasia, to cause a life-threatening epistaxis. The embolization of these symptomatic telangiectasias in the nasal mucosa is effective in preventing bleeding, provided it is regularly repeated in order to avoid a temporary occlusion. A longer lasting effect might be expected if the microparticles were previously loaded with estrogen. The juvenile angiofibroma—a benign vascular tumor in the nasopharynx of young males—sometimes exhibits aggressive forms with orbital and intracranial extensions making its resection very hazardous. A local delivery of estrogen by a microparticulate carrier might be a useful treatment for such a sex hormone–dependent lesion.

Moreover, since a number of benign or malignant tissues (liver, pancreas, colon, etc.) contain estrogen receptors (81), an adjuvant treatment might be considered on the basis of particulate carriers loaded with an antiestrogen agent. Finally, chemoembolization with an antiprogestrone agent might be appropriate for the unresectable meningioma of elderly patients, since progesterone receptors have been found in that tumor (79).

ANTICALCIC AGENTS. Vascular malformations, particularly the challenging arteriovenous malformations, present other pathological conditions that might benefit from an endovascular targeting treatment. At present they are embolized by bucrylate glue. However, the manipulation of such an agent is a problem, considering that recanalization has been reported to occur (82,83). By analogy with tumoral chemoembolization, a vascular malformation might be treated by occluding the nidus with particulate carriers loaded with an agent capable of changing the properties of the vascular endothelium. In this respect, the inhibition of the regulating effect of endothelial vasopeptides might be considered (84). The stimulated liberation of the endothelin, i.e., an endothelial contracting factor, by a humoral or mechanical effect is expected to promote the functional activity of the arteriovenous malformation. The arteriovenous dysplasia would then be self-sustained and the liberation of another vasopeptide (the relaxin, i.e., an endothelial relaxing factor) would be enhanced. The particular vascular organization and the high blood flow of the dysplasia are favorable conditions for a vicious circle of liberation of the vasopeptides to occur and a lack of balance of their activity to favor the growth of the vascular malformation. Since calcium ions are required for the endothelial vasopeptides to be active, an anticalcic agent should be released in situ in order to break the vicious cycle and to restore a balanced activity of the vasopeptides. The lesion could thus be stabilized and an aggressive arteriovenous malformation could be changed to a quiet and resectable one.

Many drugs are now available for chemoembolization. As was reported by Tomlinson (2), more than 50 drugs have been encapsulated in albumin microspheres. The potential for many additional loaded microparticle preparations is great.

IN VIVO EVALUATION

Embolization

General Requirements

The in vivo evaluation of a new embolic material is a prerequisite for any clinical application. It is thus of prime importance to assess the key requirements of an ideal occlusive agent (see Introduction), i.e., biocompatibility, effectiveness, and persistence of the embolic material. Biocompatibility means that no noxious reactions should result from the mutual interactions of both an implant and the biological environment. From an experimental point of view, the action of the implant on the tissue cannot be completely understood until we have greater knowledge of the reaction of the tissue to the implant (85). The short-term acceptance of the implant depends on interfacial phenomena, whereas long-term acceptance is mainly determined by the bulk properties of the biomaterial.

The biocompatibility of the occlusive agent has to be inferred from the toxicological, hematological, and histological points of view, respectively. In a preliminary step, toxicity and the hemostatic properties of the embolic material can be evaluated in vitro. For instance, the presence of toxic low molecular weight compounds, such as residual monomers, metals (catalysts), and solvents, can be identified by their lytic action on blood cells or cultured cells (e.g., fibroblasts). Moreover, the thrombogenicity of the biomaterial can be directly tested by means of a microscopic observation of the ability of thrombocytes to aggregate on the surface of the embolic material (28).

The investigation of the histological biocompatibility of the occlusive agent is a problem. Indeed, the biomaterial must be introduced in the vessels in such a way as to avoid any cellular and vessel reactions to anoxia. The histocompatibility can then be assessed from the direct interaction of the embolic agent with the nonperturbed biological environment. This requirement has been highlighted by Quisling et al. (86) who evaluated the histocompatibility of polyvinylalcohol "microemboli" in a rat's cerebral cortex. They injected the microemboli while avoiding the complete occlusion of the vessels. A valuable alternative consists of the embolization of organs that are irrigated by a dual vascularization, e.g., lungs and liver. However, the portal circulation flow in hepatic tissues has proved to be insufficient to meet the oxygen requirements when the arterial occlusion is complete (24). Of course, the long-term biocompatibility has still to be assessed when a persistent embolization is effective.

An ideal embolic material is supposed to trigger a complete, homogeneous, and permanent occlusion of the targeted vessels. An in vivo evaluation is thus required. Immediately after the embolization, angiograms can provide direct evidence for the selectivity and quality of the occlusion, since they display every alteration in the vascular supply and vessel diameter. The possible distal migration of the embolus can also be identified by angiograms carried out after a few days or weeks. Similarly, the recruitment of neighbor vessels by the pathological area can show evidence of a secondary vascularization.

Histological investigations are helpful in detecting a tiny recanalization of the occluded zone, which can be due to the degradation of the embolic material or to the reorganization of the thrombus itself. The direct microscopic observation has the great advantage of locating the occlusive agent in the pathological area and defining the critical diameter for a vessel to immobilize a particulate carrier. The follow-up of the injected material in the embolized vessel allows assessment of the persistence of the occlusion and an understanding of the interactions between the foreign body and the host tissue.

Results of the Experimental Evaluation

Experiments on animals testify the superiority of spherical emboli over irregularly shaped particles whenever the completeness and durability of the vessel obliteration is concerned (87). Histological observations of embolized tissues also show that no significant inflammatory response is promoted by an amorphous, deckle-edged material, either in the embolized vessels or in the surrounding tissue. Polyvinylalcohol emboli would provide such an example. They adhere to the endothelium vessels, the diameter of which is larger than the particle size. That material proves to be biologically inert and to have a purely mechanical action (86). Microscopic examinations show blood clots within interparticle interstices (88). These "thrombi" can reorganize in favor of a new vascular channel (23). Thus, two of the main drawbacks quoted, i.e., proximal obliteration and revascularization, are unambiguously proved by histological examinations.

The embolization effectiveness of spherical and irregularly shaped particles has been compared by Wright et al. (35), who injected dextran microspheres and polyvinylalcohol particles, respectively, in the splenic artery of dogs. The superiority of the microspheres was obvious, since a more homogeneous and peripheral embolization was then observed. Similarly, Dion et al. (26) reported a fast and drastic reduction in the blood flow when the renal artery of dogs was embolized with the same microspheres. Although an angiographic control concluded with total occlusion, the blood flow was partly reset up after a short while. A second injection was required to achieve a complete occlusion. A marked stretching of the vessels around the microspheres, a thinning of the internal elastic lamina, and no inflammatory reactions of the vessel wall were reported. The dextran microspheres are thus inert, acting as mechanical agents and favoring a distal and homogeneous obliteration. It must be stressed that spherical microspheres of a smooth surface have to be handled very carefully. Indeed, they are immobilized only by arteries of a smaller diameter. If that happens in a normal tissue instead of the targeted pathological tissue, then the deep anoxia and necrosis that result have a disastrous effect.

The location of spherical agents depends not only on their size but also on the elasticity of the arterial wall since a microsphere is immobilized by a twice-as-small artery. According to Madoule et al. (16), the best size range is 125–400 μm. The experimental investigation of dog kidney also showed that microspheres move in the arteries until the foreign body reaction occurs, entrapping them inside the inflammatory granuloma, i.e., on the tenth day after embolization.

Although they trigger a distal embolization, some embolic materials prove to increase the duration of the vessel obliteration. Laurent et al. (27) injected trisacryl microspheres coated with gelatin, i.e., a chemotactic agent, in the lungs of rats. An histological study showed the rapid growth of cell clumps near the microsphere surface and a more rapid formation of granuloma and vascular obliteration compared to the situation in which microspheres have only a mechanical effect. Nevertheless, the authors pointed out the difficulty in achieving a complete and permanent vascular occlusion.

Horak et al. reported that a lasting obliteration might be favored by PHEMA microspheres (29,30). That embolic material actually has a thrombogenic effect since a thrombus is observed to form itself on the surface and inside the cavities of these microspheres. The thrombus is finally replaced by a connective tissue that makes the occlusion permanent.

Chemoembolization

Since the tumoral chemoembolization aims at increasing the therapeutic drug index, two major characteristic features have to be studied by means of experimentation on animals: the local drug uptake and the systemic drug concentration. In this regard, the systemic drug injection must be compared to chemoembolization. Last but not least, the kinetics of the drug release has to be under control since it determines the local drug concentration.

Local Drug Concentration

There are very few reports on that topic in the medical literature. Hecquet et al. (45) observed that the ratio of the tissue concentration over the plasma concentration is increased by 200 times when the drug is encapsulated. According to Kato et al. (10), biologically active mitomycin C is still present in a tissue that was embolized 6 hr earlier by mitomycin C–containing microcapsules. Conversely, no active drug is detected when directly injected in the bloodstream. Surprisingly enough, Kerr et al. (56) noted only a slight increase of the adriamycin concentration in a tissue targeted by loaded microspheres. This might be caused by a problem of sensitivity in the detection of a drug that is released at a very slow rate. In addition, the drug uptake by a tumor was found to be increased when the tumor was targeted by degradable starch microspheres loaded with the drug (89).

A technical problem more likely explains why the experimentation on animals has provided very limited information on the local concentration of a drug. Indeed, the total amount of the drug present in an embolized tissue does not correspond to the actual drug uptake by the cells. The experimental methods do not actually discriminate the drug that has been released from the still encapsulated molecules. As long as the local drug uptake is unknown, no reliable connections between drug concentration and therapeutic effects can be drawn.

The efficiency of an antimitotic drug can be assessed by using a transplantable tumor, such as the model VX2 tumor implanted in a rabbit's liver (69,90). The effect of chemoembolization is in fact linked to the proliferation rate of the pathological tissue. However, the pertinence of the results might depend on how effectively the model tumor is vascularized. Nevertheless, that type of in vivo experimentation is a very convenient method for testing new antimitotic drugs. These are expected to be active in the hypoxic and hypoenergetic environment caused by the embolization.

Systemic Drug Concentration

Since the systemic concentration can be easily measured, many experimental data have been published about it. It is clear that the area under the plasma concentration vs. time curve is significantly decreased when the drug has been previously encapsulated rather than administered by a systemic injection. Reductions of 40% and 36% were reported by Kato et al. (10) and Ichihara et al. (9), respectively. However, these concordant data have to be attributed to capsules of a different nature. Juni et al. (49) used polylactide microspheres and observed that less than 10% of the drug accumulates in the venous blood. Hecquet et al. measured a 20-fold decrease in the plasma concentration with drug encapsulated prior to injection (45). Similar data were reported independently for degradable starch microspheres (36). A small but meaningful amount of drug was measured in the plasma 24 hr after the injection of drug-loaded microspheres together with an initial burst effect (38).

Kinetics of Drug Release

The fast initial release followed by a slow and sustained delivery, which occurs in vitro, has been confirmed by experimentation on animals. For instance, a burst effect has been noticed by independent research teams who used either ethylcellulose microcapsules (10,47,91) or cetylalcohol microspheres (48).

That drawback, however, is greatly reduced when protein microspheres (38–40) and PLA particles (9) are used.

In addition to the biocompatibility assessment, the time dependence of the location and the reactivity of embolic particles have been investigated in vivo. In agreement with pharmacokinetic predictions, the local release of the drug is very favorable to a decrease in the systemic side effects. Nevertheless, local pharmacological efficiency cannot be anticipated in clinical applications.

CLINICAL INVESTIGATIONS

Hyperselective Embolization by Microspheres

General Review

Since polymer microspheres have proved to be suitable and effective embolic agents, PHEMA, trisacryl gelatin, dextran, and polylactide microspheres have been tested in clinical applications (18,22,26,29). For instance, Horak et al. used microspheres to inhibit an organic haemorrhage, mainly in the lungs of 187 patients (29). They have insisted on the very easy injection of these particles in the blood vessels when a catheter is used. An immediate and permanent hemostatic effect has been promoted by PHEMA microspheres and no revascularization has occurred within 2 years. These very encouraging results might be due to the thrombogenic properties of PHEMA.

Dion et al. (26) treated several kinds of tumors and vascular malformations of 10 patients by using dextran microspheres in the size range of 40–150 μm and 100–300 μm, respectively. They also pointed out an easy handling of the microspheres, a lack of clogging, and a distal and superselective embolization. The occlusions are clearly more complete and permanent compared to those achieved by irregularly shaped particles.

Fifty-six patients affected with tumoral lesions in the head and neck area (e.g., meningiomas, nasopharyngeal angiofibromas, glomus tumors, and vascular malformations of the face, medulla, or brain) have been treated with trisacryl gelatin microspheres at the Lariboisière Hospital in Paris (22). Embolization has been successfully carried out, since no particle aggregation occurred either in the catheter or in the proximal vessel. Microspheres of a small size (200 $\pm$ 50 μm) allowed for the development of complete vascular occlusion and a deep necrosis, as observed by angiography and histological methods, respectively. The embolization has rendered the surgical excision of facial AVM without any hemostatic problem much more easy. However, the necrosis may generate unexpected complications. For instance, a meningioma has been embolized with small microspheres, and a regressive coma has been provoked by a reactional brain edema at the periphery of the deeply necrosed lesion. It is thus mandatory to carry out the embolization step by step with microspheres of increasing sizes in order to carefully control the necrosis generated in such a pathological situation. As a rule, the clinical applications of PLA microspheres have confirmed their efficiency and drawn attention to their potential risks.

Personal Experience

Recently, a series of embolizations has been carried out with PLA microspheres prepared at the University of Liège (8). The study was undertaken in order to analyze further the potentialities of that embolic material, as a preliminary step to their clinical use in chemoembolization.

The average size of the microspheres was uniform in the range 100–160 μm in order to reach the intralesional precapillary arteries. The dry microspheres were distributed in doses of 25 and 50 mg and sterilized in a 2-ml polypropylene syringe with ethylene oxide. That method of packaging allowed the embolization material to be prepared in the syringe itself. Since PLA microspheres are hydrophobic, hydroxyethyl amylum (Plasmasteril, Fresenius AG, AD-Hamburg, Germany) was used as a biocompatible surfactant in order to wet the particles. A contrast agent (Iohexol, 240 mgI/ml, Omnipaque, Nycomed, Oslo, Norway) was also added to allow the fluoroscopic control of the arterial injection. Each dose of microspheres was mixed with 2 ml of Plasmasteril and 2 ml of Omnipaque just before embolization. This mixture proved to be very suitable since the particles formed a stable suspension, which was easily injected through the microcatheter. Let us briefly describe some examples of clinical applications.

Case 1

A 46-year-old female patient was affected by a pterional meningioma on the left-hand side (Figure 10). The left middle meningeal artery was embolized by three successive injections of PLA microspheres (250 mg). This was our first clinical embolization and we were very impressed by its great easiness and the important and homogeneous reduction of the tumoral vascularization (Figure 11). Since the embolization was a preoperative one, the extratumoral arteries were sealed with a few Ivalon particles of a 150- to 250-μm size.

After a bloodless surgical excision, the anatomy of the tumor was examined. Ivalon particles were found to be in the extratumoral arteries to the exclusion of the tumor. In contrast, the PLA microspheres were selectively located in the intratumoral vessels (Figure 12). The microsphere size was thus adequate to a distal embolization. Since all the Ivalon particles were found in the extratumoral arteries, it is clear that these irregularly shaped particles of approximately the same size as the PLA microspheres agglomerate in a more proximal location.

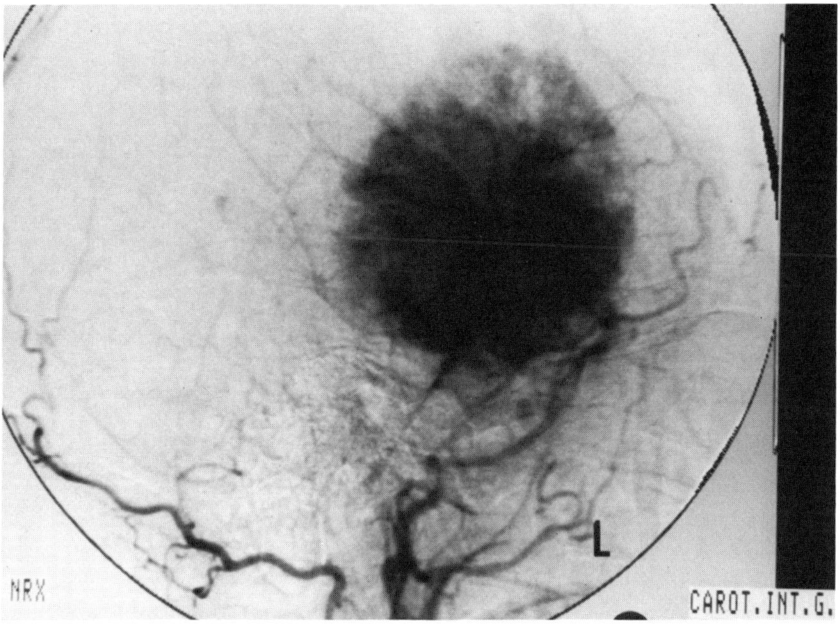

Figure 10 Lateral left external carotid artery angiogram: demonstration of a large pterional meningioma.

Case 2

A 67-year-old woman showed a left lateral sinus dural AVF fed by several meningeal arteries. The branches originating from the occipital artery were occluded with 150- to 250-μm-sized Ivalon particles. The left middle meningeal artery showed an important arteriovenous (AV) shunting (Figure 13) and was selectively injected with a 25 mg dose of PLA microspheres. The microspheres were thus gently proximally injected in the feeding artery. They were guided by a suction effect due to the AV shunts and they selectively sealed the pathological area (Figure 14). The postoperative angiogram showed the complete disappearancc of the shunts, whereas all the other normal meningeal branches were permeable. The spherical PLA particles of a smooth surface are very effective in promoting a distal embolization since they reach the lesion (second level of targeting) under the impetus of the AV shunts of the vascular malformation itself.

Case 3

This example again supports the efficiency of the microspheres in embolization but it also highlights their potential hazards. A 46-year-old female

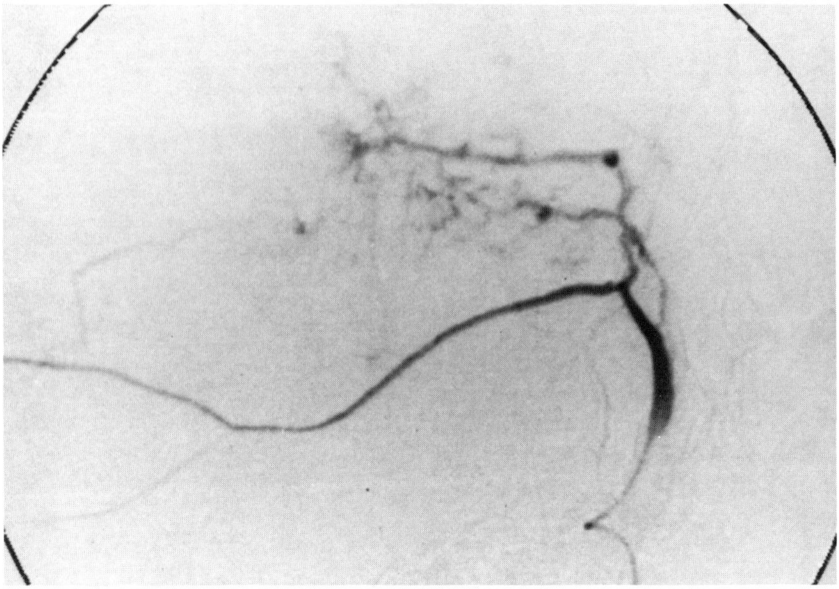

Figure 11 Lateral left middle meningeal artery angiogram shows homogeneous tumor devascularization after injection of PLA microspheres.

patient was affected by a meningioma on the left-hand side of the falx cerebri. The angiogram of the external carotid showed the meningeal vascularization of the tumor fed by the middle meningeal artery. The angiogram of the internal carotid showed a slight tumoral stain at the venous phase that was thought to be due to cortical branches originating from the anterior cerebral artery. Finally, a preembolization angiogram of the middle meningeal artery allowed the tumor fed by only one enlarged meningeal branch to be observed (Figure 15). A dose of 25 mg of PLA microspheres was delivered by repeated injection. Then the fluoroscopic control of the injection mixture showed a very transient opacification of the left ophthalmic artery (Figure 16). The embolization was immediately stopped, and the angiogram performed a few seconds later displayed the total occlusion of the lesion together with an unexpected anastomosis. The latter appeared between the branch of the middle meningeal artery and the artery of the falx cerebri coming from the ophthalmic artery. The modification of pressure in the tumoral area allowed the opening of the anastomosis and the filling of the falx cerebri artery up to the ophthalmic and the internal carotid ones. As a result, a blindness on the left-hand side occurred. As emphasized by Dion et al. (26), microspheres

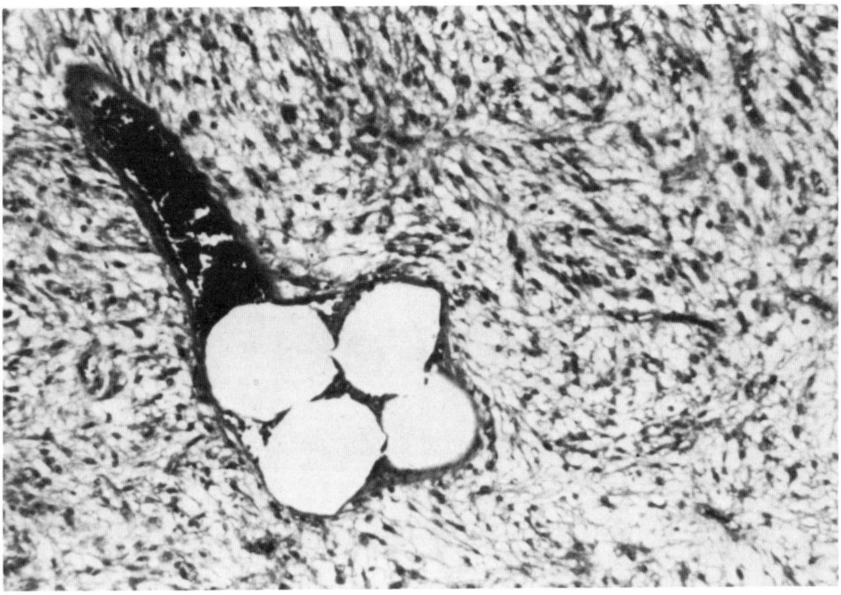

Figure 12 Anatomical examination. Microphotograph shows microspheres in the intratumoral vessels.

must be manipulated very carefully in order to avoid overembolization and reflux, but also the dangerous opening of anastomosis, particularly in the head and neck area. Indeed, smooth microspheres are stopped by vessels of an appropriate internal diameter, whatever their nature, i.e., pathological and normal arteries.

Chemoembolization of Tumors

When a tumor is treated by chemoembolization, questions about an eventual reduction in the tumoral volume, possible side effects, and the evaluation of survival time have to be addressed. The knowledge of the local drug concentration as measured on biopsied or resected tumors is also of a prime importance. Since the liver is a host organ, very appropriate for chemoembolization, it is worth reviewing the state of the art in this area.

The first clinical applications were reported by Kato et al. in the late 1970s (10). They treated tumors with ethylcellulose microcapsules loaded with mitomycin C (10,11,13,19,42,46) and cisplatin (34). A few other embolic materials were tested in human applications, such as PLA microspheres loaded with aclarubicin (9) and protein microspheres loaded with mitomycin C (14).

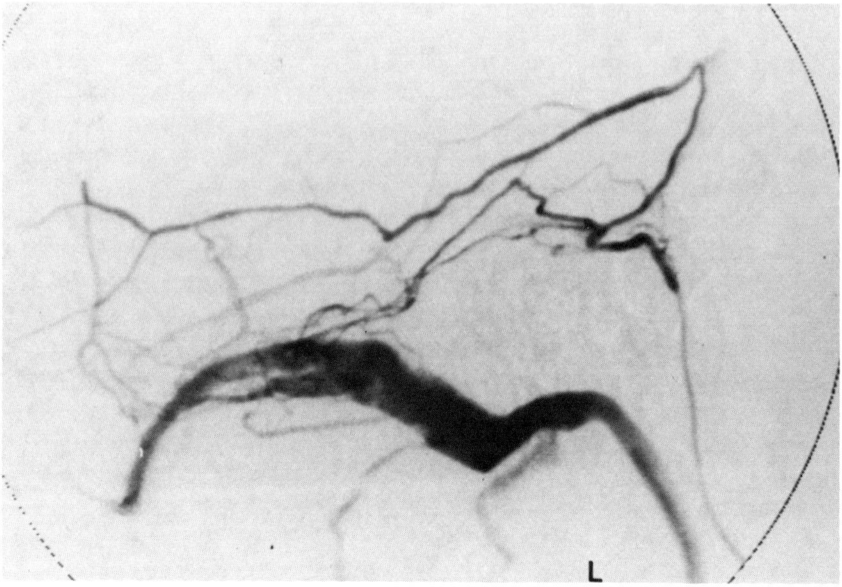

Figure 13 Lateral left middle meningeal artery hyperselective angiogram. Demonstration of numerous arteriovenous shunts within the wall of the left lateral sinus.

For the sake of comparison, results reported for degradable starch microspheres mixed with BCNU (36) and mitomycin C (92) are worth mentioning.

When microcapsules loaded with mitomycin C were used, Kato et al. reported a tumoral volume reduction exceeding 30% for 65% of the patients (12). These figures have to be compared with a reduction of more than 50% for 30% of the patients treated by Ohnishi (13). Okamoto et al. (34) noted a 50% volume reduction for 9 of 14 patients treated with ethylcellulose microcapsules loaded with cisplatin in ORL cancers. Ichihara et al. (9) found a 50% tumoral reduction for 42.6% of his patients when he used PLA microspheres containing aclarubicin. Protein microspheres loaded with mitomycin C lead to an objective tumoral response at a rate of 68.4% compared with 46.2% when the drug was delivered by an arterial infusion. As a rule, a 50% reduction of the tumoral volume seems to be independent of the embolic agent, the drug, and the location of the tumor.

When the systemic exposure is concerned, the area under the plasma concentration versus time curve is comparable with that seen in experiments on animals: a one-third to one-fourth reduction in the systemic concentration compared with the direct injection, regardless of drug and drug carrier. The

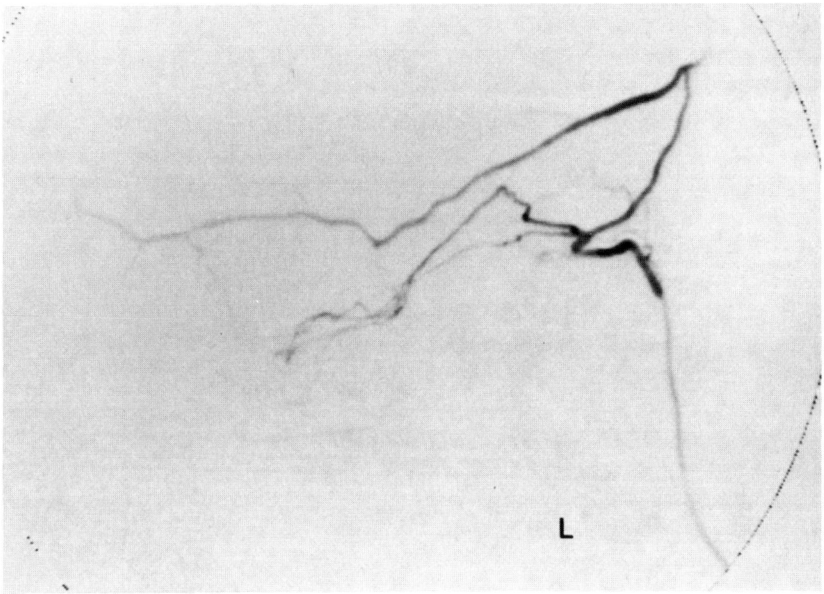

Figure 14 Lateral left middle meningeal artery after embolization with PLA microspheres shows a total occlusion of the arteriovenous shunts.

side effects are thus mild and transient; only a very limited myelosuppression can occur. In fact, local pain and fever are the most current inconveniences related to embolization. The only local side effect that is noticeable is the normal tissue necrosis. For instance, a skin necrosis in the gluteal area is typically observed when microcapsules are infused in the hypogastric artery.

It is difficult to arrive at a clear idea of the survival rate since the comparison of the data published by several authors is not straightforward. Sixty percent of the patients treated by Kato et al. (12) were still alive after 2 years. Also using microcapsules loaded with mitomycin C, Ohnishi et al. (13) observed an average survival time of 8.4 ± 5.1 months for patients affected with an unresectable hepatocellular carcinoma (HCC). By comparison, an average survival time of 7.1 months was reported for patients treated by arterial drug infusion.

Ichihara et al. (9) (PLA microspheres loaded with mitomycin C) and Fujimoto et al. (14) (protein microspheres loaded with mitomycin C) separately confronted two groups of patients treated by chemoembolization and systemic chemotherapy or conventional arterial drug infusion, respectively. In Ichihara's report, the cumulative survival rate for 59 patients with unresectable HCC tumors was 54.3% for 1 year, 24.6% for 2 years, and 19.2%

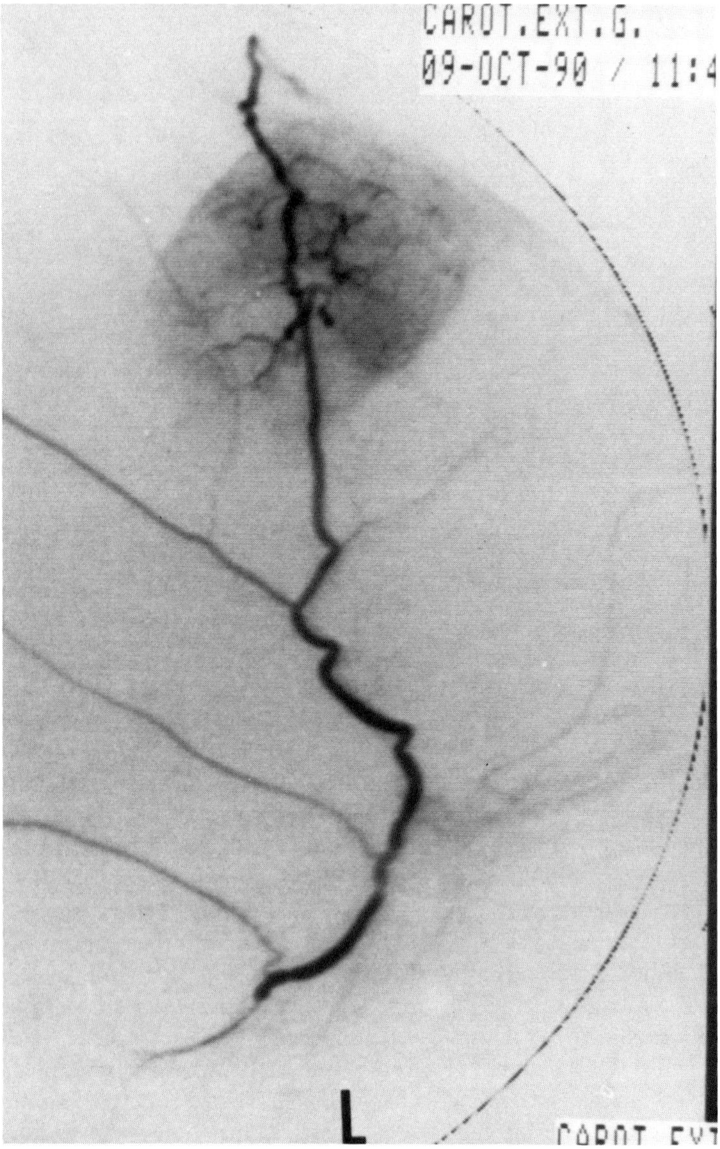

Figure 15 Lateral left middle meningeal artery angiogram reveals a meningioma situated on the left side of the falx cerebri.

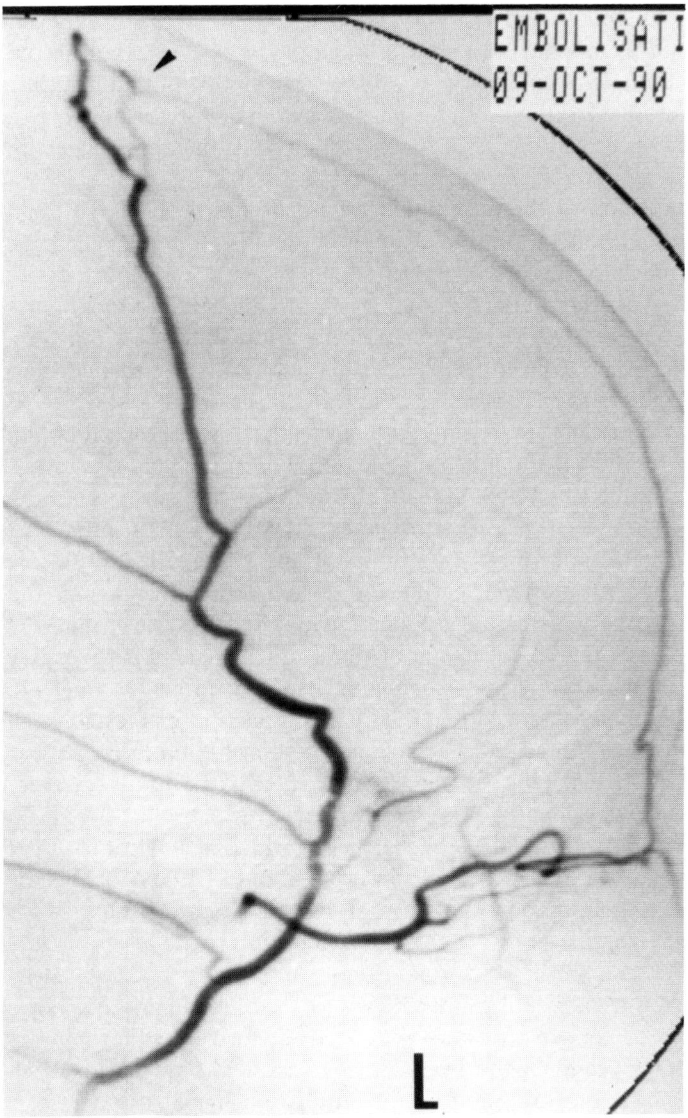

Figure 16 Lateral left middle meningeal artery angiogram after embolization with PLA microspheres. Demonstration of an anastomosis between a branch of the middle meningeal artery and the artery of the falx cerebri (arrow).

for 3 years. For the control group, the cumulative survival rate was 11.6% for 6 months and 0% after 1 year. In the Fujimoto series (14), the survival time was also significantly higher ($p = >.059$) compared with that resulting from the conventional treatment. In the particular case of unresectable HCC, chemoembolization provides at least encouraging results in comparison with conventional chemotherapy, either systemic or arterial.

The concentration of the locally delivered drug has been measured in biopsied or resected human tumors. Okamoto et al. (34) found a very large amount of platinum in the targeted tissue. Ichihara et al. (4) reported that the aclarubicin concentration in the tumor was approximately 21 times higher than in the noncancerous region. As previously mentioned, the determination of the actual local drug concentration remains dubious, since this concentration sums up the delivered as well as the still encapsulated drug.

Up to now, degradable starch microspheres mixed with an antitumoral drug have been of limited use in human application. For a group of five patients, Dakhil et al. (36) reported a 30–90% reduction in the systemic drug exposure, whereas myelosuppression was never observed. According to Ensminger et al. (92), chemoembolization was responsible for an ~30% reduction in the plasmatic drug concentration in experimentations on both human beings and animals. No objective information has been made available concerning the survival time. The potential advantage of starch microspheres is their very rapid degradation by serum amylase, which allows a repetitive treatment to be enforced. Dakhil et al. noticed that the blood circulation was restablished 15 min after embolization and was normal again after 30 min (36).

It is clear from the published reports that chemoembolization is very encouraging in the treatment of tumors of the liver. Much effort has been devoted to the chemoembolization of that organ which is a valuable target due to a dual vascularization: an arterial and a venous one (47,93–99). Hepatic tumors thrive on their blood supply mainly by the hepatic artery. That situation is most favorable since the chemoembolization of the hepatic artery is straightforward and since the normal tissue continues to be supplied by the venous route, thus preventing any necrosis from occurring. Microspheres passively injected in the hepatic artery are selectively distributed throughout the vascular compartment of the liver. However, they are unable to reach the tumoral cells to the exclusion of the intraparenchymal normal cells. Therefore, even though their kinetics of drug release and their biodegradation rate have been optimized, these microspheres will be dismissed in favor of any embolic agent – even of poor performances – able to selectively reach the tumor. The ethiodized oil (69) (Lipiodol ultrafluid, Laboratoire Guerbet, Aulnay-sous-Bois, France), for instance, is known to be selectively retained by tumoral cells of the liver and accumulated at a high concentration for a long time. The actual mechanism is still unknown but the nonvasoreactive and irregular neoplastic vessels are most likely responsible for the immobilization of the Lipiodol. Since that compound has no antimitotic activity and no ischemic effect, it has to be combined with an antimitotic drug and embolic particles. When mixed with Lipiodol, adriamycin or cisplatin is targeted to the tumor, which supports the assertion that Lipiodol is a vector agent. The Lipiodol–drug emulsion is first injected, then followed by the embolic agent which is expected to occlude the vessels and to generate necrosis of the tumor. Chemoembolization of the liver with ethiodized oil, an antimitotic drug, and embolic particles is the best treatment of unresectable hepatic tumors. The superiority of that material over targeted microspheres is supported by a 2-year follow-up of patients treated with both methods. A 2-year survival of 24.6% resulted from a treatment based on PLA microspheres loaded with aclarubicin (9), whereas Kasugai reported a 2-year survival of 45% for patients treated with a Lipiodol–drug emulsion and gelatin sponge (98).

Very recently, Yumoto et al. (100) treated HCC by chemoembolization of the hepatic artery associated with local hyperthermia. They used Lipiodol mixed with two antimitotic drugs. The survival curves showed that the combination of the two treatments was beneficial and without major side effects.

FUTURE RESEARCH AND PERSPECTIVES OF APPLICATION

Targeting: A Difficult Achievement

Basically, endovascular therapy is a targeting method. If embolization is enforced as a curative method, its efficiency and safety depends on the selectivity of the targeting. If embolization is carried out as a preoperative step, then the main arterial feeder has just to be occluded without taking the location—inside the lesion or not—and the nature and structure of the embolic agent into account.

The endovascular treatment is feasible when the damaged organ can be reached (first level of targeting). In order to avoid major side effects, a "safety point" that depends on the local vascular anatomical characteristics has to be defined (101), below which no embolic material, regardless of its nature, can be injected. The catheter is expected to deliver the embolic agent in the pathological area, which is the "ideal point" (101) for the second level of targeting. The "efficiency point" is defined at the third level and is nothing but the cell or subcellular compartment. The drug is then supposed to interact with its pharmacological receptor.

For endovascular therapy to be in a successful competition with traditional treatments, such as surgery, the entire pathological area has to be reached selectively and completely. Although endovascular devices, embolization and chemoembolization materials have been improved during the last few years, the second level of targeting is still a difficult process. It is the reason why alternatives have been explored, such as the local insertion of an implant (102,103). Furthermore, the search for new drugs specific to the systemic delivery continues in competition with that for compounds specifically active in the hypoxic conditions of embolized tissues. The future of chemoembolization as an endovascular therapy is still uncertain. Nevertheless, vascular malformations or tumor pathology are very valuable targets, for which no effective treatment is yet available.

Future of Microspheres in Embolization

Experimentation on animals and clinical treatments have proved that microspheres of a 100- to 300-μm size are the most appropriate embolic agents. Indeed, they reach the intralesional precapillary arteries, provide a dramatic reduction of the blood flow, and are responsible for a deep, permanent necrosis.

Inert spherical particles are immobilized by vessels of an appropriate inner diameter. However, they may generate side effects, even when injected by a skilled physician. The ease of injection of spherical embolic agents might thus be counterbalanced by possible complications, particularly in a preoperative treatment.

For this purpose, irregularly shaped fragments of polyvinylalcohol might be used more safely although they are less effective occluding agents. Spherical particles capable to adhere to the vessels are an alternative provided that they stabilize the occlusion during the required period.

Embolization with microspheres is thought to have great potential as a curative method. This material should be adhesive and thrombogenic and reach the intralesional arteries or arteriovenous malformation nidus. As was mentioned previously, drugs should be designed in such a way that they have a curative action under hypoxic conditions. Thus a lot of unresectable lesions or malformations could be treated by chemotherapy, provided that the kinetics of both the drug release and the polymer degradation are appropriate to the envisioned treatment. Tumoral lesions such as meningioma and vascular lesions such as dural arteriovenous fistula are potential targets for embolic particles and local sex hormone therapy, particularly in elderly patients. Epistaxis induced by nasal telangiectasia of Rendu–Osler disease and unresectable extensions of juvenile angiofibroma could also be treated by microspheres loaded with estrogen. Some types of AVM could be cured by a cocktail of microspheres loaded with an anticalcic agent and thrombin, respectively. Although more sophisticated, the following technique for an AVM eradication is expected to be very effective. The endovascular therapy should be carried out in two steps: Preliminary injection of high biodegradable microspheres of an increasing size into the malformation nidus should allow the particle size most appropriate to the therapy to be determined. Then the hyperselective injection of microspheres loaded with an anticalcic agent, exogenous estrogen, or thrombin should seal and dry the arteriovenous nidus and lead to a permanent cure. PLA-PLGA microspheres of various degradation rates and loaded with these active drugs are in preparation at the University of Liège, and it is hoped that they will be tested on animals in the near future.

Tumoral Chemoembolization

After the first enthusiastic results published by Kato and coworkers and dealing with chemoembolization as a preoperative or palliative treatment for various malignancies (10), numerous clinical reports have been made available on this subject. Most often, the authors mention preliminary clinical results relative to a limited number of patients as well as very different tumoral conditions. That situation, which makes comparisons and interpretations unreliable, is mainly due to the restriction of chemoembolization to tumors that are highly vascularized, chemoresistant to the systemic route, and nonresectable. It is worth recalling that the main drawbacks of the physical targeting are a lack of selectivity and a partial efficiency. Keeping these limitations in mind, let us summarize the different strategies that have been enforced in order to improve the efficiency of that endovascular treatment in the most favorable situation, i.e., liver tumors.

Looking for a higher selectivity in the treatment of a cirrhosis associated with HCC, Ichihara et al. (9) suggested the use of well-calibrated PLA microspheres in order to trigger embolization in the peripheral arteries. Goldberg et al. (104) also aimed at a more selective occlusion. They were interested in changing the pattern of the blood flow in the liver by the local injection of angiotensin II, a vasoactive agent. Dakhil et al. (36) were similarly concerned with the selectivity of chemotherapy. They deliberately deviated the blood flow by injecting quickly biodegraded starch microspheres in the hepatic artery.

Presently, the best strategy for achieving a selective chemoembolization relies on Lipiodol, which is retained within the HCC. Kuroda et al. (105) recently reported on the limitations of chemoembolization in the treatment of small nonencapsulated HCC. The authors suggest that the partial efficiency might result from an unaffected portal blood supply.

Although chemoembolization of liver tumors has still to be optimized, it may be considered as the best nonsurgical therapy. The last improvements of the liver therapy should now be extended to the other organs for which chemoembolization has failed. Instead of using passive embolic agents, magnetic particles might be used in order to perform an embolization as selectively as possible (106,107).

However, it must be stressed that final efficiency also depends on how fast the treatment is applied when the tumor is detected.

In conclusion, the physical targeting is certainly an elegant concept to improve the efficiency of chemotherapy. However, it remains a challenge due to a set of critical requirements: (a) a selective targeting that affects the whole tumor, (b) the selection of an appropriate drug, (c) the control of the drug release, i.e., a critical rate over a well-defined period of time that predetermines the local drug concentration accordingly. When malignant tumors can be treated with systemic chemotherapy, the question is to know whether chemoembolization can compete with it. The answer should be optimistic when the liver is concerned, but uncertain for the other organs. The chemoembolization of vascular malformations and some benign vascularized tumors is thought to deserve interest in some specific situations.

ACKNOWLEDGEMENTS

The authors are indebted to the FRSM (grant No. 3.4553.91).

REFERENCES

1. D. A. Wood, *Int. J. Pharm.*, 7, 1 (1980).
2. E. Tomlinson, *Int. J. Pharm. Techn. Prod. Mfr.*, *4*(3), 49 (1983).
3. S. Benita, J. P. Benoît, F. Puisieux, and C. Thies, *J. Pharm. Sci.*, *73*(12), 1721 (1984).
4. H. Endoh, T. Kawaguchi, T. Seki, T. Hasegawa, and K. Juni, *Chem. Pharm. Bull.*, *39*(2), 458 (1991).
5. S. S. Davis and L. Illum, *Biomaterials*, *9*, 111 (1988).
6. P. Couvreur, *CRC Crit. Rev. Ther. Drug. Carrier Syst.*, *5*(1), 1 (1988).
7. J. J. Burger, E. Tomlinson, E. M. A. Mulder, and J. G. McVie, *Int. J. Pharm.*, *23*, 333 (1985).
8. P. Flandroy, C. Grandfils, J. Collignon, A. Thibaut, N. Nihant, S. Barbette, R. Jérome, and Ph. Teyssié, *Neuroradiology*, *32*, 311 (1990).
9. T. Ichihara, K. Sakamoto, K. Mori, and M. Akagi, *Cancer Res.*, *49*, 4357 (1989).
10. T. Kato, R. Nemoto, H. Mori, and I. Kumagai, *Cancer*, *46*, 14 (1980).
11. T. Kato, R. Nemoto, H. Mori, M. Takahashi, and M. Harada, *Cancer*, *48*, 674 (1981).
12. T. Kato, R. Nemoto, H. Mori, M. Takahashi, Y. Tamakawa, and M. Harada, *JAMA*, *245*(11), 1123 (1981).
13. K. Ohnishi, S. Tsuchiya, T. Nakayama, Y. Hiyama, S. Iwama, N. Goto, M. Takashi, T. Ohtsuki, K. Kono, Y. Nakajima, and K. Okuda, *Radiology*, *152*, 51 (1984).
14. S. Fujimoto, M. Miyazaki, F. Endoh, O. Takahashi, R. D. Shrestha, K. Okui, Y. Morimoto, and K. Terao, *Cancer*, *55*, 522 (1985).
15. H.-S. G. Chen and J. F. Gross, *Cancer Treat. Rep.*, *64*(1), 31 (1980).

16. Ph. Madoule, Ph. Trampont, D. Doyon, J. Quillard, and F. Puisieux, *J. Radiol.*, *62*(8–9), 457 (1981).
17. Ph. Madoule, Ph. Trampont, and A. Roche, *J. Microencapsulation*, *1*(1), 21 (1984).
18. K. C. Lee and I. B. Koh, *Arch. Pharm. Res.*, *10*(10), 432 (1987).
19. T. L. Whateley, J. A. Goldberg, D. J. Kerr, C. S. McArdle, J. Anderson, J. G. Eley, and T. Kato, *Microencapsulated Mitomycin C for intra-arterial targeting to hepatic metastases*, Proc. Int. Symp. Controlled Release Bioact. Mater., 1990, pp. 458–459.
20. C. Marsault and A. Gatson, International course on interventional neuroradiology, oral communication, Nancy, June 1987.
21. A. J. Fox, *Am. J. Neuroradiol.*, *12*, 798 (1991).
22. R. Beaujeux, *Microsphères d'embolisation*, Mémoire de D.E.A. d'imagerie médicale, Université Paris V, 1990.
23. C. M. Strother, R. Laravuso, A. Rappe, S. L. Su, and K. Northern, *Am. J. Neuroradiol.*, *8*, 509 (1987).
24. J. R. Daniels, R. K. Kerlan, L. Dodds, P. McLaughlin, J. M. LaBerge, D. Harrington, A. M. Daniels, and E. J. Ring, *Invest. Radiol.*, *22*, 126 (1987).
25. C. R. Jack, G. Forbes, M. K. Dewanjee, M. L. Brown, and F. Earnest, *Am. J. Neuroradiol.*, *6*, 595 (1984).
26. J. E. Dion, R. N. Rankin, F. Vinuela, A. J. Fox, A. C. Wallace, and M. Mervart, *Radiology*, *160*, 717 (1986).
27. A. Laurent, M. Wassef, L. Drouet, G. Pignaud, and J. J. Merland, *Innov. Tech. Biol. Med.*, *10*(3), 358 (1989).
28. D. Horak, F. Svec, J. Kalal, K. Gumargalieva, A. Adamyan, N. Skuba, M. Titova, and N. Trostenyuk, *Biomaterials*, 7, 188 (1986).
29. D. Horak, F. Svec, J. Kalal, A. A. Adamyan, Y. D. Volynskii, O. S. Voronkova, L. S. Kokov, and K. Z. Gumargalieva, *Biomaterials*, 7, 467 (1986).
30. D. Horak, M. Metalova, F. Svec, J. Drobnik, J. Kalal, M. Borovicka, A. A. Adamyan, O. S. Voronkova, and K. Z. Gumargalieva, *Biomaterials*, *8*, 142 (1987).
31. D. Horak, F. Svec, J. Kalal, A. Adamyan, N. Skuba, M. Titova, V. Dan, B. Varava, N. Trostenyuk, O. Voronkova, K. Gumargalieva, and V. Timochina, *Biomaterials*, *9*, 367 (1988).
32. B. Chithambara Thanoo and A. Jayakrishnan, *J. Microencapsulation*, *6*(2), 233 (1989).
33. C. Chithambara Thanoo and A. Jayakrishnan, *Biomaterials*, *11*, 477 (1990).
34. Y. Okamoto, A. Konno, K. Togawa, T. Kato, Y. Tamakawa, and Y. Amano, *Br. J. Cancer*, *53*, 369 (1986).
35. K. C. Wright, S. Wallace, B. Mosier, and D. Mosier, *J. Microencapsulation*, *5*(1), 13 (1988).
36. S. Dakhil, W. Ensminger, K. Cho, J. Niederhuber, K. Doan, and R. Wheeler, *Cancer*, *50*, 631 (1982).
37. S. Kudo, K. C. Wright, V. P. Chuang, S. Wallace, S. Mir, and W. Bechtel, *Am. J. Roentgenol.*, *143*, 1069 (1984).
38. N. Willmott, J. Cummings, J. F. B. Stuart, and A. T. Florence, *Biopharm. Drug Disp.*, *6*, 91 (1985).

39. N. Willmott, J. Cummings, and A. T. Florence, *J. Microencapsulation*, *2*(4), 293 (1985).
40. N. Willmott, Y. Chen, J. Goldberg, C. McArdle, and A. T. Florence, *J. Pharm. Pharmacol.*, *41*, 433 (1989).
41. R. Nemoto, T. Kato, K. Iwata, H. Mori, and M. Takahashi, *Urology*, *17*(4), 315 (1981).
42. J. N. Bruneton, X. Dageville, G. Paleirac, E. Caramella, J. J. Manzino, J. Bourry, and G. Milano, *J. Radiol.*, *65*(12), 833 (1984).
43. K. Ohnishi, S. Tsuchiya, T. Nakayama, Y. Hiyama, S. Iwama, N. Goto, M. Takashi, T. Ohtsuki, K. Kono, Y. Nakajiama, and K. Okuda, *Radiology*, *152*, 51 (1984).
44. B. Hecquet, C. Fournier, G. Depadt, and P. Cappelaere, *J. Pharm. Pharmacol.*, *36*, 803 (1984).
45. B. Hecquet, G. Depadt, C. Fournier, and J. Meynadier, *Anticancer Res.*, *6*, 65 (1986).
46. P. Courtheoux, F. Alachkar, A. Casasco, Y. Adam, J. M. Derlon, F. Courtheoux, J. L. L'Hirondel, and J. Theron, *J. Neuroradiol.*, *12*, 151 (1985).
47. J. A. Goldberg, D. J. Kerr, R. Blackie, T. L. Whateley, L. Pettit, T. Kato, and C. S. McArdle, *Cancer*, *67*, 952 (1991).
48. P. Tilleul, G. Cavé, F. Puisieux, Ph. Trampon, and Ph. Madoule, *Labo-Pharma-Probl. Techn.*, *32*(339), 105 (1984).
49. K. Juni, J. Ogata, N. Matsui, M. Kubota, and M. Nakano, *Chem. Pharm. Bull.*, *33*(4), 1734 (1985).
50. J.-P. Benoît, in *Drug Targeting* (P. Buri and A. Gumma, ed.), Elsevier, Amsterdam, 1985, p. 95.
51. G. Spenlehauer, M. Veillard, and J.-P. Benoît, *J. Pharm. Sci.*, *75*(8), 750 (1986).
52. G. Spenlehauer, M. Vert, J. P. Benoît, F. Chabot, and M. Veillard, *J. Controlled Release*, 7, 217 (1988).
53. G. Spenlehauer, M. Vert, J.-P. Benoît, and A. Boddaert, *Biomaterials*, *10*, 557 (1989).
54. S. Benita, O. Zouai, and J.-P. Benoît, *J. Pharm. Sci.*, *75*(9), 847 (1986).
55. A. Sawaya, J. P. Benoît, and S. Benita, *J. Pharm. Sci.*, *76*(6), 475 (1987).
56. D. J. Kerr, N. Willmott, J. H. McKillop, J. Cummings, H. J. Lewi, and C. S. McArdle, *Cancer*, *62*, 878 (1988).
57. Y. Nishioka, S. Kyotani, M. Okamura, M. Miyazaki, K. Okazaki, S. Ohnishi, Y. Yamamoto, and K. Ito, *Chem. Pharm. Bull Tokyo*, *38*, 2871 (1990).
58. Ph. Dubois, R. Jérôme, and Ph. Teyssié, *Polym. Bull.*, *22*, 475 (1989).
59. Ph. Dubois, R. Jérôme, and Ph. Teyssié, *Polym. Preprint* (*Am. Chem. Soc.*, *Polym. Chem*), *31*(1), 7 (1991).
60. Ph. Dubois, R. Jérôme, and Ph. Teyssié, *Macromolecules*, *24*, 977 (1991).
61. Ph. Dubois, C. Jacobs, R. Jérôme, and Ph. Teyssié, *Macromolecules*, *24*, 2266 (1991).
62. C. Jacobs, Ph. Dubois, R. Jérôme, and Ph. Teyssié, *Macromolecules*, *24*, 3029 (1991).

63. G. E. Visscher, J. E. Pearson, J. W. Fong, G. J. Argentieri, R. L. Robinson, and H. V. Maulding, *J. Biomed. Mater. Res.*, *22*, 733 (1988).
64. R. A. Miller, J. M. Brady, and D. E. Cutright, *J. Biomed. Mater. Res.*, *11*, 711 (1977).
65. M. Vert, *More about the degradation of LA/GA-derived matrices in aqueous media*, Proceedings of the First European Symposium on Controlled Drug Delivery, Leidschendam, The Netherlands, 1990, pp. 15–16.
66. C. Grandfils, P. Flandroy, N. Nihant, S. Barbette, R. Jérome, and Ph. Teyssié, *J. Biomed. Mater. Res.*, *26*, 467 (1992).
67. P. J. Watts, M. C. Davies, and C. D. Melia, *Crit. Rev. Ther. Drug Carrier Syst.*, *7*, 235 (1990).
68. R. Bodmeier, K. H. Oh, and H. Chen, *Int. J. Pharm.*, *51*, 1 (1989).
69. D. Parienté, *Greffe hépatique et chimioembolisation de la tumeur VX2 de lapin.*, Mémoire de DEA d'imagerie médicale, Université Paris V, 1987.
70. E. C. Cadman and H. J. Durivage, in *Harrison's Principles of Internal Medicine* (J. D. Wilson and E. Braunwald, eds.), McGraw-Hill, New York, 1991, p. 1587.
71. B. A. Teicher, J. S. Lazo, and A. C. Sartorelli, *Cancer Res.*, *41*, 73 (1981).
72. K. A. Kennedy, S. Rockwell, and A. C. Sartorelli, *Cancer Res.*, *40*, 2356 (1980).
73. A. Sugitachi, Y. Takatsuka, and I. Sakamoto, *Jpn. J. Surg.*, *13*(5), 456 (1983).
74. I. Sakamoto and K. Takagi, Eur. Patent 86, 627 (1985).
75. A. V. D'Amico, A. C. Nelson, and R. K. Babayan, *J. Cancer Res. Clin. Oncol.*, *114*, 129 (1988).
76. J. M. Ruiz, B. Tissier, and J.-P. Benoît, *Int. J. Pharm.*, *49*, 69 (1989).
77. S. A. Evans, S. T. Olson, and J. D. Shore, *J. Biol. Chem.*, *257*, 3014 (1982).
78. C. Grandfils, V. Martin, N. Nihant, R. Jérome, Ph. Teyssié, and P. Flandroy, Immobilization of bovine α-thrombin by coacervation in polylactide-glycolide microspheres, Proceedings of the First European Symposium on Controlled Drug Delivery, Leidschendam, The Netherlands, 1990, pp. 97–100.
79. P. Lasjaunias and A. Berenstein, in *Surgical Neuroangiography*, Springer-Verlag, Berlin, 1987, pp. 59, 118, 383.
80. J. Suzuki and S. K. Komatsu, *Surg. Neurol.*, *16*, 438 (1981).
81. M. S. Donnell, G. A. Meyer, and W. L. Donegan, *J. Neurosurg.*, *50*, 499 (1979).
82. V. R. L. Rao, K. Ravimadalam, A. K. I. Gupta, S. Kumar, and S. Joseph, *Am. J. Neuroradiol.*, *10*, 135 (1989).
83. D. Fournier, K. Terbrugge, G. Rodesch, and P. Lasjaunias, *Neuroradiology*, *32*, 497 (1990).
84. M. Vandenabeele, *Malformations artérioveineuses cervicofaciales*, Mémoire de chirurgie plastique, UCL, Belgique, 1990.
85. G. W. Hastings, *Polymer*, *26*, 1331 (1985).
86. R. G. Quisling, J. P. Mickle, W. P. Ballinger, C. C. Carver, and B. Kaplan, *Am. J. Neuroradiol.*, *5*, 101 (1984).
87. A. Raziel, F. Puisieux, D. Terracol, P. Hoffman, G. Cave, J. J. Merland, and J. Bories, *Am. J. Roentgenol.*, *134*, 404 (1980).
88. M. D. Castaneda-Zuniga, R. Sanchez, and K. Amplatz, *Radiology*, *126*, 783 (1978).

89. E. R. Sigurdson, J. A. Ridge, and J. M. Daly, *Arch. Surg.*, *121*, 1277 (1986).
90. L. E. Lorelius, A. R. Benedetto, R. Blumhardt, H. V. Gaskill, J. L. Lancaster, and H. Stridbeck, *Invest. Radiol.*, *19*, 212 (1984).
91. W. Bechtel, K. C. Wright, S. Wallace, B. Mosier, D. Mosier, S. Mir, and S. Kudo, *Radiology*, *161*, 601 (1986).
92. W. D. Ensminger, J. W. Gyves, P. Stetson, and S. Walker-Andrews, *Cancer Res.*, *45*, 4464 (1985).
93. A. Roche, in *Radiologie et Echographie interventionnelle*, Axone, Montpellier, France, 1986, p. 495.
94. K. Nakakuma, S. Tashiro, and T. Hiraoka, *Cancer*, *52*, 2193 (1983).
95. D. Civalleri, G. Scopinaro, G. Simoni, F. Claudiani, M. Repetto, F. Decian, and U. Bonalumi, *Cancer*, *58*, 2151 (1986).
96. Y. Sasaki, S. Imaoka, H. Kasugai, M. Fujita, S. Kawamoto, S. Ishiguro, J. Kojima, O. Ishikawa, H. Ohigashi, H. Furukawa, H. Koyama, and T. Iwanaga, *Cancer*, *60*, 1194 (1987).
97. P. Courtheoux, J. L. Berthelin, Ph. Segol, P. Marchand, M. Gignoux, J. C. Verwaerde, J. Mani, and J. Theron, *J. Radiol.*, *68*(4), 247 (1987).
98. H. Kasugai, J. Kojima, M. Tatsuta, S. Okuda, Y. Sasaki, S. Imaoka, M. Fujita, and S. Ishiguro, *Gastroenterology*, *97*, 965 (1989).
99. N. Nagasue, G. Galizia, H. Kohno, Y. C. Chang, T. Hayashi, A. Yamanoi, T. Nakamura, and H. Yukaya, *Br. J. Surg.*, *76*, 871 (1989).
100. Y. Yumoto, K. Jinno, K. Tokuyama, T. Wada, H. Kobashi, T. Okamoto, H. Toki, S. Inatsuki, K. Hara, S. Moriwaki, T. Tuji, and H. Maeda, *Int. J. Hyperthermia*, 7(1), 7 (1991).
101. P. Lasjaunias, H. Alvarez, and I. Ilzuka, *Anatomic landmarks in pre-therapeutic evaluation*, Proceedings of the First Congress of the World Federation of International and Therapeutic Neuroradiology, Zurich, *Neuroradiology*, *33* (*suppl.*), p. S191 (1991).
102. H. Brem, S. M. Mahaley, N. A. Vick, H. L. Black, S. C. Schold, P. C. Burger, A. H. Friedman, I. S. Ciric, T. W. Eller, J. W. Cozzens, and J. N. Denealy, *J. Neurosurg.*, *74*, 441 (1991).
103. C. Fournier, B. Hecquet, Ph. Bouffard, M. Vert, A. Caty, M. O. Gilain, L. Vanseymortier, S. Merle, A. Krikorian, J. L. Lefebvre, A. Delubel, and L. Adonis, *Cancer Res.*, *51*, 5384 (1991).
104. J. A. Goldberg, T. Murray, D. J. Kerr, N. Willmott, R. G. Bessent, J. H. McKillop, and C. S. McArdle, *Br. J Cancer*, *63*, 308 (1991).
105. C. Kuroda, M. Sakurai, M. Monden, T. Marukawa, T. Hosoki, K. Tokunaga, K. Wakasa, J. Okamura, and T. Kozuka, *Cancer*, *67*, 81 (1991).
106. K. C. Lee, I. B. Koh, and I. J. Oh, *Arch. Pharm. Res.*, *9*(3), 145 (1986).
107. T. Kato, Y. Wada, H. Sasamuri, F. Aoki, N. Karasawa, T. Itoh, and Y. Igarashi, Experimental study on percutaneous transcatheter arterial chemoembolization under magnetic control employing newly developed ferrite microparticles coated with adriamycin, Proceedings of the *37th* International Congress of Radiology, Paris, 1989, p. 911.

13

PARTICULATE CARRIERS IN DERMAL AND TRANSDERMAL DRUG DELIVERY: MYTH OR REALITY?

Alain Rolland

GeneMedicine, Inc.
Houston, Texas

INTRODUCTION

Drug delivery to, or via, the skin presents both unique opportunities and obstacles due to the particular skin structure, physiology, and barrier properties. The skin, the largest organ of the body, may be viewed either as a natural protective barrier against penetration of toxic exogenous compounds, physical assaults, excessive loss of water and other essential compounds from the body, or as a promising portal of entry of drugs to the systemic circulation. In recent years, much work has been done to understand skin function and structure as well as the mechanisms governing the penetration of molecules through the skin. Aiming to improve dermatological formulations, which only a few years ago were mainly empirical, attempts are now being made to design new topical drug delivery systems, and even particulate carriers, that will enable a drug to reach the desired pharmacological site of action at a controlled rate and to have a sustained duration of action. Targeting of topically applied drugs to the different skin layers and appendages is becoming a major center of interest for many pharmaceutical groups working in dermatology.

Drug release from transdermal delivery systems is generally controlled by both membrane-moderated and matrix reservoir devices. However, in the transdermal drug delivery area, a new era is also emerging. Particulate carriers may find a place of choice among the plethora of devices that could be used for monitoring the percutaneous absorption of therapeutic agents. Dermal and transdermal therapies are in essence completely different means of treating either local or systemic disease via the same route of administration, i.e., the skin. Thus, although the mechanisms controlling the penetration of therapeutic molecules through the skin are identical in dermal and transdermal drug delivery, in the former case attempts to avoid the systemic absorption of the drugs will be made in order to reduce the toxic side-effects, whilst in the latter situation a controlled input of the drug in the blood circulation is aimed (Table 1). One other major difference between dermal and transdermal drug delivery is that dermatological formulations are to be applied on diseased skins, with consequently modified functions and barrier properties, whereas transdermal therapeutic systems are designed only for application on normal skin. Topical formulations for local or systemic therapy will consequently be of different designs, although common requirements still hold (Table 1).

Thus, dermal and transdermal drug delivery systems must be non-irritant, non-allergenic, stable, dermocosmetically acceptable, easy to prepare, of low cost, and they have to control drug release and sometimes improve drug bioavailability.

Table 1 Differences and Common Requirements in Dermal and Transdermal Drug Delivery

Dermal Drug Delivery	Transdermal Drug Delivery
Differences	
Diseased skin	Normal skin
Local drug delivery and effects	Systemic drug delivery and effects
Avoidance/limitation of systemic drug absorption	Controlled systemic drug absorption
Multiple sites of application	Single site of application
Classical or advanced drug delivery systems	Specific drug delivery systems (patches)
High local availability of active drug	High systemic bioavailability
Common Requirements	
Dermocosmetic acceptability	
Nonirritant formulations	
Nonallergenic formulations	
Controlled release of the drug	
Good physicochemical stability of the drug in formulation	
Improvement of drug bioavailability (e.g., penetration enhancers)	

After a brief description of the skin anatomy and physiology, this chapter will present the potential applications of particulate carriers as novel transdermal devices and as new site-specific drug delivery systems in dermatology for targeting drugs to defined skin layers and appendages. Passive targeting of topically applied drugs to hair follicles and sebaceous glands using polymeric microspheres will be illustrated in the last section of the chapter by our own experimental studies realized at CIRD GALDERMA.

SKIN PHYSIOLOGY AND BARRIER FUNCTION

The Structure of the Skin

The skin is the most accessible tissue of the body, with a surface area of 1.8 m^2, for an average 70-kg human, having per cm^2 about 10 follicles, 12 nerves, 15 sebaceous glands, 100 sweat glands, 3 blood vessels, and 3×10^6 cells (1).

The skin is a multilayered organ composed of three main layers (Figure 1): the outer, avascular, ectodermal epithelium (the epidermis); the inner, underlying mesodermal connective tissue (the dermis or corium); and below the dermis, the subcutaneous fat (subcutis). "Indeed, the *raison d'être* of the

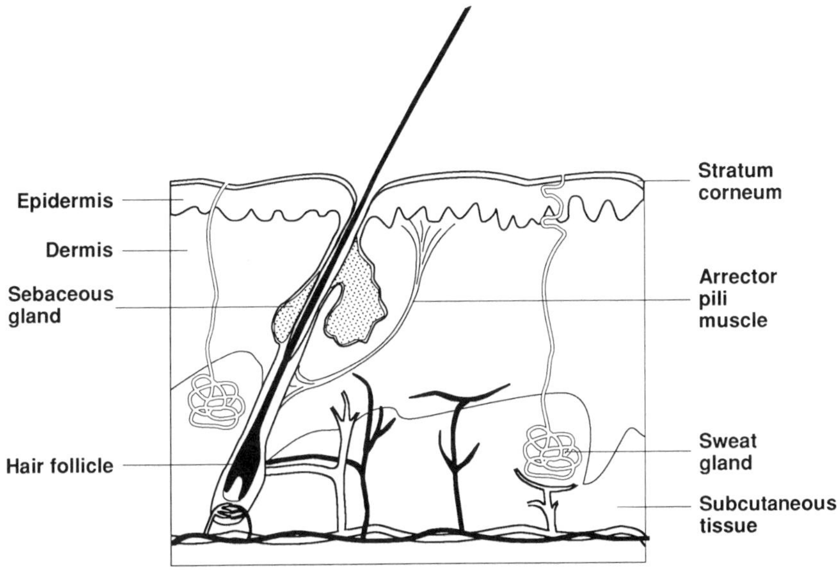

Figure 1 Schematic cross-section of full-thickness skin.

epidermis is to make the stratum corneum" (2). The stratum corneum or horny layer consists of 15–20 layers of flattened, keratinized, dead cells and it may be represented by a brick-and-mortar model (Figure 2). The "bricks" are essentially composed of anucleate corneocytes containing keratin filaments surrounded by an amorphous matrix of sulfur-rich proteins (3). The corneocyte envelope consists of a thickened inner portion of crosslinked proteins, predominantly involucrin (4,5). Involucrin becomes crosslinked through γ-glutamyl-ε-lysyl isopeptide bonds introduced by the action of the keratinocyte transglutaminase, which has apparently a calcium-dependent activity (6).

In addition to involucrin, at least six other soluble and membrane-associated proteins are present in the crosslinked protein envelope (7). The material lying outside the protein envelope is different from typical plasma membranes consisting mainly of phospholipids, cholesterol, and associated proteins. The lipid portion of the corneocyte envelope is essentially composed of ester-linked ω-hydroxyacylsphingosines (8).

The "mortar" fills the intercellular spaces of the stratum corneum and mainly consists of multiple lipid bilayers arranged in broad sheets (9,10). Unlike biological membranes, lipids of the intercellular lamellae do not contain phospholipids (9,11) but mainly ceramides, cholesterol, fatty acids, and cholesteryl esters. These intercellular membranous sheets are known to originate from the lamellar bodies found in the keratinocytes of the granular layer of the epidermis (12). The cohesive role of the corneocyte lipid envelope in the formation of lamellar sheets in the horny layer has been investigated in vitro (13). The intercorneocyte lipid bilayers, which contain predominantly

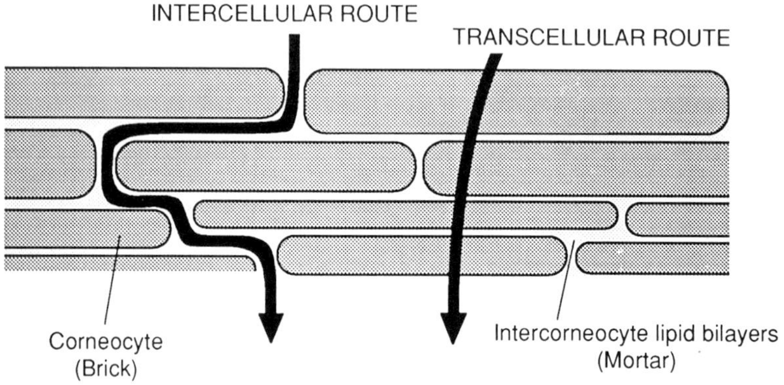

Figure 2 Schematic representation of the stratum corneum by the "brick-and-mortar" model and suggested routes of drug penetration.

saturated lipids, form an efficient barrier to the penetration of water and hydrophilic molecules.

The stratum corneum results from the continuous differentiation of keratinocytes from the basal layer. The morphologically different layers of the epidermis represent different stages in the terminal differentiation process. Thus, the epidermis, a stratified, orthokeratotic, squamous epithelium, is composed of cells that continuously change from the basal layer (stratum basale) to the stratum spinosum, stratum granulosum, and finally to the dead stratum corneum (14).

Besides keratinocytes, three other cell types are present in the epidermis: melanocytes, Langerhans cells, and Merkel cells.

The dermoepidermal junction lies below the basal cell layer and contributes to the dermal-epidermal adherence, the mechanical support of the epidermis, and the control of cells and large-molecule passage across the junction (1).

The dermis essentially consists of a network of collagen and elastin fibers embedded in an amorphous mucopolysaccharide matrix. Blood vessels, lymphatics, and nerves (Figure 1) are present in this connective tissue, which is also penetrated by skin appendages such as hair follicles, eccrine and apocrine sweat glands. The blood supply in the dermis is particularly rich and of major importance since it will readily remove molecules that have penetrated the stratum corneum and viable epidermis.

However, the normal structure of the skin may be altered in a number of diseases. In psoriasis, for instance, the epidermis becomes hyperproliferative and the total turnover time from the basal layer to shedding, which averages 28 days in healthy skin, shortens to only 4 days (1). In ichthyoses, a group of genetic diseases of keratinization, abnormal retention of cells within the stratum corneum occurs (15).

Routes of Drug Penetration Through the Skin

Owing to its multilayered structure, the skin represents a complex and relatively efficient barrier to the penetration of drugs. The stratum corneum provides a rate-limiting step in the process of percutaneous absorption. This hydrophobic membrane is thus the major obstacle to the epidermal transfer of water-soluble drugs, as illustrated by the fact that their percutaneous absorption may be 1000 times more rapid when the stratum corneum is absent or damaged than when it is present and intact (16). However, the stratum corneum is not a totally impermeable barrier hindering hydrophilic molecules from reaching the viable tissues since polar channels exist within the intercorneocyte spaces, and more precisely in between the lipid bilayers. There are indeed two possible routes of drug penetration through the stratum corneum: the transcellular and the intercellular pathway. The relative importance of one of these penetration pathways will be mainly dependent on the physicochemical characteristics of the drug molecules (17), particularly the partition and diffusion coefficients into the protein or lipid regions (18) (Figure 2).

The main transport pathway across the stratum corneum involves the tortuous, extracellular lipid domain; for very polar compounds, a "polar" route or pore pathway has been proposed (19,20). However, recent evidence suggests that aqueous channels across the stratum corneum are not necessary to explain skin permeability (21).

If a drug penetrates and does not form a reservoir within the stratum corneum, the next stage will be its partitioning, relative to its physicochemical properties, from a primarily lipid-rich domain into the viable epidermis, a predominantly aqueous-rich environment. Drugs with a suitable octanol–water partition coefficient, i.e., a reasonable solubility in both lipid and aqueous phases, will partition relatively well into the epidermis after diffusion into the stratum corneum. However, too hydrophilic or lipophilic compounds will not easily be partitioned into the stratum corneum or the underlying viable tissues, respectively.

Drugs that reach the epidermis will be able to diffuse further into the dermis and consequently enter the bloodstream. Besides, in either local or transdermal therapy, drug metabolism within the skin is of major importance, since some drugs may be inactivated by biotransformation. Conversely, prodrugs can be designed for enhancing penetration into the skin and for delivering to the viable tissues a pharmacologically active molecule.

In addition to the transepidermal pathway, a "shunt" pathway (passage through the skin appendages) may be of significance in several cases (22–24).

The transfollicular pathway in particular was recently shown to play an important role in the percutaneous absorption of some drugs such as hydrocortisone, niflumic acid, and caffeine (25). A significant appendageal diffusion may occur in spite of the small area occupied by the pilosebaceous units at the surface of the skin.

The skin appears to be a very efficient barrier to the percutaneous absorption of exogenous compounds. Drugs penetrate the stratum corneum and eventually diffuse across the viable underlying tissues, according to their physicochemical properties. Until recently, topical formulations were essentially only able to control the release of a drug but not its penetration rate or residence time in the different layers. However, over the past few years, with the advent of particulate carriers for site-specific drug delivery via parenteral administration (26–30), some research groups have focused on their potential topical applications in dermal or even transdermal therapy. Promising results show that particulate carriers may improve the penetration of some drugs, by targeting either to the stratum corneum or to the hair follicles, and control the release of drugs.

PARTICULATE CARRIERS IN TRANSDERMAL DRUG DELIVERY

Transdermal drug delivery makes use of the skin as a portal of entry of pharmacologically active molecules. Over the past 10 years, several transdermal therapeutic systems (TTS) have been marketed, which contain, for instance, nitroglycerin, scopolamine, clonidine, estradiol, and nicotine (19,31–33).

TTS present a number of advantages over classical drug delivery systems administered via other routes (oral, parenteral):

1. Ease of self-administration
2. Good patient compliance
3. Avoidance of variations in gastrointestinal absorption
4. Bypass of the hepatic first-pass metabolism
5. Production of sustained and constant plasma concentrations of drugs
6. Reduction in repeat dosing intervals
7. Reduction of potential adverse side effects
8. Removal of TTS provokes an immediate decrease of drug plasma levels
9. Substitute for oral or parenteral administration in certain clinical situations (pediatrics, geriatrics, nausea, etc.)
10. Adaptability to drugs with a short half-life
11. Suitability for drugs producing a therapeutic response at very low plasma concentrations

An ideal TTS should provide a constant plasma level of drugs so as to ensure a consistent pharmacodynamic response. The mechanisms by which drug input through the skin may be controlled are twofold:

The skin is the limiting step of percutaneous absorption by acting as an in situ rate-controlling membrane

TTS control the drug release at a prescribed rate, slower than the penetration flux through the skin

The latter situation is indeed the method of choice since the therapeutic efficacy will then be dependent on the rate of drug release and not on the skin permeability of the patient. It is therefore possible to avoid inter- and intraindividual variations due to the variability of skin permeability from one patient to another or from one site of application to another.

Transdermal delivery of drugs has so far been achieved by both membrane-moderated devices or matrix reservoir systems (19,31,34). Most of these systems can deliver the drug to the skin for prolonged periods with zero-order kinetics. Drug candidates for transdermal delivery, as represented in Figure 3, following controlled release from the TTS will have to partition into the lipophilic stratum corneum, then into the more hydrophilic viable epidermis and dermis where it will gain access to the systemic circulation (17).

Recently, particulate carriers such as liposomes have been highlighted for topical delivery of cosmetics or dermatological products (35-38). In parallel, a few experimental results have been published regarding the potential application of liposomes as part of a new generation of TTS. In that particular situation, liposomes were not designed for penetrating, say, the stratum corneum and constituting a depot controlling drug release in situ, but rather as a microreservoir at the surface of the skin controlling drug release and input via the skin. Wester et al. (39) described a liposomal transdermal system that allowed a certain degree of constant input of estradiol to the rhesus monkey in vivo over a 3-day period after application of the prototypal device.

Toulon et al. (40) also showed that encapsulation of estradiol into liposomes resulted in a delayed and reduced in vitro drug percutaneous penetration through hairless rat skin as compared to a commercial hydroalcoholic gel. Moreover, estradiol and liposomal phospholipids penetrated the skin at different rates, suggesting that estradiol did not cross the skin associated with either intact liposomes or phospholipids.

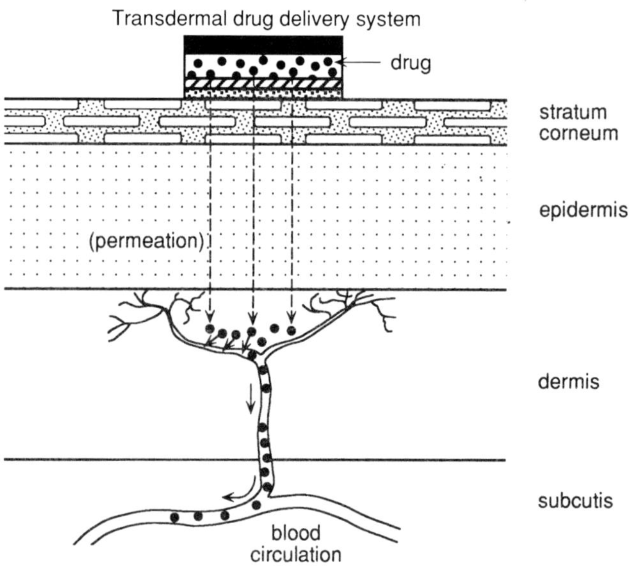

Figure 3 Diagram of transdermal drug delivery and percutaneous absorption.

Transdermal delivery of estradiol from nonionic surfactant vesicles through human stratum corneum was investigated in vitro by Hofland et al. (41). They demonstrated that, as compared to a saturated estradiol solution, incorporation of the drug into large multilamellar or small unilamellar vesicles dramatically increased the permeability of estradiol through the stratum corneum by a factor of 260 and 100, respectively.

In another study, Knepp et al. (42) investigated the potential interest of liposomes embedded in an agarose matrix as a microreservoir for controlling the release of progesterone (PG). Three different liposomal formulations containing PG were compared in vitro to free progesterone in the agarose matrix and PG associated with Intralipid. The four lipid formulations released significantly less PG than the free PG system in a pH 4.5 buffer at 37°C over 24 hr. Thus, the free PG TTS released about 90% of the dose in 24 hr as compared to 20–25% for the liposomal systems and about 50% for the Intralipid emulsion. Interestingly, these four formulations produced a zero-order PG release kinetics for at least part of the experimental time (Figure 4). Liposomal transdermal systems were able to control the release of PG over a prolonged period at a constant rate. The authors concluded that PG release was not only controlled by a diffusion mechanism through the aqueous gel network, but probably mainly by the interfacial partitioning of the drug from the lipid bilayers into the surrounding aqueous medium.

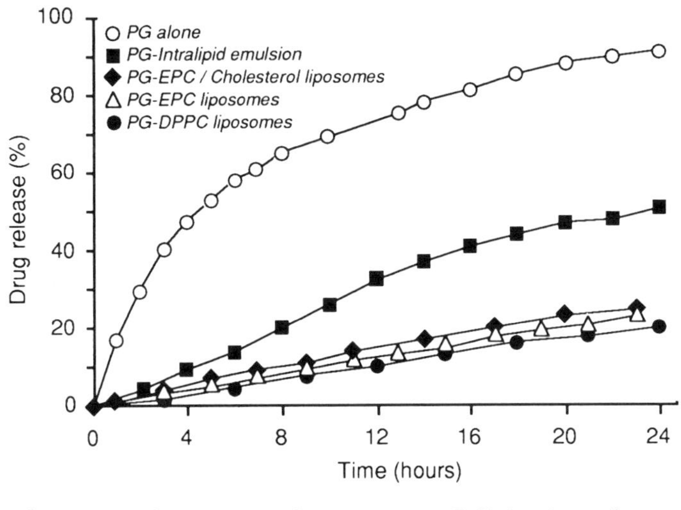

Figure 4 Comparison of progesterone (PG) in vitro release profiles from five formulations. (Adapted and redrawn with permission from Ref. 42.)

Percutaneous absorption of PG through hairless mouse skin in vitro was also investigated for the different transdermal formulations. For the free PG patch, drug percutaneous absorption was skin-controlled and not device-controlled, the skin being the rate-limiting membrane and the agarose gel a reservoir. With the liposomal formulations, the rate of PG penetration through the skin was considerably reduced. With the device containing multilamellar dipalmitoylphosphatidylcholine (DPPC) liposome-encapsulated PG, the rate of drug transport across the skin was 0.03% per hour as compared to 0.45% per hour for the free PG patch. Since the liposomes were embedded in an aqueous agarose gel, active transfer of the steroid into the intercorneocyte lipid bilayers of the stratum corneum via the liposomes is highly unlikely.

The transport of PG encapsulated into the liposomes across the full-thickness skin is even more improbable, and the absence of liposome passage for 48 hr through the skin samples was confirmed using ^{14}C-labeled DPPC liposomes. Therefore, liposomes act as a microreservoir at the surface of the skin, the rate-limiting process being PG release from the liposomes into the surrounding agarose gel. Liposomes or other particulate carriers might be of interest for controlling drug release from TTS, since they present a unique amphipathic structure allowing the encapsulation of lipophilic and/or hydrophilic drugs at high concentrations.

In a further work, Knepp et al. (43) examined the influence of the lipid composition of the liposomes on the in vitro release and percutaneous absorption of PG. Preparation of the multilamellar liposomes involved phospholipids bearing different acyl chain components, cis-unsaturated or fully saturated, which resulted in liposomes with different phase transition temperature (T_c) (Table 2). Thus, DPPC and dimyristoylphosphatidylcholine (DMPC) have fully saturated acyl chains and transition temperature higher than 20°C whereas egg phosphatidylcholine (EPC) and dioleylphosphatidylcholine (DOPC) contain cis-unsaturated acyl chains and have T_c below 0°C. In spite of the differences in the physicochemical properties of the phospholipids, there is no significant modification of the percentage of drug release from the various liposomal formulations (Table 2). Regardless of the phospholipid used, drug release from the liposomal patches is far slower than from the free PG TTS (Table 2). PG release profiles were also proved to be independent of the state of the lipids in the liposomal bilayers (gel or liquid crystalline state) (43).

In vitro percutaneous absorption studies showed that, referring to free PG patch, DPPC and DMPC liposome-encapsulated PG produced a greater reduction of the transdermal delivery rates of PG across hairless mouse skin than EPC or DOPC liposome-containing PG (Figure 5).

It is noteworthy that the presence of unsaturated fatty acids in EPC and DOPC led to an improvement of PG penetration as compared to DPPC and DMPC.

Table 2 Cumulative Amounts of Progesterone (PG), Expressed as Mean Cumulative %[± SD] of Initially Encapsulated Dose Released in 24 hr from 7 Delivery Systems into Buffer at 35°C

Formulation	T_c(°C)[a]	CCL/DB[b]	% Dose released[c]
Free PG in agarose gel	—	—	90.4 ± 5.5
PG-EPC liposomes	−5	Mixed	24.8 ± 3.8
PG-DMPC liposomes	23	14/0	24.6 ± 2.4
PG-DOPC liposomes	−22	18/1	15.9 ± 5.1
PG-DPPC liposomes	41	16/0	21.2 ± 1.1
PG − 1% OA-DPPC liposomes	41	16/0	23.1 ± 2.5
PG − 1% SA-DPPC liposomes	41	16/0	22.0 ± 2.2

[a]Phase transition temperature of phospholipid.
[b]Carbon chain length (CCL)/number of double bonds (DB).
[c]Mean ± standard deviation of four to six assays.

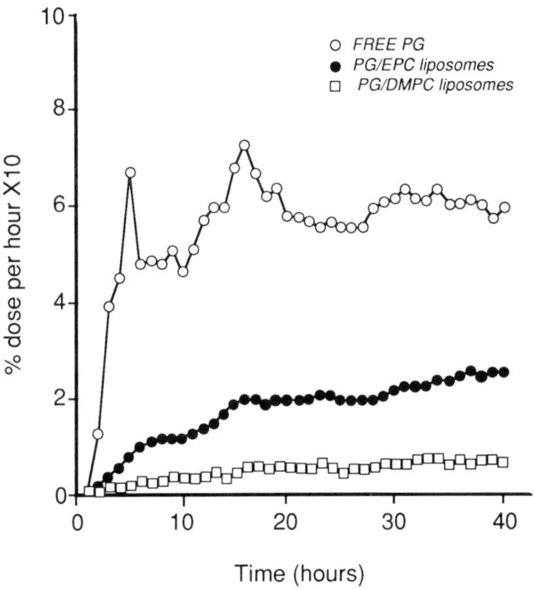

Figure 5 Transdermal delivery rate (mean ± SD; n = 6) of progesterone (PG) across hairless mouse skin in vitro from three different formulations. (Redrawn with permission from Ref. 43.)

In order to verify the influence of the unsaturated fatty acids on PG penetration, a cis-unsaturated and a fully saturated fatty acid, oleic acid (OA) and stearic acid (SA), respectively, were incorporated in multilamellar DPPC liposome-containing PG.

As observed in Table 2, drug release was not affected by the introduction of the fatty acids in the liposomes. However, although SA did not modify PG penetration rate through full-thickness hairless mouse skin to a great extent, OA dramatically enhanced drug transdermal flux to a level similar to that achieved from DPPC liposomes. cis-Unsaturated fatty acids very likely act as penetration enhancers, and consequently the PG penetration rates obtained from EPC and DOPC liposomal patches can be explained.

These studies tend to demonstrate that liposomes can be considered as very promising microreservoirs within transdermal delivery systems, allowing a controlled release of the drug and a modulation of drug penetration across the skin. In addition, effective codelivery of drug and penetration enhancer with liposomal TTS appears to be feasible.

In a recent work, Artmann et al. (44) showed that in vivo application of an aqueous suspension of antibody-loaded liposomes (oligolamellar liposomes from NAT 106 [Nattermann Phospholipid GmbH, Germany]; mean diameter, 150 nm; 100 mg phospholipid/ml dispersion) to piglet skin resulted in a homogeneous localization of the antibodies throughout the skin and even in the subcutis. Antibodies applied in buffer under the same conditions were not detected.

The nonpenetration into the skin of the antibodies, corresponding to proteins of high molecular weight (20,000–50,000), is not surprising. However, localization of antibodies in both the skin and subcutis 20 min after topical application of antibody-loaded liposomes in vivo is a surprising finding. Contamination of skin samples during the cryosection process may perhaps be questionable.

In further experiments, Artmann et al. (45) investigated quantitatively in vivo the influence of liposomes on the percutaneous absorption of labeled compounds such as [^{99m}Tc]sodium pertechnetate and heparin [^{35}S]sulfonate. Concentrations of either sodium pertechnetate or heparin remained significantly higher in skin, blood, and urine for 5 hr following topical application of liposome-entrapped substances to female piglets than after administration of free compounds. These results suggested an enhanced percutaneous absorption of sodium pertechnetate and heparin via encapsulation into the liposomes that were oligolamellar, negatively charged, with a fluid bilayer membrane and a mean size around 150 nm.

PARTICULATE CARRIERS IN DERMAL DRUG DELIVERY

In dermatology, access of the drug to the diseased tissue may be readily achieved by topical application. However, there is a need for novel drug delivery systems for the therapy of skin disorders that can precisely target the drug to the site of the disease. In the past decades, particulate carriers have been extensively investigated for delivering drugs to specific skin sites. Topical particulate carriers (liposomes, microspheres) may increase and/or regulate drug transport into the skin and consequently reduce variability in drug bioavailability from one patient to another. They can also reduce toxic side effects that may arise from undesirable high systemic drug absorption and target the drugs to definite skin compartments. Besides, particulate carriers are generally nontoxic, nonimmunogenic, nonirritant, biodegradable, and they are able to incorporate a wide range of hydrophilic and hydrophobic drugs.

Liposomes as a Topical Drug Delivery System

Experimental Studies

Liposomes, first described by Bangham et al. (46), are lyotropic liquid crystals composed mainly of amphiphilic bilayers (30). Liposomes are defined as vesicles composed of one or more lipid bilayers surrounding an aqueous space. They may be classified on the basis of two parameters:

Their mean diameter (size)
Their number of lipid bilayers

Thus, liposomes with more than one bilayer are defined as being either multilamellar or oligolamellar; they are generally large and therefore are called multilamellar large vesicles (MLV). Unilamellar vesicles, owing to their method of preparation, may be large or small and consequently are denoted large unilamellar vesicles (LUV) or small unilamellar vesicles (SUV).

Liposome classification is also sometimes based on the method of vesicle preparation, leading to abbreviations such as REV (reverse phase evaporation vesicles), SPLV (stable plurilamellar vesicles produced by hydration from organic solvents), FATMLV (frozen and thawed multilamellar vesicles), and VET (vesicles obtained by extrusion techniques). Different parameters are generally used for liposomes characterization, such as number of lipid bilayers, vesicle size, trapped volume, and distribution of solute between lamellae of MLV (30,47–49).

Although liposomes have been investigated via different routes of administration (ocular, intravenous, intramuscular, intraarticular, etc.) as new drug delivery systems for drug targeting, until recently topical administration of liposomal formulations had received little attention. Mezei and Gulasekharam first described the applicability of liposomes as drug carriers for the topical administration of triamcinolone acetonide (50,51). They applied ^{14}C-labeled triamcinolone acetonide to rabbits twice daily for 5 days in a liposomal formulation (MLV) and in a conventional ointment dosage form. As observed in Figure 6, drug concentration was more than four times higher in the epidermis and dermis with liposomes than with the ointment form. However, in spite of higher concentrations in the skin with the liposomal formulation, percutaneous absorption was greatly reduced as indicated by blood levels and urinary excretion data (Figure 6). For a therapeutic efficacy equivalent to that of conventional triamcinolone acetonide topical formulations, it would theoretically be possible to reduce the dose of drug applied in the liposomal form and thus get a much better therapeutic index. An increased absorption of triamcinolone acetonide encapsulated in liposomes was also observed after application on the forearm skin of human beings (52).

Particulate carriers, such as liposomes, might be useful for increasing the concentration of a drug in the skin, and consequently its local activity, while reducing the percutaneous absorption of the drug and its toxic side effects. Lasch and Wohlrab (53) compared the in vitro penetration of hydrocortisone through human skin 30 and 300 min after application of the drug incorporated either in egg lecithin liposomes or in an emulsion.

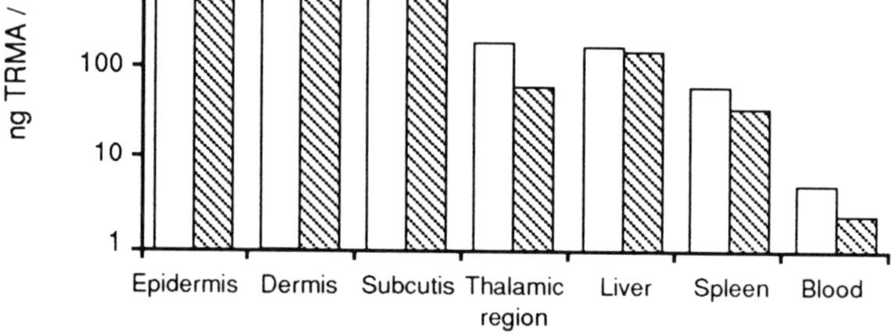

Figure 6 Triamcinolone (TRMA) concentrations after topical application of either ointment dosage form (control) or liposomal lotion twice daily for 5 days to rabbits (n = 8). (Blood concentrations are expressed in ng TRMA/ml blood.) (Redrawn with permission from Ref. 50.)

Although identical hydrocortisone concentrations were found in the stratum corneum, in the epidermis, and in the dermis, drug concentrations were dramatically enhanced with the liposomal formulations (Table 3). Hydrocortisone concentrations were thus 8- and 4-fold increased after 30 min and 14- and 9-fold increased after 300 min with the liposomal suspension as compared to the water/oil emulsion. In addition, with hydrocortisone-loaded liposomes a very small concentration gradient of the drug from the surface of the skin to deeper layers was observed, whereas the conventional emulsion led to a pronounced concentration gradient (54). In this publication, the authors assumed that the enhanced penetration of hydrocortisone was due to a loss of integrity of the liposomes, with the egg lecithin acting as a skin penetration enhancer. Jacobs et al. (55,56) examined the effect of pretreating the skin with phosphatidylcholine as a liposomal suspension on the bioavailability of corticosteroids. Using the human skin blanching assay, with a non-occluded multiple dosage regimen, they showed that at the end of the second week of corticosteroid application the blanching response on the arms pretreated with phosphatidylcholine liposomes was significantly higher than on the control-treated arms. It was suggested that phosphatidylcholine may either increase the total lipid content of the skin, thereby modifying the bioavailability of corticosteroid, or provide a thin film in intimate epidermal contact. Such a film may promote hydration of the stratum corneum and provide a fluid layer into which corticosteroids initially partition before a further, more controlled release to the underlying tissue.

Using [^{14}C]phosphatidylcholine for preparing hydrocortisone-loaded liposomes, it was shown that after in vitro application on human skin most of

Table 3 Penetration of 1% Hydrocortisone (HC) in w/o Emulsion and of 1% HC Incorporated in Liposome Suspension

Skin layer	HC concentrations (nM) after penetration times of 30 and 300 min (mean ± SD): 30 min	300 min
HC in w/o emulsion:		
Stratum corneum	15,034.40 ± 6711.61	16,137.93 ± 3984.05
Epidermis	13.79 ± 2.61	27.59 ± 4.86
Dermis	6.03 ± 1.91	11.21 ± 2.01
HC incorporated in liposome suspension:		
Stratum corneum	16,738.54 ± 8521.36	18,170.08 ± 2671.77
Epidermis	112.57 ± 92.96	103.45 ± 20.53
Dermis	83.91 ± 44.76	100.13 ± 75.26

the phospholipid was located in the horny layer (57). A marked concentration gradient existed in the stratum corneum whereas in the epidermis and dermis it was rather small. Phosphatidylcholine also penetrated rapidly into the different skin layers, the concentrations being very similar after 30 and 300 min. However, whether or not phosphatidylcholine entered the skin as intact liposomes was not elucidated. Whatever the mechanism by which phosphatidylcholine liposomes enhanced hydrocortisone penetration into human skin in vitro, liposomal hydrocortisone also showed in vivo a lower rate of percutaneous absorption as compared to a conventional ointment (57). In guinea pigs, a decreased serum concentration and urinary excretion of hydrocortisone was demonstrated after topical application of liposomal formulation (58) (Figure 7).

The selective drug delivering potential of liposomes for cutaneous application was further evidenced by Vermorken et al. (59) in a study comparing the percutaneous absorption of 5α-dihydrotestosterone encapsulated in liposomes or dissolved in acetone. The female hamster flank organs, which are sebaceous structures located one on each flank and androgen-dependent, were used as a model system. Encapsulation of 5α-dihydrotestosterone in 1-μm liposomes decreased the systemic absorption of the drug along with its biological effects, i.e., it did not stimulate flank organ growth. However, since the drug was not quantified in the different skin layers, and in particular

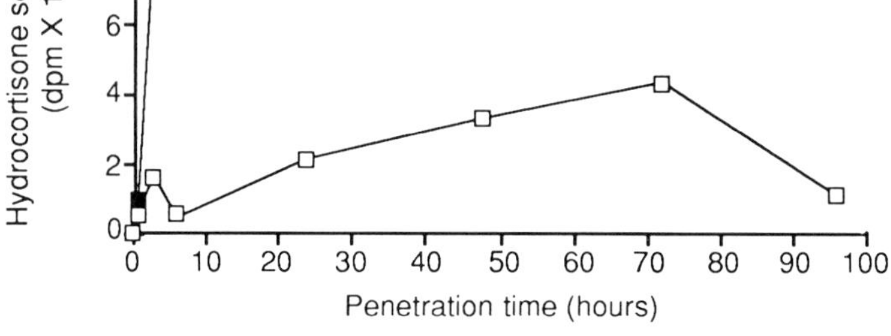

Figure 7 Serum concentrations of hydrocortisone (HC) after topical application on guinea pig ear (in dpm × 10^4/ml) of either HC ointment or HC liposomal formulation. (Redrawn with permission from Ref. 58.)

at its site of action, the sebaceous glands, it is difficult to conclude that the reduced biological effect is due to a decreased percutaneous absorption of the drug.

The in vitro percutaneous absorption through hairless mouse skin of stanozolol, an anabolic steroid that has been used orally to produce fibrinolytic enhancement in patients with lipodermatosclerosis, was evaluated from different topical formulations, e.g., solutions, oil/water creams, and liposomes (60). These topical dosage forms were designed for increasing local activity of stanozolol while reducing the systemic adverse reactions, as compared to oral therapy. Stanozolol permeability coefficients were significantly higher for the liposomal formulation than for the alcoholic and aqueous surfactant solutions, which were in turn significantly higher than for the creams.

Michniak (61) investigated the in vitro penetration of two steroids, betamethasone-17-valerate and hydrocortisone-21-acetate, through excised full-thickness male nude mouse skin. With both steroids, the flux and skin drug content were significantly increased when the drugs were encapsulated in SUV (lecithin/phosphatidic acid; 80–200 nm) as compared to various controls.

Gehring et al. (62) evaluated the effect of incorporating betamethasone valerate into different liposome formulations on blood circulation in the dermis. Although the three empty liposomal formulations (Nat 106, Nat 50, Nat 89, Nattermann Phospholipid GmbH, Germany) caused a decrease in blood circulation in the dermis, encapsulation of betamethasone valerate in the three types of liposome rapidly increased the blood circulation after topical application to the forearm of human volunteers. In these assays, no obvious paling of the skin was detected in any case and the authors questioned the relationship between an increase in the blood circulation in the middle corium and a blanching effect of corticosteroids.

Recently, the efficacy of a liposomal formulation of betamethasone has been compared to that of a commercial propylene glycol gel in a double-blind, randomized trial after application without occlusion, once daily for 14 days, to affected skin areas of patients with either atopic eczema or psoriasis vulgaris (63,64).

In eczema, betamethasone-loaded egg lecithin liposomes (SUV; mean diameter 50 nm) incorporated in a polyacrylate gel tended to reduce erythema and scaling more than a conventional gel and more rapidly. However, the commercial gel was more effective than the liposomal formulation in the treatment of psoriasis. Hence, encapsulation of betamethasone in liposomes seems to increase the anti-inflammatory but not the antiproliferative effect of the drug. Whether or not site-specific delivery of betamethasone to inflammatory cells via liposomes was achieved was not demonstrated in that work. Since betamethasone was more active as an anti-inflammatory than an antiproliferative agent when encapsulated into SUV, liposomal therapy of skin diseases may improve the benefit–risk ratio of topical glucocorticoids.

In another study, the topical bioavailability of three corticosteroid esters—hydrocortisone-21-acetate, hydrocortisone-21-hexanoate, and hydrocortisone-17-butyrate—from either a conventional cream or a phosphatidylcholine liposome suspension was evaluated in a nonoccluded topical vasoconstrictor assay in 10 volunteers (65). Blanching profiles for any corticosteroid, whether formulated in the cream or in liposomes, were very similar throughout the experiments, the daily peak blanching being observed 6 hr after steroid application. However, the cream formulations almost always gave a better vasoconstriction than the steroid-loaded liposomes. These results differ from that obtained in a previous work (55) showing the penetration enhancer-like effect of empty phosphatidylcholine liposomes on the bioavailability of corticosteroids. The authors tentatively explained the difference by the fact that the thermodynamic activity of the three steroids was different in the cream and in the liposomal suspension. They also argued that the amount of applied phospholipid was probably insufficient to supplement the lipid content of the skin and subsequently increase the steroid partitioning into that lipophilic compartment.

The treatment of four women with severe hirsutism, in both idiopathic hirsutism and polycystic ovary syndrome, by liposome-encapsulated PG was reported by Rowe et al. (66). Topically applied liposomal PG reduced the rate of hair growth in idiopathic hirsutism, this anti-androgen activity being presumably related to the inhibition of cutaneous 5α-reductase activity.

Butylparaben, one of the most effective preservatives used in topical formulations, was encapsulated in negatively charged MLV (0.1–0.5 μm in diameter) composed of egg phosphatidylcholine, cholesterol, and dicetylphosphate (67). More than 90% of butylparaben was encapsulated in the liposomes. Suspensions of liposomes-encapsulated butylparaben showed antimicrobial activity against eight kinds of microorganisms (*Escherichia coli*, *Staphylococcus aureus*, *Bacillus subtilis*, *Pseudomonas aeruginosa*, *Aerobacter aerogenes*, *Aspergillus niger*, *Penicillium citrinum*, *Candida albicans*) whereas empty liposomes were inefficient. However, the antimicrobial activity of the liposomal suspension was directly proportional to the free concentration of butylparaben in the preparation. Butylparaben was apparently encapsulated in liposomes near the surface of lipid bilayers owing to a lipid/water partition equilibrium and it seemed to have no preservative effect when incorporated into the lipid bilayers. Percutaneous penetration of liposomes containing butylparaben was studied in vivo in guinea pigs by autoradiography. ^{14}C[Butylparaben] penetrated into the body, the radioactivity being mainly detected in the gall bladder, small and large intestines, urinary bladder, and feces. ^{14}C[DPPC], used as a marker of the liposomal lipid bilayers, remained at the skin surface and did not interfere with butylparaben percutaneous absorption which was similar to that observed after topical application of butylparaben in ointment.

In another study, the percutaneous penetration of butylparaben from a liposome suspension was investigated in vitro through guinea pig skin using a flow-through diffusion cell (68). The phospholipids did not penetrate the skin to a significant extent, while the free part of butylparaben which was not encapsulated into the liposomes did enter the skin. Copenetration of butylparaben and liposomes through the skin was therefore shown to be unrealistic. Nevertheless, liposomes contributed in very small part to the penetration of butylparaben by acting as a drug carrier at the skin surface, maybe even into the stratum corneum. In addition, no alteration of the barrier function of the stratum corneum nor denaturation of skin proteins was observed as a result of the in vitro topical application of liposomes.

Encapsulation of methotrexate in liposomes was shown to greatly reduce the percutaneous absorption of the drug, as compared to the free compound applied topically (69). Moreover, methotrexate concentration in psoriatic skin was fourfold higher with the liposomal form as compared to the free form, suggesting an epidermal localization of liposomes with an in situ sustained release of the drug.

In order to improve the dermal penetration of local anesthetic agents, Foldvari et al. (70,71) entrapped tetracaine in liposomes with a high encapsulation efficiency (60–90%). The topical anesthetic effects of the liposomal tetracaine preparation was then compared to a cream formulation using the pinprick technique in adult volunteers. The liposome-encapsulated tetracaine was more effective than the cream in producing topical anesthesia, with the liposomal formulation providing long-lasting effects even at low drug concentrations.

Topical application of peptide or proteins encapsulated in liposomes was shown to be very promising in the suppression of skin lesions induced, for instance, by viruses (72).

Thus, liposomes containing interferon were prepared by the dehydration/rehydration method from egg lecithin, cholesterol, and phosphatidylserine. Twenty-four hours after inoculation of strain 148 herpes simplex virus type 1 to the back of hairless guinea pigs, free or encapsulated interferon was applied topically one to three times per day for 5 days. The efficacy of interferon-loaded liposomes was higher than that of the free drug.

Liposomal antifungal agents were developed and evaluated by Mezei (73). Two compounds, econazole base and econazole nitrate, were encapsulated in different types of MLV and their in vivo distribution was analyzed after topical application. As observed in Figure 8, referring to controls (oint-

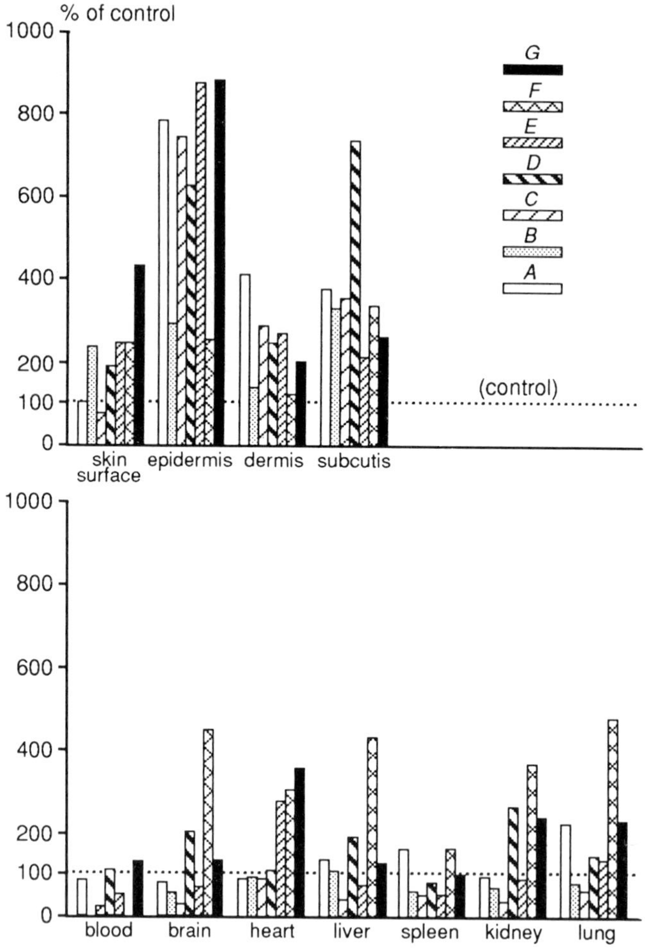

Figure 8 Comparison of drug disposition after topical application of either liposome-encapsulated econazole (A, B, C, D, E, F, G) or free econazole (control) to guinea pigs. The appropriate control (ointment, cream, suspension) for each liposomal formulation corresponds to 100%. (Adapted with permission from Ref. 73.)

ment, cream, suspension), the liposome-encapsulated econazole greatly increased drug concentration in the skin, particularly in the epidermis. Variations in the results obtained with the different liposomal formulations can be due to the physicochemical characteristics of the liposomes (size, surface charge, composition, lipid:drug ratio).

The effect of liposome-entrapped antimicrobials on *Pseudomonas aeruginosa* on colony counts of infected soft-tissue wounds in adult rats was compared with counts of animals treated with free antimicrobial agents and counts of untreated control animals (74).

One application of liposome-entrapped tobramycin or liposome-entrapped silver sulfadiazene resulted in a therapeutic effect on tissue bacterial counts that required multiple doses of free drug. The authors suggested that encapsulation of antimicrobial or antineoplastic in liposomes could offer advantages in the treatment of superficial infections or malignancies.

The in vitro percutaneous absorption of retinoic acid across hairless rat skin was evaluated after application of either free or encapsulated drug using a Franz diffusion cell (75). [^{14}C]Retinoic acid was encapsulated in [^{3}H]DPPC liposomes. Both labeled compounds were measured at various times in the receptor fluid, dermis, and epidermis, Percutaneous absorption of DPPC was negligible (0.15% after 30 hr) and increasing dissociation between retinoic acid and the liposomes was observed from the surface to deeper regions of the skin. Percutaneous absorption of retinoic acid was reduced from the liposomes as compared to a hydroalcoholic gel (mean fluxes being 22.7 ng/cm^2/hr and 42 ng/cm^2/hr, respectively). Drug concentrations in the epidermis (mainly the stratum corneum) and in the dermis were two- to threefold higher with the liposomal formulation than with the conventional gel (41% and 13% versus 18% and 8%, respectively). In vivo, steady state of retinoic acid concentrations in the skin layers was reached much more slowly with liposomes than with the gel, probably because of a controlled release of the drug from the liposomes. Higher drug amounts were found in the skin, especially below 400 μm depth, after a 12-hr application of retinoic acid–loaded liposomes than of the gel. Liposomes tend to improve the bioavailability of retinoic acid in the skin while reducing its percutaneous absorption.

Drug delivery remains the most limiting factor to the effective treatment of cutaneous virus infections such as herpes. Although systemic administration of interferon has been proven to be effective in the treatment of herpes, it is difficult to control concentration profiles of the drug in the skin and to overcome adverse reactions. Therefore, topical administration of interferon was envisaged and the drug was encapsulated in liposomes in order to get a better accessibility to the infected cells. Egbaria et al. (76) investigated the influence of liposome composition and preparation method on the distribution of interferon in the different skin layers after in vitro topical application of formulations to full-thickness hairless guinea pig skin mounted on a Franz diffusion cell.

Application of skin lipid liposomes (LUV), i.e., vesicles with a lipid composition similar to that of the stratum corneum, led to concentrations of interferon in the deeper layers of the skin almost twofold higher than with classical phospholipid liposomes. Furthermore, skin lipid liposomes prepared by the dehydration-rehydration method gave more than twice the amount of interferon into the deeper skin strata than those prepared by the reverse phase evaporation method. Encapsulation of interferon in liposomes clearly enhanced the penetration of interferon into the deeper skin strata as compared to an aqueous solution (from four to ten times, depending on lipid composition and preparation method of the liposomes).

Other polypeptides, such as the fat-soluble cyclic undeca-peptide antibiotic, cyclosporin A, were encapsulated in liposomes in order to improve their penetration in the different skin layers. In vitro diffusion studies of cyclosporin A in hairless mouse skin (77) after application of several formulations showed that drug accumulation in the skin followed the order: skin lipid MLV > phospholipid MLV >>> emulsion. Using human cadaver skin for a 24-hr in vitro penetration study, total uptake of cyclosporin A in the skin was in the order: skin lipid MLV > phospholipid MLV ≃ skin lipid LUV > phospholipid LUV >>> emulsion.

Whatever the lipid composition of the liposomal bilayers, application of MLV resulted in higher cumulative amounts of cyclosporin A in the various strata of human cadaver skin than with LUV. However, for both LUV and MLV, higher fractions of cyclosporin A were deposited in the deeper stratum corneum when the drug was encapsulated in phospholipid liposomes as compared to skin lipid liposomes.

Whatever the liposomal formulation, the total amount of cyclosporin A in the skin was higher than that obtained after topical application of an oil-in-water emulsion. Interestingly, a drug:lipid ratio close to unity was maintained in the various layers of the stratum corneum after application of LUV, suggesting association of the drug with the liposomal bilayers throughout the stratum corneum.

Liposome-encapsulated cyclosporin A, depending on the vesicle composition and preparation technique, might be useful as an alternative to systemic administration of cyclosporin A for the treatment of psoriasis, since this route of administration leads to several toxic side effects.

The pharmacological activity of a lipophilic lipoxygenase inhibitor MRZ 3/99 incorporated in either a liposome gel, a cream, or an ointment was evaluated in an arachidonic acid–induced edema model in mice (78). The MRZ 3/99 ointment did not present any anti-inflammatory activity whereas the liposomal and the cream formulations showed a significant inhibitory effect on edema formation. Although the liposome gel and the cream presented a similar efficacy in the animal model, the MRZ 3/99 amounts that penetrated into the stratum corneum 2 hr after application of both formulations on the forearm of healthy volunteers were significantly different. Thus, MRZ 3/99 penetration into the stratum corneum was about two times higher with the liposome-encapsulated drug than with the cream.

α-Tocopherol nicotinate penetration into the stratum corneum of human volunteers was also investigated by the stripping method 2 hr after application of either a liposomal formulation (MLV) or an ointment (79). Tocopherol nicotinate absorption into the stratum corneum was about three times higher with the drug-loaded liposomes than with the standard German Pharmacopoeia ointment. Moreover, drug penetration was 1.5-fold increased by applying SUV (46 ± 10 nm) instead of MLV (240 ± 100 nm). Liposomes may present an advantage over conventional topical formulations in order to overcome the skin barrier function.

Patients with the rare human genetic disease xeroderma pigmentosum, who are deficient in repair of UV-induced DNA damage, suffer from an extraordinary high incidence of skin cancer on sun-exposed skin (80).

Human DNA repair of UV damage by delivering DNA repair enzyme T4 endonuclease V encapsulated in liposomes has been investigated by Yarosh et al. (81,82). The addition of liposome-entrapped DNA repair enzymes to human fibroblasts and epidermal keratinocytes in culture stimulated DNA repair synthesis and enhanced pyrimidine dimer removal. Furthermore, topical treatment of UV-irradiated SKH-1 hairless mice with pH-sensitive liposomes containing the DNA repair enzyme increased the removal of pyrimidine dimers from the epidermal keratinocytes, whereas empty liposomes and unencapsulated T4 endonuclease V produced no effect. The authors suggested that the liposomes penetrated the stratum corneum and entered keratinocytes. Whatever the mechanism of penetration of liposome-entrapped T4 endonuclease V in vivo, the obtained results are very promising for correcting or enhancing skin cell metabolism, and thus improving human health.

Mechanisms of Topical Delivery of Liposomes-encapsulated Drugs

As reported earlier, topical application of drug-loaded liposomes may dramatically modify the bioavailability of the drug and in particular its localization in the different skin layers. However, very few studies have been done to answer to the following questions: How do liposomes modify drug penetration in the skin? Do they enter the skin? Do they cross the skin and become distributed in the bloodstream? Do they act as penetration enhancers?

In the literature, no precise answer has been given since it depends mainly on the liposome composition, the size and mode of preparation, and the physicochemical properties of the encapsulated drugs. However, in general it is known that excipients of topical drug delivery systems can affect drug absorption at two stages:

Partitioning of the drug from the dosage form to the skin
Diffusion of the drug into the stratum corneum

Liposomes have been designed for topical drug delivery as a carrier system, which means that in addition to the two precedent stages they may also act actively by transporting and releasing encapsulated drugs at specific targeted sites of the skin. One clear evidence is that after topical application, intact liposomes have never been found in the blood circulation. In vitro percutaneous absorption studies first showed that neither intact liposomes nor liposomal phospholipids were able to cross the skin and gain access to the receptor compartment (60,75,83–85). Some authors even claimed that intact liposomes could not enter the skin (67,83,84). Thus Ganesan et al. (83) using DPPC liposomes (MLV) containing model compounds (glucose, hydrocortisone, PG) developed a theory about the mechanisms involved in the in vitro skin permeation of the liposome-entrapped molecules (Figure 9). They proposed three mechanisms that are mainly dependent on the characteristics of the entrapped compound:

1. Release of the drug from the liposomes on the skin surface and penetration of the free compound according to its physicochemical properties

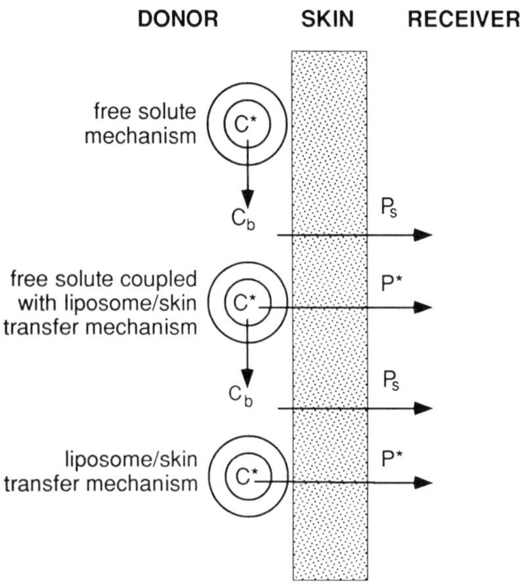

Figure 9 Schematic representation of cutaneous penetration of liposome-entrapped solutes (liposomes are not absorbed intact or fused with stratum corneum). C*: solute concentrations in the liposome, C_b: solute concentrations in the bulk aqueous medium, P*: effective permeability coefficient for the liposome/skin transfer of solute and permeation process, P_s: skin permeability coefficient for drug in aqueous solution. (Redrawn with permission from Ref. 83.)

2. Release of drug from the liposomes and skin penetration of the free molecules (1) together with direct transfer of the drug from the liposomes to the skin
3. Liposomes-skin drug transfer

Experimental and theoretical data supported the first mechanism for the skin permeation of liposome-entrapped glucose, the release rate from the MLV being the rate-limiting step. For PG a direct liposome-skin transfer and for hydrocortisone a combined mechanism was suggested. In another work (67,68) butylparaben penetration into the skin was also supposed to be dependent on free drug penetration and liposome-skin direct transfer. However, this theory was only based on in vitro experiments analyzing the fraction of compounds that had penetrated the receptor fluid. No consistent information was given regarding the amount of compound localized in the different skin strata. This theory is therefore highly criticizable and, apart from indications of potential systemic absorption of a drug via the topical route, analysis of drug concentrations in the receptor phase do not provide any information on drug localization within the skin. Analysis of drug content in the skin is, however, of major importance since drug efficacy in skin therapy will be related to drug access and retention time at the site of action.

In contrast with the previous theory, numerous studies have demonstrated that some types of liposomes were at least able to penetrate the stratum corneum. Thus, Hofland et al. (86) investigated the interactions of nonionic surfactant vesicle "niosomes" (37) with human skin in vitro. Following application of these multilamellar vesicles (0.5–2 μm in diameter) to the skin, two major types of interactions were suggested:

1. Direct penetration of intact niosomes into the stratum corneum; the vesicles consequently play the role of microreservoirs or drug carriers.
2. Fusion of niosomes with the lipid bilayers in the stratum corneum; the vesicles may act as penetration enhancers.

The application of niosomes to human skin in vitro introduced changes in the ultrastructure of the intercellular lipid lamellae of the stratum corneum down to a depth of about 10 μm. Freeze-fracture electron micrographs (Figures 10 and 11) show vesicles either on top of the first corneocyte or in the intercellular lipid bilayers of the stratum corneum. These vesicular structures were not observed in the control skin. However, using the same freeze-fracture technique, after application of a hexagonal Brij 96/water cream to the skin under occlusion in vitro, similar hexagonal structures were visualized in the intercellular lipid region of the stratum corneum. These observations may find two explanations:

1. Surfactant vesicles, and probably also liposomes, may penetrate the skin as such via the intercellular pathway of the transepidermal route.

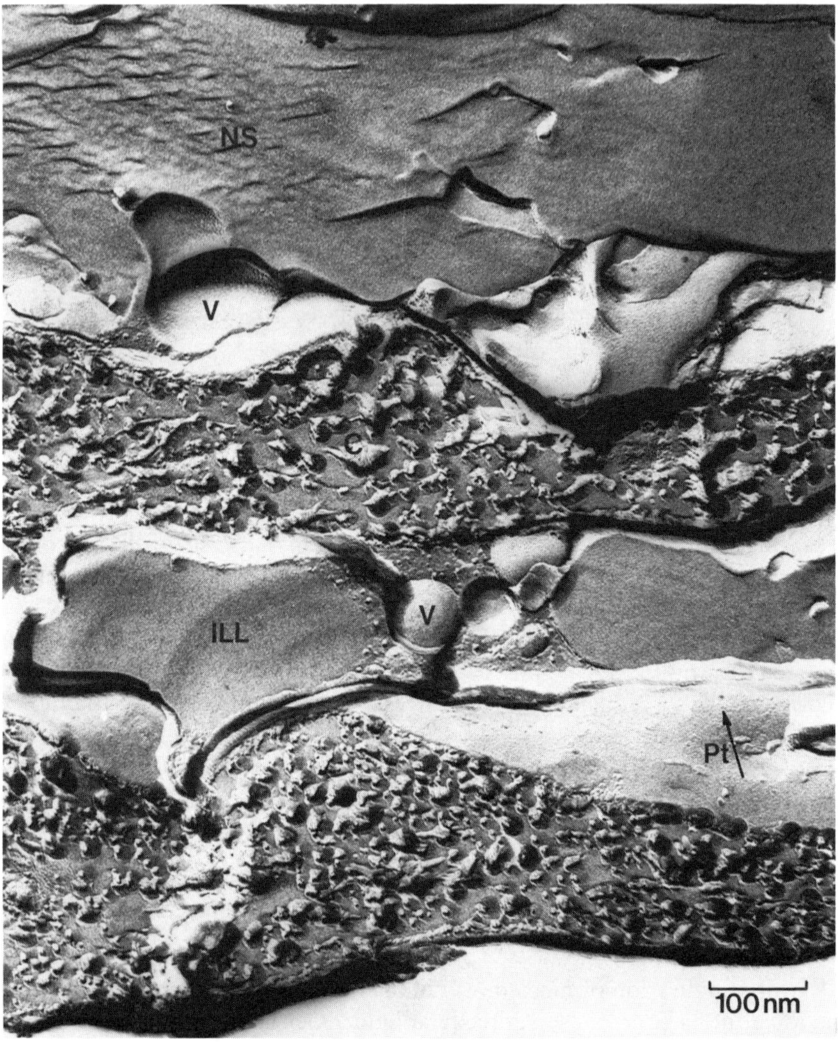

Figure 10 Freeze-fracture electron micrograph of the interface of the niosome suspension (NS) and the skin. Vesicular structures (V) can be observed on top of the first corneocyte (C) and between the intercellular lipid lamellae (ILL). The arrow indicates the direction of Pt evaporation. (Reproduced with permission from Ref. 86.)

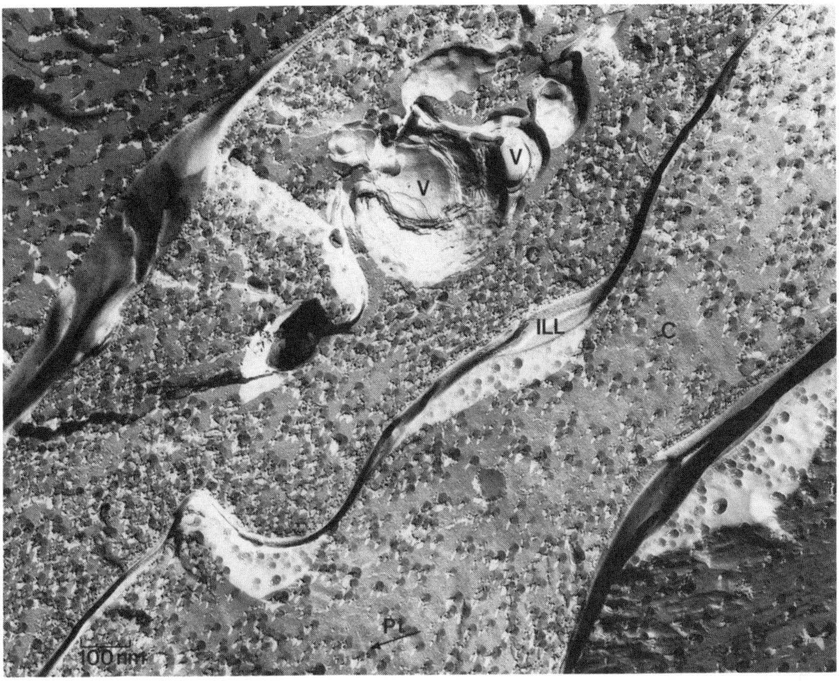

Figure 11 Freeze-fracture electron micrograph of niosome-treated human skin. A cluster of vesicular structures (V) can be seen in the region of intercellular lipid lamellae (ILL) between corneocyte (C) at a depth of approximately 10 μm. The arrow indicates the direction of Pt evaporation. (Reproduced with permission from Ref. 86.)

2. Surfactant of the vesicles may diffuse molecularly dispersed in the lipid bilayers of the stratum corneum and form in situ new vesicles.

However, depending on their physicochemical properties liposomes may either stay on the skin surface or penetrate the stratum corneum, leading then to various modifications of the lipid bilayer structure (87).

Liquid state unilamellar liposomes with an identical size (200 nm), prepared from different lipids (Nat 106, Nat 89, Nat 50, Nattermann Phospholipid GmbH, Germany), were added in vitro to stratum corneum pieces for 48 hr. Freeze-fracture electron microscopy and small-angle X-ray scattering showed that Nat 106 liposomes induced several modifications in the stratum corneum, with, in particular, the appearance of flattened, spherical, lipid structures between the corneocytes. With Nat 89 liposomes, changes in the bilayer structure were less pronounced, although the lipid structures presented a very rough fracture surface. Nat 50 liposomes did not produce any modification of the stratum corneum structures, and it was concluded that these vesicles fuse together and form lipid bilayers on the stratum corneum surface. The latter conclusion was also drawn for unilamellar phosphatidylcholine liposomes kept in contact with stratum corneum samples under occlusion at room temperature for 24 hr (88).

The different qualitative techniques mentioned above are clearly insufficient for precisely determining whether or not vesicular delivery systems penetrate the stratum corneum. The absence of spherical structures in between the corneocytes after topical application of liposomes cannot be regarded as a lack of penetration. Indeed, Abraham and Downing (13) showed that the contact in vitro between corneocytes and SUV prepared from a mixture of epidermal ceramides, cholesterol, free fatty acids, and cholesteryl sulfate (4.5:3.5:1.5:0.5) resulted in a slow transformation of the liposomes into lamellar sheets, most likely by the corneocyte lipid envelope.

In order to better understand the influence of liposomes on the kinetics and extent of uptake of different drugs in the various skin layers on topical applications of formulations either in vitro or in vivo, radiolabelled compounds have been used.

Thus, 30 min after in vitro topical application of hydrocortisone encapsulated in liposomes on human skin, drug concentrations were 8 and 14 times higher in the epidermis and dermis, respectively, as compared to a conventional ointment formulation (54). The drug concentration gradient between the surface of the stratum corneum and the deeper layers of the dermis was also very small 30 min after application of hydrocortisone-loaded liposomes whereas it was much more pronounced with the hydrocortisone ointment.

Encapsulation of retinoic acid into liposomes also led to an increase of drug concentrations in the epidermis, mainly the stratum corneum, and in the dermis 30 hr after in vitro topical application on hairless rat skin of the liposomal form as compared to a conventional gel (75).

Evaluation of the ratio of radioactivity, [^{3}H]DPPC/[^{14}C]retinoic acid, in the different skin layers demonstrated a partial dissociation between the drug and liposomal phospholipid from the skin surface to the deeper skin layers. Accumulation of cyclosporin A in the stratum corneum and deeper skin strata was far higher with either phospholipid or "skin lipid" liposomes than with oil-in-water emulsion after in vitro topical application of the different formulations on either hairless mouse skin (77,89) or human cadaver skin (85).

Interestingly, using a double labeling, i.e., [^{3}H]cyclosporin A and [^{14}C]DPPC, it was shown that a drug/phospholipid ratio close to unity was maintained in 19 successive tape strippings of the stratum corneum following application of LUV to human cadaver skin (Figure 12). After treatment with the emulsion, the drug/lipid ratio in the layers of stratum corneum largely

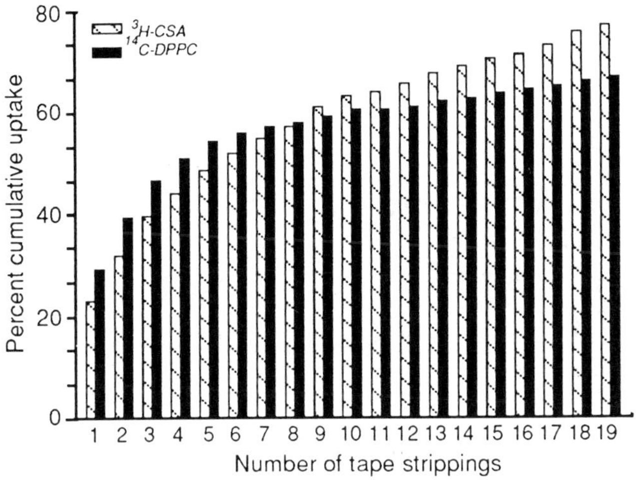

Figure 12 Cumulative uptake of [^{3}H]CSA and [^{14}C]DPPC in various strippings of stratum corneum 24 hr after in vitro topical application of phospholipid LUV to human cadaver skin (n = 4). (Reproduced with permission from Ref. 85.)

deviated from unity. These observations suggest that cyclosporin A penetrates the full thickness of the stratum corneum associated with the liposomes whereas it diffuses separately from the lipid of the emulsion.

These different quantitative in vitro experiments tend to prove that liposomes, mainly according to their size and composition, may greatly enhance the penetration of several drugs into the different skin layers (90). Dual-label studies further suggest that liposomes may act as a carrier facilitating drug passage at least into the stratum corneum. Although several studies prove that liposomes do not penetrate the skin deeper than the stratum corneum, they can act within the stratum corneum as microreservoirs that control the release and enhance the concentrations of the drug in the deeper skin strata.

In addition to in vitro studies using labeled compounds, other techniques such as one-dimensional electron paramagnetic resonance imaging (1D-EPRI) was used to evaluate quantitatively the effect of liposomes on the penetration rate of liposome-entrapped molecules into the skin (91). A paramagnetic, hydrophilic marker, ASL (*N*-(1-oxyl-2,2,6,6-tetramethyl-4-piperidinyl)-*N*-dimethyl-*N*-hydroxyethylammonium iodide) was encapsulated in either SUV or REV composed of fluid state phospholipids (lecithin and phosphatidylserine) or in MLV or REV from solid state phospholipids (dipalmitoylphosphatidylcholine and dipalmitoylphosphatidylglycerol). The diffusion of

ASL into pig ear skin was shown to be dependent on liposome size and composition (Table 4). Thus, diffusion into the skin of ASL dissolved in water or hydrogel was negligible for 60 min after application of the control formulations. Solid state liposomes did not modify the diffusion of ASL molecules into pig ear skin as compared to the control, whereas fluid state REV significantly increased ASL diffusion rate into the skin. Interestingly, penetration of ASL encapsulated in fluid state REV was compared to a mixture of free ASL and empty liposomes in the hydrogel. When ASL was not incorporated into the liposomes, no increase of the penetration was observed, proving that the liposome components had no effect on the stratum corneum barrier. These experiments suggested that fluid state REV act as an active carrier of ASL into the skin, but no precise information was given regarding liposome localization within the skin. Without any doubt, liposome size and fluidity play a major role in the penetration of hydrophilic charged compounds into the skin and 1D-EPRI might be a useful technique for

Table 4 Diffusion Constant of ASL (*N*-(1-Oxyl-2,2,6,6-tetramethyl-4-piperidinyl)-*N*-dimethyl-*N*-hydroxyethylammonium iodide) into Pig Ear Skin, from ASL-entrapped Liposomes of Various Lipid Compositions and Sizes, as well as Free ASL in Water and in Hydrogel (HG)

Lipid composition of the liposomes	Diffusion constant (*D*) ($\times 10^{-6}$)(cm^2/s)[a]
REV (EL:Chol = 7:3) in HG	0.6 ± 0.1 (2)
REV (EL:Chol = 1:1) in HG	1.8 ± 0.5 (4)
REV (EL:PS = 7:3) in HG	1.5 ± 0.1 (2)
REV (EL:PS = 7:3) in Tris	2.5 ± 0.9 (5)
REV (DPPC:DPPG = 4:1) in HG	0.0 ± 0.0 (6)
SUV (EL:Chol = 7:3) in water	1.2 ± 0.5 (5)
MLV (EL:Chol = 6:4) in HG	0.0 ± 0.0 (3)
MLV (EL:PS = 7:3) in Tris	0.0 ± 0.0 (2)
MLV (SL) in HG	0.0 ± 0.0 (3)
Free ASL in water	0.0 ± 0.0 (3)
Free ASL in HG	0.0 ± 0.0 (2)
REV (EL:Chol = 7:3) + free ASL in HG	0.0 ± 0.0 (2)

[a]Mean ± SD; EL, egg lecithin; Chol, cholesterol; PS, phosphatidylserine; DPPC, dipalmitoylphosphatidylcholine; DPPG, dipalmitoylphosphatidylglycerol.

Note: The number of measurements on different skin samples is shown in parentheses. The measurements were taken using various samples of pig ear skin and at different time intervals after killing of the animal. There were no significant differences in D when measurements were performed on fresh skin or skin which had been stored at 4°C for 10 days.

attaining a precise understanding of the mechanisms by which liposomes enhance drug penetration.

In vivo studies also support the results obtained in vitro with diffusion cells. For instance, triamcinolone acetonide was encapsulated in liposomes and its disposition was compared to that of the drug incorporated in a gel after topical application to rabbit skin (51).

Concentrations of triamcinolone acetonide were approximately fivefold and threefold increased in the epidermis and dermis, respectively, after treatment with the liposomal formulation as compared to the control dosage form. In addition, systemic absorption of the drug was considerably reduced after topical application of the liposomal formulation as supported by blood levels and urinary excretion data of triamcinolone acetonide.

Other in vivo experiments demonstrated that topical application of econazole-loaded liposomes to guinea pigs resulted in a significant increase of drug distribution in the skin, particularly in the epidermis and dermis as compared to the corresponding control formulations (73). The increase of drug concentrations in the target organ, the skin, with liposome-encapsulated econazole was concomitant with a significant reduction of drug concentrations in the nontarget, internal organs.

As stated above, liposomes have been proposed to enhance drug penetration into the skin either by direct transport into the intercorneocyte phospholipid bilayers or by a penetration enhancer effect of the liposomal lipids. In order to investigate the potential activity of phosphatidylcholine as penetration enhancer in vivo, the topical bioavailability of four corticosteroids was assessed by the human skin blanching assay after pretreating the skin with empty liposomes of phosphatidylcholine (55,56). Pretreating the skin of human volunteers with a liposomal suspension indeed increased the blanching response to all four formulations and reduced the tachyphylactic response.

The authors proposed that since liposomes cannot act as a carrier system for the corticosteroids, they probably supplement the total lipid content of the stratum corneum with phosphatidylcholine enabling a depot of steroid to be formed. They also suggested that liposomes may form a lipid film on the stratum corneum surface into which corticosteroids may first partition before entering the lipid regions of the stratum. The enhancement of diclofenac percutaneous penetration when topically applied in an aqueous gel form of hydrogenated soya phospholipids containing about 30% phosphatidylcholine and 70% phosphatidylethanolamine was demonstrated either in vitro and in vivo by Nishihata et al. (92). The application of the aqueous gel, prepared with hydrogenated soya phospholipids, increased the in vitro penetration of diclofenac through rat skin in proportion to the phospholipid concentration. The in vivo percutaneous absorption of diclofenac in rats was followed after topical application of a solution and an aqueous gel of phospholipids by measurement of drug plasma concentrations for 12 hr and evaluation of drug bioavailability. The plasma concentrations of diclofenac obtained after application of the phospholipid aqueous gel were significantly higher than in the case of the solution. The absolute bioavailability of diclofenac after topical administration of either a solution or a phospholipid aqueous gel was 5 and 25%, respectively. These results mean that hydrogenated soya phospholipids act as penetration enhancer in the percutaneous absorption process of diclofenac.

Other in vitro and in vivo experiments regarding the transdermal delivery of bunazosin HCl, a α_1-adrenoceptor–blocking drug used for the treatment of hypertension, clearly evidenced the enhancing effect of egg yolk lecithin (93). Since the improvement of bunazosin HCl transdermal delivery was shown not to be related to modification of drug solubility or thermodynamic activity, the enhancing effect of lecithin was supposed to be due to its effect on the permeability of the skin barrier.

Lipid disperse systems composed of soybean phosphatidylcholine and/or glycosylceramide also enhanced the in vitro and in vivo percutaneous penetration of flufenamic acid (94). In this case, the enhancer effect of phosphatidylcholine was proved to be mainly related to an increased solubility of flufenamic acid rather than a modification of the permeability of the stratum corneum.

Although the mode of action of phospholipids as penetration enhancers may hold true for a particular type of liposome with a defined size and composition, among the multitude of vesicular drug delivery systems other mechanisms may prevail.

Recently, penetration pathway and localization of liposomes in the skin were investigated in vivo by electron microscopy after topical application on guinea pigs of MLV prepared from DPPC and cholesterol (2:1) containing the electron-dense colloidal iron marker (71,95). Surprisingly, the electron micrographs showed the presence of intact liposomes in the dermis. Most of the liposomes found in the skin were unilamellar and had a diameter below 0.7 μm. The presence of "liposome-like" material between corneocytes suggested that liposomes penetrate the skin via the intercellular route.

As observed earlier by freeze-fracture electron microscopy, niosomes were able to penetrate the intercellular lipid region of the stratum corneum after in vitro topical application to human skin (41). The same phenomenon still occurs in vivo with solid state MLV. However, one cannot discriminate between the different possible mechanisms involved in liposome penetration and enhancement of the topical bioavailability of drugs encapsulated in vesicular systems.

Based on the different penetration studies with liposomal dosage forms, Foldvari et al. (71) suggested that in the liposome–skin interaction process more than one mechanism takes place and they proposed a cascade of events as illustrated in Figure 13:

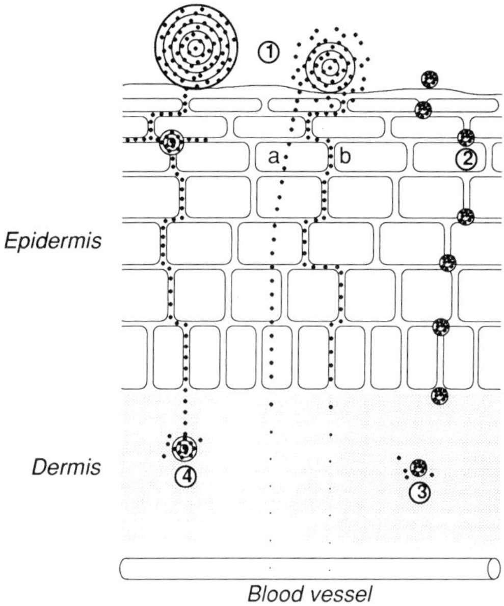

Figure 13 Hypothetical processes involved in the penetration of liposome-encapsulated drug into the skin. (From Ref. 71.)

1. Adsorption of liposomes onto the skin surface and release of the free drug by diffusion or rupture of the vesicles resulting in direct drug transfer from the liposomes to the skin. The free drug molecules consequently penetrate the skin via either the transcellular (a) or intercellular (b) pathway.
2. Penetration of small unilamellar vesicles into the intercorneocyte lipid-rich region of the stratum corneum. Intact liposomes act as drug carriers and microreservoirs that control drug localization and release. After an eventual fusion of the liposomes with the lipid bilayers in the stratum corneum, the liposomal lipids may have a penetration enhancer effect.
3. Penetration of SUV deeper into the epidermis and dermis, with a controlled release of the encapsulated drug by disruption or degradation of liposomal membranes.
4. As in 2 and 3, penetration of MLV which can lose during their transfer to the deeper skin strata one or more lipid bilayers resulting in a partial release of the encapsulated drug.

Depending on the liposome characteristics, mainly its size and composition, and on the physicochemical properties of the encapsulated drug, one or several mechanisms of either active or passive drug transfer will occur, successively or simultaneously.

All the theoretical mechanisms proposed previously may not hold true in a clinical situation where the dermatological products are to be applied to altered or diseased skin. Under certain circumstances, topically applied liposomes will be in direct contact with keratinocytes, e.g., if there is a defective or incomplete stratum corneum as in superficial wounds, in some dermatoses, and on mucocutaneous surfaces. In that situation, direct passive targeting of liposomes to keratinocytes is readily achieved, the stratum corneum barrier being inefficient. It was therefore interesting to investigate the effect of liposomes on cultured human keratinocytes.

Bonnekoh et al. (96) showed that oligolamellar liposomes (mean size 150 nm) were incorporated in vitro into HaCaT cells, an aneuploid human keratinocyte cell line with normal keratinization established from normal human adult skin. Among the different mechanisms of liposome incorporation into cells (stable adsorption with lipid transfer, membrane fusion, phagocytosis), the author suggested that liposomes were taken up by human keratinocytes by membrane fusion. A marked fluidization of keratinocytes by the liposomes was observed that may regulate cell surface receptor functions and slow down cell proliferation. The influence of empty phospholipid liposomes on cultured human epidermal keratinocytes (NHK) was recently evidenced by Schmidt et al. (97). SUV prepared from DPPC completely inhibited spontaneous formation of cornified envelopes (Figure 14) and a remarkable increase (220%) of the cellular cholesterol synthesis was observed.

Conversely, SUV enriched in cholesterol ensured cornified envelope formation even in the presence of different inhibitors of cholesterol synthesis that first suppressed terminal keratinocyte differentiation (Figure 15A). The addition of cholesterol-enriched liposomes to NHK did not antagonize the inhibitory effect of 22-ketocholesterol on cholesterol synthesis (Figure 15B) but increased the total cellular cholesterol by incorporation of liposomal cholesterol (Figure 15C).

According to their composition, liposomes can either inhibit or stimulate the formation of cornified envelope, i.e., they can modulate in vitro the differentiation of human epidermal keratinocytes. These findings are of major importance since they point out that liposomes themselves can regulate keratinocyte differentiation and consequently play a role other than drug carrier or microreservoir in the treatment of dermatological diseases.

Micro- and Nanospheres as Topical Drug Delivery Systems

About 5 years ago research interest moved in the direction of site-specific drug delivery and controlled release of drugs to the skin using nano- and

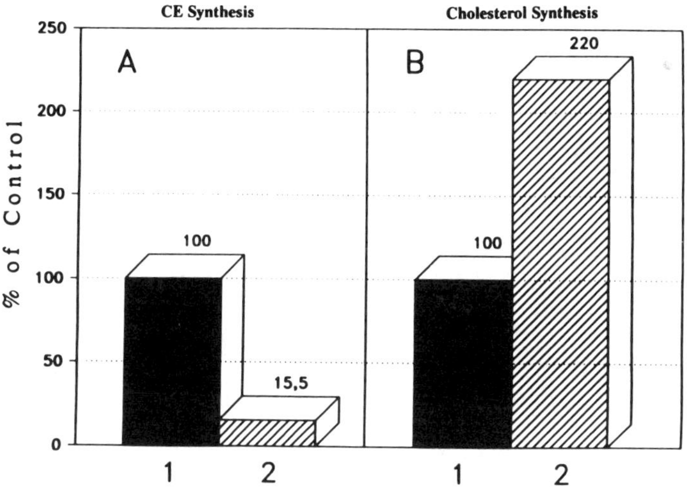

Figure 14 Influence of DPPC liposomes (bar 2) (SUV) on spontaneous cornified envelope (CE) formation (A) and on cholesterol synthesis (B) in cultured normal human keratinocytes. Results are given in % of control (bar 1). DPPC liposomes were added to the medium at confluence; the assays were performed at day 7 of confluence. (Reproduced with permission from Ref. 97.)

microcapsules. Microcapsules generally consists of a core, containing one or more drugs, surrounded by a shell; they usually have a size between 1 and 2000 μm. Capsules smaller than 1 μm are referred to nanocapsules. When no distinct shell and core structures are distinguishable, the solid particles are named either microparticles or nanoparticles, depending on their size.

Antibiotics have been incorporated into poly(DL-lactide-*co*-glycolide) microcapsules in order to improve drug efficacy in the treatment of infected wounds (98,99). Microcapsules are applied topically to open contaminated wounds so that the antibiotics are released at a controlled rate over an extended period at multiple sites within the area to be treated.

Ampicillin anhydrate–loaded microcapsules were prepared and evaluated in vitro and in vivo. In vitro release measurements were used to select an ampicillin anhydrate–loaded microcapsule formulation that would release the drug at a controlled rate over 2 weeks. Sterilized microcapsules composed of 90% DL-lactide-*co*-glycolide polymers (53:47) and 10% ampicillin anhydrate, with a diameter ranging from 45 to 106 μm, provided an initial burst of drug representing about 30% of encapsulated drug, followed by a prolonged drug release over 2 weeks.

Pharmacokinetic studies in rats corroborate the results of the in vitro release studies, the microcapsules producing a prolonged urinary excretion

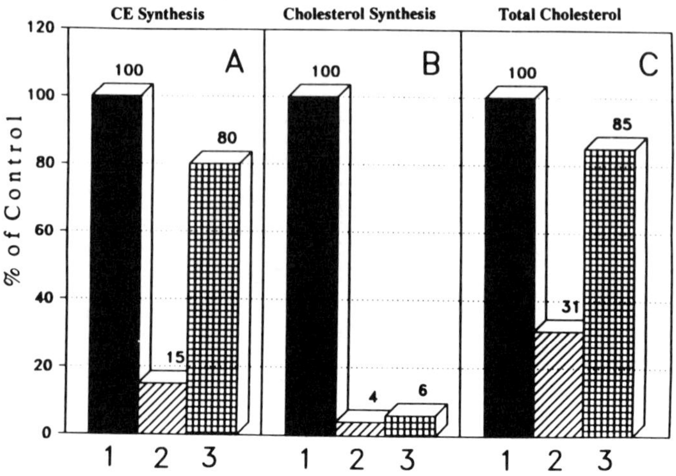

Figure 15 Effect of 25-hydroxycholesterol alone (bar 2) and in combination with DPPC/cholesterol liposomes (bar 3) (SUV) on spontaneous CE formation (A), cholesterol synthesis (B), and total cellular cholesterol (C) in cultured normal human keratinocytes. Results are given in % of control (bar 1). 25-Hydroxycholesterol was added shortly before confluence and the liposomes at confluence. The assays were performed at day 7 of confluence. (Reproduced with permission from Ref. 97.)

of drug for at least 2 weeks as compared to 4 days after subcutaneous injection of free drug. The efficacy of ampicillin anhydrate encapsulated in microcapsules was evaluated in vivo in rats presenting an artificially contaminated wound. All batches of drug-loaded microcapsules efficiently reduced bacterial counts in contaminated wounds. *Streptococcus pyogenes*, which was present in the untreated wounds after 14 days, disappeared from the microcapsule-treated wounds after only 48 hr. After 2 weeks, wounds treated with either unloaded microspheres or free ampicillin anhydrate remained infected with more than 10^5 *Staphylococcus aureus* per gram of tissue whereas 60% of the wounds treated with encapsulated drug were sterile.

Incorporation of antibiotics in microcapsules seems to be a promising approach for controlling drug release within the site of action, i.e., infected wounds, and for eliminating bacteria such as *Streptococcus pyogenes* and *Staphylococcus aureus*.

5-Fluorouracil, an anticancer drug, is used topically as 1–5% cream for the treatment of precancerous dermatosis, in particular actinic keratosis. In order to improve topical drug bioavailability, 5-fluorouracil–loaded microspheres were designed (100). Based on a solvent evaporation technique, microspheres of ethylcellulose (25–200 μm in diameter) containing 5-fluorouracil were prepared and drug release was evaluated in vitro. Drug release from the microspheres was dependent on the pH of the dissolution medium, drug-to-polymer ratio, and polymer molecular weight. In vitro release of 5-fluorouracil from microspheres incorporated in an aqueous gel was much slower than that from a gel containing free drug crystals, with an initial burst followed by a constant slow-release pattern. No experimental data have been published so far showing the potential interest of 5-fluorouracil–loaded microspheres in skin therapy.

Benzoyl peroxide is an effective antiacne drug, mainly possessing a bactericidal activity against *Propionibacterium acnes* and a mild keratolytic activity. However, a major side effect of benzoyl peroxide, related to its concentration, is skin irritation. In order to circumvent this side effect, benzoyl peroxide has been entrapped in microparticles in order to get a controlled release of the drug and a reduced irritation (101).

The spongelike microspheres (Styrene/divinylbenzene Microsponge System, Advanced Polymer Systems, USA) are stable, nontoxic, nonbiodegradable, and have a mean size of 25 μm (range 5–50 μm).

Benzoyl peroxide release in vitro through a silastic membrane from either 5% free drug in simple oil-in-water emulsion or 5% drug-loaded microparticles in o/w formulation followed an apparent zero-order kinetics. However, with the microsponges, a biphasic curve was obtained, the initial flux being similar for about 4 hr to that obtained for the free drug in an o/w emulsion and the second flux being 2.5 times slower for the next 30 hr. The first phase was explained by the release of benzoyl peroxide adsorbed onto the surface of the microspheres whereas the second flux represented the controlled release of the drug entrapped in the microporous polymer system.

In vitro percutaneous absorption of benzoyl peroxide through human skin after topical application of drug-loaded microsponges was shown to be significantly different from that obtained with freely dispersed drug in the emulsion. There was mainly less drug in the skin and in the receptor fluid with the microparticulate formulation than with the emulsion. This difference supported the concept of controlled release of benzoyl peroxide from the microparticles into and through the skin.

In vivo cutaneous absorption of benzoyl peroxide in rhesus monkey from drug-loaded microsponges (8.6%) was significantly less than from a free drug formulation (14.8%).

Entrapment of benzoyl peroxide in the microsponges reduced the drug skin irritation potential in rabbits as compared to the free drug in an o/w emulsion. This reduction of drug irritancy potential by incorporation in microparticles was further confirmed in human subjects. Thus, formulation containing entrapped benzoyl peroxide at 2.5 and 10% were significantly less irritating than their commercial counterparts. Nevertheless, formulations containing either free or entrapped drug at 2.5 and 10% showed a similar antimicrobial efficacy in vivo.

These results tend to prove that controlled release of benzoyl peroxide by incorporation in microparticles may significantly reduce drug irritancy potential while maintaining its therapeutic efficacy.

The controlled release of retinoic acid from microsponges also increased its therapeutic efficacy in hairless mice and reduced its irritancy.

Drug controlled-release via microparticulates might improve the therapeutic index of many drugs and overcome the problems related to the enormous variability in percutaneous absorption at various anatomical sites. A more precise control of topical therapy in children and elderly, whose cutaneous permeability is dramatically modified, would be feasible.

Microspheres as a Topical Site-Specific Drug Delivery System for Targeting Drugs to the Pilosebaceous Unit

The Transfollicular Pathway: A Minor Route of Drug Penetration in the Skin?

As reported above, micro- and nanoparticulate carriers have been investigated as either dermal or transdermal drug delivery systems. They mainly modulate the penetration of drugs through the stratum corneum and control drug release at different skin sites, by acting, for instance, as microreservoirs in the horny layer. Over the past few years, at CIRD Galderma we focused on the potential interest of microparticulates for delivering drugs specifically to the hair follicles at a controlled rate.

When a classical formulation is applied to the skin, drug percutaneous absorption may follow two routes: the transepidermal and the transfollicular (or "shunt") pathways. The role of skin appendages in the penetration process of drugs through the skin has been ignored for a long time. The main reason was that the surface occupied by the pilosebaceous duct orifices only represents about 0.1% of the total skin surface. However, there is increasing evidence that the transfollicular pathway plays an important role in the percutaneous absorption of some active compounds (24,25,102). Thus, Kao et al. (23) studied the influence of skin appendages on the in vitro percutaneous absorption of benzo[*a*]pyrene. For that compound, the rate and extent of in vitro skin permeation in the haired mice (7–10% of applied dose) was higher than in hairless mice (2%).

Moreover, examination of cryosections of skin samples by fluorescence microscopy 1 hr after topical application of benzo[*a*]pyrene showed intense zones of fluorescence associated with the follicular ducts and sebaceous glands. The authors concluded that for topical agents that are not well absorbed via the transepidermal route, the transappendageal pathway may contribute significantly to the overall process of percutaneous absorption.

[14C]Viprostol, a synthetic prostaglandin E_2 analog and potent peripheral vasodilator envisaged for the treatment of male pattern baldness, was applied to mice and monkeys and its distribution in the skin was followed by microscopic autoradiography (22). The drug rapidly penetrated via the transfollicular route and remained associated with the hair follicles for residence times up to 72–96 hr, in quantities apparently as great or greater than the amount in the stratum corneum.

Keister and Kasting (103) investigated the transport pathways of ibuprofen through excised human skin using mathematical analysis of transient diffusion data. They demonstrated that ibuprofen rapidly penetrated the skin via the transfollicular pathway (mean lag time = 8.6 min) as compared to the transepidermal pathway (mean lag time = 92 min) and the contribution of the shunt pathway was quite important, i.e., approximately 25% at steady state. Lamidon et al. (104) recently showed that the in vitro percutaneous absorption and in vivo distribution of a coleus extract containing forskolin and other labdanes was highly influenced by the transfollicular route. Thus, in vitro drug penetration through hairless rat skin without pilosebaceous units was reduced fourfold as compared to normal skin. Differences in the topical bioavailability of forskolin after application to skin with or without hair follicles were explained by the passage of the compound through the follicles.

The role of follicles in the in vitro percutaneous absorption of drugs such as hydrocortisone, niflumic acid, caffeine and *p*-aminobenzoic acid was estimated in appendage-free skin relative to normal hairless rat skin (25). This model for appendage-free skin, previously described by Schaefer et al. (109), preserved the barrier function of the stratum corneum as shown by morphological, histochemical, and biochemical data. For all of these compounds, the steady-state flux and the amounts diffusing in 24 or 48 hr were two to four times lower in the absence of appendages than that measured through normal skin.

These results were complemented by application of hydrocortisone to 5-day postnatal rat skin (normal skin) as compared to 1-day-old rat skin without follicles. The steady-state flux and cumulative amount of drug diffusing in 24 hr were fivefold lower in appendage-free skin.

Recent evidence from the literature tends to prove that despite the relatively small surface area of the pilosebaceous duct orifices, hair follicles may significantly contribute to the skin absorption of certain drugs.

Targeting of Adapalene to the Pilosebaceous Units by Topical Application of Polymeric Microspheres

For the treatment of pathologies associated with pilosebaceous structures, such as acne, alopecia, and other sebaceous gland dysfunctions (105) it may be important to increase further the localization of certain drugs in the hair follicles. Although some drugs may passively distribute significantly into the pilosebaceous units as reported above, most of the drugs generally enter the skin via the transepidermal pathway. Therefore, we investigated at CIRD Galderma the possibility of targeting drugs to the hair follicles using specifically designed particulate carriers (106). The objectives of follicular targeting of drugs are manifold:

Reduction or avoidance of transepidermal pathway
Decrease of drug systemic toxicity
Increase of drug concentration within the pilosebaceous units
Increase of the therapeutic index of certain drugs
Possibility of reducing the applied dose of drug and/or frequency of administration

For drug follicular targeting, microspheres were mainly envisaged as a site-specific drug delivery system for the following reasons:

Good stability of the microspheres when applied on the skin
Easy obtention of microspheres with a defined mean size in a narrow size distribution, relative to the preparation process
Protection of the drug incorporated into the microspheres against degradation in the formulation (oxidation, hydrolysis) or premature inactivation on the skin surface
Controlled release of the drug in the hair follicles from the microspheres
Possibility of incorporating either lipophilic or hydrophilic drugs into the microspheres

At CIRD Galderma, we particularly investigated the in vitro/in vivo follicular targeting of adapalene, a new naphthoic acid derivative (107) under development for the treatment of acne (108). Earlier studies demonstrated that microparticles with a diameter ranging from 3 to 10 μm selectively penetrate the follicular ducts, with particles larger than 10 μm remaining on the skin surface and microspheres smaller than 3 μm randomly distributing into the hair follicles and stratum corneum (106,109).

Adapalene [6-(3-(1-adamantyl)-4-methoxyphenyl)-2-naphthoic acid], a new modulator of cellular differentiation, was incorporated in 50:50 poly(DL-lactic-*co*-glycolic acid) microspheres (PLGA microspheres).

Polyesters such as poly(lactide-*co*-glycolide) are widely used for preparing surgical sutures, implantable prosthetic devices, and sustained-release drug delivery systems (Zoladex, ICI Pharmaceuticals, U.K.; Decapeptyl, Ipsen-Biotech., France) since these polymers are biocompatible, nontoxic, nonimmunogenic, and biodegradable by hydrolysis in an aqueous environment (110–112).

Among the different techniques for preparing PLGA microspheres (113–117), a solvent evaporation process was used. Briefly, 50:50 poly (DL) lactic-

co-glycolic acid (Medisorb 5050 DL, Dupont, USA) and 1% (w/w) adapalene (CIRD Galderma) were first dissolved in methylene chloride. The organic phase was then added to an aqueous medium (1% polyvinylalcohol), emulsified with either mechanical stirring or sonication and stirred for 4 hr at room temperature. After complete evaporation of the solvent, solid microspheres were collected, washed three times with distilled water, and freeze-dried. Depending on the stirring rate and emulsification procedure, batches of adapalene-loaded PLGA microspheres with different mean sizes were obtained, i.e., 1, 5, and 20 μm. Since adapalene is a fluorescent drug, the drug-loaded microspheres were observed by fluorescence microscopy (Figure 16A). The microspheres are homogeneously fluorescent, the drug being entrapped in the particulate carriers at a concentration of 1% as evaluated by high-performance liquid chromatography. By scanning electron microscopy, the microspheres appear to be spherical, with a smooth surface, no adapalene crystal being visible on the microparticle surface (Figure 16B).

In order to verify the concept of drug follicular targeting using microparticulates, the different PLGA microspheres containing 1% adapalene were incorporated in aqueous gels so that the final drug concentration was 0.1% (w/w). The aqueous formulations containing adapalene-loaded PLGA microspheres of different diameters were applied in vitro to hairless rat and human skin samples.

Briefly, after removing carefully subcutaneous fat of human skin obtained from surgery and excised ventral skin of female hairless rats (7–8 weeks old), circular biopsies of full-thickness skin were placed in static diffusion cells, with the dermal side in contact with the aqueous receptor phase (0.9% NaCl; 0.2% Tween 80) continuously stirred at 600 rpm and thermostated at 37°C. The skin surface was thus maintained throughout the experiment at a constant temperature of 32°C. The formulations (50 mg/cm^2) were applied to the skin surface (1.2 cm^2) and massaged for 3 min with a glass spatula. After 5 and 35 min the excess of formulation was wiped off, the diffusion cells dismantled, and the skin samples frozen at −80°C.

Frozen skin specimens (1 × 0.5 cm^2) were included in "Tissue Tek" at −20°C and longitudinal skin sections realized with a microtome were then observed by fluorescence microscopy (excitation wavelength of 313 nm and emission wavelength of 400 nm) after hydration with glycerol or by scanning electron microscopy after dehydration and gold sputter coating. Both microscopic techniques showed that the 20-μm microspheres remained on the skin surface, whereas the 1-μm microspheres randomly localized in the follicular ducts and stratum corneum (Figure 17).

After in vitro application of 5-μm adapalene-loaded microspheres to hairless rat and human skin, the fluorescence was observed solely in the hair follicles (Figure 18) and the microspheres were clearly visualized in the follicular ducts on the scanning electron micrographs (Figures 19 and 20). The

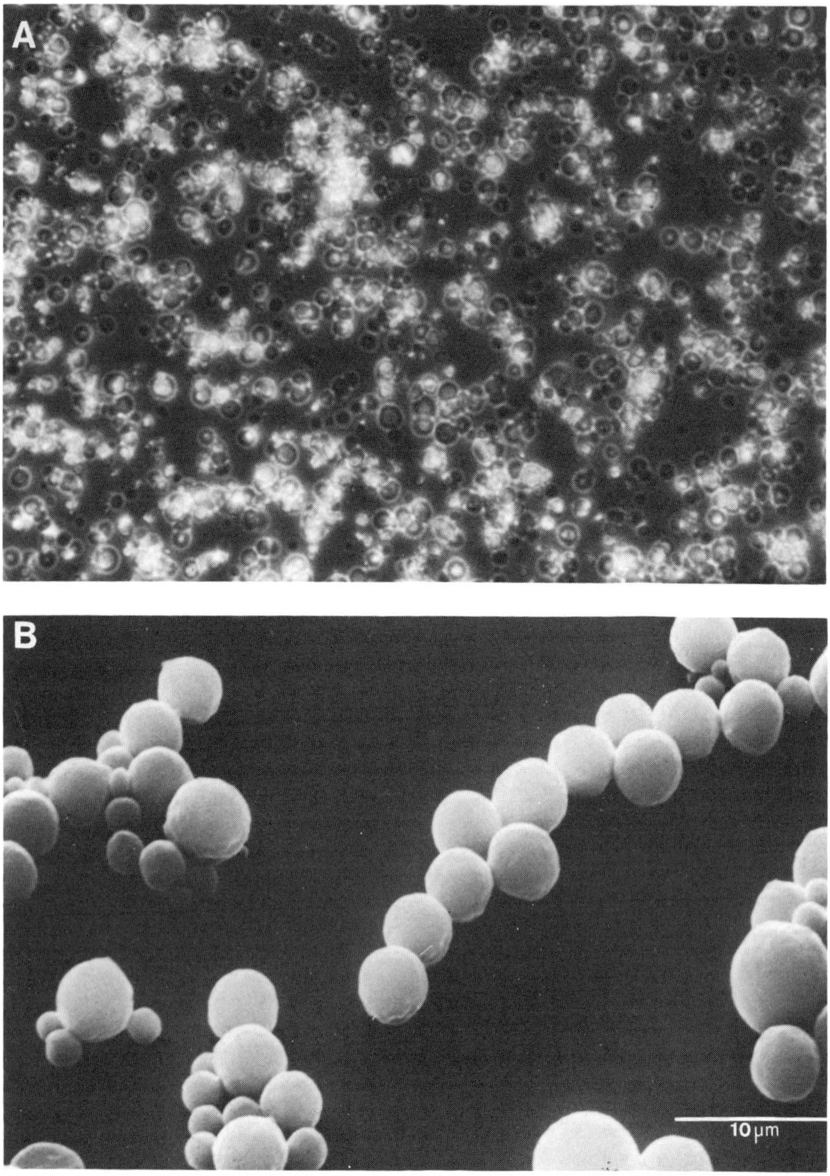

Figure 16 Adapalene-loaded PLGA microspheres. (A) Fluorescence microscopy; (B) scanning electron microscopy.

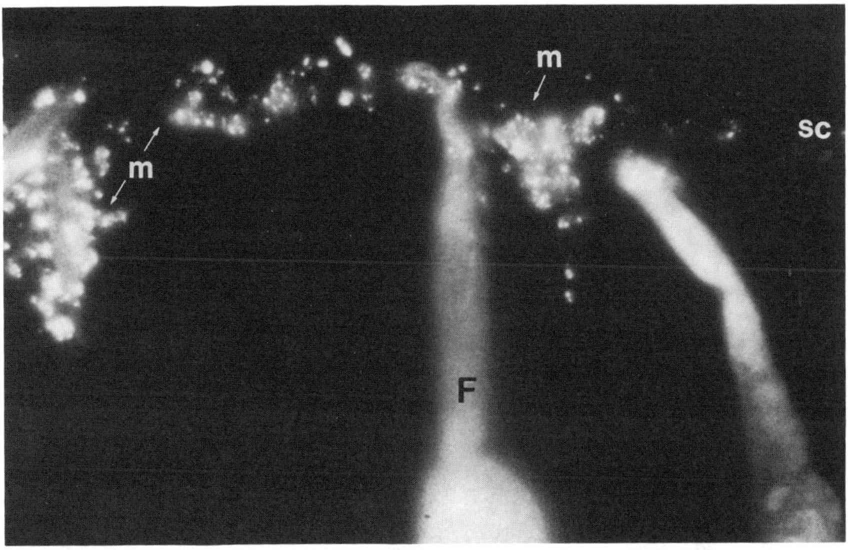

Figure 17 Cutaneous distribution of 1-μm adapalene-loaded microspheres 35 min after in vitro topical application to hairless rat skin (fluorescence microscopy). m, microspheres; SC, stratum corneum; F, hair follicle.

penetration depth of the microparticles into the pilosebaceous structures seemed to be related to the application time, with a deeper follicular penetration being observed after 35 min than after 5 min.

The principle of adapalene specific delivery to the pilosebaceous units using poly(lactide-*co*-glycolide) microspheres is illustrated in Figure 21. Although PLGA microspheres are biodegradable in an aqueous environment, in the follicular ducts that mainly contain lipophilic compounds such as sebum the microspheres are very likely stable. Therefore, to be available for producing its pharmacological effect the drug has to be released from the polymeric microspheres by diffusion. Besides, since in humans sebum excretion from the pilosebaceous structures to the skin surface is a mechanism that requires about 8 days and as the PLGA microspheres after entering the hair follicles will be eliminated via this process, drug release from the microparticles has to be much faster than 8 days.

Adapalene release from the PLGA microspheres was therefore evaluated in vitro in artificial sebum. Adapalene-loaded microspheres (5 μm) (10%, w/w) were incubated at 37°C in artificial sebum. At various periods of time, sebum aliquots were sampled and centrifuged. The supernatant was then analyzed using a UV spectrophotometer at 320 nm. As observed in Figure

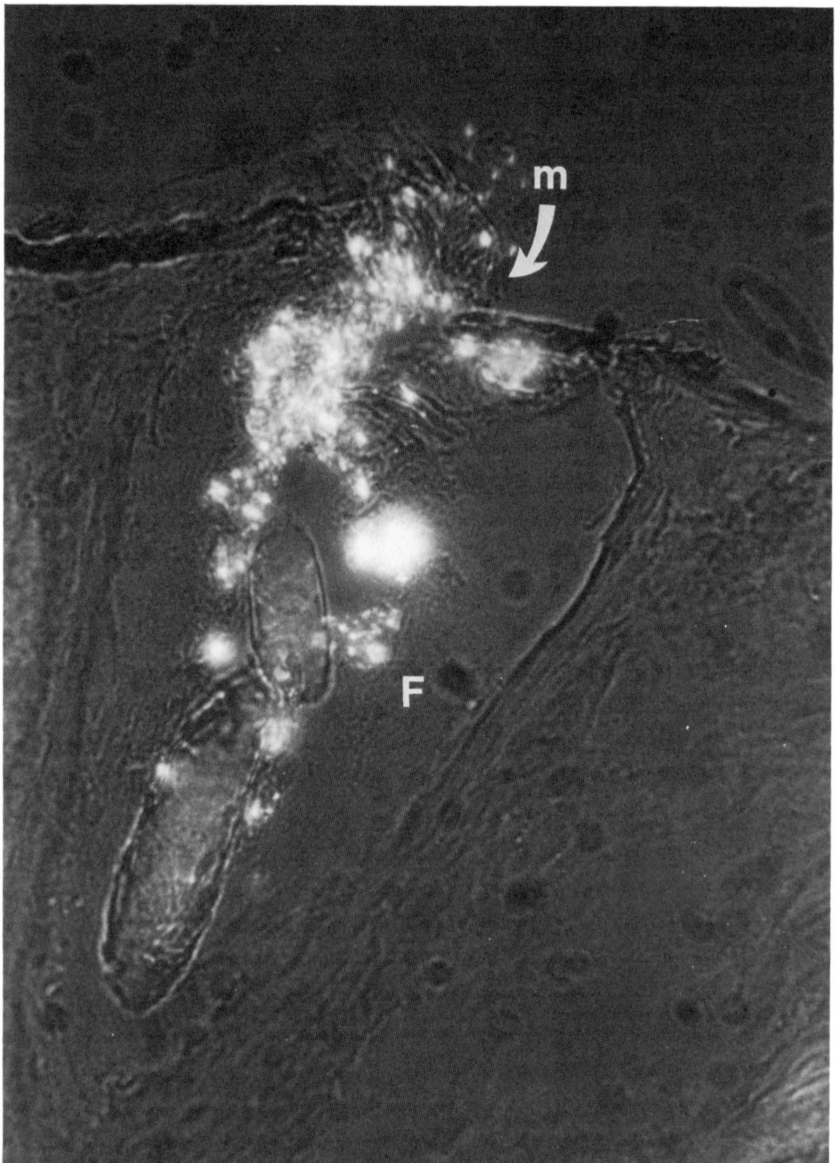

Figure 18 Follicular localization of 5-μm adapalene-loaded microspheres 300 min after in vitro topical application to human skin (fluorescence microscopy). m, microspheres; F, hair follicle.

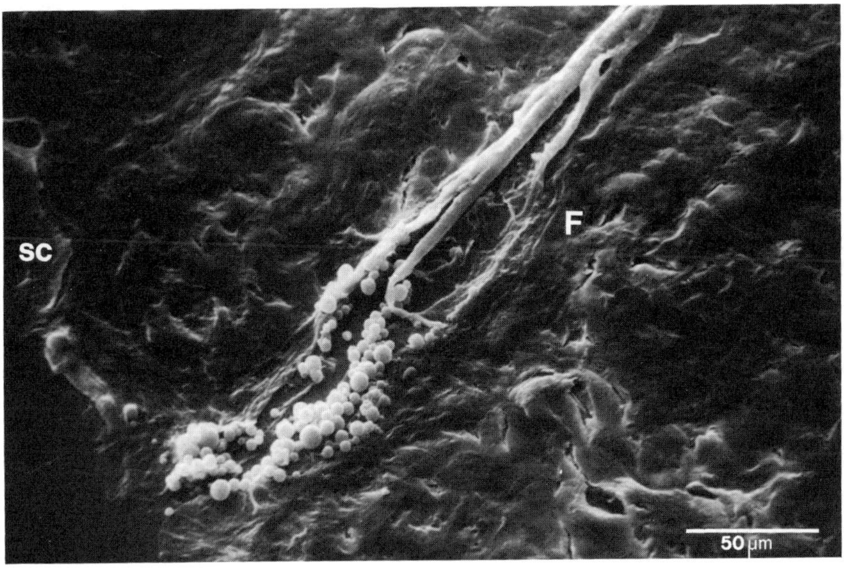

Figure 19 Follicular localization of 5-μm adapalene-loaded microspheres 35 min after in vitro topical application to hairless rat skin (scanning electron microscopy). SC, stratum corneum; F, hair follicle.

22, adapalene release from the microspheres into the sebum was a rapid phenomenon, with a half-life of approximately 60 min and a plateau of amounts of drug released was reached within 4 hr. Based on the in vitro release kinetics of adapalene from the microspheres, one may assume that in vivo the drug will be released in the hair follicles within a much shorter period than the time corresponding to sebum excretion. Following topical application of adapalene-loaded microspheres dispersed in an aqueous gel, the following steps may be proposed:

1. The microspheres are specifically delivered into the pilosebaceous structures.
2. The drug is released rapidly into the lipophilic medium contained in the follicular duct.
3. The drug diffuses to its site of action and produces its pharmacological effect.
4. The empty PLGA microspheres are eliminated from the follicular duct to the skin surface by sebum excretion.

The first two steps being verified, the third step was examined in vivo using an animal model.

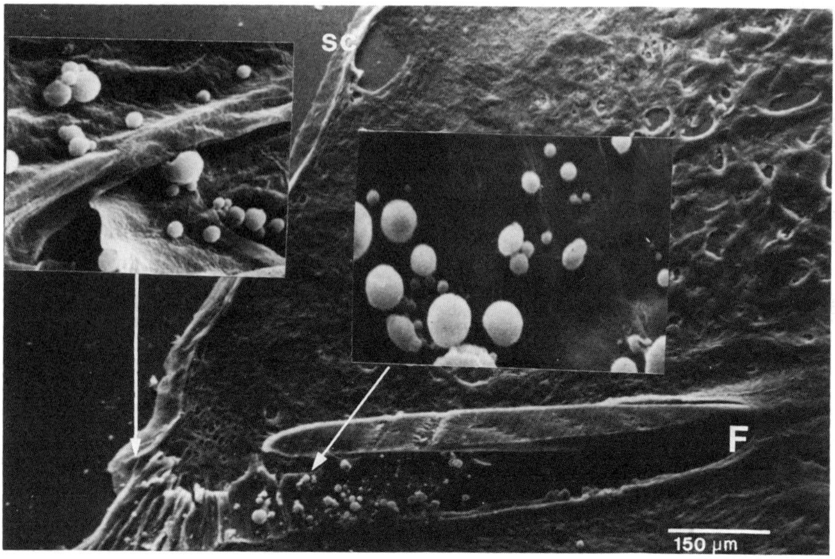

Figure 20 Follicular localization of 5-μm adapalene-loaded microspheres 35 min after in vitro topical application to human skin (scanning electron microscopy). SC, stratum corneum; F, hair follicle.

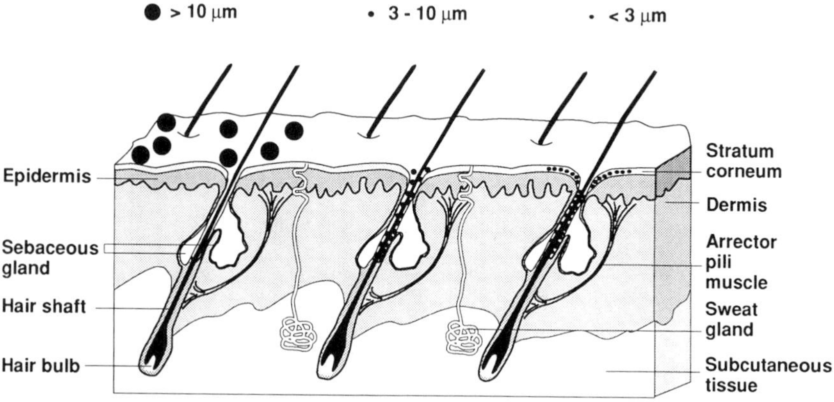

Figure 21 Schematic representation of the effect of microsphere size on cutaneous penetration pathways.

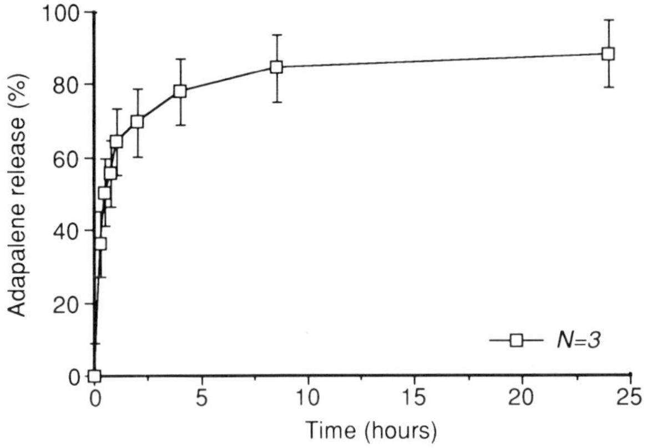

Figure 22 In vitro release of adapalene from PLGA microspheres into artificial sebum at 37°C.

The comedolytic activity of an aqueous gel containing 10% of adapalene-loaded microspheres (5 μm) was thus evaluated in the rhino mouse model (118).

Adult rhino mice are hairless mutants that carry the rhino gene, a recessive allele of the hairless gene (hr^{rh}/hr^{rh}). Prior to the end of their first hair growth cycle, a defect in catagen results in irreversible hair loss. Consequently, the follicular ducts widen, accumulate keratin, and undergo a transformation into horn-filled utriculi which resemble human noninflamed microcomedones. At CIRD Galderma, standardized histological and image analysis methods were set up to characterize and quantify the morphological effects of retinoids in rhino mouse skin.

Briefly, aqueous gels containing 10% of adapalene-loaded microspheres (5 μm) corresponding to drug final concentrations of 0.01, 0.03, and 0.1% (w/w) were administered topically for 3 weeks, on a 5-days-a week basis, to rhino mice. Comedolytic activity of the three microsphere formulations was compared to that produced by topical application of unloaded PLGA microspheres.

At the end of the treatment, skin biopsies were taken from the treated dorsal area of the rhino mice, observed with a microscope coupled to a video camera, and different parameters were measured using a semiautomatic MOP Videoplan (Kontron). The computer calculated the total number of epidermal comedones per centimeter of stratum corneum length, the epidermal thickness, and the comedo profile ($r = d/D$), where d is the width of the surface orifice and D the largest width of utricle or width at half depth. A dose-response effect was observed with the microsphere formulations for the number of epidermal comedones, with the 0.05 and 0.1% adapalene formulations producing a highly significant reduction of the epidermal comedone number (Figure 23).

A highly statistically significant and dose-related increase of the epidermal thickness and of the comedone profile was recorded in rhino mice treated with adapalene-loaded microsphere (Figures 23). This study demonstrated a dose-related comedolytic activity of the adapalene-loaded microspheres in the rhino mouse model, thus validating the third step of the proposed mechanism of action of particulate carriers in drug follicular targeting.

The in vivo experiments realized in animals will indeed have to be confirmed in patients with acne vulgaris. Since the aqueous gel containing 10% of adapalene-loaded microspheres was not irritating in a rabbit skin test, this formulation was applied in vivo to forearms of human volunteers. Following a 3-min massage, the excess of formulation was wiped off and follicular biopsy was achieved by hair shaft removal with tweezers. The inner parts of the hair shafts, corresponding to a depth of approximately 200–400 μm, were observed by fluorescence and scanning electron microscopy (Figure 24). The observed adapalene-loaded microspheres associated with the hair shafts clearly demonstrate the possibility of targeting drugs to the pilosebaceous units by using particulate carriers of defined size.

Additional tests in the rhino mouse model are in progress for optimizing this novel topical site-specific drug delivery system and clinical studies are

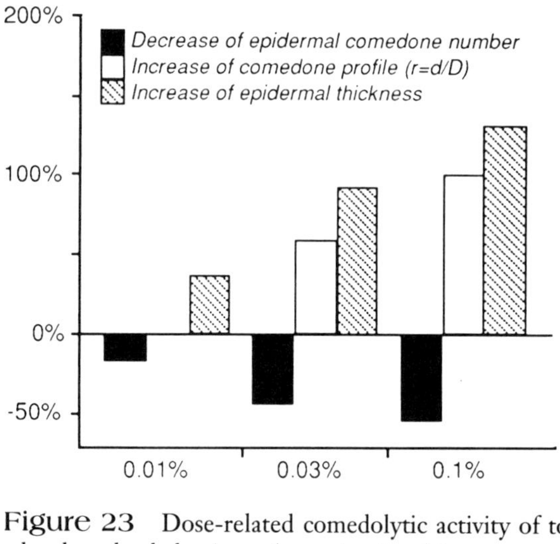

Figure 23 Dose-related comedolytic activity of topical aqueous gels containing adapalene-loaded microspheres in the rhino mouse model [values (%) relative to control (acetone)].

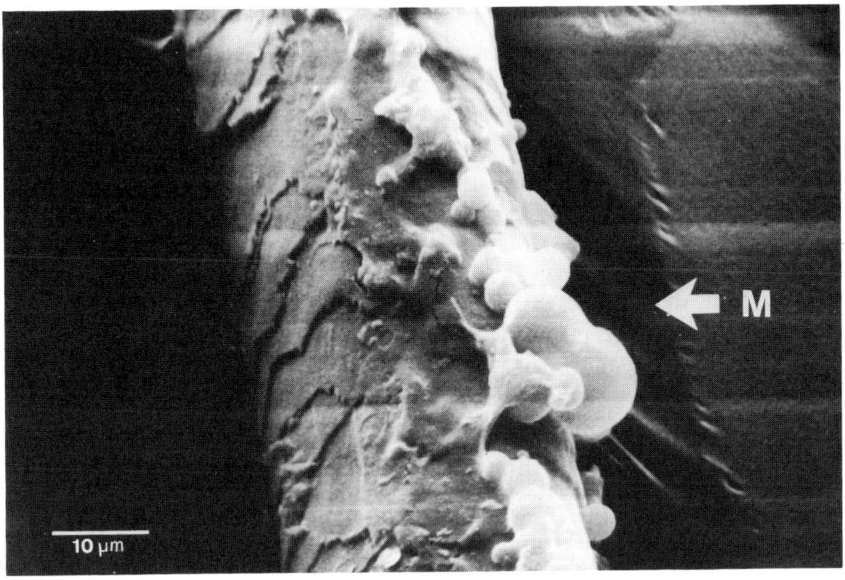

Figure 24 In vivo localization of adapalene-loaded microspheres on the inner part of hair shafts after application to forearms of human volunteers (scanning electron microscopy). *M*, microspheres.

planned in order to compare the efficacy and safety of different formulations of adapalene-loaded microspheres to classical topical drug delivery systems. Since clinical data obtained with different formulations of retinoids, in particular adapalene, indicated a good correlation between clinical efficacy in acne vulgaris and comedolytic activity in the rhino mouse, a significant therapeutic efficacy of adapalene-loaded microspheres in acne vulgaris may be anticipated.

In addition, based on the previous experiments and theoretical calculations, a reduction of either adapalene applied dose or frequency of administration might be considered using site-specific drug delivery systems such as 5-μm microspheres as compared to classical formulations.

CONCLUDING REMARKS

Over the past few years, dermal and transdermal drug delivery with particulate carriers has gained interest owing to a better knowledge of skin physiology and functions and to a more accurate understanding of drug percutaneous absorption with respect to carrier- and/or disease-related effects.

Thus, liposomes have been experimentally designed as a new transdermal drug delivery system that enables a controlled drug release and penetration across the skin. This new approach still has to be verified clinically with different types of liposomes or drugs in comparison with classical transdermal therapeutic systems.

In dermatology, drug access to specific diseased sites, either skin layers or appendages, with a controlled rate and an appropriate duration of action is becoming a major concern for improving drug therapeutic index. Over the past decade, particulate carriers such as liposomes and microspheres have been investigated for the delivery of drugs to specific skin compartments in order to control drug input, release, retention, and action into the skin and also to reduce systemic absorption and consequently adverse reactions.

Topically applied liposomes look particularly promising for providing sustained, enhanced drug levels in skin strata and for reducing percutaneous absorption. However, the mechanisms involved in skin permeation of liposome-encapsulated drugs is not yet fully understood and much work has to be done, in particular with regard to diseased skin, to elucidate these processes.

Nevertheless, in vitro and in vivo experiments have proved particulate carriers to be worthwhile for increasing cutaneous bioavailability and decreasing systemic absorption of many drugs, e.g., corticosteroids, antifungal agents, local anesthetics, antibiotics, retinoids, anti-inflammatory agents, peptides, etc. In the near future, additional particulate carrier–based drug delivery systems will likely follow up the first commercially available topical liposome-based formulation containing econazole.

Finally, particulate carriers not only represent an effective means of modulating drug penetration and/or release into the skin layers, mainly the stratum corneum, but they may also be used for targeting drugs in a specific way to the pilosebaceous structures. Drug follicular targeting using microparticulates has proved so far to be feasible and successful in the case of retinoids, and such an approach may be very promising for treating patients with acne, alopecia, or other disorders associated with the pilosebaceous structures, with a concomitant reduction of either applied dose or frequency of administration of drug.

The basic idea of dermal and transdermal drug delivery using particulate carriers, considered as a myth a few years ago, is now clearly becoming a reality.

ACKNOWLEDGMENTS

The authors thank Mrs. Catherine Gueston for typing the manuscript and Mrs. Dominique Poisson for artwork. We appreciate the collaboration of

Mrs. Nathalie Wagner who generated much of the experimental data presented in this chapter on drug follicular targeting with polymeric microspheres. We are also indebted to Pr. Hans Schaefer and Dr. Braham Shroot for valuable discussion and comments.

We also wish to thank Dr. J. A. Bouwstra (Center for Bio-Pharmaceutical Sciences, The Netherlands) for donation of freeze-fracture electron micrographs and the following publishers for granting permission to reprint published material: Elsevier Science Publishers, Amsterdam (Figures 4, 5, 8, 9, 10, 11 and Tables 2, 4); S. Karger AG, Basel (Table 3 and Figure 12); Taylor & Francis Ltd., London (Figure 13); Pergamon Press PLC, Oxford (Figure 6); Johann Ambrosius Barth, Leipzig (Figure 7); Elsevier Science Publishing Company, New York (Figures 14 and 15).

REFERENCES

1. B. W. Barry, *Dermatological Formulations: Percutaneous Absorption*, Marcel Dekker, New York, 1983.
2. A. M. Kligman, in *The Epidermis* (W. Montagna and W. C. Lobitz, Jr., eds.), Academic Press, New York, 1964, p. 387.
3. P. M. Steinert and J. S. Cantieri, in *Biochemistry and Physiology of the Skin*, Vol. 1 (L. A. Goldsmith, ed.), Oxford University Press, New York, 1983, p. 135.
4. R. H. Rice and H. Green, *Cell*, *11*, 417 (1977).
5. F. M. Watt and H. Green, *J. Cell Biol.*, *90*, 738 (1981).
6. M. Simon and H. Green, *Cell*, *40*, 677 (1985).
7. M. Simon and H. Green, *Cell*, *36*, 827 (1984).
8. P. W. Wertz and D. T. Downing, *Biochim. Biophys. Acta*, *917*, 108 (1987).
9. M. L. Williams and P. M. Elias, *CRC Crit. Rev. Ther. Drug Carrier Syst.*, *3*, 95 (1987).
10. K. C. Madison, D. C. Swartzendruber, P. W. Wertz, and D. T. Downing, *J. Invest. Dermatol.*, *88*, 714 (1987).
11. P. W. Wertz, D. C. Swartzendruber, K. C. Madison, and D. T. Downing, *J. Invest. Dermatol.*, *89*, 419 (1987).
12. L. Landmann, *J. Ultrastruct. Res.*, *72*, 245 (1980).
13. W. Abraham and D. T. Downing, *Biochim. Biophys. Acta*, *1021*, 119 (1990).
14. E. Christophers, C. Schubert, and M. Goos, in *Pharmacology of the Skin*, Vol. 1 (M. W. Greaves and S. Schuster, eds.), Springer-Verlag, Berlin, 1989, p. 3.
15. M. L. Williams, *Pediatr. Dermatol.*, *1*, 1 (1983).
16. R. L. Bronaugh and R. F. Stewart, *J. Pharm. Sci.*, *74*, 1062 (1985).
17. R. H. Guy and J. Hadgraft, in *Transdermal Drug Delivery: Developmental Issues and Research Initiatives* (J. Hadgraft and R. H. Guy, eds.), Marcel Dekker, New York, 1989, p. 59.
18. B. W. Barry, *J. Controlled Release*, *15*, 237 (1991).
19. K. A. Walters, in *Topics in Pharmacy, Vol. 2, Routes of Drug Administration* (A. T. Florence and E. G. Salole, eds.), Wright, London, 1990, p. 78.

20. G. L. Flynn, in *Principles of Route-to-Route Extrapolation for Risk Assessment* (T. R. Gerrity and C. J. Henry), Elsevier, New York, 1990, p. 93.
21. R. O. Potts and R. H. Guy, *Proc. Int. Symp. Controlled Release Bioact. Mater.*, *18*, 175 (1991).
22. G. Nicolau, R. A. Baughman, A. Tonelli, W. McWilliams, J. Schiltz, and A. Yacobi, *Xenobiotica*, *17*, 1113 (1987).
23. J. Kao, J. Hall, and G. Helman, *Toxicol. Appl. Pharmacol.*, *94*, 93 (1988).
24. A. Bernd, M. Breuer, K. Dold, and H. Holzmann, *Arzneim.-Forsch./Drug Res.*, *40*, 413 (1990).
25. B. Illel, H. Schaefer, J. Wepierre, and O. Doucet, *J. Pharm. Sci.*, *80*, 424 (1991).
26. A. Rolland and D. Bourel, in *Targeted Therapeutic Systems* (P. Tyle and B. P. Ram, eds.), Marcel Dekker, New York, 1990, p. 259.
27. G. Lopez-Berestein and I. J. Fuller (Eds.), *Liposomes in the Therapy of Infectious Diseases and Cancer*, Alan R. Liss, New York, 1989.
28. E. Tomlinson, *Adv. Drug Deliv. Rev.*, *1*, 87 (1987).
29. E. Tomlinson, in *Drug Delivery Systems: Fundamentals and Techniques* (P. Johnson and J. G. Lloyd-Jones, eds.), Ellis Horwood Ltd., Chichester, 1987, p. 32.
30. *Liposomes: From Biophysics to Therapeutics* (M. J. Ostro, ed.), Marcel Dekker, New York, 1987.
31. A. Rolland and H. Schaefer, in *Biologie de la Peau* (J. Thivolet, M. Faure, and D. Schmitt, eds.), Editions INSERM, Vol. 214, 1991, p. 209.
32. R. H. Guy and J. Hadgraft, *J. Controlled Release*, *4*, 237 (1987).
33. Y. W. Chien, in *Transdermal Delivery of Drugs*, Vol. 1 (A. F. Kydonieus and B. Berner, eds.), CRC Press, Boca Raton, 1987, p. 81.
34. R. W. Baker, and J. Heller, in *Transdermal Drug Delivery: Developmental Issues and Research Initiatives* (J. Hadgraft and R. H. Guy, eds.), Marcel Dekker, New York, 1989, p. 293.
35. H. Lautenschlager, *Cosmet. Toiletries*, *105*, 89 (1990).
36. A. A. Siciliano, *Cosmet. Toiletries*, *100*, 43 (1985).
37. R. M. Handjani-Vila, in *New Approaches to Controlled Drug Delivery* (A. A. Hincal, H. S. Kas, and M. Sumnu, eds.), Editions de Santé, Paris, 1990, p. 213.
38. H. Lautenschläger, *J. Appl. Cosmetol.*, *8*, 1 (1990).
39. R. C. Wester, F. C. Szoka, D. A. W. Bucks, and H. I. Maibach, American Pharmaceutical Association, *Proc. 37th National Meeting*, *14*, 229 (1984).
40. A. Toulon, M. Besnard, B. Leclerc, C. Weingarten, D. Salin-Drouin, and G. Couarraze, *S. T. P. Pharm. Sci.*, *1*, 76 (1991).
41. H. E. J. Hofland, J. A. Bouwstra, H. E. Bodde, F. Spies, R. van der Geest, and H. E. Junginger, *Proc. Int. Symp. Controlled Release Bioact. Mater.*, *18*, 307 (1991).
42. V. M. Knepp, R. S. Hinz, F. C. Szoka, and R. H. Guy, *J. Controlled Release*, *5*, 211 (1988).
43. V. M. Knepp, F. C. Szoka, and R. H. Guy, *J. Controlled Release*, *12*, 25 (1990).
44. C. Artmann, J. Röding, M. Ghyczy, and H. G. Pratzel, *Arzneim.-Forsch./Drug Res.*, *40*, 1363 (1990).
45. C. Artmann, J. Röding, M. Ghyczy, and H. G. Pratzel, *Arzneim.-Forsch./Drug Res.*, *40*, 1365 (1990).

46. A. D. Bangham, M. M. Standish, and J. C. Watkins, *J. Mol. Biol.*, *13*, 238 (1965).
47. P. R. Cullis, L. D. Mayer, M. B. Bally, T. D. Madden, and M. J. Hope, *Adv. Drug Deliv. Rev.*, *3*, 267 (1989).
48. D. Lichtenberg and Y. Barenholz, in *Methods of Biochemical Analysis* (D. Glick, ed.), John Wiley and Sons, New York, 1988, p. 337.
49. P. Machy and L. Leserman (Eds.), *Liposomes in Cell Biology and Pharmacology*, John Libbey and Co. Ltd, London, 1987.
50. M. Mezei and V. Gulasekharam, *Life Sci.*, *26*, 1473 (1980).
51. M. Mezei and V. Gulasekharam, *J. Pharm. Pharmacol.*, *34*, 473 (1982).
52. L. Krowczynski and T. Stozek, *Pharmazie*, *39*, 627 (1984).
53. J. Lasch and W. Wohlrab, *Biomed. Biochim. Acta*, *45*, 1295 (1986).
54. W. Wohlrab and J. Lasch, *Dermatologica*, *174*, 18 (1987).
55. M. Jacobs, G. P. Martin, and C. Marriott, *J. Pharm. Pharmacol.*, *38*, 69P (1986).
56. M. Jacobs, G. P. Martin, and C. Marriott, *J. Pharm. Pharmacol.*, *40*, 829 (1988).
57. W. Wohlrab, U. Lachmann, and J. Lasch, *Dermatol. Mon. Schr.*, *175*, 344 (1989).
58. W. Wohlrab and J. Lasch, *Dermatol. Mon. Schr.*, *175*, 348 (1989).
59. A. J. M. Vermorken, M. W. A. C. Hukkelhoven, A. M. G. Vermeesch-Markslag, C. M. A. A. Goos, P. Wirtz, and J. Ziegenmeyer, *J. Pharm. Pharmacol.*, *36*, 334 (1984).
60. J. G. McGurk, G. Sandhu, D. W. Lendrem, C. J. Potter, and A. R. Ross, *J. Pharm. Pharmacol.*, *40*, 47P (1988).
61. B. Michniak, *Pharm. Res.*, *5*, S-85 (1988).
62. W. Gehring, M. Ghyczy, M. Gloor, Ch. Heitzler, and J. Röding, *Arzneim.-Forsch./Drug Res.*, *40*, 1368 (1990).
63. H. C. Korting, M. Schäfer-Korting, H. Zienicke, and O. Braun-Falco, *Skin Pharmacol.*, *3*, 206 (1990).
64. H. C. Korting, H. Zienicke, M. Schäfer-Korting, and O. Braun-Falco, *Eur. J. Clin. Pharmacol.*, *39*, 349 (1990).
65. M. Jacobs, G. P. Martin, and C. Marriott, *J. Pharm. Pharmacol.*, *39*, 142P (1987).
66. T. C. Rowe, M. Mezei, and J. Hilchie, *Prostate*, *5*, 346 (1984).
67. H. Komatsu, K. Higaki, H. Okamoto, K. Miyagawa, M. Hashida, and H. Sezaki, *Chem. Pharm. Bull.*, *34*, 3415 (1986).
68. H. Komatsu, H. Okamoto, K. Miyagawa, M. Hashida, and H. Sezaki, *Chem. Pharm. Bull.*, *34*, 3423 (1986).
69. H. M. Patel, U.K. Patent 2143-433-A (1985).
70. A. Gesztes, and M. Mezei, *Anesth. Analg.*, *67*, 1079 (1988).
71. M. Foldvari, A. Gesztes, and M. Mezei, *J. Microencapsulation*, 7, 479 (1990).
72. N. Weiner, N. Willams, G. Birch, C. Ramachandran, C. Shipman, and G. Flynn, *Antimicrob. Agents Chemother.*, *33*, 1217 (1989).
73. M. Mezei, in *Topics in Pharmaceutical Sciences* (D. D. Breimer and P. Speiser, eds.), Elsevier, Amsterdam, 1985, p. 345.
74. C. I. Price, J. W. Horton, and C. R. Baxter, *J. Surg. Res.*, *49*, 174 (1990).

75. V. Masini, F. Bonte, A. Meybeck, and J. Wepierre, *Proc. Int. Symp. Controlled Release Bioact. Mater.*, *17*, 425 (1990).
76. K. Egbaria, C. Ramachandran, D. Kittayanond, and N. Weiner, *Antimicrob. Agents Chemother.*, *34*, 107 (1990).
77. K. Egbaria, C. Ramachandran, and N. Weiner, *Skin Pharmacol.*, *3*, 21 (1990).
78. C. Michel, T. Purmann, E. Mentrup, G. Michel, and J. Kreuter, *Proc. Int. Symp. Controlled Release Bioact. Mater.*, *18*, 485 (1991).
79. T. Purmann, C. Michel, E. Mentrup, and J. Kreuter, *Proc. Int. Symp. Controlled Release Bioact. Mater.*, *18*, 489 (1991).
80. K. Kraemer, M. M. Lee, and J. Scotto, *Arch. Dermatol.*, *123*, 241 (1987).
81. D. B. Yarosh, *J. Photochem. Photobiol.*, *6*, 445 (1990).
82. D. B. Yarosh, J. Tsimis, and V. Yee, *J. Soc. Cosmet. Chem.*, *41*, 85 (1990).
83. M. G. Ganesan, N. D. Weiner, G. L. Flynn, and N. F. H. Ho, *Int. J. Pharm.*, *20*, 139 (1984).
84. N. F. H. Ho, M. G. Ganesan, N. D. Weiner, and G. L. Flynn, *J. Controlled Release*, *2*, 61 (1985).
85. K. Egbaria, C. Ramachandran, and N. Weiner, *Skin Pharmacol.*, *4*, 21, (1991).
86. H. E. J. Hofland, J. A. Bouwstra, M. Ponec, H. E. Boddé, F. Spies, J. Coss Verhoef, and H. E. Junginger, *J. Controlled Release*, *16*, 155 (1991).
87. J. A. Bouwstra, H. E. J. Hofland, G. S. Gooris, F. Spies, and H. E. Junginger, *Proc. Int. Symp. Controlled Release Bioact. Mater.*, *18*, 305 (1991).
88. F. Spies, H. E. Boddé, A. Meybeck, and F. Bonté, *Proc. Int. Symp. Controlled Release Bioact. Mater.*, *18*, 529 (1991).
89. K. Egbaria and N. Weiner, *Adv. Drug Deliv. Rev.*, *5*, 287 (1990).
90. M. Schäfer-Korting, H. C. Korting, and O. Braun-Falco, *J. Am. Acad. Dermatol.*, *21*, 1271 (1989).
91. V. Gabrijelcic, M. Sentjurc, and J. Kristl, *Int. J. Pharm.*, *62*, 75 (1990).
92. T. Nishihata, K. Kotera, Y. Nakano, and M. Yamazaki, *Chem. Pharm. Bull.*, *35*, 3807 (1987).
93. A. Kato, Y. Ishibashi, and Y. Miyake, *J. Pharm. Pharmacol.*, *39*, 399 (1987).
94. T. Kimura, N. Nagahara, K. Hirabayashi, Y. Kurosaki, and T. Nakayama, *Chem. Pharm. Bull.*, *37*, 454 (1989).
95. M. Foldvari, G. T. Faulkner, and M. Mezei, *J. Microencapsulation*, *5*, 231 (1988).
96. B. Bonnekoh, J. Röding, G. R. F. Krueger, M. Ghyczy, and G. Mahrle, *Br. J. Dermatol.*, *124*, 333 (1991).
97. R. Schmidt, E. J. Parish, V. Dionisius, C. Cathelineau, S. Michel, B. Shroot, A. Rolland, A. Brzokewicz, and U. Reichert, *J. Invest. Dermatol.*, *97*, 771 (1991).
98. J. A. Setterstrom, T. R. Tice, and W. E. Meyers, in *Recent Advances in Drug Delivery Systems* (J. M. Anderson and S. W. Kim, eds.), Plenum Press, New York, 1984, p. 185.
99. J. A. Setterstrom, T. R. Tice, W. E. Meyers, J. W. Vincent, and G. C. Battistone, *Polym. Mater. Sci. Engl.*, *53*, 620 (1985).
100. M. M. Ghorab, H. Zia, and L. A. Luzzi, *J. Microencapsulation*, *7*, 447 (1990).
101. R. C. Wester, R. Patel, S. Nacht, J. Leyden, J. Melendres, and H. Maibach, *J. Am. Acad. Dermatol.*, *24*, 720 (1991).

102. D. Yeung, W. P. Smith, and S. Nacht, in *Transdermal Delivery of Drugs*, Vol. 2 (A. F. Kydonieus and B. Berner, eds.), CRC Press, Boca Raton, 1987, p. 19.
103. J. C. Keister and G. B. Kasting, *J. Controlled Release*, *4*, 111 (1986).
104. M. C. Lamidon, F. Bonte, A. Meybeck, and J. Wepierre, *Proc. Int. Symp. Controlled Release Bioact. Mater.*, *18*, 425 (1991).
105. F. J. G. Ebling, and W. J. Cunliffe, In *Textbook of Dermatology*, 5th Ed., Vol. 3 (R. H. Champion, J. L. Burton, and F. J. G. Ebling, eds.), Blackwell Scientific, Oxford, 1992, p. 1699.
106. H. Schaefer, F. Watts, C. Papantoniou, and C. Mahieu, Eur. Patent 0,375,520 (1989).
107. B. Shroot, J. M. Bernardon, and J. Eustache, Eur. Patent, 199,636 (1986); U.S. Patent 4,717,720 (1988).
108. M. Verschoore, A. Langner, H. Wolska, S. Jablonska, J. Czernielewski, and H. Schaefer, *Br. J. Dermatol.*, *124*, 368 (1991).
109. H. Schaefer, F. Watts, J. Brod, and B. Illel, in *Prediction of Percutaneous Penetration: Methods, Measurements, Modelling* (R. C. Scott, R. H. Guy, and J. Hadgraft, eds.), IBC Technical Services, London, 1990, p. 163.
110. J. P. Benoît, S. Benita, F. Puisieux, and C. Thies, in *Microspheres and Drug Therapy: Pharmaceutical, Immunological, and Medical Aspects* (S. S. Davis, L. Illum, J. G. McVie, and E. Tomlinson, eds.), Elsevier, Amsterdam, 1984, p. 91.
111. M. Nakano, N. Wakiyama, T. Kojima, K. Juni, R. Iwaoku, S. Inoue, and Y. Yoshida, in *Microspheres and Drug Therapy: Pharmaceutical, Immunological, and Medical Aspects* (S. S. Davis, L. Illum, J. G. McVie, and E. Tomlinson, eds.), Elsevier, Amsterdam, 1984, p. 327.
112. H. B. Rosen, J. Kohn, K. Leong, and R. Langer, in *Controlled Release Systems: Fabrication Technology* (D. Hsieh, ed.), CRC Press, Florida, 1988, p. 83.
113. B. W. Bruhn and B. W. Müller, *Proc. Int. Symp. Controlled Release Bioact. Mater.*, *18*, 668 (1991).
114. R. Arshady, *J. Controlled Release*, *17*, 1 (1991).
115. C. Thomasin, B. Gander, and H. P. Merkle, *Proc. Int. Symp. Controlled Release Bioact. Mater.*, *18*, 646 (1991).
116. A. Kishida, J. B. Dressman, S. Yoshioka, Y. Aso, and Y. Takeda, *J. Controlled Release*, *13*, 83 (1990).
117. R. Jalil and J. R. Nixon, *J. Microencapsulation*, *6*, 473 (1989).
118. M. Bouclier, A. Chatelus, J. Ferracin, C. Delain, B. Shroot, and C. N. Hensby, *Skin Pharmacol.*, *4*, 65 (1991).

Index